Sustainable Distributed Energy Resources

Academic Advisory Group (AAG) of the Section on Energy, Environment, Resources, and Infrastructure Law (SEERIL) of the International Bar Association (IBA)

General Editors

TITLES IN THE SERIES

Human Rights in Natural Resource Development: Public Participation in the Sustainable Development of Mining and Energy Resources **(2002)**

Energy Security: Managing Risk in a Dynamic Legal and Regulatory Environment **(2004)**

Regulating Energy and Natural Resources **(2006)**

Beyond the Carbon Economy: Energy Law in Transition **(2008)**

Property Law in Energy and Natural Resources **(2010)**

Energy Networks and the Law: Innovative Solutions in Changing Markets **(2012)**

The Law of Energy Underground: Understanding New Developments in Subsurface Production, Transmission, and Storage **(2014)**

Sharing the Costs and Benefits of Energy and Resource Activity **(2016)**

Innovation in Energy Law and Technology **(2018)**

Energy Justice and Energy Law: Distributive, Procedural, Restorative and Social Justice in Energy Law **(2020)**

Resilience in Energy, Infrastructure, and Natural Resources Law **(2022)**

Net Zero and Natural Resources Law: Sovereignty, Security, and Solidarity in the Clean Energy Transition **(2024)**

Sustainable Distributed Energy Resources

Law and Policy

Edited by

ÍÑIGO DEL GUAYO, *LLM, PHD*
Professor of Administrative Law, University of Almería, Spain

DAMILOLA S. OLAWUYI SAN, *LLM, PHD*
Professor and UNESCO Chair on Environmental Law and Sustainable Development, Hamad Bin Khalifa University, Qatar

LOUIS DE FONTENELLE, *PHD, HDR*
Associate Professor in Public Law, Université de Pau et des Pays de l'Adour, France

MILTON FERNANDO MONTOYA, *PHD*
Director, Mining and Energy Law Institute, Externado University of Colombia, Colombia

JOSÉ JUAN GONZÁLEZ, *LLM, PHD*
Full Time Professor of Environmental Law, Metropolitan Autonomous University, Mexico

OXFORD
UNIVERSITY PRESS

OXFORD
UNIVERSITY PRESS

Great Clarendon Street, Oxford, OX2 6DP,
United Kingdom

Oxford University Press is a department of the University of Oxford.
It furthers the University's objective of excellence in research, scholarship,
and education by publishing worldwide. Oxford is a registered trade mark of
Oxford University Press in the UK and in certain other countries.

Published in the United States of America by Oxford University Press
198 Madison Avenue, New York, NY 10016, United States of America.

British Library Cataloguing in Publication Data

Data available

Library of Congress Control Number: 2025950212

ISBN 9780198991915

DOI: 10.1093/9780198991946.001.0001

Printed and bound by
CPI Group (UK) Ltd., Croydon, CR0 4YY

MIX
Paper | Supporting
responsible forestry
FSC® C013604

The manufacturer's authorised representative in the EU for product safety is
Oxford University Press España S.A. of Parque Empresarial San Fernando de Henares,
Avenida de Castilla, 2 – 28830 Madrid (www.oup.es/en or product.safety@oup.com).
OUP España S.A. also acts as importer into Spain of products made by the manufacturer.

Preface

This is the thirteenth book within a fruitful and long-standing collaboration among the International Bar Association's Section on Energy, Environment, Natural Resources, and Infrastructure Law (IBA SEERIL), the Oxford University Press (OUP), and the Academic Advisory Group (AAG) of the IBA SEERIL. The International Bar Association is the world's largest association of lawyers and bar associations. As an active section of the IBA, the IBA-SEERIL spearheads public interest initiatives to keep members informed on significant developments relating to the Section's six committees: Environment, Health and Safety Law, International Construction Projects, Mining Law, Oil and Gas Law, Power Law, and Water Law. The AAG is a network of academics that carry out research and teaching in the areas pertinent to SEERIL.

This latest book is designed to inform a worldwide audience of practitioners, policy-makers, scholars, researchers, and students on the legal tools and techniques for advancing sustainability in the development, financing, implementation, and integration of distributed energy resources (DERs) in electricity systems. DERs refer to the small-scale generation, transmission, and storage of electricity through microgrids, backup batteries, fuel cells, and virtual power plants that are often at, or closer to, the point of consumption (for example, solar panels in the consumer's home). DERs, of course, centrally include self-generation installations. Over the last decades, the ongoing global energy transition and the quest for innovative ways to promote energy efficiency; and to reduce the environmental footprints, infrastructure gaps, and costs of electricity from centralized energy systems, have resulted in an exponential rise in initiatives aimed at advancing DERs. It is globally accepted that DERs have crucial roles to play in advancing the four vectors for energy transition, namely decarbonization, decentralization, digitization, and democratization (4Ds). At the same time, despite the promise and potential of DERs, a wide range of environmental, social, and economic challenges arise in their implementation, which if not holistically addressed could limit the full potential of DERs to advance a secure and sustainable energy future.

This book aims to contribute to the analysis of the legal and energy policy challenges facing DERs in a wide range of jurisdictions. The common thread throughout the analysis, both internationally and nationally, is the achievement of sustainability in its three dimensions: environmental, social, and economic in DER design and implementation. The book is the result of an effort in planning, co-ordination, and the need to approach this subject through a sustainability lens, which is the common thread across the chapters. Through thematic chapters and case studies, the book examines the latest developments of law and policy on sustainable DERs in Africa, Asia, Europe, Australasia, and North and South America. Second, it evaluates how legal and regulatory systems can better enhance sustainable DERs by addressing the wide range of environment and land tenure concerns, technology gaps, network regulation, gender justice, human rights, financing, data and consumer protection, and other broad sustainability challenges in current implementation and deployment. Third, it provides practical recommendations on how legal, governance, and contractual frameworks, including energy entrepreneurship education, can better enhance decentralized energy communities, energy entrepreneurship, and co-operative business models, to promote secure and sustainable energy.

The analyses incorporated into the chapters delve into the granularity of DER law and policy, and do not limit themselves to reiterating well-known principles. The book therefore offers a comprehensive and authoritative account of the guiding principles of sustainable DERs law and policy, unpacking the interplay of the 4Ds. It analyses and offers legal pathways for addressing the sustainability problems that arise with the emergence and expansion of DERs.

In May 2025, the AAG members met at the *Université de Pau et des Pays de l'Adour*, in Pau (France), for an intensive three-day workshop to present draft chapters and refine the book's focus. We express our most sincere appreciation to our marvelous hosts, Professor Louis de Fontenelle, Director of the Energy Justice & the Social Contract (EJ&SC), *Chair de la Unité Mixte de Recherche (UMR) Transitions Energétiques et Environnementales (TREE),* at the University of Pau and Pays de l'Adour, France; and Alice Moulene Daba, Stéphanie Roposte, and Isabelle Montin, members of the Chair, for their efficient help and dedication in making that mid-term workshop very successful and memorable.

We are also very thankful to the IBA SEERIL for their continued support for the work of the AAG. Over the last forty-one years, the AAG has proudly benefited from active collaboration and a productive circle of ideas, practices, joint publications, and public events, with IBA-SEERIL committees, which have remained instrumental to shaping the continued development of law and practice in this rapidly evolving field.

We would also like to thank the editorial staff of Oxford University Press, especially Fay Gibbons and Paulina dos Santos Major, for the thoughtful, constructive, and professional review process which guided this book through to its timely completion. Darío Moya García (University of Almería) deserves special appreciation for providing remarkable and thoughtful research assistance. He assisted greatly at the editing stage. We also acknowledge the dedication and collegiality of all the contributing authors, whose substantial research and commitment to leading-edge scholarship in this field has been pivotal to the production of this book.

The Editors

Contents

Abbreviations

2019 ED	Directive (EU) 2019/944 of the European Parliament and of the Council of 5 June 2019 on common rules for the internal market for electricity and amending Directive 2012/27/EU
AAOIFI	Accounting and Auditing Organization for Islamic Financial Institutions
AC	Collective Self-Generators
ACO	Alberta Government Aboriginal Consultation Office
AD	Anaerobic Digestion
AfDB	African Development Bank
ADR	Alternative Disputes Resolution (mechanisms)
AEMC	Australian Energy Markey Commission
AER	Alberta Energy Regulator
AERC	Special Areas of Collective Reincorporation
AETCR	Territorial Training and Reincorporation Spaces
AETP	African Energy Transition Programme
AGGE	Large-Scale Self-Generation
AGPE	Small-Scale Self-Generation
AMI	Advanced Metering Infrastructure
APSR	Authority for Public Services Regulation (Oman)
AREI	Africa Renewable Energy Initiative
ASSELLMU	Association of Environmental Law Lecturers in Middle East and North African Universities
AUC	Alberta Utilities Commission
BCEAO	Central Bank of West African States (*West African Economic and Monetary Union*)
BESS	Battery Energy Storage System
BM	Balancing Market
BOP	Bottom of the Pyramid
BRP	Balancing Responsible Party
BSC	Balancing and Settlement Code
BTM	Batteries Behind-the-meter
CALISO	California Independent System Operator
CAPEX	Finance Capital Expenditures
CARES	(Scotland's) Community and Renewable Energy Scheme
CBAM	Carbon Border Adjustment Mechanism
CBN	Central Bank of Nigeria
CfD	Contracts for Difference
CDG	Clean Distributed Generation
CDS	Closed Distribution System
CEPAL	Comisión Económica para América Latina y el Caribe
CEREEAC	Centre for Renewable Energy and Energy Efficiency for Central Africa
CET	Common External Tariff
CFE	Federal Commission of Electricity

CHP	Combined Heat and Power
CITA	Companies Income Tax Act
CNSC	Canadian Nuclear Safety Commission
COP	Conference of Parties
COP26	26th Annual United Nations Climate Change Conference of the Parties
COP28	28th Conference of the Parties to the United Nations Framework Convention on Climate Change
CPUC	California Public Utility Commission
CREG	Energy and Gas Regulation Commission
CSR	Corporate Social Responsibility
CUP	Cambridge University Press
Ds	Decarbonization, Decentralization, Digitalization, and Democratization
DA	Day-Ahead
DAM	Day-Ahead Market
DE	Distributed Energy
DERs	Distributed Energy Resources
DERs	Distributed Energy Resources
DERMS	Smart Distributed Energy Resource Management Systems
DES	Distributed Energy System
DEWA	Dubai Electricity and Water Authority
DFS	Demand Flexibility Service
DNV	Det Norske Veritas (consulting/assurance company)
DR	Demand Response
DRC	Democratic Republic of the Congo
DSF	Demand-Side Flexibility
DSO	Distribution System Operator
DSR	Demand-Side Response
EAB	Alberta Environmental Appeals Board
EAN	European Article Number
EAP	Energizing Agriculture Programme
ECs	Energy communities
ECB	Electricity Control Board
ECOWAS	Economic Community of West African States
ECREEE	Centre for Renewable Energy and Energy Efficiency
EDB	Electricity Distribution Business
EDF	Électricité de France
EEG	(Germany's) Renewable Energy Sources Act
EEI	Energizing Economies Initiative
EEP	Energizing Education Programme
EIB	European Investment Bank
EIL	Electricity Industry Law
EMS	Energy Management System
EPSL	Electricity Public Service Law
ERA	Electricity Regulation Act 4 of 2006 (*South Africa*)
ERAA	Electricity Regulation Amendment Act 38 of 2024
ESG	Environmental, Social, and Governance
ESIA	Environment and Social Impact Assessment

ESL	Energy Sector Law
ESO	Electricity System Operator (*UK*)
ESTs	Environmentally Sustainable Technologies
ETC	Energy Transition Council
EU	European Union
EV	Electric Vehicle
EY	Ernst & Young (consulting/financial services firm)
FENOGE	Non-Conventional Energy and Efficient Energy Management Fund
FERC	Federal Energy Regulatory Commission
FITs	feed-in-tarrifs
FtM	Batteries In-front-of-the-meter
FNCER	Non-conventional Renewable Energy Sources
FOTEASE	Fund for Energy Transition and the Use of Renewable Energy
FPIC	Free, Prior and Informed Consent
FPSC	Florida Public Service Commission
FRA	Financial Regulatory Authority *(Egypt)*
FSA	Financial Services Act *(Malaysia)*
GATT	General Agreement on Tariffs and Trade 1994
GB	Great Britain
GCCIA	Gulf Cooperation Council Interconnection Authority
GD	Distributed Generation
GDC	Collective Distributed Generators
GDP	Gross Domestic Product
GDPR	General Data Protection Regulation
GHG	Greenhouse Gas(es)
GNWT	Government of Northwest Territories
Gt/GT	Gigatonne (billion tonnes)
GW	Gigawatt
GW(e)	Gigawatt electric
GWh	Gigawatt-hour
HS Code	Harmonized System Code
HWR	Heavy Water Reactor
IAEA	International Atomic Energy Agency
ICCPR	International Covenant on Civil and Political Rights
ICJ	International Court of Justice
IDB	Islamic Development Bank
IEA	International Energy Agency
IEP	Integrated Energy Plan, 2016 (*South Africa*)
IFC	International Financial Corporation
IFIs	Islamic Financial Institutions
IFSA	Islamic Financial Services Act (*Malaysia*)
IFSB	Islamic Financial Services Board (*International, based in Malaysia*)
IICM	Integrated Islamic Crowdfunding Microfinance
ILO	International Labour Organization
INTSA	Innovation and Non-Traditional Solutions Allowance
IOUs	Investor-Owned Utilities
IPP(s)	Independent Power Producer(s)

IREI	Innovative Renewable Energy Initiative
IRENA	International Renewable Energy Agency
IRP	Integrated Resource Plan (*South Africa*)
IRP2019	Integrated Resource Plan of 2019 (*South Africa*)
ISES	Integrated Sustainable Energy Strategy (Egypt)
ISMO	Independent System and Market Operator (*South Africa*)
ISO(s)	Independent System Operator(s)
ITO	Independent Transmission Operator
ITSMO	Independent Transmission System and Market Operator (*South Africa*)
IUCN	International Union for Conservation of Nature
IVEKA	Intercommunale Vereniging voor de Elektriciteitsdistributie in de Kempen en het Antwerpse
JET-IP	Just Energy Transition Investment Plan (*South Africa*)
JETP	Just Energy Transition Partnership
JRC	Joint Research Centre (European Commission)
kW	Kilowatt
LAEP	Local Area Energy Planning
LCOE	Levelised Cost of Electricity
LGP	Limited Generation profile
LMP	Locational Marginal Price
LWR	Light Water Reactor
MDAG	Market Development Advisory Group
MoERE	Ministry of Electricity and Renewable Energy (Egypt)
MME	Ministry of Mines and Energy
MPI	Multidimensional Poverty Indicator
MPRDA	Mineral and Petroleum Resources Development Act 28 of 2002 (*South Africa*)
MS	Member State(s)
MSCF	Market Scale-up Challenge Fund
MW	Megawatt
MWe/MW(e)	Megawatt electric
MWh	Megawatt-hour
MWth/MW(th)	Megawatt thermal
NABE	National Energy Security Agency (*Poland*)
NAFTA	North American Free Trade Agreement
NBET	Nigeria Bulk Electricity Trading Company
NCCRP	National Climate Change Response Policy
NCEP	National Clean Energy Access Programme (Ghana)
NDC(s)	Nationally Determined Contribution(s)
NDP	National Development Plan
NDRC	National Development and Reform Commission
NEA	National Energy Administration
NEA (OECD)	Nuclear Energy Agency (OECD)
NEBEF	Notification d'Échange de Blocs d'Éffacement (*France's demand response scheme*)
NEC	National Energy Commission
NEM	Net Energy Metering

NEP	Nigeria Electrification Project
NERC	Nigerian Electricity Regulatory Commission
NERSA	National Energy Regulator of South Africa (*South Africa*)
NESO	National Energy System Operator (Great Britain)
NETP	Nigeria Energy Transition Plan
NGO	Non-Governmental Organisation
NIMBY	Not In My Backyard
NIPC	Nigeria Investment Promotion Commission
NM	Net Metering
NRA	National Regulatory Authority
NRCB	Natural Resources Conservation Board
NTCSA	National Transmission Company of South Africa
NTPC	Northwest Territories Power Corporation
NURIN	Unique National Recognition and Identification Number
NWAs	Non-wires alternatives
NWT	Northwest Territories
NZ	New Zealand
OECD	Organisation for Economic Co-operation and Development
OETC	Oman Electricity Transmission Company
OUP	Oxford University Press
O&M	Operations & Maintenance
PACE	Property-Assessed Clean Energy
PDET	Programa de Desarrollo con Enfoque Territorial
PESTEL	Political, Economic, Social, Technological, Environmental and Legal
PETL	Planning and Energy Transition Law
PLATEASE	Energy Transition and Sustainable Energy Use Plan
PM-KUSUM	Pradhan Mantri Kisan Urja Suraksha evam Utthaan Mahabhiyan (*India*)
PPA(s)	Public Purchase Agreement(s)
PPA/PPAs	Power Purchase Agreement(s)
PPP	Public–Private Partnership
PSC	Public Service Commission (generic)
PUA	Public Utilities Act
PV	Photovoltaic
RCE	Energy Communities Register
R&D	Research and Development
RE	Renewable Energy
REA	Rural Electrification Agency
RECs	Renewable Energy Communities
RED III	Directive (EU) 2023/2413 of the European Parliament and of the Council of 18 October 2023 amending Directive (EU) 2018/2001, Regulation (EU) 2018/1999 and Directive 98/70/EC as regards the promotion of energy from renewable sources and repealing Council Directive (EU) 2015/652
REF	Rural Electrification Fund
REIPPPP	Renewable Energy Independent Power Producer Procurement Programme (*South Africa*)
REMA	Review of Electricity Market Arrangements Programme
RES	Renewable Energy Source

RESP	Regional Energy Strategic Plan
RPS	Renewable Portfolio Standard(s)
RTAs	Regional Trade Agreements
RTO	Regional Transmission Operator (*US*)
SA	South Africa
SAC	Shariah Advisory Council (*Malaysia*)
SACREEE	Southern African Development Community Centre for Renewable Energy and Energy Efficiency
SDG(s)	Sustainable Development Goal(s)
SEB	State Electricity Boards (*India*)
SEG	Small Export Guarantee
SEPAP	Solar Energy for Poverty Alleviation Program
SESP	Sustainable Energy Support Policy
SGIP	Self-Generation Incentive Program
SIN	National Interconnected System
SIRC	State Islamic Religious Council (*Malaysia*)
SMEs	Small and Medium-scale Enterprises
SMEEs	Small and medium energy and eco-entrepreneurs
SMR(s)	Small Modular Reactor(s)
Solar PV	Solar Photovoltaic
SPN	Solar Power Naija
SRI	Socially Responsible Investment
SSDP	South Sudan Development Plan (*South Sudan*)
STEEPLED	Social, Technological, Economic, Environmental, Political, Legal, Ethical, and Demographic
STEER	Socio-cultural, Technological, Economic, Ecological, and Regulatory
STEM	Science Technology Engineering and Mathematics
SWOT	Strengths, Weaknesses, Opportunities and Threats
TFEU	Treaty on the Functioning of the European Union
TRA	Tanzanian Revenue Authority (*Tanzania*)
TSO	Transmission System Operator
UAE	United Arab Emirates
UK	United Kingdom
UN	United Nations
UNCTAD	UN Trade and Development
UNDP	United Nations Development Programme
UNDRIP	United Nations Declaration on the Rights of Indigenous Peoples
UNECA	United Nations Economic Commission for Africa
UNFCCC	United Nations Framework Convention on Climate Change
UNGPs	United Nations Guiding Principles on Business and Human Rights
UNIDO	United Nations Industrial Development Organization
UPME	Mining and Energy Planning Unit
US	United States
USD	United States Dollar
USMCA	United States-Mexico-Canada Agreement
UoT	University of Toronto
VAT	Value Added Tax

VIC	Vertically Integrated Company
VLP	Virtual Lead Party (*UK*)
VPP(s)	Virtual Power Plant(s)
WAEMU	West African Economic and Monetary Union (*West Africa*)
WHO	World Health Organization
WMSCM	Waqf-Muzara'ah-Supply Chain Model (*Zanzibar, Tanzania*)
YDC	Yukon Development Corporation
ZNI	Non-Interconnected Areas
ZOMAC	Zonas Más Afectadas por el Conflicto Armado de Colombia (*Most Affected Area by the Armed Conflict in Colombia*)

Contributors

Godswill Agbaitoro, PhD, Lecturer in Energy and Environmental Law, University of Essex, United Kingdom.

Nadia B Ahmad, JD, LLM is an Associate Professor of Law at Barry University School of Law, USA.

Adenike Akinsemolu, IAS Vanguard Fellow, University of Birmingham, United Kingdom and Senior Research Fellow, Institute for Oil, Gas, Energy, Environment and Sustainable Development, Afe Babalola University, Ado Ekiti, Nigeria.

Elena I Athwal, JD, LLM, is a Research Fellow, UNESCO Chair on Environmental Law and Sustainable Development, at the Hamad bin Khalifa University (HBKU) Law School, Doha.

Barry Barton, Professor of Law, University of Waikato, Hamilton, New Zealand.

Evangeline Fletcher, Juris Doctor from Barry University, School of Law, USA and a Research Fellow of the Just Energy Transitions and Place project.

Louis de Fontenelle, Associate Professor at Pau University, France.

Lee Godden, Professor of Law, Melbourne Law School, The University of Melbourne, Australia.

José Juan González, Professor of Law at the Metropolitan Autonomous University, Azcapotzalco, México and the Director of the Mexican Institute for Environmental Law Research, Mexico.

Íñigo del Guayo, Professor of Administrative Law, at the University of Almería, Spain.

Kaisa Huhta, Associate Professor of European Law at the University of Eastern Finland Law School and Part-time Professor at the European University Institute.

Tasniem Ahmed Elyass Hussain, LL.M candidate at Yale Law School, United States. She holds a Juris Doctor degree in law from the Hamad Bin Khalifa University, Doha Qatar.

Josephine Ilesanmi, BSc (Aberdeen), MSc (Aberdeen), PhD Candidate, Durban University of Technology, South Africa and Research Fellow, Institute for Oil, Gas, Energy, Environment and Sustainable Development (OGEES Institute), Afe Babalola University, Ado Ekiti, Nigeria.

Allan E Ingelson, Professor, Faculty of Law and Haskayne School of Business, Strategy and Global Management (SGMA), Executive Director, Canadian Institute of Resources Law.

Jaccques Jacobs, Research Project Manager, at the DST/NRF SARChI Research Chair: Mineral Law in Africa, University of Cape Town.

Wikus Kruger, Director: PowerFutures Lab, Graduate Business School, University of Cape Town.

Alastair R Lucas KC, Professor Emeritus and Senior Research Fellow, Canadian Institute of Resources Law, Faculty of Law, University of Calgary, Alberta, Canada.

Emma Merikanto, Doctoral Researcher, Law School, Faculty of Social Sciences and Business Studies, University of Eastern Finland.

Ivett Montelongo Buenavista, PhD in Environmental Law, Director of Environmental Affairs at the firm González y Asociados, a consulting firm specializing in Environmental Law.

Milton F Montoya, Director of the Mining and Energy Law Institute, at the Externado de Colombia University, Bogotá, Colombia.

Hanri Mostert, Professor of Law, University of Cape Town, South Africa and DST/NRF SARChI Research Chair: Mineral Law in Africa.

Pedi Obani, PHD, Associate Professor of Law, University of Bradford, United Kingdom.

Kehinde Olaoye, Assistant Professor of Business and Commercial Law, Hamad Bin Khalifa University, Doha. Qatar.

Damilola S. Olawuyi SAN, Professor of Law and UNESCO Chair on Environmental Law and Sustainable Development at HBKU Law School, at the Hamad Bin Khalifa University, Qatar and a Chancellor's Fellow and Director at the Institute for Oil, Gas, Energy, Environment and Sustainable Development, at the Afe Babalola University, Nigeria.

LeRoy C Paddock, Distinguised Professorial Lecturer in Environmental Law at the George Washington University Law School, Washington DC, United States.

Kierra Peoples, Graduate Research Assistant at Barry University School of Law, USA.

Brianna Pruitt, Research Assistant at Barry University School of Law, USA.

David Robinson, Analyst at the Oxford Institute for Energy Studies.

Martha M. Roggenkamp, Professor Emeritus of Energy Law, Faculty of Law, University of Groningen, the Netherlands and Honorary Professor of Energy Law, University of Pau, France.

Ana Laura Silva Lopez, Professor of Law at the Metropolitan Autonomous University, Azcapotzalco, México.

Hugo Meyer van den Berg, Adjunct Associate Professor of the Faculty of Law, University of Cape Town, affiliated with the DST/NRF SARCHI Research Chair: Mineral Law in Africa and a lecturer at the University of Namibia.

Juan Manuel Zamora Velázquez, Professor of Law at the Metropolitan Autonomous University, Azcapotzalco, México.

José V Zapata, partner at Holland & Knight (Bogotá).

Hao Zhang, Associate Professor of Law, Faculty of Law, The Chinese University of Hong Kong.

1

Introduction

Distributed Energy Resources as the New Electricity Paradigm: Legal Context and Principles

Íñigo del Guayo

I. Introduction

This chapter analyzes the legal principles and the regulatory context of distributed energy resources (DERs), a key component of future electricity systems worldwide. It provides foundational information on both the contours and significance of DERs, and the role of law in addressing the fundamental sustainability questions that DER technologies and initiatives raise.

Traditionally and for several decades, the generation and distribution of electricity depended on large, centralized energy systems and power infrastructure, distant from consumption centres, which required the construction of transmission and distribution networks.[1] However, over the last decades, the ongoing global energy transition and the quest for innovative ways to promote energy efficiency; and to reduce the environmental footprints, infrastructure gaps, and costs of electricity from centralized energy systems, have resulted in an exponential rise in initiatives aimed at advancing DERs.[2]

DERs refer to the small-scale generation, transmission, and storage of electricity through microgrids, backup batteries, fuel cells, and virtual power plants that are often at, or closer to, the point of consumption (for example, solar panels in the consumer's home).[3] That is why DERs were usually referred to as *distributed generation*,[4] but currently the wider expression DER is used instead. The concept has now been expanded to include not only generation, but also other resources used on-site, such as storage. Electricity generated out of renewable energy sources and stored and/or consumed where it has been generated is the basis for several electricity businesses of a new electricity system, such as individual or

[1] MM Roggenkamp and others (eds), *Energy Networks and the Law: Innovative Solutions in Changing Markets* (OUP 2012) 1–10.

[2] E Noonan and E Fitzpatrick, 'Will Distributed Energy Resources Change How We Get our Energy?', European Parliamentary Research Service at <https://www.europarl.europa.eu/RegData/etudes/ATAG/2020/651944/EPRS_ATA(2020)651944_EN.pdf> accessed 10 November 2024.

[3] T Bin Nadeem and others, 'Distributed Energy Systems: A Review of Classification, Technologies, Applications, and Policies' (2023) 48 Energy Strategy Reviews 101096.

[4] Directive (EU) 2019/944 of the European Parliament and of the Council of 5 June 2019 on common rules for the internal market for electricity and amending Directive 2012/27/EU (OJ L 158, 14 June 2019, p 125) (hereinafter referred as to 2019 ED) uses several times the expression *distributed generation*, which means 'generating installations connected to the distribution system': art 2, 32 of the 2019 ED (the expression is used in arts 8, 3; 31, 1 and 58, letter d). The 2019 ED only twice uses the expression 'distributed energy resources': at Recital 61, to state that DSO should be enabled to use services from DER such as demand response and energy storage, and art 16, 3, when dealing with citizen energy communities.

Íñigo del Guayo, *Introduction* In: *Sustainable Distributed Energy Resources*. Edited by: Íñigo del Guayo, Damilola S. Olawuyi San, Louis de Fontenelle, Milton Fernando Montoya, and José Juan González, Oxford University Press.
 DOI: 10.1093/9780198991946.003.0001

collective self-generation (or self-consumption),[5] renewable energy communities, citizen energy communities, and aggregation. Microgrids, insofar as they consist of a distribution network and do not involve transmission, can also be classified among DERs. A further refining of the expression DER is needed, since the concept not only excludes electricity transmission (transport), but sometimes it excludes even electricity distribution. That is the case of self-generation, in any case, and although some energy communities may have their own distribution network, there are communities without one.

In addition to reducing the costs and environmental impacts associated with centralized electricity generation, transmission, and storage, DERs aim to promote energy security and sustainable development by increasing the availability, affordability, accessibility, reliability, and energy self-sufficiency of the end-users. DERs also aim to foster energy democracy and diversification by advancing citizen-led energy solutions that address electricity deficits through innovative renewable energy technologies. The projected rise of small and medium energy and eco-entrepreneurs (SMEEs) involved in the DER value chain could significantly advance economic diversification while also generating livelihood and decent employment for the communities. DERs are already fundamentally reshaping the global energy landscape.[6]

However, despite the promise and potential of DERs, a wide range of environmental, social, and economic challenges arise in their implementation, which if not holistically addressed could limit the full potential of DERs to advance a secure and sustainable energy future.[7] This book discusses legal and policy developments and reforms needed to address social, economic, and environmental challenges associated with DERs, in a manner that promotes harmonized and sustainable DER systems.

To lay a foundation for the rest of the book, this chapter examines the concept of DERs and their key role in the emergence of a new electricity system. The central argument is that the legal order must help DERs to emerge and be spread in a sustainable manner. The more DERs there are in the system, the more changes the legal order will experience, from a centralized to a decentralized one, to a legal order based in supply to a new one based on demand, and from a rigid legal order to a flexible and changing one, while keeping the exigencies of certainty and predictability. Since DERs are to take the place of the primitive industrial electricity model, there is the need to examine the change of the traditional premises of electricity Law.

This chapter proceeds in five sections. After this introduction, section II examines how DERs are at the heart of each of the four energy transition vectors, the place of DERs within a quadrinomial electricity sector, together with demand response, flexibility, and active consumers. At the center of the chapter there is an analysis of the benefits for sustainability of distributed energy resources (section III). After examining the content of the book and the coherence among chapters (section IV), the chapter concludes with several reflections about the nature and scope of the law related to distributed energy resources (section V).

[5] In EU Law, the common expression to refer to the situation where a consumer produces the electricity it consumes is 'self-consumption': see, for example, art 21 of the 2018 RED, under the title 'Renewable self-consumers'.

[6] Noonan and Fitzpatrick (n 2).

[7] See Chapter 2 of this book. See also I Bantekas, 'The Off-Grid Revolution and the "Promise" of Energy Equality Under the Paris Climate Change Agreement' (2024) 23(1) Northwestern Journal of Human Rights 45–72.

II. The Gradual Change in the Traditional Premises of Electricity Law and the Emergence of a New Electricity System Based on Distributed Energy Resources

Electricity law was built during the twentieth century on several economic and technological premises, some of which are currently under review. Among the premises that are being challenged is the non-storable nature of electricity, which must be consumed at the time it is produced. In contrast to this old premise, some techniques have recently emerged that will allow electricity storage, at least for domestic and commercial uses.

The process of decentralization of the electricity system and the widespread use of self-consumption (or self-generation) poses a challenge to these networks, whose functions will change. The interconnected grid will continue to be necessary, but its importance will diminish in quantitative terms as consumers disconnect and meet their electricity needs themselves, without needing to be connected to the grid. The proliferation of consumers capable of providing flexibility techniques to the system (in the form of storage, for example, or in the form of making their electricity available when they do not need it) will transform the grid: it will cease to be a mere vehicle for the transmission of electricity and become a platform for various electricity transactions between market participants (producers, storage, system operators, grid managers, aggregators, and customers). The businesses of the new electricity system will rely on smart, digitalized grids that utilize blockchain.

The third relevant premise is that in the electricity system, there is someone who produces and supplies, distinct from the consumer. The dichotomy between the company that produces (and supplies, through a grid) and the consumer who needs that electricity to live, largely shaped electricity law into what it is today: a special consumer protection regime. This situation came to define the supply of electricity as a public service. The widespread use of self-consumption (individual and collective) is breaking down this otherness and will imbue electricity law with completely different content.

A. The Four Vectors of Energy Transition

Energy transition means that the future of the electricity system will be reliant on renewable sources, and that the system will be governed by the following four vectors of transformation (the four "Ds"): Decarbonization, Decentralization, Digitalization, and Democratization.[8]

1. Decarbonization

Renewable sources of energy, such as wind or solar power, do not emit GHGs and, therefore, they decarbonize the system. Legislation ensuring that electricity is predominantly generated by renewable energy sources will advance decarbonization. The identification between 'energy transition' and 'electrification' is subject to review, since the original intent, which must be reinstated, is decarbonization. DERs are mostly based on the use of

[8] JD Makholm, 'The New "Three Ds" in Regulation—Decarbonization, Decentralization, and Digitization' (2018) 35 Natural Gas & Electricity 18–22. https://doi.org/10.1002/gas.22075 accessed 17 October 2025.

renewable energies, which emit lesser or no GHGs. This means DERs are a main source of decarbonization.

2. Decentralization

Since electricity generated with renewable sources of energy, due to its lower scale in comparison to traditional thermal sources, tends to be consumed where it has been produced, the system will become decentralized. This is having a big impact on networks, which were designed to transport electricity over long distances, to which distribution networks connect to deliver electricity to consumers. This means that networks will no longer be unidirectional, but rather bidirectional. This also means that networks will no longer be only a cable to transmit electricity, but rather a trade platform of electricity and attached services. In this context, DERs are at the heart of a decentralized system.[9]

3. Digitalization

Those two vectors of transformation (decarbonization and decentralization) are based on the new digital technologies invading the sector, which are disrupting the economic and technological premises that shaped the electricity paradigm in the twentieth century. Therefore, the electricity system will be digitalized. DERs are based in the use of digital technologies. Consider, for example, aggregation of demand. Aggregators will offer consumers the chance to sell their surplus electricity in the market. In doing so, aggregators will sum up the surplus electricity of thousands of consumers. This activity can only be based on new techniques which allow aggregators to manage a huge amount of data. Another example is storage: batteries will store electricity when the price is low and will sell its stored electricity when the price is high. For doing that, operators of batteries need to deal with digital technologies having the proper information.

The benefits of DERs are, on the one hand, that the electricity is generated with renewable installations, and, on the other hand, that there is no need to build a network or extend it. Digitalization (new information and communication technologies) is allowing to extend the benefits of more traditional distributed generation to new DERs, where proximity is not of so much relevance. The need for proximity is no longer an essential requirement for citizen energy communities,[10] although the fact that they can also be engaged in the production of electricity with renewable installations may work as a signal of the local character of said type of energy community. Geographical proximity is neither a component of the definition nor operation of aggregation and storage. There is also the example of the public purchase agreements (PPA), where the generated electricity can be sold to the consumers located miles away. This means that the increase of renewable energies has been working towards decentralization, and digitalization is helping to extend the benefits of decarbonization from renewable to other electricity businesses involving agents separated geographically from one another.

[9] See AJ Wieczorek and others, 'Citizen-Led Decentralised Energy Futures: Emerging Rationales of Energy System Organisation' (2024) 113 Energy Research & Social Science 103557; and R Leal-Arcas, N Akondo, and J Alemany Rios, 'Energy Decentralization in the European Union' (2019) Queen Mary School of Law Legal Studies Research Paper No 307/2019, available at SSRN: https://ssrn.com/abstract=3333694.

[10] As defined, for example, in art 16 of the 2019 ED.

4. Democratization

As a result of all that process, the aim is to ensure that electricity systems become more democratic, since consumers (not a bunch of oligopolistic producers) will have the power to decide when and how to produce and consume electricity. Legislation enhancing access to new distributed and local generation digital technologies and digital tools enables the energy consumers themselves to become energy citizens that can actively produce energy for their own use, while being able to sell and infuse the grid with any excess electricity as prosumers. DERs are a source of democratic values, also because they can on a voluntary basis join other consumers to jointly generate and trade electricity. DERs are the result of shifting the accent of the legal order from supply to demand.[11]

The increased recognition of the values implicit in the four 'Ds' is already resulting in transformations to energy law and practice, especially the design and financing of energy infrastructure projects.

B. A Quadrinomial Electricity System: Distributed Energy Resources, Demand Response, Flexibility, and Active Consumers

The future electricity system can be described as the result of the inextricable interaction of four elements: (i) distributed energy resources, (ii) demand response, (iii) demand and system flexibility, and (iv) active consumers. Understanding each of those elements implies understanding its relationship with the other three. The future electricity system will rely upon DERs, but also upon demand response (which actually is, for the 2019 ED, an example of DER[12]). The new electricity system will be flexible in three main regards: first, there will be network flexibility, meaning that networks will be bidirectional and thus allowing consumers to sell surplus electricity (self-generated); second, there will be demand-side flexibility thanks to new digital technologies, allowing consumers to decide when they want to consume, store, or sell their surplus electricity, depending on the price of electricity in the market; third, digital technologies will make it possible for consumers to take those decisions. This flexibility is as much a cause and a consequence of the use of DERs. Simultaneously, consumers of that future electricity system are active consumers, making use of the system flexibility and contributing themselves to increase the flexibility. That is why sometimes they are referred to as prosumers (subjects who produce and consume).

These four components are analyzed in subsequent subsections B.1 (DERs), B.2 (Demand response), B.3 (flexibility), and B.4 (active consumers).

1. DERs

As noted earlier, DERs are small-scale units of local generation or storage connected to the electricity distribution network or *behind-the-meter*[13] (on the customer's side). According

[11] See D Olawuyi, 'From Energy Consumers to Energy Citizens: Legal Dimensions of Energy Citizenship' in K Huhta and others (eds), *Sustainable Energy Democracy and the Law* (Brill 2021) 101–23; and A Savaresi, 'The Rise of Community Energy from Grassroots to Mainstream: The Role of Law and Policy' (2019) 31 Journal of Environmental Law 487.

[12] Recital 61 of the 2019 ED states that DSO should be enabled to use services from DER such as demand response and energy storage.

[13] For example, *Batteries Behind-the-meter* (BTM) are those connected behind the customer or group of customer meters, whose primary purpose is to manage energy for a consumer or group of consumers. These batteries are not part of the distribution or transmission system, although they could constitute controllable assets if

to the International Energy Agency, DERs include various technologies: distributed PV, energy storage,[14] electric vehicles (EVs), and smart buildings, to name just a few.[15] The IEA compares the definitions of DERs given by the European Commission, the US Federal Energy Regulatory Commission (FERC), and the Australian Energy Markey Commission (AEMC). Both the FERC and the European Commission include resources on both sides of the meter, but the AEMC pays more attention to resources *behind-the-meter*. Whereas FERC considers energy efficiency a type of DER, the European Commission does not.[16] *Behind-the-meter* resources pose many more challenges to the existing grid, and it is more difficult for system operators to monitor the energy system impacts of emerging resources.

The IEA makes a further distinction between dispatchable DERs, such as EVs and other DERs, which must be forecast in advance to identify their potential impact on the grid (such as distributed PVs).

Connection to the distribution network (as opposed to the transmission or transport network) is an essential component of the definition of DERs. The 2019 ED states that 'distributed generation' means 'generating installations connected to the distribution system".[17] One may conclude that big photovoltaic or wind generation units connected to the transmission system cannot be classified as DERs. That conclusion, in turn, seems to be phased out by the definition of 'distribution' of the same 2019 ED, which means 'the transport of electricity on high-voltage, medium-voltage and low-voltage distribution systems with a view to its delivery to customers, but does not include supply".[18] This would seem to mean that big renewable generation units connected to the transmission system could not be excluded from the concept of DER. However, they can be excluded from the concept of DERs, since the electricity poured into the transmission line is to be transported, rather than delivered to a consumer. In other words, DERs avoid the construction and use of transmission lines, since the electricity generated is consumed where it is produced or within its vicinity . . . Generation units to be classified under the concept of DER produce, store, or manage electricity close to the point of consumption rather than at large, centralized power plants.

DERs therefore refer to the small-scale generation, transmission, and storage of electricity through one or several of the following means: (i) microgrids, (ii) solar panels (especially rooftop PV), (iii) storage (or batteries), (iv) fuel cells, (v) electric vehicles (especially with bidirectional charging),[19] (vi) small wind turbines, (vii) smart appliances and

consumers are under an aggregation scheme in which providing services to the system can be rewarded. On the contrary, *In-front-of-the-meter* (FtM) batteries are those interconnected in distribution or transmission networks or directly connected to a generation asset on the grid.

[14] In EU electricity law, 'energy storage' means deferring the final use of electricity to a moment later than when it was generated, or the conversion of electrical energy into a form of energy which can be stored, the storing of such energy, and the subsequent reconversion of such energy into electrical energy or use as another energy carrier; and 'energy storage facility' means a facility where energy storage occurs: arts 2, 59, and 67 of the 2019 ED.

[15] International Energy Agency, *Unlocking the Potential of Distributed Energy Resources. Power System Opportunities and Best Practices*, Paris 2022, at https://www.iea.org/reports/unlocking-the-potential-of-distributed-energy-resources accessed 11 November 2025.

[16] ibid, p 14. For a comprehensive analysis of DERs in the USA, see SAW Fitts and FK Davis (eds), *Distributed Generation Law: A Guide to Regulations, Policies, and Programs* (American Bar Association 2020).

[17] Art 2(32) of the 2019 ED.

[18] Art 2(28) of the 2019 ED. The 2019 ED definitions are sometimes divergent from the definitions given by national laws. Said divergences are acceptable up to the extent they do not mean a contradiction between EU and national laws. They impose the need of an extra effort to properly construe the applicable law. For example, under Spanish law, 'a high voltage line' cannot be classified under 'distribution' under no circumstances.

[19] EVs can also serve as energy storage devices and contribute to solar electricity self-consumption, if parked within the premises of the owner or user. Linking the EV's consumption at home while recharging away from

thermostats (such as heat pumps), and (viii) combined heat and power (CHP) systems.[20] DERs can operate independently or be aggregated to provide services to the grid, such as peak shaving, voltage support, or frequency regulation. Aggregation includes not only independent aggregators, commonly referred to as Virtual Power Plants (VPPs) that are often at, or closer to, the point of consumption, but also aggregation in collaboration with suppliers, energy communities, the system operator, and other industry stakeholders.

Some countries and subnational authorities have already enacted DER legislation, and some jurisdictions include a definition of DERs within their legislation. This is the case of Colombia, which defines DERs as energy resources installed near consumption centres, connected to the distribution grid, and capable of dynamically injecting energy, consuming energy, or providing complementary services to the grid. DERs include Demand Response (DR), electric vehicles, distributed generation, storage, and small- and large-scale self-generation, among others.[21]

DERs have multiple benefits for the electricity system. Besides the reduction of the costs and environmental impacts associated with centralized electricity generation, transmission, and storage, DERs aim to promote energy security and sustainable development by increasing the availability, affordability, flexibility, resilience, accessibility, reliability, and self-sufficiency of electricity for the end-users. All those benefits of DERs lead to a decentralized electricity system.

DERs also aim to foster energy democracy and diversification by advancing citizen-led energy solutions that address electricity deficits through innovative renewable energy technologies.[22] The projected rise of small and medium energy and eco-entrepreneurs (SMEEs) involved in the DER value chain could significantly advance economic diversification while also generating livelihood and decent employment for the communities. DER are already fundamentally reshaping the global energy landscape. Studies show that global development of DER currently outstrips the development of centralized electricity projects fivefold with more growth expected as the energy transition grows in scale and ambition.[23]

home, for instance through the same electricity supplier, can potentially contribute to a more dynamic system integration of distributed solar energy assets. This may also allow owners and users to use the same contract and data-sharing agreement for their recharging need: Communication from the Commission to the European Parliament, the Council, the European Economic and Social Committee, and the Committee of the Regions, *EU Solar Energy Strategy*, COM (2022) final, 18 May 2022, p 11. See, also, Communication from the Commission to the European Parliament, the Council, the European Economic and Social Committee, and the Committee of the Regions, *Industrial Action Plan for the European Automotive Sector*, COM (22025) 95 final, 5 March 2025

[20] T Bin Nadeem and others, 'Distributed Energy Systems: A Review of Classification, Technologies, Applications, and Policies' (2023) 48 Energy Strategy Reviews 101096.

[21] Resolution No 40283 of 3 August 2022 of the Colombian Ministry of Energy and Mines, which establishes guidelines for the incorporation of distributed energy resources. See also Decree 1403 of 22 November 2024 which modifies Decree 1073 of 26 May 2015 (Single Regulatory Decree of the Mining and Energy Administrative Sector), in relation to the energy policy guidelines for self-generation and marginal production. It has raised many controversies. The latter has been developed by Resolution No 24 of 13 March 2015 of the CREG, which regulates large-scale self-generation activity in the national interconnected system (SIN) and establishes other provisions, and Resolution No 174 of 7 October 2021 of the CREG, regulating small-scale self-generation and distributed generation activities in the National Interconnected System. I am thankful to Professor Juan José Arango, from *Universidad del Externado de Colombia*, for having given to me these legislative references.

[22] See W Gao (ed), *Distributed Energy Resources: Solutions for a Low Carbon Society* (Springer 2023).

[23] E Noonan and E Fitzpatrick, 'Will Distributed Energy Resources Change How We Get our Energy?', European Parliamentary Research Service at <https://www.europarl.europa.eu/RegData/etudes/ATAG/2020/651944/EPRS_ATA(2020)651944_EN.pdf> accessed 10 November 2024.

At the EU level, for example, the EU published an *Action Plan for Affordable Energy*, which foresees a set of reforms directed to enhance flexibility, electrification, demand response, efficient use of the network, energy efficiency,[24] in line with the efforts to decarbonize the European industry.[25] Authorities must make an effort to accelerate the permitting procedures for grid, storage, and clean energy projects and for infrastructure providing flexibility, for example EVs recharging points. Grid connection requests to distribution networks are growing and create long queues, slowing down renewables, electrification, and the establishment of industrial clusters. Following the 2023 *Grid Action Plan* and the 2023 DER III,[26] the *Accele-RES* strategy aims to help to implement streamlined permitting procedures, crucial, in turn, for meeting the 2030 renewable energy targets.[27]

2. Demand response

New energy technologies and new information and communication technologies are placing consumers and their demand at the center of the system.[28] New demand-side management policies must now gain unprecedented strength and centrality. With the help of these techniques, it will be necessary to constantly modulate demand and take advantage of the flexibility provided by DERs. These resources include storage, self-consumption, and local energy communities. The background atmosphere of DERs is based on demand flexibility and demand response. In closed distribution networks (in conjunction with the special status of electro-intensive consumers), a certain DER can also be identified. These distributed energy resources make it easier for consumers to self-supply with decarbonized energy and offer flexibility services to the central system. To these distributed energy resources, we must add aggregation, which is destined to play a central role in demand flexibility.

This is how the phenomenon of demand response emerges. It is a demand-side management instrument for consumers to alter their consumption by increasing or decreasing the amount of electricity purchased, based on the fluctuation of prices. Consumers will consume more when electricity is cheaper and will consume less when electricity is more

[24] Communication from the Commission to the European Parliament, the Council, the European Economic and Social Committee, and the Committee of the Regions, *Action Plan for Affordable Energy. Unlocking the true value of our Energy Union to secure affordable, efficient and clean energy for all Europeans*, Brussels, COM (2025) 79 final, 26 February 2025.

[25] Communication from the Commission to the European Parliament, the Council, the European Economic and Social Committee, and the Committee of the Regions, *The Clean Industrial Deal: A joint roadmap for competitiveness and decarbonization*, COM (2025) 85 final, Brussels, 26 February 2025.

[26] Communication from the Commission to the European Parliament, the Council, the European Economic and Social Committee, and the Committee of the Regions. *Grids, the missing link—An EU Action Plan for Grids*, COM (2023) 757 final, Brussels, 28 November 2023.

[27] Commission Recommendation (EU) 2024/1343 of 13 May 2024 on speeding up permit-granting procedures for renewable energy and related infrastructure projects, C/2024/2660 (OJ L 2024/1343 of 21 May 2024) and Commission Staff Working Document, *Guidance on designating renewables acceleration areas. Accompanying the document 'Commission Recommendation on speeding up permit-granting procedures for renewable energy and related infrastructure projects'*, SWD (2024) 333 final, Brussels, 13 May 2024; and Communication from the Commission to the European Parliament, the Council, the European Economic and Social Committee, and the Committee of the Regions. *A simpler and faster Europe. Communication on implementation and simplification Communication on implementation and simplification European Commission 2024-2029*, COM (2025) 47 final, 11 February 2025, p 3.

[28] There is however a long path to go before consumers get full information: '*However, the lack of real-time or near real-time information provided to consumers about their energy consumption has prevented them from being active participants in the energy market and the energy transition*': 2019 ED, Recital no 10. For a review of the EU laws and policies putting the consumer at the center, see Part III of L Meeus, *The Evolution of Electricity Markets in Europe* (Edward Elgar 2020).

expensive. In doing so, they benefit from direct involvement in the electricity market, thanks to price signals, whether alone or through aggregation.[29]

There is an incentivized demand response, whereby consumers who reduce their consumption are compensated, upon request from the Transmission System Operator (TSO) or the Distribution System Operator (DSO) (this is a more generalized interruptibility, not limited to industrial customers, as is currently the case). There is another type of demand response, which consists of an end-consumer offering to increase or decrease their consumption in exchange for a price set by an organized market.

Demand response entails reducing generation needs, lowers electricity prices, improves efficiency, and contributes to the security and reliability of the system by reducing supply outages. Therefore, demand response must be remunerated. Demand response is made possible by new digital techniques (smart grids and smart metering) and the automation of production and consumption processes. Due to these new techniques, consumers know the price of electricity at every moment of the day. All of the above facilitates the emergence of the central concept of Virtual Power Plants (VPPs), because, for example, the new demand flexibility, combined with storage, to the extent that it saves new generation, can be considered and regulated as a particular (virtual) power generation plant.[30]

Demand response to electricity has always been important to reduce the need for investment in grids and generation (to meet seasonal peaks). What is new is, first, the frequency of the need for flexibility; it will almost always be necessary to respond to changes in intermittent renewable generation and demand from electric vehicles, heat pumps, and storage. A second novelty is that flexibility needs are not limited to the entire system; they are now local.[31] A third major change is that demand response used to consist of reducing demand at peak times; now it is about increasing demand (including storage) when there is excess renewable production; and reducing demand or increasing local production when renewables are in short supply.

3. Demand and system (network) flexibility: Digitalization facilitating flexibility and going beyond proximity

In all electricity systems, supply and demand must be balanced at all times. In the electricity model developed over the last 100 years, the flexibility essential for matching demand and supply always lay on the supply side (generation or production). On the supply side, flexibility consisted of the ability to increase (or reduce) production—depending on increases or decreases in consumption—using traditional sources of flexible production: natural gas (combined cycle) or coal-fired power plants, and large hydroelectric plants (with tank storage and/or pumped storage plants). These power plants are well-suited to provide flexibility on the supply side. The premise was that electricity demand is inelastic or inflexible because, as a basic need, consumers do not change their consumption even if the price of

[29] I Herrera-Anchústegui, 'Is EU Competition Law a Hindrance for Aggregated Demand Response?', in P Perales and others (eds), *The Digital Economy Regulatory Contractual and Competition Aspects* (Tirant Lo Blanch 2021). See Recital 37 and Art 2(20) of the 2019 ED.

[30] TG Werner and R Remberg, 'Technical, economical and regulatory aspects of Virtual Power Plants', in Institute of Electrical and Electronics Engineers, *2008 Third International Conference on Electric Utility Deregulation and Restructuring and Power Technologies*, Nanjing, China (2008), pp 2427–33, doi: 10.1109/DRPT.2008.4523819.

[31] For a discussion of engineering challenges of DERs, see G Graditi and M Di Somma (eds), *Distributed Energy Resources in Local Integrated Energy Systems: Optimal Operation and Planning* (Elsevier 2021).

electricity increases. This does not mean that electricity consumption is invariable. On the contrary, electricity demand is seasonal, increasing during periods of hot and cold weather. Inflexibility or inelasticity means that, in times of rising or falling demand, the necessary balance with supply must always come from the production side, as consumers cannot fail to meet their cooling or heating needs.

The necessary flexibility in the electricity system was normally provided by some flexible generation plants, such as coal-fired plants, natural gas plants, and large hydroelectric plants (both reservoir storage and pumped storage). Flexibility was provided by supply (production). In the context of decarbonization, coal plants should have no future and must be closed. They are being replaced by natural gas combined cycles. In the future, there will be a massive penetration of intermittent renewable energies (mainly wind and solar) for electricity generation. Although renewable energies are the necessary solution from a sustainability perspective, these are unpredictable and weather-dependent generation sources. The growing penetration of intermittent and unmanageable renewable energy is raising the need for flexible backup. Recourse to combined gas cycles and hydroelectric power must be ruled out as the sole and permanent solution (even though combined cycles are temporarily a good instrument for modulating supply (production) in the system).

Consequently, given that generation will become increasingly uncertain (intermittent renewable energy and the closure of flexible fossil fuel plants), the electricity system must gain flexibility on the demand side. This new policy should not consist solely of an increase in traditional demand-side management policies. The scope of this demand-side management was limited in electricity regulations, as it was limited to action on consumption to reduce it and/or make it more efficient but did not consider the negotiability of this reduced demand by consumers.[32]

DERs are a source of flexibility for the system, and distribution networks must facilitate the exercise of that role.[33] In accordance with EU law Member States must provide the necessary regulatory framework to allow and provide incentives to DSOs to procure flexibility services, including congestion management in their areas, in order to improve efficiencies in the operation and development of the distribution system. In particular, the regulatory framework must ensure that DSOs are able to procure such services from providers of distributed generation, demand response, or energy storage and must promote the uptake of energy efficiency measures, where such services cost-effectively alleviate the need to upgrade or replace electricity capacity and support the efficient and secure operation of the distribution system. DSOs must procure such services in accordance with transparent, non-discriminatory, and market-based procedures unless the regulatory authorities have established that the procurement of such services is not economically efficient or that such procurement would lead to severe market distortions or to higher congestion.[34]

One of the tasks of TSOs, under European law, is to manage electricity flows on the system, taking into account exchanges with other interconnected systems. To that end, the

[32] For example, within the Spanish Electricity Act no 24/2013 of 26 December, this limited approach is still evident.

[33] Current discussions for the approval of a demand response network code works in that direction: Recommendation no 01/2025 of the ACER of 7 March 2025 on reasoned proposal for the establishment of the Network Code on Demand Response according to art 59(1)(e) of Regulation (EU) 2019/943 of the European Parliament and of the Council of 5 June 2019 on the internal market for electricity: https://www.acer.europa.eu/sites/default/files/documents/Recommendations/ACER_Recommendation_01-2025_Demand_Response_Network_Code.pdf accessed 26 October 2025.

[34] Art 31, 1 of the 2019 ED.

TSOs is responsible for ensuring a secure, reliable, and efficient electricity system and, in that context, for ensuring the availability of all necessary ancillary services, including those provided by demand response and energy storage facilities, insofar as such availability is independent from any other transmission systems with which its system is interconnected.[35]

Digitalization (ie, progress of digital communication and control technologies), are empowering DSOs and TSOs, customers, aggregators, trading firms, and other stakeholders to manage grid demand more effectively, including at the residential level.[36]

For the EU, flexibility should be in the hands of consumers, for whom new opportunities have come into the scene thanks to new technologies and DER. It is of the opinion that technological progress in grid management and the generation of renewable electricity has unlocked many opportunities for consumers. Healthy competition in retail markets is still essential.[37]

4. Active consumers

An active consumer is a consumer who consumes, or stores electricity generated within its premises located within confined boundaries or within other premises, or who sells self-generated electricity or participates in flexibility or energy efficiency schemes, provided that those activities do not constitute its primary commercial or professional activity.[38]

In other words, active consumers are those engaged in DERs. Within the EU Renewable Energy Directive (RED), self-consumers (or, reversely, self-generators) are only *renewable self-consumers*.[39] The 2019 ED, when regulating active consumers and when referring to self-generated energy, does not expressly refer to electricity generated with renewable energy. However, for several reasons, it seems that the 2019 ED is aligned with the RED in that regard. Even if they were not, it would be important to stress that the only relevant self-generation and active consumption for this Book is renewable self-generation and renewable active consumption.

Final customers must be entitled to act as active customers without being subject to disproportionate or discriminatory technical requirements, administrative requirements,

[35] Art 40, 1, d of the 2019 ED.

[36] There is an increasing number of scientific contributions on the influence of new technologies in the promotion of renewable energies and DERs, among which some must be highlighted: T Gómez San Román and J Pablo Chaves, *Integration of Renewable and Distributed Energy Resources in Power Systems* (MDPI 2020); P Verma and others, *Digitalization: Enabling the New Phase of Energy Efficiency* (UNECE 2020); E Barlett, C Gimeno, and C Majó, 'Digitalisation at the core of energy networks: the backbone of the energy transition. A brief assessment on the infrastructure investment regulatory framework', paper presented at the 10th FSR Annual Conference (10-11 June 2021), at https://fsr.eui.eu/digitalisation-at-the-core-of-energy-networks-the-backbone-of-the-energy-transition/ accessed 25 October 2025; M Andoni and others, 'Blockchain Technology in the Energy Sector: A Systematic Review of Challenges and Opportunities' (2019) 100 Renewable and Sustainable Energy Reviews 100, 143–74.

[37] Recital 10 of the 2019 ED. Empowering consumers to actively participate in the green transition is also one of the main conclusions of the Report from the Commission to the European Parliament, the Council, the European Economic and Social Committee, and the Committee of the Regions, *Report on energy prices and costs in Europe*, COM (2025) 72 final, Brussels, 26 February 2025.

[38] Art 2(8) of the 2019 ED.

[39] Art 2(14) of the 2018 RED. In this Chapter, the expression RED refers to RED II, which is Directive (EU) 2018/2001 of the European Parliament and of the Council of 11 December 2018 on the promotion of the use of energy from renewable sources (OJ L 328 of 21 December 2018, p 82), which repealed RED I of 2009. RED III refers to Directive (EU) 2023/2413 of the European Parliament and of the Council of 18 October 2023 amending Directive (EU) 2018/2001, Regulation (EU) 2018/1999 and Directive 98/70/EC as regards the promotion of energy from renewable sources and repealing Council Directive (EU) 2015/652 (OJ L 2023/2413 of 31 October 2023).

procedures and charges, and to network charges that are not cost-reflective.[40] They must be entitled to operate either directly or through aggregation, to sell self-generated electricity, including through PPA, to participate in flexibility schemes and energy efficiency schemes, and to delegate to a third party the management of the installations required for their activities, including installation, operation, data handling, and maintenance, without that third party being considered to be an active customer.[41] Active consumers must be subject to cost-reflective, transparent, and non-discriminatory network charges that account separately for the electricity fed into the grid and the electricity consumed from the grid, ensuring that they contribute in an adequate and balanced way to the overall cost sharing of the system.[42] Active consumers are financially responsible for the imbalances they cause in the electricity system; to that extent they shall be balance-responsible parties or shall delegate their balancing responsibility.[43]

III. From DERs to Sustainable DERs: Benefits for the Sustainability of an Electricity System Based on Distributed Energy Resources

Sustainability has three dimensions: environmental, economic, and social, as discussed in Chapter 2 of this book. With regard to the environmental dimension, DERs play a significant role in the transition to decarbonization and, thus, to the sustainability of the energy sector. Insofar as the basis for the development of DERs is renewable energies, they are a source of decarbonization of the electricity system and, thus, a source of sustainability in environmental terms. DERs, such as individual and collective self-generation and renewable energy communities are based in the use of renewable energies. Consequently, they are key components in the evolution of the electricity sector towards sustainability. Several factors are promoting DERs, such as the growing electrification of energy consumption. The increasing penetration of renewable generation sources (in particular, wind and sun) allows DERs to develop rapidly. Demand (consumption) aggregation helps sustainability since adding up electricity demand avoids the need for new generating installations as well as the need for constructing new transmission and/or distribution lines.

As to the economic dimension, renewable generation sources are lower carbon and also competitive with traditional fossil fuel-based power generation in terms of cost, and that makes DERs sustainable, from an economic point of view. The spread of PPAs contracts and also Contracts for Difference (*CfD*) fights volatility and makes the supply of electricity more stable pricewise, and thus more economically sustainable. DERs may help fighting energy poverty since DERs technologies allow consumers to organize themselves and generate and consume the electricity they produce. This cannot be achieved by vulnerable consumers themselves and there is the need of public support, also because some of the DERs technologies are unaffordable for some consumers. That support, in turn, raises issues of fair competition insofar as some consumers may be financing others. The spread of DERs

[40] Art 15, 1 of the 2019 ED.

[41] Art 15, 2, letters a, b, c, and d of the 2019 ED.

[42] Art 15, 2, letter e of the 2019 ED. This must be done in accordance with art 59(9) of the 2019 ED and art 18 of Regulation (EU) 2019/943 of the European Parliament and of the Council of 5 June 2019 on the internal market for electricity (OJ L 158, p 54).

[43] Art 15, 2, letter f of the 2019 ED. This must be done in accordance with art 5 of Regulation (EU) 2019/943.

has an immediate economic benefit since some traditional costs, such as the costs of networks, is avoided.

There are a number of social aspects associated to DERs. The more DERs there are in the electricity system, the more security there is for the entire system since DERs are flexible by nature and allow the use of a remarkable variety of electricity sources to satisfy demand. Security does not increase by the increased use of DERs, it must go hand by hand with a sound digitalization and better regulation of grids. This is one of the warnings of the International Energy Agency, since it invites regulators to incentivize DERs owners 'to align their equipment use and location with the needs of the grid' and encourages digitalization to achieve that alignment.[44] In other words, there is the need to strike a balance between the decentralization benefits of DERS and a certain degree of centralization provided by networks. The evident benefits of DERs in terms of decarbonization are counterbalanced by a social tendency to reject DERs installations in the vicinity of dwellings. This poses a challenge for social acceptance of DERs. This is, for example, the case of new *stand-alone* electricity storage.

IV. Structure and Methodology of the Book

This book offers a sustainability assessment of the design, financing, and implementation of DERs in major world regions. It unpacks social, economic, and environmental gaps in emerging laws and policies on DERs, and innovative regulatory reforms needed to address such gaps.

Given the sheer number of laws, regulations, policies, and contracts around the world on DERs, this book focuses on a select number of case studies of countries with significant uptake of DERs, and with publicly accessible legal instruments and information on their DER programs. In Africa, comparative examples are drawn from Ghana, Nigeria, South Africa, Sudan, and Kenya, countries with growing penetration of DERs, while China's role as the world's largest producer and consumer of solar panels makes it an ideal case study in Asia. Case studies from Colombia and Mexico in Latin America, as well as discussion of distributed energy storage in the Middle East and North Africa region offer additional perspectives from developing countries. On the other hand, substantive chapters on the United States, United Kingdom, Canada, and Australia, Global North countries with ambitious DER programs as well as the EU, offer balanced and representative perspectives from different legal traditions (common law, civil law, and mixed jurisdictions) on how legal systems are responding to the wide range of sustainability questions posed by DERs.

All the chapters adopt a comparative and lessons learned approach aimed at unpacking regulatory responses to advance sustainable DERs. Part I of the book includes thematic chapters that examine key principles, theories, and tenets that underpin sustainable DERs. After this introductory Chapter, Chapter 2 examines the nature, scope, and dimension of sustainable DERs under a general framework of international law (Olawuyi), followed by a very relevant policy aim, which is that of decarbonization of electricity systems (Mostert). DERs pose several environmental questions, such as those related to land use (Van den

[44] See n 12 p 8.

Berg and Agbaitoro) or raise sustainability challenges, such as small modular reactors (SMR: De Fontenelle). These thematic chapters lay a foundation for evaluating the sustainability dimensions of DERs.

The rest of the book (Parts II and III) then offers national and regional case studies that demonstrate how these concepts are being translated to legislation, contracts, financial tools, industry guidelines, and entrepreneurship-focused educational tools on DERs in Africa, Asia, Europe, Australasia, and North and South America. DERs also challenge the traditional central grid model, as in the case of sustainable cross border distributed distribution systems that could at the same time play a part in achieving one internal energy market in the EU (Roggenkamp). Most jurisdictions are engaged in finding a prominent role to DERs as we read in the cases of New Zealand (Barton) and China (Zhang). The relevance of DERs is so paramount that they have direct implications on human rights (González, Montelongo, Silva, and Zamora). DERs benefit from tax and fiscal incentives (Paddock) and financing them has peculiarities, particularly in Africa (Tasneem Elias). Storage is one of the prominent DERs and is extensively analyzed (Athwal). Demand aggregation is also a key type of DER, providing flexibility (Robinson). One of the key drivers of DERs is digitalization, which, in turn, poses questions of data protection (Godden). DERs are, simultaneously, a source of democratization of the energy system, and a result of such democratization. This is shown in the cases of energy communities (Huhta and Merikanto) and community-centric models (Ahmad), for which there are several examples in Colombia (Montoya and Zapata). Democratization, in turn, has to do with gender justice (Obani and Akinsemolu) and eco-entrepreneurship law, distributed energy resources (Ilesanmi and Olaoye), and entrepreneurial legal education (Lucas and Ingelson). Each case study chapter includes a recommendation section that extracts how lessons learned from legal developments in the jurisdiction could inform the development of legislation on sustainable DERs in other countries and regions.

Finally, the concluding chapter of the book provides a synthesis and sets of recommendations on how lessons from the diverse jurisdictions in the preceding chapters may be applied to ensure good and sustainable governance of DERs.

V. Conclusion: Granularity of the Law of Sustainable Distributed Energy Resources

If there is any future new electricity paradigm, it is the electricity system dominated by DERs. The four 'Ds' (decarbonization, digitalization, decentralization, and democratization) are intertwined, and together they play in favour of that new electricity system, decarbonization and distributed generated renewable electricity being the most relevant key drivers. This is reflected in several jurisdictions, for example in the EU third legislative package (2018–19).[45]

Despite their decarbonization potential, DERs have been linked to adverse environmental impacts, including land-use concerns. Concerns arise as to how the high cost of DER technologies may lead to exclusion and inequitable access of low income and vulnerable

[45] The 2019 ED and the 2018 Renewable Directive (RED II).

groups. Questions also arise about grid integration and balancing, especially how to accelerate investments in smart grids, to ensure that DERs do not damage existing electricity infrastructure designed for centralized generation. Furthermore, questions on how the rights of consumers with respect to data protection, billing, and metering, will be addressed in a decentralized supply system will also require careful consideration. The lack of supportive legal and policy frameworks that promote energy entrepreneurship and citizenship may hinder the abilities of energy citizens and prosumers to benefit from the energy democratization objectives of DERs. Without a sustainability perspective, the implementation of DERs risks exacerbating environmental pollution, energy poverty, social exclusions, as well as complex human rights, legal, and commercial risks in energy markets worldwide.

The key point between the interaction of renewable energy, electrification, and information and communications technologies (digitalization) is that the presence and management of DERs can often balance end-user net demand with supply at lower costs than centralized alternatives, particularly large-scale electricity storage and attempts to decarbonize dispatchable fossil fuel generation in the context of short-term flexibility needs within the daily cycle.

If we are asking questions about the nature of law in this field, it becomes obvious that it is very granular. Advances need many adjustments to small technical rules. Grand statements of principle are not of much benefit. Principles are essential, but clearly insufficient in this area. Unlike the relative simplicity of the traditional model (a handful of large power plants on one side and millions of consumers connected to those plants via a grid on the other), the DER-based model is much like a gigantic puzzle, with many pieces to fit together. All the elements are interconnected: for example, storage must be regulated, but it is of little use if, at the same time, the aggregation that allows the stored electricity to be added and traded is not regulated. Ideally, in a hypothetical future, each consumer (or group of consumers) produces their own electricity and sells the excess electricity on the market. For reasons of scale, there will have to be agents (aggregators) who aggregate all this excess electricity for sale on the market. The commercial and contractual aspects of this future model are complex, but there is a prior and more serious complexity, which is maintaining grid stability given that several of the technologies that support DERs are not manageable. That is why we need a granular law that advances by breaking down barriers while maintaining the necessary levels of security and economic efficiency. Regulation must help DERs to be integrated within the electricity system and they should be able to participate in all electricity markets, including balancing and congestion management markets, in a non-discriminatory and homogeneous manner. The granularity of sustainable DERs law and policy requires perfecting the principles of participation by all stakeholders and cooperation among all stakeholders involved in the implementation of a new decarbonized, decentralized, digitalized, and democratic electricity model.

PART I

DERS AND THE SUSTAINABLE DECARBONIZATION OF ELECTRICITY SYSTEMS

2
International Law on Distributed Energy Resources

Damilola S. Olawuyi

I. Introduction

This chapter examines interrelated and mutually reinforcing bodies of international law that govern the sustainable design, financing, and implementation of distributed energy resources (DERs). It discusses international law principles and norms underpinning sustainable DERs; the gaps to their implementation; and how such gaps can be coherently addressed to advance sustainability in the implementation of DERs.

With the adoption of the Paris Agreement, and the United Nations Sustainable Development Goals (SDGs) in 2015, the last decade has witnessed a significant rise in policies and projects aimed at achieving two mutually reinforcing objectives. The first objective is to address the energy poverty crises ('defined as the inability of households to access electricity and modern energy services at an affordable cost') facing many parts of the world, especially in the Global South.[1] Consequently, investments in DER projects, aimed at decentralizing and expanding energy access beyond traditional large and centralized energy systems, are on the increase across the world to bridge current electricity deficits and to advance a sustainable energy future.[2] A central aim of DERs is to advance SDG 7 on clean, reliable, and affordable energy for all by the year 2030.[3] For example, though not legally binding, the Joint G20 Energy-Climate Ministerial Communiqué released in 2021 expressed support for DERs 'as concrete means to facilitate energy affordability, reliability, viability, accessibility, and sustainability'.[4] Global annual investments in DERs are already projected to increase by 75 per cent by 2030, with more growth expected as the energy transition grows in scale and ambition.[5]

[1] See D Olawuyi, 'From Energy Consumers to Energy Citizens: Legal Dimensions of Energy Citizenship' in K Huhta and others (eds), *Sustainable Energy Democracy and the Law* (Brill 2021) 101–23; also I Del Guayo and others (eds), *Energy Law and Energy Justice* (OUP 2020) 1–15, 254–72.

[2] S Oguntola, 'Energy Crisis: Teaching Hospitals opt for Solar Power' (*Nigerian Tribune*, 5 January 2025), also G Elizondo and R Poudineh, 'Harnessing the Power of Distributed Energy Resources in Developing Countries: What Can Be Learned from the Experiences of Global Leaders?' (Oxford Institute for Energy Studies 2023).

[3] United Nations, 'Transforming our World: the 2030 Agenda for Sustainable Development' UNGA Res 70/1 (25 September 2015).

[4] See The Group of 20 (G-20), 'Joint G20 Energy-Climate Ministerial Communiqué' (para 55) <http://g20italy.org/wp-content/uploads/2021/07/2021_G20-Energy-Climate-joint-Ministerial-Communique.pdf> accessed 10 March 2025.

[5] See Frost & Sullivan, 'Nearly $1 Trillion in Investments Estimated by 2030 as Power Sector Transitions to a More Decarbonized and Flexible System' <https://www.frost.com/news/press-releases/nearly-1-trillion-in-investments-estimated-by-2030-as-power-sector-transitions-to-a-more-decarbonized-and-flexible-system/> accessed 10 March 2025; see also European Parliament, 'Will Distributed Energy Resources Change How We Get

Damilola S. Olawuyi, *International Law on Distributed Energy Resources* In: *Sustainable Distributed Energy Resources*.
Edited by: Íñigo del Guayo, Damilola S. Olawuyi San, Louis de Fontenelle, Milton Fernando Montoya, and José Juan González,
Oxford University Press. © International Bar Association 2026. DOI: 10.1093/9780198991946.003.0003

The second objective is to address the climate change emergency that is already threatening human existence in many parts of the world.[6] Efforts to address the climate change emergency and achieve net zero energy systems have resulted in increased investment in the decarbonization of energy systems, especially by phasing out electricity generation from coal and natural gas, while increasing investment in cleaner renewable energy sources, including solar, wind, hydro, and battery technologies and infrastructure.[7] As the G20 rightly notes, 'the progressive deployment of renewable energy and technologies may enable consumers to become simultaneously producers (i.e. prosumers)'.[8]

However, as highlighted in Chapter 1 of this book, despite its prospects, the off-grid revolution and the growing use of DERs and non-wires alternatives (NWAs) raises a wide range of questions.[9] Questions arise about grid integration and balancing, especially how to accelerate investments in smart distributed energy resource management systems (DERMs) to ensure that integrating electricity from DERs does not overwhelm or damage extant distribution and storage infrastructure.[10] In addition to addressing cybersecurity risks in DERs, implementing DERMs will require new contractual relationships and partnerships with aggregators, virtual power plants (VPPs), and other digital third-party service providers.[11] Furthermore, as DER projects ramp up in scale and ambition, global demand for cobalt, lithium, and other transition minerals needed to power the digitalization of electricity infrastructure and systems are projected to significantly increase.[12] Human rights questions associated with the mining, processing, and distribution of energy transition minerals have intensified calls for a just transition that ensures that renewable energy projects do not exacerbate human rights violations, modern slavery, and forced labour in the extractive sector.[13] Infusing the design, financing, and implementation of DERs with sustainability safeguards will be essential to address the crosscutting social, environmental, and human rights questions in their current implementation across the world.

Despite the rise in infrastructure investments and commitments on distributed energy resources, corresponding legal and governance framework to advance sustainability in

Our Energy?' <https://www.europarl.europa.eu/RegData/etudes/ATAG/2020/651944/EPRS_ATA(2020)651944_EN.pdf> accessed 10 November 2024.

[6] See PR Shukla and others (eds), 'Climate Change 2022: Mitigation of Climate Change: Working Group III Contribution to the Sixth Assessment Report of the Intergovernmental Panel on Climate Change' <https:// www.ipcc.ch/ rep ort/ar6/wg3/downloads/rep ort/IPCC_ A R6_ W GIII-Ful lReport.pdf> accessed 1 May 2025.

[7] See D Olawuyi and others, *Net Zero and Natural Resources Law: Sovereignty, Security, and Solidarity in the Clean Energy Transition* (OUP 2024) 1–15.

[8] G20 (n 4).

[9] DERs are examples of the broad category of non-wires alternatives (NWAs) in electricity generation. NWAs include electricity grid investment or projects 'that use non-traditional transmission and distribution (T&D) solutions, such as distributed generation (DG), energy storage, energy efficiency (EE), demand response (DR), and grid software and controls'. See Navigant Research, *Non-Traditional Transmission and Distribution Solutions: Market Drivers and Barriers, Business Models, and Global Market Forecasts* (Navigant 2017); see also I Bantekas, 'The Off-Grid Revolution and the "Promise" of Energy Equality Under the Paris Climate Change Agreement' (2024) 23(1) Northwestern Journal of Human Rights 45–72.

[10] See eg, G Chazan, 'Bavarian village reaps renewable energy: Success of Bavarian "prosumers" creates dilemma for German electricity system' https://www.ft.com/content/4bd1ad70-793e-11e3-b381-00144feabdc0 accessed 27 October 2025; P Basak and others, 'A Literature Review on Integration of Distributed Energy Resources in the Perspective of Control, Protection and Stability of Microgrid' (2012) 16(8) Renewable and Sustainable Energy Reviews 5545–56.

[11] L Strezoski, 'Distributed Energy Resource Management Systems—DERMS: State of the Art and How to Move Forward' (2023) 12(1) WIREs Energy and Environment e460.

[12] See Olawuyi (n 7); also United Nations Working Group on Business and Human Rights (UN WGBHR), 'Extractive Sector, Just Transition and Human Rights' (2023) UN General Assembly Report A/78/155.

[13] UN WGBHR (n 12).

DER projects have yet to be exhaustively analysed and communicated. This chapter aims to fill a gap in this regard. It examines interrelated and mutually reinforcing bodies of international law that harness, incentivize, and govern the sustainable design, financing, and implementation of DERs.

This chapter is organized into five sections. Following this introduction, section II explores the scope and dimensions of sustainable DERs. It discusses international law principles and norms underpinning sustainable DERs, especially sustainable development, energy security, energy citizenship, and energy/environmental justice. Section III explores barriers that limit the coherent integration of these sustainability principles in DERs, most notably: the lack of international modalities and frameworks on distributed energy resources; the impact of unilateral trade tariffs and isolationism on the security of supply of minerals and materials needed to advance DER systems; the barriers to the transfer and absorption of grid decarbonization and digitization technologies; and the seemingly intractable issue of addressing human rights risks in the implementation of energy distribution projects. Section IV discusses integrative legal frameworks that are required to address extant barriers to sustainable DERs and to advance a coherent, holistic, and whole-society approach to DER implementation.

II. Scope and Dimensions of Sustainable DERs

Sustainable DERs has its roots in principles and concepts in international energy and environmental law, international climate change, and international human rights law that aim to advance the safe, orderly, sustainable, and efficient development and management of decentralized electricity generation, distribution, and storage for the benefit of current and future generations. Even since the Brundtland Report defined sustainable development as 'development which meets the needs of the present without compromising the ability of future generations to meet their own needs', the need to integrate the three pillars of sustainable development—social, economic, and environmental—in all aspects of planning has been crucial.[14] In the context of DERs, recent studies have underlined the importance of sustainable DERs as a framework for coherent, holistic and integrated planning to address crosscutting adverse environmental, social, governance, and human rights impacts of DER projects in a manner that avoids overlap, constraints, and tradeoffs.[15]

Thus, as discussed in Chapter 1 of this book, sustainable DERs refers to 'decentralised energy systems that balance social, economic, and environmental objectives, including: decarbonization, digitalization, and democratization—in a coherent and holistic manner'. Sustainable DERs is about developing net-zero aligned electricity systems that leave no one behind in terms of availability, affordability, accessibility, and citizen empowerment.[16] At the same time sustainable DERs emphasize the importance of integrating the three pillars

[14] J Mensah, 'Sustainable Development: Meaning, History, Principles, Pillars, and Implications for Human Action: Literature Review' (2019) 5(1) Cogent Social Sciences 1653531.

[15] S Gawusu, R Afriyie Mensah, and O Das, 'Exploring Distributed Energy Generation for Sustainable Development: A Data Mining Approach' (2022) 48 Journal of Energy Storage 104018.

[16] C Dadzie and others, 'Scaling Utility-Enabled Distributed Energy Resources in Nigeria: A Roadmap to Boost Distribution Company Revenues and Improve Power Availability and Reliability' (RMI, June 2024) 17 <https://rmi.org/insight/scaling-utility-enabled-distributed-energy-resources-in-nigeria/>, accessed 23 December 2024.

of sustainable development in the design, approval, financing, planning, and implementation of DERs to anticipate and address adverse impacts.[17]

The environmental pillar *(planet)* emphasizes the protection of natural capital (ie, land, air, water, minerals, etc) in DER design and implementation.[18] DERs can play a critical role in advancing the goals of the 2015 Paris Agreement which expressly sets the target to hold 'the increase in the global average temperature to well below 2°C above pre-industrial levels . . . and to limit the temperature increase to 1.5°C above pre-industrial levels'.[19] A 1.5°C aligned electricity sector will require: leveraging DER systems, particularly wind and solar, for decarbonization and GHG emission reductions to achieve net zero;[20] reducing all forms of environmental pollution and land degradation in DER implementation; and utilizing green supply chains in the procurement of goods and services for DER projects. Sustainable DERs should therefore reduce GHG emissions and environmental pollution in electricity generation, distribution, and storage while minimizing adverse impacts on lands and ecosystems. By so doing, DERs could play significant roles in advancing the right to a clean, healthy, and sustainable environment, a right recognized by the United Nations General Assembly, and by the constitutions of many countries.[21]

The social pillar *(people)* emphasizes the need to address social exclusions and human rights violations that may result from DER design and implementation.[22] For example, while DER implementation is gaining momentum in large cities and urban centres, residents of rural and peripheral areas including Indigenous peoples, and racialized and vulnerable communities face a high risk of exclusion from DER due to a lack of information, resources, and technologies needed to generate their own energy.[23] To be sustainable, the off grid revolution and energy transformation must address social exclusions that may result from location, gender, age, race, poor population density, and low income.[24] The need for states, business enterprises, and all other actors to address adverse human rights impacts in all economic activities and sectors has gained significant momentum since the adoption of the United Nations Guiding Principles on Business and Human Rights (UNGPs) in 2011.[25] The UNGPs build upon core international human rights instruments that emphasize the need for public participation, accountability, non-discrimination and equality, empowerment, and legality (the PANEL principles) in all development activities and projects.[26]

[17] Mensah (n 14).

[18] United Nations Environment Programme (UNEP) 'Nature Risk Profile: A Methodology for Profiling Nature Related Dependencies and Impacts' (UNEP 2023); see also DS Olawuyi and E Athwal, 'Net Zero, Human Rights, and ESG Risk Management' in Olawuyi (n 7).

[19] United Nations Framework Convention on Climate Change (UNFCC), The Paris Agreement <https://unfccc.int/sites/default/files/english_paris_agreement.pdf> accessed 1 March 2025.

[20] See International Renewable Energy Agency, *World Energy Transitions Outlook 2022*<https://www.irena.org/-/media/Files/IRENA/Agency/Publication/2022/Mar/IRENA_World_Energy_Transitions_Outlook_2022.pdf> accessed 1 March 2025.

[21] UN Resolution (A/76/L.75) on the human right to a clean, healthy, and sustainable environment.

[22] Olawuyi and Athwal (n 18); also P Śleszyński, 'How to Recognize the Threat of Social Energy Exclusion? Demographic and Settlement Determinants of Distributed Energy Development in Polish Towns and Cities' (2023) 98 Sustainable Cities and Society 10478.

[23] Śleszyński (n 22).

[24] ibid; also Bantekas (n 9).

[25] United Nations, 'Guiding Principles on Business and Human Rights: Implementing the United Nations "Protect, Respect and Remedy" Framework' (UNGPs) <www.ohchr.org/documents/publications/guidingprinciplesbusinesshr_en.pdf> accessed 12 May 2025.

[26] Olawuyi and Athwal (n 18), see also D Olawuyi, 'Corporate Accountability for the Natural Environment and Climate Change' in I Bantekas and MA Stein (eds), *Cambridge Companion to Business and Human Rights* (CUP 2021) 237–42.

With the growing adoption of national action plans and legislation across the world that highlight rights-based implementation of energy projects, environmental, social, and governance (ESG) risks relating to DERs must be carefully considered by DER operators. Sustainable DERs must evaluate, assess, and address risks of human rights violations and social exclusions in the penetration of DERs in order to better understand the resources and tools needed to support areas not covered. Social inclusion in DERs also includes addressing concerns of child labour and modern slavery in the sourcing of minerals and raw materials used across the DER value chain, especially in conflict affect regions. Also important are ensuring citizen empowerment as prosumers, consumer protection, gender justice, land justice, fair and equitable wages in DER operations, and eradicating fraudulent or inaccurate reporting, also known as greenwashing, in all aspects of the DER project.[27]

The economic pillar *(profit)* emphasizes the vital roles of DERs as tools for economic diversification, especially by advancing citizen-led and cost-saving energy solutions that address electricity deficits. Studies already indicate the potential of solar and wind DERs to provide cheaper energy when compared to traditional electricity generation technologies.[28] Furthermore, the projected rise of small and medium energy and eco-entrepreneurs (SMEEs) involved in the DER value chain could significantly generate livelihood and decent employment for the communities, while also empowering SMEEs to commercialize and market their energy inventions and technologies. However, the economic potential of DERs will depend on a wide range of factors including legal and regulatory protection for intellectual property rights associated with DER innovation, access of SMEEs to financing, tax incentives, and subsidy schemes to reduce the cost of installation of solar, wind, and other DERs, attractive compensation schemes, especially net metering (NM) and feed-in-tariffs (FIT) to stimulate more investment in DERs, and clear interconnection rules and regulations to integrate DERs.[29] Sustainable DERs should provide economic incentives to support the cost-effective implementation of DER programs and projects, including the protection of intellectual property rights associated with DER innovation, while also streamlining regulatory requirements for SMEE engagement in DER.

As seen in Table 2.1, sustainable DERs are underpinned by a wide range of international law instruments that aim to advance these social, environmental, and economic pillars of sustainable development. The core principles that underpin sustainable DERs are discussed next.

III. Principles and Norms Underpinning Sustainable DERs

Advancing sustainable DER is about integration, ie, integrating principles of sustainable development, energy citizenship, energy security, and energy justice in the decentralization, decarbonization, digitilization, and democratization of electricity generation, distribution, and storage in a manner that leaves no one behind. Despite some debates on the

[27] See UN WGBHR (n 12).

[28] E Oliver, 'The Economics of Integrating Distributed Energy Resources into the Electricity System. Distributed Generation—Pioneering the Future of Decentralized Energy' (2025) IntechOpen.

[29] ibid; see also R Valova and G Brown, 'Distributed Energy Resource Interconnection: An Overview of Challenges and Opportunities in the United States' (2022) 2 Solar Compass 100021.

legal status of some of these principles they are widely recognized and have become the bedrock of legislation in several jurisdictions.

A. Sustainable Development

As a foundational principle of international environmental law, sustainable development envisions a symbiotic balance between three essential and mutually reinforcing pillars: economic development, maintenance of a sound environment, and social development.[30] Building on the widely cited definition of sustainable development in the Brundtland report, Principle 3 of the Rio Declaration states that 'the right to development must be fulfilled so as to equitably meet developmental and environmental needs of present and future generations'.[31] Sustainable development therefore includes *intragenerational equity*, which requires us to balance decisions on electricity decentralization with decarbonization, democratization, and digitalization objectives in a manner that safeguards the energy needs of the current generation.[32] The *intergenerational* equity principle on the other hand requires us to balance decisions on electricity decentralization in a manner that safeguards the environment and advances a sustainable energy future for generations yet unborn.

Advancing sustainable DERs therefore requires a comprehensive process that places sustainability screening at the heart of the design, approval, financing, and implementation of DERs. Such screening will include environmental impact assessments to mitigate adverse environmental impacts and human rights due diligence to address adverse human rights impacts that may result from DER implementation. Achieving inter and intra generational equity will also require addressing barriers to the participation of women, youth, Indigenous peoples, and other marginalized groups in DER projects to ensure that current and future generations play lead roles in decision-making processes on DERs.[33]

B. Energy Security

One of the primary aims of international energy law is to advance energy security, ie the availability, affordability, and accessibility of energy resources, and reducing the vulnerability of vital energy infrastructure systems to natural and ecological disasters.[34] Given the crucial importance of energy to human life and activities, meeting the demand for energy resources and addressing disruptions in energy supply has been a crucial driver of energy law and policy across the world. In response to energy security threats, DERs have emerged

[30] See MC Cordonier Segger and D Olawuyi, *Sustainable Development Law: Principles, Practices, and Prospects* (2nd edn, OUP 2025) 1–10.

[31] United Nations, Rio Declaration on Environment and Development, UN Doc.A/CONF.151/26 (Vol I), 12 August 1992.

[32] M Aryanpour, 'Sources and Principles of Natural Resources Law and Policy in a Net Zero Era' in Olawuyi (n 7).

[33] ibid.

[34] T Morgandi and JE Viñuales, 'Energy Security in International Law' in R Geiß and N Melzer (eds), *The Oxford Handbook of the International Law of Global Security* (OUP 2021); also International Energy Agency ('IEA'), 'Energy Security' <https://www.iea.org/topics/energysecurity/> accessed 12 January 2020, defining energy security as 'uninterrupted availability of energy resources at an affordable price'.

as pivotal instruments that allow energy diversification by expanding energy generation and distribution beyond centralized systems.

By creating supportive legal and institutional frameworks to support the investments in alternative electricity generation, DERs can significantly contribute to bridging electricity deficits and addressing energy poverty. However, the potential contributions of DERs to energy security will to a great extent depend on putting in place policies and laws to address demand and supply sides of DER implementation. First, to encourage sustainable investments in DERs, legislation that introduce renewable energy performance standards will be essential. In India for example, the amendment of the Electricity Act to introduce mandatory Renewable Purchase Obligations (RPOs) and targets for the use of renewable energy has been key in driving investments in DERs.[35] Similarly, in Europe, the introduction of the Renewable Energy Directive (2009/28/EC), updated by the Revised Directive EU/2023/2413, has contributed to the increase in the share of renewable energy sources across EU countries from 12.5 per cent in 2010 to 24.5 per cent in 2023.[36] The Directive sets 'an overall renewable energy target of at least 42.5% binding at EU level by 2030—but aiming for 45%'.[37] Adopting legal frameworks to increase demand for electricity from renewable energy resources can significantly increase interest and investment in DER projects.

Second is the need to incentivize DER investments. Sustainable DERs must move beyond just legislating on DERs to providing actual deployment incentives that would make the initiative attractive from a cost and economic perspective. Such incentives can be in the form of grants or other direct financial support to reduce the initial expenditure and burden of DER infrastructure, or in the form of investment tax credits or reversed taxes to provide equivalent funding for the upfront capital investment in DER initiatives. As regulatory authorities develop legal and policy frameworks on DERs, such regulatory developments should be underpinned by dedicated fiscal incentives that will enhance the broad appeal of such incentives from practical and economic perspectives. A good example is the EU Innovation Fund which provides fiscal incentives for clean infrastructure projects.[38] Similarly, in the United States, the Energy Policy Act, 1992 established tax and other financial incentives such as the Renewable Electricity Production Tax Credit and the Business Energy Investment Tax Credit for facilities investing in renewable energy generation.[39] Such incentives will not only drive interest in DERs, but could also significantly lower the costs and enhance the affordability of DERs.

As legislation and policies to incentivize and support DERs grow in scale and ambition, balancing and managing demand and supply of electricity from DERs and their sustainable integration to existing electricity grids will be crucial. Surplus electricity, if not properly managed, could undermine the stability and reliability of the grid.[40] As earlier noted,

[35] Morgandi (n 34), see also Ministry of Power of India, Tariff Policy, Gazette of India, Part I, Section I, Resolution No 23/2/2005-R&R (Vol III) (6 January 2006) <http://www.orierc.org/documents/National%20Electricity%20Tariff%20Policy.pdf> accessed 15 March 2025.

[36] European Commission, Renewable Energy Directive <https://energy.ec.europa.eu/topics/renewable-energy/renewable-energy-directive-targets-and-rules/renewable-energy-directive_en> accessed 15 March 2025.

[37] ibid.

[38] European Union, Innovation Fund, <https://climate.ec.europa.eu/eu-action/eu-funding-climate-action/innovation-fund_en> accessed 12 May 2025.

[39] United States Environmental Protection Agency, 'Renewable Electricity Production Tax Credit Information' <https://www.epa.gov/lmop/renewable-electricity-production-tax-credit-information> accessed 12 May 2025.

[40] See for example, Chazan (n 10).

sustainable DERs would require investments in DERMS to ensure the safe, orderly, and sustainable sale and integration of electricity from decentralized sources into existing electricity infrastructure designed for centralized generation, distribution, and storage.

C. Energy Citizenship

A central driver of DERs is the need to advance energy citizenship. Until recently, policies aimed at addressing energy poverty tended to focus primarily on centralized electricity generation and distribution driven by governments and large energy conglomerates.[41] However, over the last decade studies have shown the importance of energy citizenship—'the active engagement of citizens, especially small and medium scale innovators and entrepreneurs', in promoting the availability, reliability, and affordability of energy resources—through innovative and decentralized solutions.[42] Underpinned by notions of participatory development and localism, energy citizenship advocates the development of citizen-led and locally made technologies, tools, and products that advance domestic energy policies and goals.[43] Energy citizenship therefore centres around citizen and consumer-driven energy generation, distribution, and supply initiatives, as opposed to a passive dependence on centralized energy systems controlled by governments and large corporations.[44]

The importance of energy citizenship has assumed greater significance and urgency with the adoption of the SDGs. SDG 17 specifically underscores the need for enhanced partnerships by various actors, including individuals, local community associations, and businesses to actively contribute towards achieving the various SDGs.[45] Energy citizenship places the citizens at the heart of the SDGs. In the context of advancing SDG 7 on energy for all, sustainable DERs aim to advance energy security by enabling decentralized and small-scale renewable energy projects that allow citizens, in individual or collective capacities, to supply electricity generated from renewable energy sources to extant electricity grid structures. The sustained flow of citizen-led energy generation and distribution will require putting in place supportive policies, legislation, and incentives that streamline the process of developing decentralized energy projects, as well as their financing and commercialization.[46] Through microgrids and community DER projects, citizens can move from being mere energy consumers to active electricity producers (prosumers) and energy entrepreneurs.

As Chapter 20 of this book demonstrates, sustainable DERs will therefore require addressing legal barriers that limit the ability of energy entrepreneurs and individuals to develop and launch DER solutions and ventures including regulatory complexities and delays in project approval and registration, as well as financial constraints. Advancing sustainable energy

[41] See P Devine-Wright, 'Community Versus Local Energy in a Context of Climate Emergency' (2019) 4 Nature Energy 894–96.
[42] See Olawuyi (n 1) 101–23; also Devine-Wright (n 41).
[43] Olawuyi (n 1) 101–23; see also O Renn, F Ulmer, and A Deckert, *The Role of Public Participation in Energy Transitions* (Elsevier 2020) 1–10.
[44] Olawuyi (n 1) 101–23.
[45] United Nations, Transforming Our World (n 3).
[46] Devine-Wright (n 42); Olawuyi (n 1).

access for all through DERs requires clear and coherent legal frameworks that advance energy citizenship and prosumerism. It would be difficult, if not impossible, to achieve the energy democratization objectives of DERs without such supportive legal and policy frameworks.

D. Energy and Environmental Justice

While energy decentralization through DERs aims to accelerate decarbonization and democratization, studies already highlight how the implementation of some DERs have increasingly been linked with compulsory acquisition of lands as project sites; forced displacements; social exclusions; concentration of projects in poor and vulnerable communities; governmental repression; and other complex human rights violations in many countries.[47] Debates on sustainable DERs therefore aim to ensure the fair, inclusive and equitable implementation of DERs in a manner that upholds the human rights of impacted peoples and communities and ensures that no one is left behind.[48]

The social pillar of sustainable DERs therefore draws on longstanding principles relating to climate justice, energy justice, and environmental justice which emphasize the rights-based design and implementation of energy policies with meaningful participation of all people regardless of income, race, colour, creed, or national origin.[49] By integrating human rights considerations in the implementation of DERs, DER activities and projects achieve wider legitimacy, while advancing a just and inclusive energy transition.[50] The imperatives for a just transition is already resulting in the integration of just transition clauses in energy and electricity contracts, legislation and policies.[51] Sustainable DERs must emphasize the responsibilities of investors, participants, and stakeholders in DERs to commit to just transition principles and to coherently anticipate and address gender justice, environmental justice, ESG, and other justice risks associated with DERs.

The foregoing principles and norms underline the importance of coherent and rights-based design, financing and implementation of DERs in a manner that leaves no one behind. Yet, as the next section demonstrates, piecemeal and unco-ordinated approaches to the policies and legislation on sustainable DERs have not provided the necessary regulatory clarity to drive such coherence.

IV. The Search for Coherence in Sustainable DERs Implementation

While the elements of sustainable DERs are robustly underpinned by a wide range of international instruments, several gaps from regulatory and enterprise governance standpoints

[47] Aryanpour (n 32); United Nations, 'Renewable Energy and the Right To Development: Realizing Human Rights For Sustainable Development' <https://www.ohchr.org/sites/default/files/2022-05/KMEnergy-EN.pdf> accessed 12 March 2025. See also C Gonzalez, 'Racial Capitalism, Climate Justice, and Climate Displacement' (2021) 11(1) Oñati Socio-Legal Series 108–47.

[48] Aryanpour (n 32); also UN WGBHR (n 12).

[49] Aryanpour (n 32); see also Guayo (n 1) 1–15.

[50] IRENA, 'A just and inclusive energy transition in emerging markets and developing economies: Energy planning, financing, sustainable fuels and social dimensions', International Renewable Energy Agency (Abu Dhabi, IRENA 2024).

[51] Aryanpour (n 32).

remain at international and domestic levels which will need to be addressed to effectively mainstream sustainability in the design, approval, financing, and implementation of DERs. This section explores barriers that limit the coherent integration of these sustainability principles in DERs, and the role of international law in addressing such gaps.

A. Lack of International Framework on DERs

As indicated in Table 2.1, the Paris Agreement expressly recognizes the need for countries to take actions, including mitigation, adaptation, finance, technology transfer, and capacity-building, to achieve the objective of reducing GHG emissions and limit global temperature increase to 1.5 °C above pre-industrial levels. While this recognition provides impetus for the decentralization of energy sources to integrate renewable energy, the lack of clear and comprehensive guidelines or a roadmap at the international level for its sustainable implementation remains a practical gap.[52] The lack of an international framework on DERs means that countries and sub-regional authorities are implementing energy decentralization programs according to their own local priorities and contexts, especially in the context of their nationally determined contributions (NDCs).[53] Given the different levels of ambition and commitments in the NDCs, coherent global standards on DERs are yet to emerge. The implication is the proliferation of implementation standards with respect to grid interconnection, interoperability, and digitalization such that it requires sifting through a maze of what represents the international best practice on DERs. For example, in the United States, industry bodies have come forward with technical guidelines that could guide the standardization of DER implementation, including interconnection with distribution grids.[54] However, such standards mainly focus on technical and interconnection aspects, with little or no elaboration of the social and environmental dimensions of DERs. While these efforts can help streamline DER methodologies and requirements, there is also a need for international standard-setting under the international climate change framework to promote a common language and multijurisdictional methodology on DER implementation, verification, and reporting.

To promote transparency in the use of DERs for climate mitigation under the Paris Agreement, it is imperative to provide a set of common guidelines and modalities on DERs. Such international guidance will go a long way to provide a common understanding on the potential and benefits of DERs to GHG emission reductions; energy efficient DER technologies and the technical standards for their application; social, environmental, and economic aspects and risks of DER implementation; and risk mitigation strategies including

[52] IRENA (n 50); also, H Carr, 'Distributed Energy Resources: what we learned from regulators about managing the energy transition in Africa' (24 October 2024) <https://crossboundaryenergy.com/regulators-managing-the-energy-transition-in-africa/> accessed 12 March 2025.

[53] See for example, Ontario Energy Board (OEB), *Framework for Energy Innovation: Setting a Path Forward for DER Integration* (30 January 2023) <https://engagewithus.oeb.ca/fei/news_feed/the-oeb-has-released-the-framework-for-energy-innovation-setting-a-path-forward-for-der-integration-2> accessed 12 March 2025.

[54] T Basso, *IEEE 1547 and 2030 Standards for Distributed Energy Resources Interconnection and Interoperability with the Electricity Grid* (US National Renewable Energy Laboratory, December 2014) <https://www.nrel.gov/docs/fy15osti/63157.pdf> accessed 12 March 2025; see also National Renewable Energy Laboratory, 'Using Distributed Energy Resources: A How-To Guide for Federal Facility Managers' <https://www.nrel.gov/docs/fy02osti/31570.pdf> accessed 12 March 2025.

Table 2.1. Core international instruments on sustainable DERs

Pillars of Sustainable DERs	Core International Instruments	Key provisions
The Environmental Pillar Sustainable DERs should reduce GHG emissions and environmental pollution in electricity generation, distribution and storage, while minimizing adverse impacts on lands and ecosystems.	United Nations Framework Convention on Climate Change 1992	The ultimate aim of the Convention, as set out in Article 2, is to reduce the emission of greenhouse gases (GHGs) that cause climate change.
	Paris Agreement, 2015	The Paris Agreement sets the target of 'holding the increase in the global average temperature to well below 2°C above pre-industrial levels and pursuing efforts to limit the temperature increase to 1.5°C above pre-industrial levels.' Article 6 (8) emphasises the need for 'integrated, holistic and balanced non-market approaches' in advancing climate objectives, which will include innovative DERs.
	<u>Soft law instrument</u> UN Resolution (A/76/L.75) on the human right to a clean, healthy and sustainable environment	This landmark UN Resolution 'calls upon States, international organizations, business enterprises and other relevant stakeholders to adopt policies' and measures to ensure a clean, healthy and sustainable environment for all.
	United Nations Sustainable Development Goals	SDG 7: Ensure 'access to affordable, reliable, sustainable and modern energy for all.' Target 7.2: 'By 2030, increase substantially the share of renewable energy in the global energy mix.' SDG 13: 'Take urgent action to combat climate change and its impacts' by 2030.
The Social Pillar Sustainable DERs should address ESG risks, particularly social exclusions and human rights violations that may result from DER design and implementation	Paris Agreement, 2015	The Paris Agreement notes in its preamble, that 'Parties should, when taking action to address climate change, respect, promote and consider their respective obligations on human rights.'
	Nine core international human rights instruments, as well regional instruments, recognize the need to address all forms of social exclusions in development activities:	These international human rights instruments emphasize the need for public participation, accountability, non-discrimination and equality, empowerment and legality (the PANEL principles) in all development activities and projects, which will include DER projects.

(*continued*)

Table 2.1. Continued

Pillars of Sustainable DERs	Core International Instruments	Key provisions
	International Covenant on Civil and Political Rights (ICCPR), Dec. 19, 1966	
	International Covenant on Economic, Social and Cultural Rights, (ICESCR) Dec. 16, 1966	
	International Convention on the Elimination of All Forms of Racial Discrimination, Mar. 7, 1966, 660 U.N.T.S. 195 (CERD)	
	Convention on the Elimination of All Forms of Discrimination against Women (CEDAW), Dec. 18, 1979	
	Convention against Torture and Other Cruel, Inhuman or Degrading Treatment or Punishment (CAT) Dec. 10, 1984, 1465 U.N.T.S. 85	
	Convention on the Rights of the Child, (CRC) Nov. 20, 1989, 1577 U.N.T.S. 3	
	International Convention on the Protection of the Rights of All Migrant Workers and Members of Their Families (CMW), Dec. 18, 1990, 2220 U.N.T.S. 3	
	Convention on the Rights of Persons with Disabilities (CRPD), Dec. 13, 2006	
	International Convention for the Protection of All Persons from Enforced Disappearance (CED) 23 December 2010	
	Regional level	
	Regional Agreement on Access to Information, Public Participation and Justice in Environmental Matters in Latin America and the Caribbean (Escazú Agreement) (entered into force 22 April 2021)	

Table 2.1. Continued

Pillars of Sustainable DERs	Core International Instruments	Key provisions
	African Charter on Human and Peoples Rights	
	American Convention on Human Rights, the 'Pact of San Jose, Costa Rica' Soft law instrument The United Nations Guiding Principles on Business and Human Rights (UNGPs) 2011	The UNGPs a set of 31 principles that provide a framework for states and businesses to prevent and address human rights abuses linked to business activity, including in the energy and extractive sectors.
The Economic Pillar Sustainable DERs should unlock economic diversification, especially by advancing the integration of citizen-led energy solutions that address electricity deficits while promoting trade and investments, green economic opportunities, decent employment, cost savings and revenue for the public.	Marrakesh Agreement Establishing the World Trade Organisation (WTO)	Preamble to the Marrakesh Agreement recognizes that the primary objectives of trade and economic relations shall be to raise standards of living, 'while allowing for the optimal use of the world's resources in accordance with the objective of sustainable development'
	Trade-Related Aspects of Intellectual Property Rights (TRIPS) (1 January 1995)	Article 7 of TRIPS emphasizes the need for the protection and enforcement of intellectual property rights to enhance innovation.
	Regional level The Energy Charter Treaty (ECT), and the *Energy Charter Protocol on Energy Efficiency* and Related Environmental Aspects (PEEREA)	The ECT, and Article 1 (2) of the PEEREA emphasize energy efficiency and sustainable development as core objectives.
	Agreement Establishing the African Continental Free Trade Area (AfCFTA) ECOWAS Energy Protocol 1986 Agreement on ASEAN Energy Cooperation	These regional agreements recognize the importance of regional energy cooperation and trade to advance sustainable development. DERs will play key roles in boosting regional trade in critical materials needed for clean energy projects.
	Soft law instruments Rio Declaration on Environment and Development, 1992	Principle 10 emphasizes the importance of localism and public participation, noting that 'environmental issues are best handled with the participation of all concerned citizens, at the relevant level.'

Source: Compiled by Author.

the requirements for human rights due diligence, environmental impact assessments, and gender vulnerability screening. Such modalities should also elaborate on the responsibilities of parties, especially developed countries, to take the lead in increasing access to DER technologies and financing required to promote the use of DERs.

B. Trade Isolationism and the Weakening of the Multilateral Trade System

The implementation of DERs will to a large extent depend on the availability of raw materials, environmentally sustainable technologies (ESTs), and goods, including transition minerals needed to manufacture solar panels, batteries, and other key components of DERs.[55] For many years, international trade law and the multilateral trading system have been widely developed to address barriers to the seamless flow of goods, services, and ESTs across borders. However, the rise of unilateral trade tariffs, resource nationalism, wars and conflicts, and the increased weakening of multilateralism over the last decade pose great challenges to the free flow of transition minerals and technologies required to drive DER projects.[56] A good example is the implication of the unilateral announcement of trade tariffs by US President Donald Trump and its reverberating impacts for clean energy transition, which is projected to cut the US and other countries off from technologies needed to drive energy efficiency and DERs.[57] Given China's dominant role as the manufacturer on nearly all the components of solar panels and cells, such isolation and any retaliatory tariffs by China and other countries will not only limit access to DER technologies and transition minerals, but could significantly weaken international solidarity and collaborative efforts needed to advance sustainable DERs.[58]

Article II of the General Agreement On Tariffs And Trade 1994 (GATT 1994) requires that members of the WTO must accord to the goods of other Member's 'treatment no less favourable than that provided for' in its Schedule.[59] As the WTO Appellate Body found in the *Argentina—Footwear* case, 'the application of customs duties in *excess* of those provided for in a Member's Schedule, inconsistent with the first sentence of Article II:1(b), constitutes "less favourable" treatment under the provisions of Article II:1(a)'.[60] The unilateral imposition of trade tariffs in excess of the agreed tariff concession limits under GATT is therefore a potential violation of WTO rules, which is expected to result in a flurry of cases before the WTO's Dispute Settlement Body (DSB). The ability of the WTO SDB and the WTO to clarify these questions in a timely, transparent, and effective manner will be

[55] Aryanpour (n 32).

[56] S Aiyar and others, *Geo-economic Fragmentation and the Future of Multilateralism* (International Monetary Fund 2023) 1–10; Aryanpour (n 32).

[57] C Edwards, 'US tariffs will make global trade shrink, says WTO' (BBC News, 16 April 2025); see also B Arris, 'What impact will Trump's tariffs have on US' Green Transition Progress?' (21 April 2025) https://www.intellinews.com/what-impact-will-trump-s-tariffs-have-on-us-green-transition-progress-377305/ accessed 18 October 2025.

[58] N Gordon, 'How the U.S.-China Trade War Could Derail the Energy Transition' (17 April 2025) https://carnegieendowment.org/emissary/2025/04/us-china-trade-war-tariffs-critical-minerals-clean-energy-impacts accessed 18 October 2025.

[59] Article II of the General Agreement on Tariffs and Trade 1994 (GATT 1994) https://www.wto.org/english/res_e/publications_e/ai17_e/gatt1994_art2_oth.pdf accessed 18 October 2025.

[60] Appellate Body Report on *Argentina—Measures Affecting Imports of Footwear, Textiles, Apparel, and Other Items* (hereafter 'Argentina—Footwear'), adopted 22 April 1998, WT/DS56/AB/R, para 47.

crucial given the time sensitive nature of energy transition related disputes.[61] Debates over the effectiveness and future relevance of the WTO SDB means that diplomatic solutions will also be required to address tariff barriers to the flow of clean technologies in an area of unilateralism.[62]

The increasing fragmentation and weakening of multilateral trade calls for a new form of trade partnership that prioritizes the urgency of climate change and fosters the required collaboration to streamline the flow of ESTs across borders. With the US announcing its decision to freeze financial contributions to the WTO,[63] there is an urgent need for the WTO, UNCTAD, and other trade related organizations to embark on comprehensive trade diplomacy and reforms aimed at reinstilling confidence and trust in the multilateral trading system.

C. Legal Barriers to Technology Absorption

While DERs aim to advance energy citizenship, one of the key barriers to DER implementation in many parts of the world, especially in developing countries is the local unavailability of technologies and materials, including smart digital solutions required to advance DER projects. In Nigeria for example, the lack of homegrown facilities to develop solar panels has resulted in a continuous dependence on the importation of solar panels, at above market rates. Estimates indicate that in 2023 alone, Nigeria imported over four million solar panels at the cost of more than $200 million.[64] Import-related costs hike the prices of solar panels making it less affordable to businesses and households, especially in poor and underserved communities. Furthermore, solar panels designed for other countries and climates may not meet local specifications and requirements, especially weather conditions, which may result in their sub-optimal performance and quality control challenges in local contexts.[65]

Advancing sustainable DERs will require a transformational shift from a one-track focus on technology importation to technology absorption, that is 'the process of learning to understand, utilise and replicate technology, including the capacity to choose it and adapt it to local conditions and to integrate it with indigenous technologies'.[66] In addition to a lack of sustained investments in clean technology entrepreneurship to promote home grown development of DER solutions, legal barriers to technology absorption must also be carefully addressed. First is the weak legal protection for intellectual property rights (IPRs)

[61] Generally, a WTO dispute takes two to three years for adoption and consequently another 300 days for securing compliance after the adoption, raising concerns over delays especially given the time sensitive nature of the energy transition. See A Reich, 'The Effectiveness of the WTO Dispute Settlement System: A Statistical Analysis' EUI Working Papers LAW 2017/11, 18.

[62] ibid. See also G Dimitropoulos, 'Investor-State Dispute Settlement Reform and Theory of Institutional Design' (2018) 9(4) Journal of International Dispute Settlement 535–69.

[63] J Bacchus, 'Another Misguided US Attack on the World Trade Organization' https://www.cato.org/blog/another-misguided-us-attack-world-trade-organization accessed 18 October 2025.

[64] Premium Times, 'Reducing solar panel importation: A path towards sustainable energy in Nigeria' (11 February 2025) https://www.premiumtimesng.com/promoted/773588-reducing-solar-panel-importation-a-path-towards-sustainable-energy-in-nigeria.html accessed 18 October 2025.

[65] Premium Times (n 64).

[66] See D Olawuyi, 'From Technology Transfer to Technology Absorption: Addressing Climate Technology Gaps in Africa' (2018) 36(1) Journal of Energy & Natural Resources Law 61–84.

in many countries which serves as a barrier to technology deployment and absorption.[67] Previous studies document how the lack of high quality patent and IPR systems continues to hinder clean technology innovation in Africa.[68] Similarly, the prohibitive cost of patent registration, especially as seen in many African countries, continues to serve as a barrier to patent registration and innovation resulting in almost total dependence on technology importation.[69]

Furthermore, the prevalence of legislative provisions that allow state authorities to appropriate or make use of a patented product or process without authorization or compensation to the patent rights holder remains a key challenge in many countries. At the international level, Article 31 of the Agreement on Trade-Related Aspects of Intellectual Property Rights (TRIPS) clearly provides some safeguards against the use of a patented product or process without authorization or compensation to the patent rights holder.[70] However, lack of clear recognition of such safeguards in domestic legislation has raised concerns over the potential for unfair use of patented renewable energy technologies and inventions in countries with lax or incomprehensive intellectual property (IP) protection regimes.[71] A comprehensive legal framework on IP is required to address these and other legal barriers to the development and commercialization of DER technologies and innovation in order to accelerate homegrown development. In addition to strengthening the level of protection accorded to IPR and addressing the prohibitive costs of patent registrations, there is also a need to modernize IP institutions and regulatory structures to make them fit for purpose, and to 'give confidence to inventors that transferred technology will be protected from arbitrary confiscation or abuse'.[72]

Considering the importance of technology for climate mitigation and adaptation under the Paris Agreement, the Technology Mechanism has a key role to play in providing technical assistance and support to developing countries to enhance their absorptive and deployment capacity for solar, wind, batteries, and other DER technologies.[73] Article 10(4) of the Paris Agreement mandates the Technology Mechanism to promote and facilitate enhanced action on technology development and transfer. Achieving this mandate will require providing technical knowledge and financial assistance, especially to developing countries, to reform their patent, IPR, and technology development systems in line with international best practices.[74]

[67] United Nations Environment Programme, Patents and Clean Energy Technologies in Africa (United Nations Environment Programme, Division of Environmental Law and Conventions 2013) 7–8; also A Abdel-Latif, 'Intellectual Property Rights and the Transfer of Climate Change Technologies: Issues, Challenges, and Way Forward' (2015) 15 Climate Policy 103.

[68] See Olawuyi (n 66); also United Nations, *The Role of Intellectual Property Rights in Promoting Africa's Development: Overview of IPR in Africa*, <https://www.un.org/osaa/sites/www.un.org.osaa/files/final_policy_paper_on_iprs_in_africa_fin_en_230822_v56883.pdf> accessed 21 May 2025.

[69] United Nations (n 68).

[70] WTO, *Agreement on Trade-Related Aspects of Intellectual Property Rights* https://www.wto.org/english/docs_e/legal_e/27-trips.pdf accessed 11 September 2025.

[71] Olawuyi (n 66).

[72] ibid. See also I Mgbeoji, 'African patent offices not fit for purpose' in J DeBeer and others (eds), *Innovation & Intellectual Property: Collaborative Dynamics in Africa* (University of Cape Town Press 2014).

[73] See arts 10(2)–(4) of the Paris Agreement.

[74] See art 11 of the Paris Agreement noting that capacity building should facilitate technology development, dissemination, and deployment in developing countries.

D. Weak Legal and Regulatory Support for Clean Technology Entrepreneurship

The fundamental promise of DERs is to promote energy democratization by empowering energy citizens and SMEEs to be actively involved in the manufacture of solar panels, wind turbines, as well as in electricity generation and commercialization processes. To achieve this objective, a clear and comprehensive legal framework to incentivize and promote clean technology entrepreneurship is essential.[75]

Yet, regulatory uncertainty and the lack of supportive policies and legislation to promote clean technology entrepreneurship remains a barrier in developing countries.[76] For example, in the Global Entrepreneurship Index (GEI) virtually all countries in Africa and the Middle East rank very low in terms of providing supportive legal and regulatory frameworks on clean technology entrepreneurship, with a few exceptions such as Oman and Qatar.[77] The GEI reflects challenges in these and other regions in terms of creating a supportive legal order that simplifies the licensing and registration of energy ventures.[78] Similarly, the lack of dedicated financing and incentives, such as innovation funds, grants, and low-interest loans, to alleviate the high upfront transaction costs of DERs remains a barrier to DER implementation.[79] An important starting point, therefore, is for countries to address these and other pre-existing barriers to financing, formalization, and registration faced by private energy producers.

In addition to establishing robust legal protection for innovation and intellectual property, the legal and regulatory elements of a supportive entrepreneurial environment include 'establishing simplified processes for starting or registering a business venture; open, predictable, and business friendly legislation or policies; tax incentives; flexible entry and exit options for investors; and access to fair and timely dispute resolution mechanisms'.[80] This will include addressing delays, costs, and administrative barriers associated with registering or commercializing DER innovation.[81] For example, in 2023, South Africa streamlined the approval requirements for DER projects, while completely removing the requirement for licensees to operate a generation facility.[82] Relaxing the licensing rules has helped to simplify and incentivize the DER approval processes.[83] The South African experience can provide

[75] H Cao and others, 'Does Legislation Promote Technological Innovation in Renewable Energy Enterprises? Evidence from China' (2024) 188 Energy Policy 114111. See also Global Entrepreneurship and Development Institute (GEDI) 2019, 'Global Entrepreneurship Index 2019' <http://thegedi.org/global-entrepreneurship-and-development-index/> accessed 10 March 2025.

[76] Olawuyi (n 1); see also O Agu and K Onasoga, 'Navigating the regulatory landscape for distributed solar power projects in Nigeria' (International Bar Association, 6 November 2024) <https://prod-bo.ibanet.org/regulatory-landscape-solar-power-nigeria> accessed 27 October 2025; see also P Onuh and others, 'Challenges and Opportunities in Nigeria's Renewable Energy Policy and Legislation' (2024) 23(2) World Journal of Advanced Research and Reviews 2354–72.

[77] Global Entrepreneurship Index 2019 (n 75). See also Olawuyi (n 1).

[78] ibid, see also World Bank, 'Doing Business 2020: Middle East and North Africa', which notes significant reform and progress in some countries such as Qatar and the UAE, but administrative delays and regulatory burdens in several others, available at https://espanol.doingbusiness.org/content/dam/doingBusiness/media/Profiles/Regional/DB2020/MENA.pdf accessed 9 May 2025.

[79] Agu and Onasoga (n 76); see also Chapter 20 of this book.

[80] Global Entrepreneurship Index (n 75); see also Olawuyi (n 1).

[81] Olawuyi (n 1).

[82] M Richards, 'South Africa Exempts Private Generators from Generation Licence Requirements' (White & Case 2023) <https://www.whitecase.com/insight-alert/south-africa-exempts-private-generators-generation-licence-requirements> accessed 18 October 2025.

[83] Carr (n 52).

lessons for other countries on the legal approaches required to streamline DER implementation, and to incentivize and support clean technology innovation.

E. Capacity Gaps

Implementing sustainable DERs will depend largely on the abilities of individuals, businesses, and energy entrepreneurs to effectively anticipate and address social, environmental, and economic risks associated with DER projects. Furthermore, public awareness and education on DERs will be essential to avoid local community opposition to DER projects. This includes training and capacity development on DER technologies, environmental regulations, pollution management, land ownership, human rights safeguards, and ESG risk management in DER implementation.[84] It is equally essential to empower national authorities and monitoring entities with the skills, resources, and tools necessary for enforcing sustainability and human rights standards in project implementation.

Due to a lack of training, equipment, and data-gathering technologies, many DER operators and regulators, especially in developing countries, struggle to properly anticipate and address sustainability concerns in DER implementation.[85] For example in Nigeria, the lack of capacity to manufacture, assemble, and maintain solar facilities has increased the transaction costs of DERs making it unaffordable to the most underserved communities.[86] Similarly, across Africa, studies have highlighted how due to a lack of training, equipment, and tools, regulators struggle to monitor compliance with sustainability standards in DER implementation.[87] Furthermore, the lack of statistical and data-gathering technologies and tools weaken the abilities of regulators in a number of countries to transparently collate, evaluate, and process data relating to DER implementation in a manner that can instil public confidence on its overall contributions and effectiveness.[88]

Closely related is the lack of sustained investment in research and development (R&D) on renewable energy, especially with respect to DERs.[89] Given the crucial importance of homegrown innovation to reducing the deployment costs of DERs, it is imperative for countries, higher education institutions and business enterprises to devote more resources to R&D efforts on DERs and renewable energy technologies. To bridge capacity gaps, higher education institutions have key roles to play in designing innovative training and research programs that provide skills and knowledge acquisition opportunities for innovators, financial institutions, regulators, and other stakeholders involved in DER implementation.

The next section discusses integrative and good governance frameworks through which state and non-state actors can better advance sustainable DERs needed to accelerate a low carbon and sustainable energy future.

[84] Olawuyi and Athwal (n 18).
[85] Carr (n 52).
[86] Onuh (n 79); also D Abdullahi and others, 'Key Barriers to the Implementation of Solar Energy in Nigeria: A Critical Analysis (2017) IOP Conf Ser Earth Environ Sci 83 012015.
[87] Carr (n 52).
[88] ibid; also Abdullahi (n 86); Agu and Onasoga (n 76).
[89] Abdullahi (n 89); Agu and Onasoga (n 76).

V. Integrative Legal Frameworks for Advancing Sustainable DERs

As earlier discussed, sustainable DERs is about systemic integration: integrating and harmonizing institutions, information sharing, policy, and investment decision-making relating to DER through a co-ordinated and holistic approach that minimizes environmental, social, and economic risks. As the case studies in this book demonstrate, the current piecemeal implementation of DERs has not fostered coherence in terms of anticipating, balancing, and addressing social, economic, and environmental risks in DER projects. Dynamic legal innovation is required at international, regional, and national levels to promote greater coherence in the design, approval, and implementation of DERs.

At the international level, there is a need for greater clarity on the guiding principles and modalities for the sustainable and cost-effective implementation of DERs in the context of tackling the climate emergency and promoting a sustainable energy future. Article 13 of the Paris Agreement recognizes the need for building transparency-related capacity to support climate action, especially in developing countries. Providing technical support and guidance on the effective deployment, integration, and use of DERs for climate action will be crucial to promoting transparent implementation in accordance with the Paris Agreement. In addition to providing a working definition of DERs in the context of climate action, such technical guidance should clarify issues of interconnectivity, reliability, and verification of GHG reductions; measuring the costs and benefits of DER use relative to traditional electricity distribution; financing and incentivizing DERs, as well as addressing adverse environmental and human rights impacts through human rights due diligence tools. Establishing guidance on appropriate reporting requirements for generators, distributors, and other stakeholders in DER projects will also enhance predictability and clarity on the pace of DER implementation. Guidance on how countries and sub-national authorities can monitor and report on issues relating to cost-effectiveness, consumer confidence and satisfaction, including supportive regulations to promote clean technology entrepreneurship, and providing a level playing field for accessing finance, support, and knowledge on DER will be crucial. Such clarity will go a long way in promoting greater coherence while also promoting greater public confidence that DER integration and use will progress in a just and rights-based manner.

Furthermore, there is a need to address the increasing fragmentation and weakening of multilateral trade systems in order to foster the trade partnerships needed to advance DERs. Reform of the WTO and its dispute settlement system to promote greater transparency and effectiveness will be crucial for rebuilding trust in trade multilateralism.[90] Also, regional trade agreements (RTAs) can help like-minded countries build on existing trade co-operation to promote horizontal and collaborative policies targeted at addressing barriers to environmental goods and services and ESTs required to accelerate DERs.[91] For example, the growing focus on regional energy interconnectivity in Asia-Pacific Economic Cooperation (APEC) and Gulf Cooperation Council (GCC) countries can provide impetus for promoting regional trade agreements that simplify the exchange of goods, services, and technologies that are crucial to DERs. Regional knowledge exchange could also

[90] See D Olawuyi, 'Environment' in I Van Damme and others (eds), *Oxford Handbook of International Trade Law* (2nd edn, OUP 2022) 692–93.
[91] ibid.

enable experience sharing on DERs, thereby fostering trade co-operation and solidarity for DER implementation. Such a regional approach could gradually help to reopen greater interest in trade multilateralism in an era of isolationism.

Second is the need for comprehensive and supportive legislation on DERs at the domestic level. The lack of standardization in DER implementation and reporting is directly linked to the lack of clear national strategies and regulatory standards on DERs. While a few countries have announced national strategies and legislation aimed at regulating DERs, such efforts are not widespread especially across Africa, Asia, Latin America, and the Middle East.[92] The absence of supportive and comprehensive laws on DERs hinders clean technology entrepreneurship and innovation needed to drive DERs. DER legislation is vital to providing financial incentives to support DERs and at the same time establish performance standards relating to DER use, especially in public institutions and facilities while stipulating penalties and sanctions in case of non-performance. Such performance standards will greatly drive demand for DERs, providing compensation and reward for DER investments. Such legislation should also place DER project planners under obligation to integrate human rights into natural investments and projects. Such legislation can standardize key measures for project design and approval, including permitting, licensing, efficiency standards, and financing options, as well as standardize reporting requirements to promote transparent, accessible, and timely reporting.[93] It should also clarify the integration of DERs in electricity distribution and establish incentives and support for utilities, Independent System Operators (ISOs), distributors, and customers to effectively implement all aspects of DERs and other NWAs.[94] Legislative clarity could enable efficient evaluation of DER programs and progress tracking based on set timelines. Such legislation should also clarify the roles and responsibilities of different government entities and institutions to avoid role duplication and promote greater synergy across the environment, human rights, social development, and justice institutions.

Third is the need for DER investors, operators, and industry bodies to incorporate sustainability standards in all aspects of project planning and risk management. As earlier noted, while some technical industry guidelines on DER are emerging, focus on sustainability has been lacking.[95] It is imperative for industry associations and bodies, such as the International Organization for Standardization, International Energy Agency, and World Energy Council amongst others to revise and update guidance documents on DERs to specifically address issues of mitigating social, environmental, and economic risks in implementation. Sustainability aligned standards for DER projects should focus on material risk

[92] Examples include South Africa's amendments to s 3 of Sch 2 of the Electricity Regulation Act 4 of 2006, specifically focused on DER implementation; Argentina's Law no 27424 creating the Promotion Regime for Distributed Generation of Renewable Energy Integrated in the Public Electricity Grid; Ghana's License Manual for Service Providers in the Renewable Energy Industry 2012 and Guidelines for Distributed Energy Generation 2023; Philippines' Energy Regulatory Commission, 'Governing Distributed Energy Resources', ERC Resolution No 11, Series of 2022; Portugal's Decree-Law No 15/2022, of 14 January 2022. For examples of DER legislation in Latin America, see A Bernal and others, 'Bottom-up energy transitions: Managing the rise of energy communities in Latin America' <https://www.iea.org/commentaries/bottom-up-energy-transitions-managing-the-rise-of-energy-communities-in-latin-america> accessed 8 August 2025.

[93] See Olawuyi and Athwal (n 18).

[94] See for example, Ontario Energy Board (OEB), *Framework for Energy Innovation: Setting a Path Forward for DER Integration* (30 January 2023) <https://engagewithus.oeb.ca/fei/news_feed/the-oeb-has-released-the-framework-for-energy-innovation-setting-a-path-forward-for-der-integration-2> accessed 12 March 2025.

[95] Basso (n 54); National Renewable Energy Laboratory (n 54).

identification, risk control, and information dissemination with respect to how actual or potential adverse sustainability risks are anticipated and addressed from DER design to decommissioning stages.[96] Such industry guidance should also establish benchmarks and standards for monitoring and tracking the success of DERs in the context of addressing energy poverty and advancing climate action, using high-quality statistical data and verification methods. Furthermore, tailored training programs and knowledge exchange on due diligence, risk mitigation, and financing of DER technologies and infrastructure should be actively designed and promoted by energy industry regulators and industry bodies to address current capacity gaps with respect to DERs. Professional associations, such as the International Bar Association and the Association of Environmental Law Lecturers in Middle East and North African Universities (ASSELLMU), in collaboration with higher education institutions, have key roles in promoting courses, certifications, and knowledge exchange platforms to provide awareness on embedding sustainability in DER implementation.

VI. Conclusion

As DERs grow in scale and ambition across the world, an integrative framework and approach will be central to the sustainable design and implementation of DERs. Advancing sustainable DERs will require fostering international and regional trade co-operation, promoting clean technology entrepreneurship and technology absorption at domestic levels, and establishing national plans, legislation, industry benchmarks and standards to address social, human rights, and environmental risks in DER design and implementation.

Legal and institutional barriers that stifle the development of a coherent DER market must be comprehensively addressed to ensure risk anticipation and mitigation in DER projects. The lack of an international framework and guidance on DERs in exacerbated by the absence of robust legal and institutional frameworks on DERs at domestic levels. The slow pace of standardization has not fostered the sharing of best practices and capacity development tools needed to coherently integrate DERs in policy and practice. Furthermore, regulatory clarity on the interconnection of DER with distribution grids is required to incentivize DER penetration and growth.

The regional sharing of expertise, knowledge, and best practices on DERs could foster the exchange of knowledge, resources, and best practices to advance regulatory development on sustainable DERs. For example, lessons from countries such as South Africa that have introduced legislative changes on DERs and the decarbonization of the electricity system can inform other countries on the legal reforms needed to achieve a safe, orderly, and sustainable shift towards sustainable DERs. Industry associations, regional bodies, and higher education institutions have key roles to play in fostering such knowledge exchange, capacity development, and further research on the integration of sustainability standards in all aspects of DER planning and implementation.

[96] Olawuyi and Athwal (n 18).

3

From Coal to Distributed Energy

Sustainable Decarbonization and Decentralization of Electricity Systems

Hanri Mostert, Jacques Jacobs, and Wikus Kruger

I. Introduction

As the global energy landscape shifts towards sustainability, it becomes crucial to understand the legal reforms necessary to facilitate the transition from coal-generated energy to renewable energy sources. For coal-reliant economies in particular, resistance is to be expected, not just from incumbents in the energy markets, but also from institutional habits, entrenched infrastructural interests, and vested loyalties to national energy icons. These habits, interests, and loyalties are likely as emotional as they are political and economic. This chapter focuses on repurposing and replacing the legislative frameworks that support coal-fuelled energy. It entails a fundamental shift in legislative, political, and institutional logics, as well as a commitment from a centralized system to one based on distributed energy resources (DER).

Energy transitions must address both infrastructure and the cultural narratives that sustain coal's hold.[1] Across diverse jurisdictions, gaps in the legal framework—alongside economic and political interests, institutional path dependencies, and the entrenched stability of incumbent energy regimes—continue to slow the decarbonization and decentralization imperatives.

Some countries (eg, France and parts of the US) exhibit stronger legal frameworks and cultural or normative commitment to the transition. However, they still face economic or political resistance, especially from industrial actors or grid incumbents.[2] On the other hand, countries like South Africa, Indonesia, and Russia are characterized by deeper political and emotional entanglements with centralized or fossil-fuel-based energy.[3] For clarification, 'emotional' attachment to coal means more than mere nostalgia. It reflects how

[1] BK Sovacool and MH Dworkin, 'Energy Justice: Conceptual Insights and Practical Applications' (2015) 142 Applied Energy 435.

[2] G Bang, 'The United States: Conditions for Accelerating Decarbonisation in a Politically Divided Country' (2021) 21(1) International Environmental Agreements: Politics, Law and Economics 43–58; S McDaniel and D Bailey, 'Winning the Transition: Strategic State Action in France and Germany Amid the Low Carbon Transition and Energy Shock' (2025) 29(2) Competition & Change 225, 232.

[3] I Overland and J Loginova, 'The Russian Coal Industry in an Uncertain World: Finally Pivoting to Asia?' (2023) 102 Energy Research & Social Science 103150; R Clark, N Zucker, and J Urpelainen, 'The Future of Coal-fired Power Generation in Southeast Asia' (2020) 121 Renewable and Sustainable Energy Reviews 109650; L Baker and J Phillips, 'Tensions in the Transition: The Politics of Electricity Distribution in South Africa' (2019) 37(1) Environment and Planning C: Politics and Space 177–96.

Hanri Mostert, Jacques Jacobs, and Wikus Kruger, *From Coal to Distributed Energy* In: *Sustainable Distributed Energy Resources*. Edited by: Íñigo del Guayo, Damilola S. Olawuyi San, Louis de Fontenelle, Milton Fernando Montoya, and José Juan González, Oxford University Press. © International Bar Association 2026. DOI: 10.1093/9780198991946.003.0004

fossil fuels are embedded in national identity and state-building.[4] Such entanglements are often linked to historical narratives of sovereignty, development, or national control, as well as very real economic interests and impacts that reflect their higher levels of economic vulnerability—both at a macro-level and due to the energy intensity of their economies.[5]

Somewhere between these opposite poles are countries like Poland and India, where structural economic dependencies and political cautions curtail regulatory ambitions in favour of decarbonization and decentralization.[6] China presents a complex case. It is expanding coal production to ensure short-term energy security and regional stability, while also rapidly scaling up renewable energy sources to enhance energy sovereignty and to position itself as a global leader in clean energy manufacturing. This dual strategy reflects a blend of security concerns and industrial policy aimed at long-term economic and strategic competitiveness.[7] Australia, which is technically advanced and well-resourced for decarbonization and decentralization, illustrates how federal politics and regional economies can nevertheless stall decentralization.[8]

What follows is a thematic exploration of the resistance to transition pathways and the twists and turns that can achieve motion in an otherwise stuck system. The case of South Africa provides a particularly helpful illustration, not because its challenges are unique—they are not. But they are concentrated and interwoven in ways that amplify the tensions inherent in crossing the edge from a centralized, fossil-fuels-based system towards one of distributed energy resources. The South African case reveals the tight entanglement between state power and fossil energy. It offers a compelling vantage point from which to explore the structural and affective contradictions that many energy systems must confront on their path to decentralized, decarbonized energy. Among several coal-dependent jurisdictions, Poland, India, and Indonesia may serve as mirrors and foils to the South African experience. The analysis below traces a trajectory from entrenched institutional patterns and protective policy behaviours in these jurisdictions, through moments of destabilization and political reconsideration, towards the cautious emergence of alternative energy solutions and reform pathways.

II. Coal's Structural Grip

Countries that build their economies on coal often find themselves not only reliant on the resource, but also bound to it through their infrastructure and institutional design. Once coal becomes the backbone of the national energy supply, it often becomes entrenched in

[4] P Newell and D Mulvaney, 'The Political Economy of the "Just Transition"' (2013) The Geographical Journal <https://doi.org/10.1111/geoj.12008> accessed 7 August 2025.

[5] M Hafner and S Tagliapietra, 'The Global Energy Transition: A Review of the Existing Literature' in M Hafner and S Tagliapietra (eds), *The Geopolitics of the Global Energy Transition* (Springer 2020) 1–24, 17.

[6] R Chandra and others, 'India's Coal Conundrum: Decarbonisation Amidst A Developmental Legacy' (2025) 16(1) Wiley Interdisciplinary Reviews: Climate Change 928; M Antosiewicz and others, 'Pathways for the Transition of the Polish Power Sector and Associated Risks' (2020) 35 Environmental Innovation and Societal Transitions 271–91.

[7] O Odgaard and J Delman, 'China's Energy Security and Its Challenges Towards 2035' (2014) 71 Energy Policy 107–17.

[8] L Godden, 'Energy Law and Regulation in Australia' in G Bellantuono and others (eds), *Handbook of Energy Law in the Low-Carbon Transition* (Walter de Gruyter GmbH & Co KG 2023) 369–86, 370.

the broader economic system.[9] It becomes crucial for job creation in coal-mining regions and for financing energy utilities.[10] It can even become politically symbolic.[11] This makes decarbonization more than simply a redesign exercise: it is an undertaking with structural, societal, and psychological dimensions. In South Africa[12] and Poland, resistance to phase-out stems not only from economic concerns but also from perceived threats to legitimacy.[13]

As the world's seventh-largest coal producer[14] and a top emitter of greenhouse gases (GHG),[15] South Africa's dependence on coal is deeply structural.[16] The coal sector employs nearly 95,000 workers, pays some of the highest wages in the formal sector, and supports extensive procurement networks.[17] For the longest time, coal exports have generated significant revenue, and they continue to do so.[18]

South Africa's vertically integrated system of generating, transmitting, and distributing predominantly coal-generated electricity[19] now grapples with the legal, governance, and logistical aspects of an inevitable shift to decarbonization. Eskom, the national utility, generates nearly 90 per cent of the country's electricity, primarily from coal.[20] It owns the national transmission grid and supplies electricity to most municipalities and directly to large users.[21] This vertical integration, backed by long-term coal contracts[22] and government

[9] WR Freudenburg, 'Addictive Economies: Extractive Industries and Vulnerable Localities in a Changing World Economy' (1992) 57(3) Rural Sociology 305–32.

[10] RT Perdue and G Pavela, 'Addictive Economies and Coal Dependency: Methods of Extraction and Socioeconomic Outcomes in West Virginia, 1997–2009' (2012) 25(4) Organisation & Environment 368.

[11] B Brown and SJ Spiegel, 'Coal, Climate Justice, and the Cultural Politics of Energy Transition' (2019) 19 Global Environmental Politics 149.

[12] P Mirzania and others, 'Barriers to Powering Past Coal: Implications for a Just Energy Transition in South Africa' (2023) 101 Energy Research & Social Science 1; R Cramer and H Mostert, 'Chapter 14: Losing or Exiting Negatively Valued Ownership in a Climate-Distressed World: The Options in South Africa' in M Habdas and others (eds), *Rethinking Expropriation Law IV: Takings for Climate Justice and Resilience* (11th edn, Den Haag 2024) 374.

[13] B Fine and Z Rustomjee, *The Political Economy of South Africa: From Minerals-Energy Complex to Industrialisation* (Wits University Press 1996); M Kuchler and G Bridge, 'Down the Black Hole: Sustaining National Socio-technical Imaginaries of Coal in Poland' (2018) 41 Energy Research & Social Science 136.

[14] US Energy Information Administration, 'Country Analysis Brief: South Africa' (2023) BP Energy Outlook 5 <https://www.eia.gov/international/content/analysis/countries_long/South_Africa/pdf/SAfrica_CAB_FY24_final.pdf> accessed 15 April 2025.

[15] E Nel, L Marais, and Z Mqotyana, 'The Regional Implications of Just Transition in the World's Most Coal-Dependent Economy: The Case of Mpumalanga, South Africa' (2023) 4 Frontiers in Sustainable Cities 1–2; Department of Forestry, Fisheries & the Environment, *Africa's 5th Biennial Update Report (BUR-5) to the United Nations Framework Convention on Climate Change (UNFCC)* (2023) 6–7.

[16] K Mathu and R Chinomona, 'South African Coal Mining Industry: Socio-economic Attributes' (2013) 4(14) Mediterranean Journal of Social Sciences 347.

[17] Minerals Council South Africa, 'Facts and Figures Pocketbook 2023' (2024) 21 <https://doi.org/10.1007/978-3-662-62217-9_13> accessed 15 April 2025; N Makgetla and M Patel, 'The Coal Value Chain in South Africa' (2021) Trade & Industrial Policy Strategies 9–13.

[18] Minerals Council South Africa, 'Comprehensive Facts and Figures 2023' (2024) 19 <https://www.mineralscouncil.org.za/industry-news/publications/facts-and-figures> accessed 2 August 2025.

[19] L Baker and J Phillips, 'Tensions in the Transition: The Politics of Electricity Distribution in South Africa' (2019) 37(1) Environment and Planning C: Politics and Space 177, 181.

[20] IP Morosele and KE Langerman, 'The Impacts of Commissioning Coal-fired Power Stations on Air Quality in South Africa: Insights from Ambient Monitoring Stations' (2020) 30(2) Clean Air Journal 1 <https://doi.org/10.17159/caj/2020/30/2.8833> accessed 15 April 2025; Eskom, 'Generation Plant Mix' (2024) <https://www.eskom.co.za/wp-content/uploads/2024/07/GX-0001-Generation-Plant-Mix-Rev-29.pdf> accessed 15 April 2025.

[21] Eskom, 'Transmission and Distribution of Electricity Fact Sheet' (2021) 1 <https://www.mineralscouncil.org.za/industry-news/publications/facts-and-figures> accessed 2 August 2025; Minerals Council South Africa (n 18) 20; *Platinum Mile Inv 442 (Pty) Ltd and Another v Chief Director, the Department of Forestry, Fisheries and the Environment and Others* (17469/2024) [2025] ZAWCHC 129 (20 March 2025) <https://www.saflii.org/za/cases/ZAWCHC/2025/129> accessed 15 April 2025.

[22] Some pertaining to mines that are Eskom-owned or -built. Morosele and Langerman (n 20) 1.

protection, creates a powerful lock-in, so strong that even the national imperative to reform state institutions (including those in the energy sector) for inclusivity, equity, and accountability[23] could not turn the tide towards unbundling monopolies, encouraging competition, and allowing independent power producers (IPPs).[24] Reform efforts such as the 1998 White Paper on Energy Policy[25] and the 2011 Renewable Energy Independent Power Producer Procurement Programme (REIPPPP)[26] envisioned a competitive, market-based electricity industry. However, attempts to unbundle Eskom through the Independent System and Market Operator (ISMO) Bill[27] failed, due to Eskom's resistance to unbundling and relinquishing control over transmission assets, compounded by legislative and policy inertia, as well as political and ideological challenges, particularly concerning ownership of the transmission system.[28]

This lock-in dynamic is not unique to South Africa. It was also observed in Poland, where coal has long been framed as a guarantor of sovereignty, employment, and industrial stability.[29] It has been an essential part of livelihoods, supply chains, and local economies in coal-producing regions of Central, Southern, and Southwestern Poland.[30] However, although South Africa and Poland share the common challenge of coal dependency, their responses to it differ.

Poland still relies on coal and is reluctant to relinquish it altogether as a source of energy.[31] Still, it is pursuing a managed decline, supported by EU funds, evolving markets, and policy frameworks.[32] A key step has been the separating of coal assets from renewable

[23] S Essex and J de Groot, 'Understanding Energy Transitions: The Changing Versions of the Modern Infrastructure Ideal and the "Energy Underclass" in South Africa, 1860–2019' (2019) 133 Energy Policy 110937, 4–6 <https://doi.org/10.1016/j.enpol.2019.110937> accessed 15 April 2025.

[24] Power Futures Lab, University of Cape Town, *Restructuring Eskom: Why the ERA Amendment Act is so Important* (GSB/Power Futures Lab Policy Brief 1, 2024).

[25] Department of Minerals and Energy (South Africa), *White Paper on the Energy Policy of the Republic of South Africa* (1998) <https://www.gov.za/sites/default/files/gcis_document/201409/whitepaperenergypolicy19980.pdf> accessed 16 December 2025.

[26] Launched in response to three converging pressures: (i) the 2007/2008 power crisis; (ii) the Government's failure to implement the 1998 White Paper's vision, especially the unbundling of Eskom and enabling of private generation; and (iii) South Africa's commitment to international climate obligations, including reducing its carbon emissions, as part of its Nationally Determined Contributions (NDCs) under the UNFCCC. A Eberhard and N Raine, 'The South African Renewable Energy Independent Power Producer Procurement Programme: A Review and Lessons Learned' (2016) 27(4) Journal of Energy in Southern Africa 1–14 <https://doi.org/10.17159/2413-3051/2016/v27i4a1483> accessed 15 April 2025.

[27] Department of Energy, 'Independent System and Market Operator Bill' (Gazette Number 290, Government Gazette 34289, 13 May 2011 <https://www.gov.za/sites/default/files/gcis_document/201409/b9b-2012-130716a.pdf> accessed 15 April 2025.

[28] Department of Energy, 'Independent System and Market Operator Bill (B9-2012)' <https://pmg.org.za/bill/151/> accessed 15 April 2025. South African Institute of International Affairs (SAIIA), 'State of the Nation: Follow-through now imperative' <https://saiia.org.za/research/state-of-the-nation-follow-through-now-imperative/> accessed 15 April 2025.

[29] H Brauers and P-Y Oei, 'The Political Economy of Coal in Poland: Drivers and Barriers for a Shift Away from Fossil Fuels' (2020) 144 Energy Policy 111621, 2 <https://doi.org/10.1016/j.enpol.2020.111621> accessed 15 June 2025

[30] A Nowakowska, A Rzeńca, and A Sobol, 'Place-Based Policy in the "Just Transition" Process: The Case of Polish Coal Regions' (2021) 10(10) Land 1072, 6–9 <https://www.mdpi.com/2073-445X/10/10/1072> accessed 1 August 2026.

[31] Brauers and Oei (n 29) 2.

[32] Reported figures show a quite rapid decline in coal dependency over the past five years, from 76.9 per cent in 2022, to 68 per cent in 2023 and most recently reaching levels below 50 per cent. S Tokarski and A Tajduś, 'Challenges Facing the Polish Energy Sector in 2024 and Upcoming Years' (2024) 69 Archives of Mining Sciences 127, 135; A Ostrovsky, 'Poland Striving for a Future Powered by Renewables' (*CurationCorp*, 11 June 2025) <https://www.curationcorp.com/blog/poland-striving-for-a-future-powered-by-renewables> accessed 15 June 2025.

investments. Previously, both were held within a de facto vertically integrated system of state-owned companies, which created barriers to financing clean energy. The creation of the National Energy Security Agency (NABE) was intended to overcome these barriers.[33] However, NABE now controls over 70 per cent of domestic electricity generation, which means state control over fossil-based capacity has been effectively recentralized.[34] This raises concerns about slowing the coal phase-out.

Poland's unbundling of its electricity industry into separate generation, transmission, and distribution companies, along with creating a wholesale electricity market, has improved competition and market access for renewables. Yet, conflicts of interest remain where distribution companies are owned by firms with generation assets, creating potential bottlenecks for the energy transition.[35]

Declining dependency rates suggest that economic and infrastructural entrenchment have eased, but prevalent symbolic and emotional inertia may explain Polish resistance to alternative (non-coal) electricity sources.[36] Green energy is perceived as externally imposed and unreliable.[37] Also, coal is officially positioned as a 'strategic reserve'.[38] Coal's embeddedness, hence, is not just a legacy in Poland; it is an ongoing planned insurance which echoes the cultural and emotional attachment to it from some parts of the Polish constituency.

From both these jurisdictional scenarios, it is clear that coal's dominance in the energy mix is just a part of the problem. The real entrapment relates to the system's structure: as long as the need to decentralize the energy system is not prioritized, countries like Poland and South Africa will remain trapped in a structure that debilitates national and local government, inadvertently sabotages industry and business, and limits individual and communal opportunities for self-agency and entrepreneurship.[39]

III. Reasons to Quit Coal

Across the world, coal-reliant countries face mounting pressure to pivot away from the fossil fuels that underpinned their industrialization. Disentangling from coal is neither simple nor easy, and yet it is necessary. The economic, environmental, and social costs of continued dependence are too high. Besides, global decarbonization frameworks are gaining momentum. Coupled with the emergence of cost-competitive renewable energy sources, quitting coal is becoming more than just desirable: it is becoming inevitable.

In coal-dependent systems with entrenched centralized energy utilities, it has become necessary to address old, frail infrastructure, grid fragility, and other structural weaknesses. Each jurisdiction's trajectory is different, but there are some similar underlying tensions.

[33] Tokarski and Tajduś (n 32) 127, 131.

[34] ibid.

[35] E Bayer, *Report on the Polish Power System: Version 2.0* (Agora Energiewende 2018) <https://www.agora-energiewende.de> accessed 6 August 2025.

[36] Brauers and P-Y Oei (n 29) 2; A I Baruk and A Goliszek, 'The Ways of Interpreting Green Energy by Young Polish Individual Recipients vs. Their Gender' (2023) 50 Energy Strategy Reviews 101212, 5–6 https://doi.org/10.1016/j.esr.2023.101212 accessed 15 June 2025.

[37] Brauers and P-Y Oei ibid 2; Baruk and Goliszek (n 36) 5–6.

[38] Tokarski and Tajduś (n 32) 137–38.

[39] A Lawrence, 'Energy Decentralisation in South Africa: Why Past Failure Points to Future Success' (2020) 120 Renewable and Sustainable Energy Reviews 1, 1–2, 6.

South Africa offers a particularly concentrated example of the dynamics at play. India, which faces similar pressures, offers additional insight into how these tensions interact.

A. Systemic Costs of Continued Coal Dependence

In South Africa, over-reliance on a centralized, coal-fired energy monopoly created an unsustainable energy model.[40] Eskom's dominance and its commitment to coal have hindered the country's adoption of energy system decentralization[41] that accompanies the decarbonization agenda introduced by the Paris Agreement[42] and the Just Energy Transition initiative.[43] It resulted in a system characterized by bureaucratic rigidity and systemic grid fragility.[44] Some of Eskom's coal-fired power plants are decades old and operating well beyond their intended lifespans, making them prone to breakdowns. Mounting maintenance backlogs increased the frequency of scheduled power outages (colloquially known as 'load-shedding'), which severely affected business, industry, and practically all aspects of daily living.

The cost implications are profound. Eskom's debt burden now exceeds R400 billion,[45] largely attributable to procurement corruption,[46] operational mismanagement,[47] and delays and cost escalations[48] in constructing new coal mega-plants, such as Medupi and

[40] In particular, lack of competitors to Eskom has resulted in bureaucratic inefficiency, lack of innovation, and resistance to change. A Bowman, 'Parastatals and Economic Transformation in South Africa: The Political Economy of the Eskom Crisis' (2020) 119 (476) African Affairs 395–431; L Mondi, 'State, Market and Competition—Can Eskom be rescued?' (2018) 3 Viewpoints 1, 7–8.

[41] MB Ting and R Byrne, 'Eskom and the Rise of Renewables: Regime-resistance, Crisis and the Strategy of Incumbency in South Africa's Electricity System' (2020) 60 Energy Research & Social Science 1 <https://doi.org/10.1016/j.erss.2019.101333> accessed 15 April 2025.

[42] A global pact, between nearly all countries in the world. United Nations Framework Convention on Climate Change <https://unfccc.int/files/essential_background/background_publications_htmlpdf/application/pdf/conveng.pdf> accessed 29 June 2024.

[43] E Atkins, *A Just Energy Transition: Getting Decarbonisation Right in a Time of Crisis* (Bristol University Press 2023) 18, 19 <https://bristoluniversitypress.co.uk/a-just-energy-transition> accessed 15 April 2025.

[44] G Ndubuisi, EK Avenyo, and R Asiama, 'Dancing on the Grid: Electricity Crises, Manufacturing Energy Vulnerability, and Jobs in South Africa' UNU-WIDER Working Paper 2024/41, 5 (United Nations University World Institute for Development Economics Research 2024).

[45] P Zondi and Z Robinson, 'The Relationship Between Government Debt and Economic Growth in South Africa with Specific Reference to Eskom' University of South Africa Working Papers 14 (2021); K Dlamini, 'An Investigation into the Relationship Between Debt Sustainability and Fiscal Support for Eskom' (MCom thesis, University of the Witwatersrand 2019) 1 <https://wiredspace.wits.ac.za/server/api/core/bitstreams/3b79127a-574e-4482-972c-87d20c742dea/content> accessed 15 April 2025.

[46] Exposed through state capture inquiries. *President of the Republic of South Africa v Office of the Public Protector and Others* (91139/2016) [2017] ZAGPPHC 747, 2018 (2) SA 100 (GP), [2018] 1 All SA 800 (GP), 2018 (5) BCLR 609 (GP) (13 December 2017) <https://www.saflii.org/za/cases/ZAGPPHC/2017/747.html> accessed 15 April 2025.

[47] Summarized in *United Democratic Movement and Others v Eskom Holdings SOC Ltd and Others* (005779/2023;003615/2023;022464/2023) [2023] ZAGPPHC 1949 (1 December 2023) <https://www.saflii.org/za/cases/ZAGPPHC/2023/1949.html> accessed 14 April 2025. And see Eskom, *Integrated Report for the Year Ended 31 March 2024* (2024) 21 <https://www.eskom.co.za/wp-content/uploads/2024/12/Eskom-integrated-report-2024.pdf> accessed 15 April 2025.

[48] Originally projected to cost around R80 billion, the costs have ballooned to over R450 billion. See generally F Tshidavhu and N Khatleli, 'An Assessment of the Causes of Schedule and Cost Overruns in South African Megaprojects: A Case of the Critical Energy Sector Projects of Medupi and Kusile' (2020) 27 Acta Structilia 119, 128 <https://www.ajol.info/index.php/actas/article/view/208209> accessed 15 April 2025; H Labuschagne, '18 years and R467 billion later—Most expensive mistakes in the history of Eskom' (*My Broadband*, 31 March 2025) <https://mybroadband.co.za/news/energy/588945-18-years-and-r467-billion-later-most-expensive-mistakes-in-the-history-of-eskom.html> accessed 15 April 2025.

Kusile.[49] Electricity prices have surged, disproportionately affecting poor households.[50] Investor confidence has eroded. Municipalities are struggling to meet payment obligations.[51] The broader economy has had to absorb the shocks,[52] straining especially low- and middle-income households:[53] productivity losses, job insecurity,[54] and damage to critical infrastructure from erratic power supply[55] have all become par for the course. Moreover, pollution from South Africa's coal plants causes at least 2,680 premature deaths per year and 1.28 million days of work absence.[56]

Similar vulnerabilities are apparent in India's experience.[57] The world's second-largest coal producer,[58] India's coal-fired plants face similar operational inefficiencies and supply chain challenges.[59] India's federated state-run utilities often operate at financial losses, undercut by politically sensitive tariffs and cross-subsidies.[60] Rural areas, in particular, suffer from inadequate electricity supply and frequent power outages. Then there are also the public health costs: India's Air Quality Index often reach hazardous levels in the coal-intensive areas.[61]

As can be seen from both the South African and Indian experiences, the true cost of coal dependency extends beyond coal infrastructure and coal-fired electricity generation. The

[49] Tshidavhu and Khatleli (n 48) 119–43.

[50] B Qeqe, F Kapingura, and B Mgxekwa, 'The Relationship between Electricity Prices and Household Welfare in South Africa' (2022) 15 (20) Energies 2 <https://doi.org/10.3390/en15207794> accessed 15 April 2025.

[51] Fiscal Brief, '2022/23 Financial Year Analysis' (2023) at 8 <https://www.parliament.gov.za/storage/app/media/PBO/Quarterly_Economic_Briefs/2023/2-june/30-06-2023/June_2023_Fiscal_Brief_15_June_2023.pdf> accessed 15 April 2025. Municipalities provide 'end user' consumer services to its communities, thus selling the electricity on to communities but not paying ESKOM as a supplier: *Emfuleni Local Municipality and Another v Eskom Holdings SOC Ltd and Others* (104254/2024) [2024] ZAGPPHC 1138 (11 November 2024) <https://www.saflii.org/za/cases/ZAGPPHC/2024/1138.html> accessed 15 April 2025. H Mostert and T Naude, 'State Protection of Energy Consumers: Between Human Rights and Private Sector Regulation' in I del Guayo and others (eds), *Energy Justice and Energy Law* (OUP 2020) 139–59.

[52] Public Servants Association South Africa 'Impacts of Loadshedding on Economic Growth and Unemployment' (2022) <https://www.psa.co.za/docs/default-source/psa-documents/psa-opinion/impact-of-loadshedding.pdf?sfvrsn=49f278de_0> accessed 12 April 2025.

[53] H Bhorat and T Köhler, 'Power cuts have hit South Africa's labour market hard: the biggest toll has been on jobs—new research' (*The Conversation*, 27 May 2024) <https://theconversation.com/power-cuts-have-hit-south-africas-labour-market-hard-the-biggest-toll-has-been-on-jobs-new-research-230793?utm> accessed 29 May 2024; Ndubuisi (n 44) 5; M Masoga and others, 'Do Energy Prices Drive Inflation in South Africa? An Unrestricted VAR Approach' (2022) 6 Journal of Applied Economics and Business Studies 1–14 <https://doi.org/10.34260/jaebs.621> accessed 15 April 2025.

[54] Power shortages have forced businesses to cut operational hours, reduce staff, or shut down completely. See JL Erero, 'Impact of Loadshedding in South Africa: A CGE Analysis' (2023) 10(2) *Economics and Political Economy* 79.

[55] M Lusková and B Leitner, 'Societal Vulnerability to Electricity Supply Failure' (2021) 20(1) Interdisciplinary Description of Complex Systems 16–26.

[56] Centre for Research on Energy and Clean Air (CREA), *Health Impacts of Delaying Coal Power Plant Decommissioning in South Africa* (October 2023) https://energyandcleanair.org accessed 6 August 2025.

[57] K Agrawal and others, 'Just Transition Away from Coal: Vulnerability Analysis of Coal Districts in India' (2024) 108 Energy Research & Social Science 103355 <https://doi.org/10.1016/j.erss.2023.103355> accessed 2 August 2025.

[58] Energy Institute, *Statistical Review of World Energy: 2024* (473rd edn, 2024) 47 <https://www.energyinst.org/__data/assets/pdf_file/0006/1542714/684_EI_Stat_Review_V16_DIGITAL.pdf> accessed 2 August 2025.

[59] Y Ding, D Mallapragada, and RJ Stoner, 'The Role of Coal Plant Retrofitting Strategies in Developing India's Net-Zero Power System: A Data-Driven Sub-National Analysis' (2025) 86 Energy for Sustainable Development 101687 <https://www.sciencedirect.com/science/article/abs/pii/S0973082625000377> accessed 2 August 2025.

[60] MK Verma and others, 'Indian Power Distribution Sector Reforms: A Critical Review' (2020) 144 Energy Policy 111672 <https://doi.org/10.1016/j.enpol.2020.111672> accessed 2 August 2025.

[61] J Kopas and others, 'Environmental Justice in India: Incidence of Air Pollution From Coal-Fired Power Plants' (2020) 176 Ecological Economics 106711 <https://doi.org/10.1016/j.ecolecon.2020.106711> accessed 2 August 2025.

deeper costs, including health issues, financial strain on middle- and low-income households, and systemic rigidity, limit innovation and economic opportunities. Electricity supply constraints also dampen the economies of these coal-reliant countries.

Beyond these realities, the times are also changing: financial institutions are increasingly reluctant to fund new coal projects due to climate commitments and Environmental, Social, and Governance (ESG) considerations, as well as stranded asset risks.[62] This is increasingly turning the transition to a low-carbon future into a present-day reality.

B. Beyond Decarbonization: Sustainability and Energy System Transformation

The Paris Agreement and the national commitments flowing from it[63] have rendered coal less tenable, economically and diplomatically, as a long-term strategy. Renewable energy sources are not only becoming more cost-competitive, but they are also more favourable choices in terms of international climate commitments and domestic environmental pressures in coal-reliant countries.[64]

For South Africa, the $8.5 billion international commitment in the Just Energy Transition Partnership (JETP)[65] signals an ambition that goes deeper than simply decarbonizing energy: it provides an opportunity to reimagine the structure of the energy system altogether. Its implementation roadmap, the Just Energy Transition Investment Plan (JET-IP),[66] emphasizes not only decarbonization, but also the decentralization of power generation, ownership, and access.[67] It represents a desire to break with the Eskom-centred monopoly model and a move towards an inclusive, democratized, and community-empowering system of distributed energy (DE).[68]

[62] C Elsner and M Neumann, 'Caught Between Path-Dependence and Green Opportunities: Assessing the Impetus for Green Banking in South Africa' (2023) 18 Earth System Governance 100194, 108 <https://doi.org/10.1016/j.esg.2023.100194> accessed 15 April 2025.

[63] Paris Agreement, Art 4. South African Government, Updated Nationally Determined Contribution, <https://www.dffe.gov.za/sites/default/files/docs/southafricasINDCupdated2021sept.pdf> accessed 15 April 2025; see also Presidential Climate Commission, *South Africa's NDC Targets for 2025 and 2030: Technical Analysis to Support Consideration of the Emissions Trajectory in South Africa's NDC* (Technical Report No 2, 2023) <https://www.climatecommission.org.za/publications/south-africas-ndc-targets-for-2025-and-2030> last accessed 12 April 2025; Government of India, *India's Updated First Nationally Determined Contribution under the Paris Agreement (2021–2030)* <https://unfccc.int/sites/default/files/NDC/2022-08/India%20Updated%20First%20Nationally%20Determined%20Contrib.pdf> accessed 4 August 2025.

[64] H Wrinkler and others, 'Just Transition Transaction in South Africa: An Innovative Way to Finance Accelerated Phase-Out and Fund Social Justice' (2021) 3 Journal of Sustainable Finance & Investment 1228, 1243.

[65] T Zprava, 'France, Germany, UK, US and EU launch ground-breaking International Just Energy Transition Partnership with South Africa' (2 November 2021) <https://ec.europa.eu/commission/presscorner/detail/cs/ip_21_5768> accessed 15 April 2025. France, Germany, the UK, and the EU—as originally pledged for the $8.5 billion South Africa JETP—remain committed despite the US withdrawal. N Ferris, 'Don't write off the Just Energy Transition Partnership just yet' (Energy Monitor, 12 October 2023) <https://www.energymonitor.ai/just-transition/dont-write-off-the-jetp-just-energy-transition-partnership-just-yet/> accessed 12 April 2025.

[66] Presidential Climate Commission, 'South Africa's Just Energy Transition Investment Plan (JET IP) for the Initial Period 2023–2027' (2022) 129–33 <https://www.climatecommission.org.za/south-africas-jet-ip> accessed 29 July 2024. Overseen by the South African Presidential Climate Finance Task Team, established in 2022. For more information, see Presidential Climate Commission, 'Towards a Just Transition' <https://www.climatecommission.org.za> accessed 15 April 2025.

[67] Presidential Climate Commission (n 66) 48–52, 106.

[68] Lawrence (n 39) 3.

Indonesia, like South Africa, is a JETP recipient, with a $20 billion pledge from international partners.[69] While it has announced a coal retirement plan and shown high-level support for the energy transition, progress remains slow. The state-owned utility PLN dominates a highly centralized power sector,[70] with little movement towards decentralization or inclusive ownership. There is a risk that the JETP will focus on emissions targets, addressing the deeper structural changes needed for lasting reform.[71]

Both South Africa and Indonesia are at an inflection point: the drive for an energy transition raises the chance not only to transition their reliance on fossil fuels such as coal, but also to rethink and reform their energy systems, from centralized and carbon-heavy to flexible and inclusive. The pace and designs for each jurisdiction differ, but the direction and the momentum towards low-carbon can no longer be denied.

IV. Policy and Legal Response: Distributed Energy as 'Recovery Plan'

In any systemic shift, recognizing the problem and identifying the need to change are only the first steps. The real challenge lies in building the scaffolding to enable something new to emerge: creating new patterns, rules, and responsibilities that support the departure from an entrenched paradigm, and the move towards a new one. For transitioning energy systems, this means reimagining the structure of generation, transmission, ownership and control. Systemic shifts are multifaceted and complicated. It is not just a matter of swapping coal for renewables. It entails redesigning structural and institutional logistics. Law and policy are the incubators of such design.

Articulating climate ambitions and clean energy goals is one important step towards effecting the systemic transition, but implementing them may still be complicated, uneven, and dependent on the ability of a regulatory reform initiative to address deep-seated structural lock-in and dependencies. Yet even in systems where coal dependency is tenacious, reconfiguring of legal and policy frameworks can begin to enable DE. South Africa's evolving approach is presented below, with reference to parallel efforts and obstacles in India and Poland. The discussion that follows shows how the regulatory scaffolding is being installed, where gaps remain, and how institutional and political resistance could still compromise the transition.

A. Law and Policy Architecture Emerging

At the heart of the transition is the understanding that South Africa's vertically integrated, coal-dependent electricity model is aged, brittle, and increasingly unsustainable.[72] DE offers a way out: it will directly address the root problems of the current coal-dominated vertical

[69] Reuters, 'Global JETP plans to help developing nations clean up power sectors' (25 September 2024) https://www.reuters.com/sustainability/climate-energy/global-jetp-plans-help-developing-nations-clean-up-power-sectors-2024-09-25/> accessed 7 August 2025.

[70] W Kruger and others, '*Just Energy Transitions in South Africa and Indonesia: Institutional Location, Political Power, and the Role of the JETP*' (under review, *Energy Policy*, 2025).

[71] ibid.

[72] Power Futures Lab (n 24) 1–2.

system, by decentralizing electricity generation and ownership, improving resilience, and creating opportunities for more equitable participation.[73] Unlike coal's mega-projects that take years to implement, cost more than foreseen, and often reproduce old inequalities,[74] DE allows for immediate, scalable interventions that (literally) bring power closer to the people.[75] It is also most often financed, built, and owned by private owners, shifting risk away from electricity consumers and lightening the fiscal burden on an already constrained Eskom and National Treasury. It provides a possible pathway out of energy precarity for marginalized communities and overburdened municipalities.[76] Section B below discusses how legal and policy instruments are beginning to support this shift, albeit unevenly.

The logic of DE is also gaining traction in other coal-dependent systems. In India, where coal still accounts for nearly 70 per cent of electricity generation, the government's approach is to decentralize the grid while pursuing large-scale renewable targets in tandem.[77] India's legal framework[78] is enabling captive power generation, open access to the grid and distributed solar development.[79] India also has a modest solar rooftop programme suggesting a growing legal uptake on decentralized participation.[80]

Poland's energy law has also begun to accommodate DE, even if more reluctantly. The Polish energy system has historically dominated by vertically integrated utilities and coal-fired generation,[81] but reforms aim to liberalize the markets and promote renewables. Poland's 2015 Renewable Energy Sources Act (as amended) enables prosumer schemes and community energy models,[82] but permitting processes are still slow and entrenched interests continue to hamper decentralization.[83]

B. Phase-out Plans for the Complex Path to Decarbonization

The sections below outline the early tentative steps towards systemic change, the recognitions that preceded systemic disruption, and the supporting framework that may accompany the future evolution of a distributed energy system in various jurisdictions.

[73] Lawrence (n 39) 2, 5 explores three ways in which the relationship between energy decentralization and participatory energy governance, in the form of community ownership, has progressed.

[74] A Andreoni, 'How Can South Africa Advance A New Energy Paradigm? A Mission-oriented Approach to Megaprojects' (2022) 38(2) Oxford Review of Economic Policy 237–38.

[75] Lawrence (n 39) 1.

[76] ibid 3.

[77] A Shankar and M Bukya, 'Decentralisation to Decarbonise the Indian Economy' (2023) 19 Visions for Sustainability art 7153 https://doi.org/10.13135/2384-8677/7153 accessed 4 August 2025.

[78] Eg, India's Electricity Act, 2003 (Act No 36 of 2003) s 2(8), s 9, s 42, s 49, s 86(1)(e). S Mukherjee, T Dhingra, and A Sengupta, 'Electricity Act 2003: Review of Select Indicators' (9 August 2016) SSRN <https://ssrn.com/abstract=2820492 or http://dx.doi.org/10.2139/ssrn.2820492> accessed 4 August 2025.

[79] P Yadav, P J Davies, and D Palit, 'Distributed Solar Photovoltaics Landscape in Uttar Pradesh, India: Lessons for Transition to Decentralised Rural Electrification' (2019) 26 Energy Strategy Reviews 100392.

[80] V Kumar, JK Jethani, and L Bohra, 'Combating Climate Change Through Renewable Sources of Electricity—A Review of Rooftop Solar Projects in India' (2023) 60 Sustainable Energy Technologies and Assessments 103526.

[81] D Chwieduk, W Bujalski, and B Chwieduk, 'Possibilities of Transition from Centralised Energy Systems to Distributed Energy Sources in Large Polish Cities' (2020) 13(22) Energies 6007 <https://www.mdpi.com/1996-1073/13/22/6007> accessed 4 August 2025.

[82] *Act on Renewable Energy Sources* (RES Act, Dz.U. 2015 poz. 478).

[83] H Kryszk and others, 'Barriers and Prospects for the Development of Renewable Energy Sources in Poland during the Energy Crisis' (2023) 16(4) Energies 1724 https://www.mdpi.com/1996-1073/16/4/1724 accessed 15 April 2025.

1. Tentative early steps

South Africa's coal phase-out is foreseen to be gradual, staggered, and politically negotiated.[84] Its path towards decentralization, however, is still under construction and fraught with unevenness in implementation. The 1998 White Paper was an early step in the process. Following it, South African policy acknowledged that coal dependence is more than just a supply issue: Eskom's heavy investment in coal creates a co-dependency, manifested in the vertical model. However, the early reforms were modest. The 2008 National Energy Act,[85] Energy Efficiency Strategy,[86] and the 2011 National Climate Change Response Policy[87] (NCCRP) introduced goals on efficiency, diversification, and emissions reduction, but decentralization remained marginal, leaving Eskom's dominance largely unchallenged.

The 2011 launch of the Renewable Energy IPP Procurement Programme (REIPPPP)[88] was a breakthrough:[89] its transparent, competitive bidding process[90] attracted substantial investment in solar and wind energy projects,[91] embedding local content, community ownership, and socio-economic upliftment into renewable energy procurement.[92] Early rounds demonstrated that renewables could be delivered quickly and cost-effectively, while supporting development goals.[93] However, later political and institutional resistance, particularly from Eskom in delaying or refusing power purchase agreements,[94] curtailed momentum.

India's early progress shared similarities. Along the same lines as South Africa's REIPPPP, India's reforms initially favoured centralized renewables. The Jawaharlal Nehru National Solar Mission, from 2010, kick-started growth in utility-scale renewables, most often procured through federal or state-level competitive bidding programmes.[95] India quickly

[84] As evidenced by the activities of the Presidential Climate Commission that has published various policy position papers and public participation. See Presidential Climate Commission, 'Publications' <https://www.climatecommission.org.za/publications> accessed 15 April 2025.

[85] National Energy Act 34 of 2008.

[86] DME, *Energy Efficiency Strategy of the Republic of South Africa* (2005) <https:/www.gov.za/sites/default/files/gcis_document/201409/energy-efficiencystrategy051.pdf> accessed 15 April 2025.

[87] Department of Environmental Affairs, 'White Paper on the National Climate Change Response' (Gazette Notice 757, Government Gazette 34695, 19 October 2011).

[88] Department of Mineral Resources and Energy, *Renewable Energy Independent Power Producer Procurement (REIPPP) Programme* <https://www.dmre.gov.za/energy-resources/reippp-programme> accessed 15 April 2025.

[89] A Sonday, *The Impact of the South African Renewable Energy Independent Power Producers Procurement Programme on South African Communities: A Case Study* (MPhil thesis, University of Cape Town, Graduate School of Business 2021) 8–9 <http://hdl.handle.net/11427/36126> accessed 15 April 2025.

[90] Eberhard and Raine (n 26) 1.

[91] ibid.

[92] F Müller and S Claar, 'Auctioning a "Just Energy Transition"? South Africa's Renewable Energy Procurement Programme and its Implications for Transition Strategies' (2021) 48 (169) Review of African Political Economy 333–51 <DOI: 10.1080/03056244.2021.1932790> accessed 15 April 2025; DME, 'White Paper on the Energy Policy of the Republic of South Africa' (Gazette Notice 3007, Government Gazette 19606, 17 December 1998) and the DME, 'White Paper on the Renewable Energy Policy of the Republic of South Africa' (Gazette Notice 513, Government Gazette 26169, 14 May 2004). Statistics South Africa, *2023 Sustainable Development Goals (SDGs) Country Report: South Africa* (2023) 112ff <http://www.statssa.gov.za/MDG/SDG_Country_report.pdf> accessed 15 April 2025.

[93] A Eberhard, J Kolker, and J Leigland, *South Africa's Renewable Energy IPP Procurement Program: Success Factors and Lessons* (University of Cape Town, Graduate School of Business 2014) 4 <https://www.gsb.uct.ac.za/files/ppiafreport.pdf> accessed 15 April 2025.

[94] Around Bid Window 4 (2015); A Eberhard and R Naude, *The South African Renewable Energy IPP Procurement Programme: Review, Lessons Learned & Proposals to Reduce Transaction Costs* 27 <https://www.gsb.uct.ac.za/files/EberhardNaude_REIPPPPReview_2017_1_1.pdf> accessed 15 April 2025; A Moyo, 'Renewable energy delays in 2016 haunt SA today' (14 June 2022) ITWeb <https://www.itweb.co.za/article/renewable-energy-delays-in-2016-haunt-sa-today/wbrpO7g2z8zvDLZn> accessed 15 April 2025.

[95] International Energy Agency, *Jawaharlal Nehru National Solar Mission (Phase I, II and III)* (last updated 24 August 2021) <https://www.iea.org/policies/4916-jawaharlal-nehru-national-solar-mission-phase-i-ii-and-iii> accessed 2 August 2025.

became one of the largest solar markets in the world, with an installed capacity more than 90 GW by 2024. However, state-level constraints, inconsistent procurement and transmission bottlenecks remain challenges to the country's 500 GW 2030 renewable energy target.

Poland might have been quicker to legislate, but it was slower to act. It passed its Energy Law in 1997 and a Renewable Energy Sources Act in 2015, but uptake of decentralized renewables remained marginal due to licensing complexity, limited grid access, and local opposition.[96] Decentralization has only begun to gain support more recently, through EU funding schemes and the pressure of climate obligations.[97]

2. Recognition of the need to change

South Africa's signing[98] of the Paris Agreement[99] and submission of its Nationally Determined Contributions (NDCs)[100] marked a formal acknowledgement that its coal-dominated vertically integrated model is unsustainable. The 2019 Integrated Resource Plan (IRP)[101] reduced future coal allocations, expanded renewables and created opportunities for distributed, and embedded generation.[102] As such, it was an early signal that continued reliance on coal was neither sustainable nor affordable.

By contrast, the broader 2016 Integrated Energy Plan (IEP),[103] intended to address structural energy dependencies more systemically, was never adopted,[104] leaving it without policy effect.[105] Its absence weakens strategic alignment and long-term planning.[106] The narrower, electricity-focused Integrated Resource Plan (IRP)[107] remains the planning tool.

[96] M Serowaniec, 'Sustainable Development Policy and Renewable Energy in Poland' (2021) 14(8) *Energies* 2244, 6 <https://www.mdpi.com/1996-1073/14/8/2244> accessed 3 August 2025.

[97] ibid.

[98] Signed 22 April 2016 and ratified 1 November 2016. See UN Treaty Collection, Chapter XXVII.7.d Paris Agreement <https://treaties.un.org/Pages/ViewDetails.aspx?src=TREATY&mtdsg_no=XXVII-7-d&chapter=27&clang=_en accessed> 2 August 2025.

[99] Note 42 above.

[100] Republic of South Africa, *South Africa's First Nationally Determined Contribution (Updated Submission, September 2021)* (UNFCCC 2021) <https://unfccc.int/sites/default/files/NDC/2022-06/South%20Africa%20updated%20first%20NDC%20September%202021.pdf> accessed 15 April 2025.

[101] Department of Mineral Resources and Energy, *Integrated Resource Plan 2019* (Government Notice 1360, Government Gazette 42784, 18 October 2019) 11 <https://www.gov.za/documents/electricity-regulation-integrated-resource-plan-irp2019-october-2019-18-oct-2019-0000> accessed 15 April 2025. A 2024 draft update to the IRP2019 aimed to realign electricity planning with current energy realities, including updated demand projections, cost declines in renewable technologies, and South Africa's climate obligations through 2050. Department of Mineral Resources and Energy, *Draft Review of Integrated Resource Plan 2019* (Government Notice 4238, Government Gazette 49974, 4 January 2024) <https://www.gov.za/sites/default/files/gcis_document/201401/49974gon4238.pdf> accessed 15 April 2025. However, these updates are not in force yet.

[102] Note 101 above.

[103] Department of Energy, *Integrated Energy Plan* (Government Notice 1430, Government Gazette 40445, 25 November 2016) <https://www.gov.za/sites/default/files/gcis_document/201611/40445gon1430reduced.pdf> accessed 15 April 2025.

[104] Although Government Gazette 40445 from November 2016 contains a draft Integrated Energy Plan (IEP), it was released only for public comment and has not been finalized or formally adopted. T Creamer, 'Timeline for inaugural IEP announced as govt sets year-end target for updated IRP' Engineering News 13 September 2024 available at < https://www.engineeringnews.co.za/article/timeline-for-inaugural-iep-announced-as-govt-sets-year-end-target-for-updated-irp-2024-08-23>. Instead, the Integrated Resource Plan (IRP), which focuses only on electricity generation, has become the de facto planning document. Presidential Climate Commission 'The State of Climate Action in South Africa: Priorities for Action for the Government of National Unity' <https://pccommissionflo.imgix.net/uploads/images/24_REP_State-of-Climate-Action-in-South-Africa_v4.pdf> p 51.

[105] Despite being a legally mandated planning instrument under s 6, National Energy Act 34 of 2008.

[106] K Akom, T Shongwe, and M K Joseph, 'South Africa's Integrated Energy Planning Framework, 2015–2050' (2021) 32(1) Journal of Energy in Southern Africa 68, 78 <https://doi.org/10.17159/2413-3051/2021/v32i1a8517> accessed 15 April 2025.

[107] The Integrated Resource Plan (IRP) (n 101), which currently guides electricity planning, was finalized in 2010 and updated through a 2013 version (never published), a 2018 draft, and the most recent official version in

Fragmented, reactive planning continues to undermine the strategic consistency needed for a long-term transition.[108]

India reflects a similar pattern: although national commitments through NDCs and long-term planning acknowledge the necessity of low-carbon growth, these are undermined by ongoing coal expansion and fragmented policy-making—including simultaneous promotion of renewables and continued coal investments.[109]

3. System disruption and withdrawal

Between 2020 and 2022, legal reforms in South Africa began dismantling Eskom's monopoly.[110] The Licensing Threshold Reforms[111] allowed private companies (eg, mines, manufacturers, and commercial farms) to generate or procure their own electricity without a license,[112] removing a long-standing barrier to investment and competition. This shift, driven by the worsening electricity supply crisis, opened an opportunity for rapid growth in the private generation sector.

Eskom's functional unbundling also commenced[113] with the creation of the National Transmission Company of South Africa (NTCSA) to improve grid access for renewables.[114] The enactment of the Electricity Regulation Amendment (ERAA) Act in 2024[115] went further, enabling a competitive multi-market electricity system,[116] mandating an *independent* transmission system operator (outside of Eskom) and ensuring open grid access.[117] These steps could reduce Eskom's dominance[118] and support new technologies and roles,[119] enabling a diversified, competitive electricity sector. However, progress is slow, and resistance from both Eskom and the government casts doubt on the depth of reform.

2019—marking nearly a decade since the last formal adoption. The IRP reflects a delayed acknowledgement of the need for diversification and decarbonization.

108 Akom (n 106) 76–78.

109 Chandra (n 6).

110 National Treasury, *Operation Vulindlela Phase 1 Review: Reflecting on Progress in Driving South Africa's Economic Reform Programme 2020–2024* (20 May 2024) <https://www.gov.za/sites/default/files/gcis_document/202405/operationvulindlelaphase1review.pdf> accessed 15 April 2025.

111 Schedule 2 Amendment, 2021. Initially lifting the licensing exemption threshold for embedded generation projects from 1 MW to 10 MW (2020), and later to 100 MW (2021), this policy shift radically reconfigured the playing field for energy generation.

112 As long as they registered their project with the National Energy Regulator of South Africa (NERSA).

113 Preamble to the Electricity Regulation Amendment Act 38 of 2024: 'to provide for an open market platform that allows for competitive electricity'.

114 Section 34(1)(b) of the Electricity Regulation Act 4 of 2006. 'Determination under Section 34(1)(b) of the Act: New Electricity Transmission Infrastructure' published in GN 6070 in GG 52418 of 28 March 2025 <https://www.nersa.org.za/wp-content/uploads/bsk-pdf-manager/2020/10/Appendix-A-S34-Determination-New-generation-1.pdf> accessed 15 August 2025. The NTCSA is still under Eskom Holdings, as an interim entity to facilitate fair access to the grid and increased integration of renewables while full reforms are being implemented.

115 Electricity Regulation Amendment Act (ERAA) 38 of 2024; Government Notice 5139, Government Gazette 51100 Vol 710 (20 August 2025).

116 Section 1(a) ERAA.

117 ibid, s 1(h).

118 Portfolio Committee on Mineral Resources and Energy, *Report on Public Hearings on the Electricity Regulation Amendment Bill [B23–2023] Held from 26 September 2023 to 29 February 2024* (9 February 2024) para 3.4 <https://pmg.org.za/tabled-committee-report/5645/> accessed 15 April 2025.

119 Such as an Independent Transmission System Operator (ITSO), which could buy electricity from multiple generators (public and private) and ensure open access to the grid. Section 34A(1) of the Electricity Amendment Act 38 of 2024.

These steps mirror India's reforms, particularly its efforts to unbundle state electricity boards (SEBs) and create competitive electricity markets. India's Electricity Act (2003) created a legal basis for open access, captive generation, and third-party sales.[120] However, political interference and weak enforcement have limited impact, especially in financially stressed state utilities.[121]

Poland's transition has also faced institutional roadblocks. While transmission unbundling was achieved earlier,[122] genuine market competition remains limited. DE is still largely peripheral, and coal continues to benefit from legacy subsidies.[123] Recent energy crisis pressures have seen a reversion to coal in some areas, highlighting relapse risks.[124]

4. Building systems of support

The 2022 publication of the Just Energy Transition Investment Plan (JET-IP)[125] marked a decisive step for South Africa towards recovering from its coal dependency. The JET-IP articulates a vision for inclusive, decarbonized energy grounded in infrastructure expansion, skills development, and local industrialization.[126] Importantly, it committed to ensuring that historically disadvantaged and coal-dependent communities are not 'left behind', reinforcing equity in the transition process.[127] The JET-IP is more than just investment planning. It is a blueprint for new governance which is less dominated by Eskom and more open to DE. A key regulatory advance made since its introduction is the South African government's emphasis on expanding the national transmission grid, which is a critical bottleneck to integrating renewables at scale.[128] Simultaneously, regulatory and institutional changes were introduced to streamline embedded generation and enable municipalities to procure energy independently, further decentralizing control.[129] These shifts symbolize the unclenching of the monopolistic hold once exercised by Eskom, and an opening of space for diverse actors to co-create a more adaptive, distributed energy future.

India's parallel can be found in its PM-KUSUM scheme, which supports decentralized solar for farming,[130] and in its Green Energy Open Access Rules (2022),[131] which facilitate grid access for clean generators. Yet state-level regulatory capacity remains a key

[120] Electricity Act 2003 (India), preamble.

[121] SKN Nair, 'Electricity Regulation in India: Recent Reforms and Their Impact' (2008) 2(1) Margin: The Journal of Applied Economic Research 87–144.

[122] B Nowak, 'Challenges of Liberalisation: The Case of Polish Electricity and Gas Sectors' (2009) 2(2) Yearbook of Antitrust and Regulatory Studies 141–68.

[123] A Marciniuk-Kluska and M Kluska, 'Transformation of the Energy Market in Poland in the Context of the European Union Over the Last 20 Years' (2025) 18(13) Energies 3410, 6 <https://www.mdpi.com/1996-1073/18/13/3410> accessed 2 August 2025.

[124] J Bijańska and K Wodarski, 'Hard Coal Production in Poland in the Aspect of Climate and Energy Policy of the European Union and the War in Ukraine: Investment Case Study' (2024) 88 Resources Policy 104390, 2.

[125] Presidential Climate Commission (n 66) 31, 129–33.

[126] ibid, 35, 129–33.

[127] ibid, 33, 129–33.

[128] ibid, 81.

[129] ibid, 249.

[130] Ministry of New and Renewable Energy (India), *Pradhan Mantri Kisan Urja Suraksha evam Utthaan Mahabhiyan (PM-KUSUM)* <https://mnre.gov.in/en/pradhan-mantri-kisan-urja-suraksha-evam-utthaan-mahabhiyaan-pm-kusum/> accessed 2 August 2025.

[131] The Green Open Access Rules is a formal set of regulations issued by India's Ministry of Power under the Electricity Act 2003, aimed at promoting non-discriminatory access to renewable energy for small consumers and ensuring open access to the transmission and distribution network. Electricity (Promoting Renewable Energy through Green Energy Open Access) Rules 2022 (India) <https://mnre.gov.in/en/centre-rules/> accessed 2 August 2025.

bottleneck.[132] Legal frameworks exist, but enforcement and administrative coherence remain weak in many areas.[133]

Poland's transition plan is still taking shape. The updated Energy Policy of Poland to 2040[134] acknowledges the need to expand renewables and improve energy efficiency, but critics argue that institutional inertia and political ambivalence continue to slow progress.[135] Decentralization is nominally supported but often lacks corresponding legal or financial instruments to scale participation.[136]

C. Navigating Institutional and Political Resistance

Despite progress made, entrenched structures still impede South Africa's transition.[137] Policy incoherence persists: the Just Energy Transition Investment Plan (JET-IP) promotes renewables and advances climate commitments, yet the 2019 Integrated Resource Plan[138] still allocates new coal capacity.[139] Municipalities are encouraged to procure power independently, but remain constrained by Eskom's debt collection requirements,[140] which creates a tug-of-war over control.

Grid limitations, opaque access processes, and slow regulatory reform hamper the REIPPPP.[141] Meanwhile, the Electricity Regulation Amendment Bill[142] gestures towards a competitive market, but the institutional capacity needed is not in place.[143] Eskom adds to the paradox. It aligns publicly with transition goals through its JET Strategy,[144] but coal-fired

[132] R Debnath, V Mittal, and A Jindal, *A Review of Challenges from Increasing Renewable Generation in the Indian Power System* (2020) 27–28 <https://www.jstor.org/stable/pdf/resrep30482.pdf> accessed 2 August 2025.

[133] SK Sinha, R Srikanth, and S Mahalingam, 'Regulatory Framework for India's Energy Security and Sustainability' (2022) 162 Energy Policy 112815, 4 <https://www.sciencedirect.com/science/article/pii/S0301421522000404> accessed 2 August 2025.

[134] Ministry of Climate and Environment 'Energy Policy of Poland until 2040 (EPP2040)' https://www.gov.pl/web/climate/energy-policy-of-poland-until-2040-epp2040 (2 February 2021).

[135] A Tajduś and S Tokarski, 'Risks Related to Energy Policy of Poland Until 2040 (EPP 2040)' (2020) 65(4) *Archives of Mining Sciences* <https://psjd.icm.edu.pl/psjd/element/bwmeta1.element.oai-journals-pan-pl-118180/c/oai-journals-pan-pl-118180_full-text_Archiwum-65-4-11-Tajdus.pdf-1> accessed 1 August 2025.

[136] O Juszczyk and others, 'Barriers for Renewable Energy Technologies Diffusion: Empirical Evidence from Finland and Poland' (2022) 15(2) Energies 527, 7 <https://www.mdpi.com/1996-1073/15/2/527> accessed 2 August 2025.

[137] Presidential Climate Commission, *The State of Climate Action in South Africa: Priorities for Action for the Government of National Unity*, 3–4 <https://pccommissionflo.imgix.net/uploads/images/24_REP_State-of-Climate-Action-in-South-Africa_v4.pdf> accessed 2 August 2025; S Bega, 'SA's progress on climate and just transition hindered by incoherent policies, weak governance and inconsistency' (*Mail & Guardian*, 30 July 2024) <https://mg.co.za/the-green-guardian/2024-07-30-sas-progress-on-climate-and-just-transition-hindered-by-incoherent-policies-weak-governance-and-inconsistency> accessed 29 July 2024.

[138] Note 101 above.

[139] Presidential Climate Commission (n 66) 9, 249.

[140] P Lutchman, *Incorporating Embedded Generation into Municipal Networks* (Association of Municipal Electricity Utilities 2023) <https://www.ameu.co.za/Embedded_Generation_update_-_P_Lutchman.pdf> accessed 16 April 2025.

[141] V Obisie-Orlu, *Maximising the Impact of the Renewable Energy Independent Power Producer Procurement Programme (REIPPPP)* (Good Governance Africa, Policy Briefing 2, 2023) <https://gga.org/maximising-the-impact-of-the-renewable-energy-independent-power-producer-procurement-programme-reipppp/> accessed 16 April 2025.

[142] Electricity Regulation Amendment Bill B 23B-2023.

[143] ibid, ss 34A, 34B, and 35C read with the Objects.

[144] Eskom, *Just Energy Transition Fact Sheet* (2021) <https://www.eskom.co.za/wp-content/uploads/2021/10/JET_Factsheet13Oct2021.pdf> accessed 15 April 2025.

power station life extensions continue,[145] with environmental compliance challenges often deprioritized in the face of immediate supply pressures.[146] Overcoming these contradictions will require coherent policy and stronger institutions in moving away from coal.

India faces parallel dilemmas. Tariff regulation, cross-subsidies, and political control of state utilities create disincentives for decentralized energy.[147] Even when national law permits open access or embedded generation, state regulators may delay approvals or impose restrictive charges.[148] This mismatch between legal enablement and political-economic practice mirrors South Africa's experience.

Poland's resistance is more symbolic and cultural in nature. Coal retains its political salience as a source of national energy security and employment.[149] Even when legal pathways exist for decentralization, political rhetoric and institutional habits remain focused on coal.[150] EU pressure has prompted legal adaptation, but implementation remains ambivalent.[151]

V. Conclusion: Redefining the Energy System

Experiences in South Africa, India, Poland, and Indonesia show that decarbonization cannot succeed without deeper structural and institutional decentralization. Coal's hold on a system is often more than just infrastructural. It can be institutional, political, and emotional. Dislodging it requires coherent reforms that dismantle vertical monopolies and align incentives to enable newcomers to participate in the sector.

From the comparative analysis, progress seems to have been uneven. South Africa's reforms, which included a liberalization of licensing, opening of the market under the ERA and the JET-IP, are undermined by policy contradictions and grid constraints.[152] India's national targets coexist with state-level bottlenecks. Poland's managed coal decline risks slowing under recentralized control. Indonesia's JETP illustrates how high-level commitments can stall in centralized, utility-dominated systems. Across these cases, insufficient alignment between decarbonization and decentralization (alongside and limited democratization) limits the pace and inclusiveness of change.

The transition lens offered in this chapter aligns with the Four Ds outlined in Chapter 1: decarbonization, decentralization, digitalization, and democratization. Each of these is essential to sustainable distributed energy resources (DERs), which are systems that balance the environmental, social, and economic pillars of sustainability. Environmentally,

[145] Most notably at Kendal and Grootvlei.
[146] Morosele and Langerman (n 20) 6–7.
[147] Verma (n 60) 6 (Figure 5).
[148] V Garg and others, *India's Renewable Energy Open Access Market: Trends and Outlook* (IEEFA and JMK Research, New Delhi 2022) 3 <https://jmkresearch.com/wp-content/uploads/2022/11/India-Renewable-Energy-Open-Access-Market.pdf> accessed 2 August 2025.
[149] F Černoch, L Lehotský, and A Konvalinová, 'Navigating Russia's War and Energy Transition: Poland's Coal Challenge' (2024) 113 Energy Research & Social Science 103548, 1–2 <https://www.sciencedirect.com/science/article/pii/S2214629624001397> accessed 2 August 2025.
[150] ibid.
[151] JB Skjærseth, 'Implementing EU Climate and Energy Policies in Poland: Policy Feedback and Reform' (2018) 27(3) Environmental Politics 498–518, 506 <https://www.tandfonline.com/doi/pdf/10.1080/09644016.2018.1429046> accessed 2 August 2025.
[152] Mirzania (n 12) 1.

the shift from coal to renewables must achieve a predetermined reduction in emissions, with the broader aims of minimizing ecological harm and building climate resilience. Socially, embodying the just transition principles requires addressing energy poverty and expanding energy citizenship to ensure that coal-region communities, municipalities, small businesses, and marginalized groups can participate and benefit. Economically, the shift from coal must enable cost-competitive, resilient power systems that support diversification and local industry.[153] Aligning the Four Ds with the three pillars of sustainability provides a framework to guide law and policy, enabling DERs to fulfil their promise of a cleaner, fairer, and more resilient energy future.

To 'leave no one behind,' as emphasized in Chapter 2, legal and governance frameworks for distributed energy must be coherent, rights-based, and sustainability-centred. For one, ensuring a just transition means preventing socio-economic collapse in coal-dependent towns, where large-scale unemployment is an imminent existential threat. Moreover, decommissioning coal plants in an environmentally responsible way, while avoiding economic shocks, demands clear decommissioning policies, rehabilitation plans, and funding mechanisms for repurposing old sites.[154] Regulatory frameworks must embed safeguards to this effect in their requirements, for example, for planning, finance, and project approvals. They must also promote open grid access, transparent market rules, and investment in transmission to unlock decentralized generation. Importantly, enabling local ownership, innovation, and skills development must be prioritized in regulatory frameworks.

The information offered by studying coal-dependent jurisdictions offers a clear insight: without dismantling vertical monopolies, decarbonization will remain partial and precarious. A just transition is not only about replacing fossil fuels with renewables; it is about redefining power: electrical and political. The path forward requires institutional unbundling and emotional disentanglement from systems of control. With co-ordinated action, even the most co-dependent energy systems can recover. The opportunity remains to convert a history of entanglement into a future of resilience, where power (both electrical and political) is more evenly shared.

[153] C Strambo, M Patel, and S Maimele, *Taking Stock of the Just Transition from Coal in South Africa* (Stockholm Environment Institute 2024) <https://doi.org/10.51414/sei2024.034> accessed 15 April 2025.

[154] H Winkler and others, 'Just Transition Transaction in South Africa: An Innovative way to Finance Accelerated Phase-out of Coal and Fund Social Justice' (2021) 13(3) Journal of Sustainable Finance & Investment 1228–51 <https://doi.org/10.1080/20430795.2021.1972678> accessed 15 April 2025.

4

Land Tenure, Project Siting, and Environmental Aspects of DERs

Godswill Agbaitoro and Hugo Meyer van den Berg

I. Introduction

This chapter examines the legal and policy challenges affecting the use of distributed energy resources (DERs) to support decentralized energy systems and promote decarbonization. DERs play a vital role in enhancing energy access, decarbonization, and energy efficiency, which are three key objectives of modern energy systems. While the use of DER is regarded as a valuable tool for realizing the three objectives, there are specific challenges associated with its deployment. The challenges manifest in the form of competing interests in DER projects, which are linked to issues surrounding project sitings without due consultation and participation from Indigenous communities, human and environmental rights concerns, and land access and tenure systems. While energy development, including DERs, is significantly linked to all aspects of human life, enhancing our ability to work, live, survive, and carry out various tasks,[1] its development broadly poses multiple challenges to human and environmental rights.

As has been mentioned in Chapter 2, DERs have the capacity to increase energy access by enhancing grid capacity for more people, offering significant advantages for rural areas by providing a reliable and sustainable energy supply.[2] Notwithstanding its potential, there are specific challenges that may affect its success in advancing decarbonization. For example, land tenure and security systems (including ownership rights) are often challenges associated with the deployment of DERs for sustainable transition.[3] In practice, customary land ownership systems frequently create conflicts between energy project developers and local communities. This is sometimes attributed to environmental and social justice paradigms, which are part of broader environmental and social debates that affect energy projects.[4] Additionally, there are other associated challenges stemming from human and environmental rights violations resulting from DER projects that are likely to affect successful deployment. These issues make the use of DERs as a strategy for achieving decentralization and decarbonization a challenge.

[1] RJ Heffron, *Energy Law: An Introduction* (Springer 2021) 1–2.
[2] See Chapters 1 and 2 of this book.
[3] Natural Resource Institute, 'COP26: Mitigating and Adapting to Climate Change where Land Tenure is Insecure' https://www.nri.org/latest/news/2021/cop26-mitigating-and-adapting-to-climate-change-where-land-tenure-is-insecure accessed 28 February 2025.
[4] B El Houda Lamhamedi and W Timo de Vries, 'An Exploration of the Land-(Renewable) Energy Nexus' (2022) 11(6) Land 767.

Godswill Agbaitoro and Hugo Meyer van den Berg, *Land Tenure, Project Siting, and Environmental Aspects of DERs* In: *Sustainable Distributed Energy Resources*. Edited by: Íñigo del Guayo, Damilola S. Olawuyi San, Louis de Fontenelle, Milton Fernando Montoya, and José Juan González, Oxford University Press. © International Bar Association 2026. DOI: 10.1093/9780198991946.003.0005

Beyond the conflict over land access and tenure security, the absence of comprehensive and effective DER legal and policy frameworks creates confusion among investors, project developers, and host communities. Many countries lack comprehensive DER regulations, leaving gaps in licensing, grid interconnection, and compliance requirements.[5] Furthermore, the inconsistencies between national and local policies complicate implementation and slow down project approvals. Even where there are legal and policy frameworks they often fail to integrate human rights and environmental sustainability principles into the use of DER for sustainable energy transitions. Therefore, we argue that there is a need to reframe DER legal and policy frameworks to adopt an approach that considers the human rights and environmental implications in the implementation of projects. Ultimately, this approach will present all stakeholders with the opportunity to advance a just and rights-based approach to the implementation of DERs.

With the growing energy demand and the urgent need for climate action, developing strong legal and policy frameworks for DER is critical. This approach will entail providing theoretical insights and practical legal solutions to some of the most pressing challenges related to the use of DER, including addressing issues around access to Indigenous lands, tenure security, and human rights and environmental sustainability issues. This chapter makes a substantive contribution to scholarly discourse by exploring how law and policy frameworks can be used to advance a just and rights-based implementation of DERs towards sustainable energy transitions. In doing so, it offers strategic approaches for balancing competing interests in DERs, with a focus on the protection of human rights, enhancing environmental sustainability, and ensuring social equity in DER deployment. By identifying legal and policy gaps in DER implementation, the chapter aligns with the book's broader aim of evaluating how legal and policy systems can better support the use of DERs for sustainable energy transitions.

The chapter is divided into five sections, including this introduction. Section II explores the intersections of DER with land tenure systems, environmental sustainability, and human rights. In this section, we highlight potential concerns related to these areas in the context of DER projects. Section III examines the legal and policy framework for DER deployment, pointing out gaps in achieving sustainable energy transitions. It also discusses how international, regional, and domestic legal and policy frameworks often overlook human and environmental rights issues in project implementation when approaching DER utilization. Section IV offers recommendations to address the challenges and risks associated with DER. Here, we argue through the energy justice lens to balance the competing interests in implementation, together with promoting a just and rights-based approach that considers various human, environmental, and social rights issues in DER deployment. Section V is the conclusion.

[5] Presently, the global shift to adopt clean energy has spotlighted DERs, with solar panels, wind turbines, battery storage, and electric cars all seen as examples of small-scale decentralized energy resources that offer potential for achieving sustainability, stability, and resilience. However, the biggest challenge for many countries is how to integrate DERs into existing energy infrastructure through policy and regulatory means. The reality is that existing regulations and policies were designed for centralized generation where power flows in a single direction. See also DM López González and J Garcia Rendon, 'Opportunities and Challenges of Mainstreaming Distributed Energy Resources Towards the Transition to More Efficient and Resilient Energy Markets' (2022) 157 Renewable and Sustainable Energy Reviews 112018.

II. DER's Intersection with Land Tenure Systems, Human and Environmental Rights Issues

As mentioned earlier, DER deployment offers opportunities to drive the decarbonization of energy systems. However, its proliferation presents some external challenges for regulators and policymakers.[6] Before the introduction of a structured system for managing land for economic purposes, land management was formerly based on nation-states, city-states, kingdoms, and chieftains fulfilling their territorial ambitions.[7] As socio-economic changes unfolded, there were drastic changes driven by centres of urban economic production seeking land,[8] increasing demand for natural resources and labour, and growing markets for selling their products.[9] These changes made land tenure systems and security among the challenges hindering the successful development of energy projects, including DER for sustainable energy transitions. As the implementation of DER projects requires the use of land, conflicts often arise with land tenure and security systems, resulting in injustices that may hinder the advancement of DERs.

The conflict associated with DER implementation begins with issues related to project siting. The siting of DER projects raises various concerns related to human and environmental rights that impact members of host communities. Energy decisions regarding the use of DERs to promote sustainable energy systems and transitions often lack proper safeguards for protecting these rights. Consequently, the deployment of DERs as a strategy to address energy access and security challenges and to facilitate sustainable energy transition can lead to issues when human and environmental considerations are overlooked. The following sub-sections examine the implementation challenges associated with DER in terms of land tenure systems, human rights, and environmental factors.

A. Land Access and Tenure Security

The concept of land is multi and trans-dimensional.[10] It is part of an environmental unit, an economic asset, a productive resource, a territory, an accommodation, a sense of place, an identity, and a symbolic scenery.[11] Additionally, land is an important natural resource, both economically, socially, and politically, yet it is a finite resource requiring sustainable utilization.[12] The importance of land to economic growth and development makes the associated issues such as land access, tenure security, and ownership concerning in modern

[6] A Akimov and others, 'Are Australian Regulator and Policy Frameworks ready for a Distributed-Energy-Resources-Led Energy Transition?' (2023) 82 Australian Journal of Public Administration 529–56.

[7] L Royston, 'Security of Urban Tenure in South Africa: Overview of Policy and Practice' in A Durand-Lasserve and L Royston (eds), *Holding Their Ground: Secure Land Tenure for the Urban Poor in Developing Countries* (Earthscan 2002).

[8] ibid. See also JD Unruh, 'Urbanization in the Developing World and the Acutely Tenure Insecure' (2007) 11 City Debates 115–20.

[9] A Cattaneo, 'Deforestation in the Brazilian Amazon: Comparing the Impacts of Macroeconomic Shocks, Land Tenure, and Technological Change' (2001) 77(2) Land Economics 219–40; K Firmin-Sellers, 'Custom, Capitalism, and the State: The Origins of Insecure Land Tenure in West Africa' (2000) 156(3) Journal of Institutional Theoretical Economics 513–30.

[10] Lamhamedi and Timo de Vries (n 4) 767.

[11] ibid 767.

[12] F Kariuki and R Ng'etich, 'Land Grabbing, Tenure Security and Livelihoods in Kenya' (2016) 9 African Journal of Legal Studies 79–99.

society. For example, legal issues concerning land may take various forms—such as ownership and property rights, tenure systems, Indigenous and customary claims, land reforms, and disputes—all of which pose significant legal and policy challenges to effective land use. With population growth, high food prices, the impact of climate change, trade regimes, global consumer and corporate-driven food systems, and growing demand for agrofuels and feed, there remains fierce competition for land access and ownership, thus causing high pressures on tenure systems.[13]

Access to land and tenure security are crucial factors for sustainable development.[14] In practice, tenure insecurity is both a primary and underlying issue, exacerbating other pressures that create conflict and uncertainty and hindering sustainable use.[15] One potential problem often faced by countries in the deployment of DERs is the use of land located at a certain distance from the consumption point for a facility designed to meet self-consumption needs.[16] Access to and the use of land close to DER infrastructure is crucial to the successful advancement of DER projects, along with addressing any challenges to tenure security. Just as DER development has the potential to increase energy access, it could equally limit other economic opportunities, such as subsistence farming and hunting for Indigenous communities.[17] Without access to land or rights to use it, DER projects such as onshore wind farms and solar panels cannot be implemented for sustainable energy transitions. Therefore, access to land and the applicable tenure system are crucial in DER implementation. This is in addition to the role landowners play in the deployment of DER projects, which facilitates the energy transition.[18] For example, large-scale energy projects carried out by corporations or states often require the acquisition of land through lease, allocation, concession, purchase, or compulsory acquisition for public interest. Such projects may lead to negative impacts on the rights of members of host communities, including the displacement of hundreds of families and individuals.

Furthermore, lands used for energy projects are often those that local and Indigenous communities have historically utilized to enhance their livelihoods. For example, lands in Indigenous and host communities and the way they are utilized may be connected to the realization of their economic, social, and cultural rights, as access to land remains vital for achieving these rights. With the implementation of DER projects, there could be an impact on their right to food, housing, health, and an adequate standard of living—all of these are key features of economic, social, and cultural rights and are essential for securing better livelihoods. Access to land is essential for achieving a decent standard of living, and this

[13] International Fund for Agricultural Development (IFAD), 'Improving Access to Land Tenure Security Policy' https://www.ifad.org/documents/d/new-ifad.org/land_e-pdf accessed on 20 March 2025.

[14] YJ Masuda, BE Robinson, and MB Holland, 'Property Rights, Tenure Form and Tenure Security' in MB Holland, YJ Masuda, and BE Robinson (eds), *Land Tenure Security and Sustainable Development* (Palgrave Macmillan 2022) 3.

[15] ibid.

[16] M Feridun İzgi and B Selçuk Ardıyok Keki, 'Navigating the Global Landscape of Distributed Energy Resources: Opportunities, Challenges, and Legal Considerations' https://www.ibanet.org/distributed-energy-resources-opportunities-challenges-legal accessed 20 March 2025.

[17] T Kaime and G Agbaitoro, 'An Energy Justice Approach to Resolving the Conflict between the Development of Energy Access Projects and Human Rights Risks and Violations in Africa: Can a Balance be Struck?' (2022) 3(1) Global Energy Law and Sustainability 39–71.

[18] R Wade and D Rudolph, 'Making Space for Community Energy: Landed Property as Barrier and Enabler of Community Wind Projects' (2024) 79 Geographica Helvetica 35–50 <https://doi.org/10.5194/gh-79-35-2024> accessed 18 October 2025.

need can be affected by the implementation of large-scale DER projects. A notable example of this challenge is the Kinangop Wind Power Project in Kenya, which was planned to generate 61 MW and faced significant opposition from local farmers. They voiced concerns about protecting their ownership rights to the land designated for the project, emphasizing that it should be used for food production instead of energy generation.[19]

On the other hand, land tenure systems include the rights and institutions that regulate access to and use of land. This involves a framework of rights, duties, and responsibilities related to the use, transfer, alienation, and ownership of land. Land tenure security remains a fundamental element of many global agendas, such as the United Nations Sustainable Development Goals ('SDGs') and the Paris Agreement. This is due to its crucial role in promoting actions aimed at sustainable development. The importance of land makes tenure security an explicit element of many policies and programmes, including those designed to tackle climate change. For example, initiatives such as Reducing Emissions from Deforestation and Degradation and the establishment of protected areas, as well as sustainable agricultural programmes, all depend on land use for their success. Therefore, when energy decisions concerning the implementation of DERs fail to recognize or consider land tenure systems, conflict between project developers and host communities is likely to arise.

Furthermore, similar to carbon-based energy projects, one of the most oppressive aspects of the injustice caused by DER projects is the compulsory acquisition of land (also known as land grabbing) by corporations and governments for the public interest.[20] Land grabbing manifests, among other ways, through the development of energy projects, including the exploration and extraction of mineral deposits, as well as the development of renewable energy. This approach often leaves a lasting impact on land tenure systems, particularly since its implementation frequently requires large areas of land, resulting in numerous conflicts with Indigenous communities. For instance, many energy projects aimed at increasing access through renewable energy development have been marred by human rights violations such as land dispossession and grabbing. This has been evidenced in Uganda, where the construction of a hydroelectric power plant resulted in a series of land dispossessions and grabs to accommodate the project.[21]

B. Environmental Aspects

DER's primary purpose is to enhance energy security, improve efficiency, and reduce the environmental impact of energy systems. However, sometimes the goals of environmental

[19] I Renkens, 'The Impact of Renewable Energy Projects on Indigenous Communities in Kenya: The Cases of the Lake Turkana Wind Power Project and the Olkaria Geothermal Power Plants' (2019) International Work Group for Indigenous Affairs Report <https://iwgia.org/en/resources/publications/3535-the-impact-of-renewable-energy-project-on-indigenous-communities-in-kenya.html> accessed 5 July 2025.

[20] FA Ndi, 'Land Acquisition, Renewable Energy Development, and Livelihood Transformation in Rural Kenya: The Case of the Kipeto Wind Energy Project' (2024) 112 Energy Research & Social Science 103530.

[21] The government of Uganda approved the Bujagali hydropower dam on the River Nile in Uganda in 1994 as the lowest-cost option to increase power production and increase access to electricity in the country. It was completed in 2012 after eighteen years of controversy that delayed the dam's construction. The project involved the development of a 200–50MW hydropower plant on the Nile, about 8 kilometres downstream of the existing Kiira Power Station. It also included construction of a 100-kilometre transmission line, substations, and other associated works.

public policies hinder the achievement of DERs; particularly regarding the implementation of large-scale DERs, such as solar and wind farms. This is because their implementation often requires significant land areas, which can potentially lead to habitat loss and other environmental effects. While DER is frequently promoted as a green alternative to fossil fuels, it is vital to recognize that a full understanding of their environmental impacts can only be gained through their life cycle assessment. This involves evaluating the environmental impacts throughout the entire life cycle of the project, from raw material extraction to manufacturing, operation, and disposal at the end of life. For example, the construction, operation, and decommissioning of offshore installations, such as wind turbines, have various impacts on the marine environment, including the displacement of marine species and the alteration of their growth and development.

Furthermore, DER environmental impacts can also be ascertained when considering the production of equipment used for implementing projects, including solar panels, wind turbines, as well as battery storage equipment and hydro schemes. It has been reported that their production involves resource extraction and energy consumption, thereby contributing to environmental footprints.[22] In practice, most DER systems require technical components for production and construction, resulting in a heavy reliance on various materials, including rare earth elements, which are subject to resource depletion. As mentioned earlier, other environmental aspects of DER include the disposal of end-of-life equipment, which can create environmental problems if not properly recycled and disposed of. For example, solar panels contain materials that are harmful, such as lead and cadmium, and batteries used in energy storage contain hazardous chemicals, including lithium and lead.[23] Additionally, offshore wind farms can have a significant impact on the marine environment due to the constant noise and vibration generated during operation, which can be disruptive or damaging to fish and other marine species.[24] The same applies to hydropower projects, which can impact local ecosystems, particularly as they involve the construction of dams that disrupt fish migrations and create reservoirs that emit significant amounts of greenhouse gases.[25]

Another environmental aspect of DER projects is the lack of environmental and social impact assessments (ESIA), which are not always mandatory. The lack of ESIA is largely attributed to the government's desire to enhance energy access and security, as well as promote sustainable energy transitions. Proponents of DER projects, including governments, often neglect or disregard the importance of conducting an effective ESIA due to the broader climate goals that the project aims to achieve, thereby overlooking the potential of ESIA to provide an opportunity to ascertain environmental sustainability. When they fail to conduct an effective ESIA, this allows Indigenous communities to raise environmental concerns and issues regarding the siting of such projects. This misalignment between DER deployment for energy access and security and broader climate goals presents challenges to the feasibility and long-term success of DER. For example, the implementation of DER projects in some African countries is often executed without a detailed environmental impact

[22] G Scandrett, 'Environmental Impacts of Renewable Energy Sources' (2017) https://www.adecesg.com/resources/blog/environmental-impacts-of-renewable-energy-sources/ accessed 2 June 2025.
[23] ibid.
[24] ibid.
[25] ibid.

assessment.[26] DER technologies that involve combustion, particularly burning fossil fuels, can produce many of the same types of impacts as larger fossil-fuel-fired power plants, such as air pollution.[27] While the environmental impacts of DER may be smaller in scale than those from large power plants, they are often located closer to populated areas due to the plant's capacity to serve the local area. Consequently, making certain that DER systems impact air emissions, leaving a negative environmental footprint due to their proximity to the end-user, particularly those that are byproducts of fossil fuel combustion.

C. Human Rights Issues

Just as we have observed human rights risks linked to the deployment of carbon-based energy resources—covering impacts on the right to life, subsistence, food, and health—the same concerns also apply to the deployment of DERs.[28] Although DERs are regarded as self-supporting distributed energy systems that do not rely solely on electricity supplied by large-scale power stations, the introduction of new operators of various sizes often raises local concerns. These concerns include issues related to human rights, safety, disaster prevention, the impact on landscapes and the environment, and the future disposal of facilities.[29] Additionally, their implementation presents other specific human rights challenges, such as forced displacement, marginalization, exclusion of Indigenous communities ie through lack of consultation and participation, and the concentration of energy projects in vulnerable communities.

As with all energy projects, there are broader human rights issues including DER projects known to have caused fatalities in Indigenous communities where such projects are situated. For instance, the development of renewable energy projects can negatively impact the rights to life, water, health, and food if adequate health and safety measures are not implemented. Additionally, large-scale renewable projects can potentially lead to massive displacement of Indigenous communities and environmental refugees. This is evident in India, where the country's energy transition agenda aims to generate 500 GW of power from renewable sources, such as solar energy, by 2030. The ongoing massive renewable energy projects in the Oran land at Thar Desert in India continues to generate mixed reactions due to systematic disregard for the human rights of Indigenous communities that rely on traditional lands for their livelihoods, sustenance, and cultural identity.[30]

Furthermore, the challenge of siting DER projects in residential areas, especially in poor and vulnerable Indigenous communities, raises concerns of a lack of consultation and participation in the decision-making process leading to their implementation. This is

[26] S Obaji, 'Environmental impact assessments don't work in Nigeria: here's why' The Conversation <https://theconversation.com/environmental-impact-assessments-dont-work-in-nigeria-heres-why-188796> accessed 21 March 2025.

[27] United States Environmental Protection Agency, 'Distributed Generation of Electricity and its Environmental Impacts' https://www.epa.gov/energy/distributed-generation-electricity-and-its-environmental-impacts# accessed 20 March 2025.

[28] Kaime and Agbaitoro (n 17) 40–41.

[29] K Yokoi and H Wada, 'Initiatives and Challenges for the Introduction of Distributed Energy Systems' (2024) https://www.ibanet.org/initiatives-challenges-distributed-energy-systems accessed 8 August 2025.

[30] S Misra and G Agbaitoro, 'Mapping Human Rights Violations Connected to Renewable Energy Development in India: A Case Study of the "Oran Land" at Thar Desert and Energy Transition' (2025) 10(1) Business and Human Rights Journal 249–59.

particularly concerning, as some of the DER generation systems are visually unappealing to Indigenous communities. The role of consultation and participation while developing energy projects cannot be overemphasized. Under international law, including the Aarhus Convention 1998 and Escazu Agreement 2018, the concept of participatory democracy and prior informed consent procedures in environmental matters is well recognized.[31] For instance, Article 25(a) of the ICCPR stipulates that every citizen shall have the right and opportunity to participate, without distinction and unreasonable restrictions, in the conduct of public affairs, directly or through freely chosen representatives.[32] This makes it important for all energy project proponents to allow affected communities, through consultation and participatory mechanisms, to have input in energy decision-making processes.

Another human rights issue associated with the implementation of DER projects is the failure of project proponents to secure a social licence from Indigenous host communities. This presents a significant challenge in situations where the process leading to the siting of the project may not have adhered to established policies. For example, India's renewable energy projects, which involve installing extensive large-scale solar panels for power generation have not been accepted by Indigenous communities where the projects are situated.[33] Due to its geographical position, the Thar Desert, a traditional land heavily relied on by Indigenous communities, was earmarked as a suitable site for hosting solar and wind energy projects to drive India's energy transition agenda.[34] Consequently, the project developers acquired Indigenous land without conducting prior environmental, social, or human impact assessments, thus failing to use such opportunity to secure a social licence. Reports indicate that the project proponents restricted Indigenous communities from accessing their land and cut down trees and vegetation to create space for the solar panels, including installing power transmission lines and building ancillary infrastructure.[35]

The failure to secure a social licence for developing energy projects through consultation and participatory mechanisms has historically caused conflicts between project proponents and Indigenous communities. Furthermore, it has led to an increase in legal claims in courts and posed a consistent threat to the social acceptance of renewable energy projects through protests and, in some cases, violent responses from Indigenous communities. It seems that from most national governments' perspective, the benefits of renewable energy projects for sustainable transitions significantly outweigh the negative impacts on human and environmental rights.[36]

[31] M Fitzmaurice, 'Some Reflections on Public Participation in Environmental Matters as a Human Right in International Law' (2002) 2(1) Non-State Actors and International Law 12.

[32] International Covenant on Civil and Political Rights (adopted 22 November 1969, entered into force 23 March 1976) 999 UNTS 171 (ICCPR).

[33] The Energy Resource Institute India (TERI), 'Addressing Land Issues for Utility Scale Renewable Projects in India' (2018) https://shaktifoundation.in/wp-content/uploads/2018/01/Study-Report-Addressing-Land-Issues-for-Utility-Scale-Renewable-Energy-Deployment-in-India.pdf accessed 28 October 2025.

[34] Misra and Agbaitoro (n 30).

[35] ibid 251; see also A Parvaiz, 'Feature—Rare Birds, Forest Protectors Clash with India's Clean Energy Vision' Thomson Reuters Foundation (13 February 2022) https://www.reuters.com/article/india-solar-just-transition-idAFL8N2TL3I6/ accessed 18 October 2025.

[36] International Institute for Sustainable Development, 'Benefits of Renewables Outweigh Negative Impacts REN21 Report Finds' https://sdg.iisd.org/news/benefits-of-renewables-outweigh-negative-impacts-ren21-report-finds/ accessed 7 August 2025.

III. Legal and Policy Gaps in DER Initiatives

Research to better understand the contentious legal and policy issues surrounding DERs, such as rooftop solar panels, wind turbines, and small-scale battery storage, has largely focused on specific case studies.[37] Notwithstanding, the legal and policy framework for advancing DERs remains at a level that provides limited support. This is due in part to the absence of comprehensive strategies and mandates that clearly promote energy access, decarbonization, and efficiency through the deployment of DERs. Furthermore, existing frameworks either fail to recognize competing interests, such as land access, community rights, and ownership, tenure security conflicts, and environmental sustainability objectives, or do not address them at all. These challenges are not unique to any one region. Across the globe, the success of DER initiatives often hinges on whether legal systems can accommodate decentralization while balancing environmental, social, and economic priorities. While there has been some resilience in DER investment, it remains insufficient to meet the scale of universal energy access goals.[38] The challenge now is to translate global ambition into regulatory clarity, social inclusion, and environmental safeguards on the ground.

A. International Framework for DERs

The global policy framework for energy transition is shaped by a range of international instruments, including the United Nations SDGs[39] and the Paris Agreement.[40] Additional guidance is provided by international organizations such as the International Renewable Energy Agency ('IRENA')[41] and the International Energy Agency ('IEA'),[42] which play influential roles in shaping discourse and best practices. However, these bodies issue non-binding recommendations and do not create enforceable legal obligations.[43] While these frameworks reflect a broad consensus on the need for clean, affordable, and reliable energy, DERs are not explicitly addressed in binding international law. As a result, the legal

[37] Akimov (n 6 above).

[38] International Renewable Energy Agency Report 2024, 23 https://www.iea.org/reports/world-energy-outlook-2024 accessed 21 July 2025.

[39] United Nations Sustainable Development Goals https://sdgs.un.org/goals accessed 22 July 2025.

[40] United Nations Climate Change https://unfccc.int/process-and-meetings/the-paris-agreement accessed 22 July 2025.

[41] International Renewable Energy Agency https://www.irena.org accessed 22 July 2025.

[42] International Energy Agency https://www.iea.org accessed 26 July 2025.

[43] Pursuant to art IV of its Statute, IRENA functions as a centre of excellence and facilitator for renewable energy by systematizing best practices, providing non-binding policy advice, building capacity, promoting technology transfer, supporting financing mechanisms, and fostering research collaboration. In carrying out these activities, IRENA seeks to raise public awareness, operates in accordance with the purposes and principles of the United Nations, and co-operates with other international and regional bodies to avoid duplication of effort, with particular regard to the needs of developing countries and remote regions. See IRENA, 'Statute of the International Renewable Energy Agency' (2009) art 4 https://www.irena.org/-/media/Irena/Files/Official-documents/IRENA-Statute/IRENAstatuteenIRENAFCStatutesignedinBonn26012009incldeclarationonfurtherauthenticversions.pdf accessed 11 August 2025. The IEA, on the other hand, is a global forum providing authoritative analysis, data, and non-binding policy recommendations across the full spectrum of energy sources and technologies. Adopting an all-fuels, all-technology approach, it engages with both member and emerging economies to advance energy security, affordability, and sustainability through co-operation on energy policy, market analysis, efficiency measures, clean energy deployment, and system reliability. See https://www.iea.org/about/mission accessed 11 August 2025.

and regulatory environment for DERs remains largely dependent on regional or domestic policy and implementation.

SDG 7 commits states to ensuring 'access to affordable, reliable, sustainable and modern energy for all' by 2030.[44] Its sub-targets—expanding the share of renewables,[45] improving energy efficiency,[46] and expanding infrastructure and technology in developing countries[47]—align closely with the benefits offered by DERs, particularly in underserved or remote areas where grid extension is not viable. However, SDG 7 does not prescribe binding obligations or regulatory standards specific to DERs, leaving their promotion and regulation to national governments. According to the 2025 SDG Progress Report, electrification must accelerate by at least 1.2 per cent annually to meet global access targets, yet an estimated 645 million people may still lack electricity by 2030. Off-grid solar solutions remain the most cost-effective means of reaching these populations, with the potential to provide electricity to up to 389 million people.[48]

Similarly, the Paris Agreement obliges parties to pursue 'economy-wide' emissions reductions[49] and to promote technology development and transfer to improve resilience to climate change and reduce greenhouse gas emissions.[50] DERs play an essential role in this context, offering a pathway to decarbonize energy systems while enhancing resilience and local energy security. Nonetheless, the Agreement remains technology-neutral and non-prescriptive. It relies on nationally determined contributions (NDCs), which vary widely in how they address decentralized renewable energy systems, if at all.

There is, at present, no binding international legal regime specifically tailored to the governance of DERs. This absence contributes to fragmented policy environments and regulatory uncertainty at the national level, where DER deployment often intersects with complex questions of land access, grid interconnection, environmental impact, and community participation. Strengthening the alignment between global commitments and domestic DER policy frameworks remains a key challenge for ensuring a just and effective energy transition worldwide.

B. Regional and Domestic Approaches to DERs

At the regional level, responses to global policy signals have varied significantly. While many jurisdictions recognize the potential of DERs to advance energy access, decarbonization, and resilience, regional frameworks differ in regulatory maturity, legal enforceability, and alignment with the specific requirements of DER deployment. Some regions have established institutional frameworks to support decentralized energy systems, but these efforts often lack coherence and fall short of integrating essential issues such as land tenure, project siting, environmental safeguards, and community participation.

[44] United Nations Sustainable Development Goals https://sdgs.un.org/goals/goal7 accessed 22 July 2025.
[45] Target 7.2 of SDG 7.
[46] Target 7.3 of SDG 7.
[47] Target 7b of SDG 7.
[48] United Nations, 'The Sustainable Development Goals Report 2025' 22 https://unstats.un.org/sdgs/report/2025/The-Sustainable-Development-Goals-Report-2025.pdf accessed 22 July 2025.
[49] Article 4.4 of the Paris Agreement 2015.
[50] Article 10 of the Paris Agreement 2015.

In the European Union, the Clean Energy for All Europeans Package establishes a clear legal basis for supporting consumers, enabling community energy initiatives, and integrating DERs through smart grids and demand-side flexibility.[51] Article 8.2 of Directive 2019/944 specifically requires member states to determine the criteria for the grant of authorizations for the construction of generating capacity in their territory. Several factors must be considered when determining the criteria, including the protection of the environment,[52] land use, and siting.[53] Furthermore, member states are obliged to establish specific, simplified, and streamlined authorization procedures for small-scale decentralized and distributed generation, taking into account their limited size and potential impact.[54] Article 8.4 requires the authorization procedures and criteria to be made public by the member states.

In Africa, high-level policy commitments to energy transition are reflected in frameworks such as the African Union's Agenda 2063,[55] the Africa Renewable Energy Initiative (AREI),[56] and the African Energy Transition Programme (AETP),[57] to mention a few.[58] However, while these initiatives acknowledge the importance of achieving universal energy access and goals relating to combating climate change, they lack specific commitments or enforceable mandates tailored to DERs. Moreover, these continental strategies do not directly address the complex interplay between DER expansion, land tenure security, community rights, and environmental commitments—factors central to equitable and sustainable energy development. Regional bodies like the ECOWAS,[59] Centre for Renewable Energy and Energy Efficiency (ECREEE), in West Africa, the East African Centre of Excellence for Renewable Energy and Energy Efficiency (EACREEE), the Southern African Development Community Centre for Renewable Energy and Energy Efficiency (SACREEE), and the Centre for Renewable Energy and Energy Efficiency for Central Africa (CEREEAC), promote renewable energy co-operation and efficiency, but their frameworks remain

[51] In particular, Directive (EU) 2019/944 of the European Parliament and of the Council of 5 June 2019 on common rules for the internal market for electricity; see preambles 10, 43, and 61 of this Directive, which in broad terms recognize the role of DERs by emphasizing the importance of consumer participation in enabling system flexibility (Preamble 10), the potential of community energy as a cost-effective and inclusive model (Preamble 43), and the need to empower system operators to utilize DER-based services through market mechanisms to optimize network efficiency and avoid costly expansion (Preamble 61). See also art 22.4(d) of Directive (EU) 2018/2001 of the European Parliament and the of the Council of 11 December 2018 on the promotion of the use of energy from renewable sources,

[52] Article 8.2(c) of the Directive.

[53] Article 8.2(d) of the Directive.

[54] Article 8.3 of the Directive.

[55] 'Agenda 2063: The Africa We Want' (2015) https://au.int/en/agenda2063/overview accessed 3 April 2025.

[56] Africa Renewable Energy Initiative https://www.arei.info/infoandstrategy accessed 3 April 2025.

[57] African Energy Transition Programme https://au-afrec.org/energy-transition-programme accessed 3 April 2025.

[58] The key aspiration of Agenda 2063 in this regard is Aspiration 1, which seeks a 'prosperous Africa based on inclusive growth and sustainable development'. Member states commit, under Agenda 2063, to speed up action to connect Africa through world-class infrastructure, which includes 'harnessing all African energy resources to ensure modern, efficient, reliable, cost-effective, renewable and environmentally friendly energy to all African households, businesses, industries and institutions, through building the national and regional energy pools and grids, and [Programme for Infrastructure Development in Africa] energy projects'. Mandated by the AU and endorsed by the Committee of African Heads of State and Government on Climate Change (CAHOSCC), the AREI seeks to promote inclusive and sustainable energy development. Its targets include the deployment of at least 10GW of new and additional renewable energy generation capacity by 2020 and the long-term mobilization of Africa's potential to generate at least 300GW of renewable energy by 2030. See https://au-afrec.org/node/788 accessed 3 April 2025.

[59] Economic Community of West African States (ECOWAS).

underdeveloped when it comes to addressing the legal complexities of DER deployment, including land use conflicts, tenure insecurity, and environmental safeguards. As in other regions, Africa's regulatory landscape reflects growing ambition but lacks a coherent and justice-oriented legal architecture that fully integrates land access, tenure systems, and community rights into DER implementation.

On the African continent, the deployment of DERs has gained momentum to fill the gaps where centralized systems cannot reach.[60] Africa contributes more than 16 per cent of the world's decentralized renewable energy capacity, with off-grid solar solutions playing a pivotal role in delivering electricity to millions of people across sub-Saharan Africa.[61] The most widely adopted DER technologies include solar home systems, community mini-grids, and battery energy storage solutions. Some countries have seen a surge in off-grid solar installations, such as Kenya,[62] Nigeria,[63] and Rwanda,[64] driven partly by falling technology costs[65] and the widespread use of mobile payment systems that support pay-as-you-go models.[66] In Sierra Leone[67] and Zambia,[68] donor-supported mini-grid projects have emerged as community-level energy solutions for households, schools, and small businesses. Meanwhile, South Africa[69] and Morocco[70] are increasingly exploring battery storage integration, particularly to support renewable energy intermittency and enhance grid reliability.

While the growth of DERs across Africa signals a promising shift towards decentralized and inclusive energy systems, the successful implementation of these projects depends not only on technology and finance, but also on the siting of these projects. DERs are often deployed close to communities, farms, and ecologically sensitive areas, which raises a host of challenges, including land rights, environmental degradation, and lack of community consultation or consent.[71] These concerns are becoming more prevalent, with several DER

[60] African Energy Chamber 'The State of African Energy: 2025 Outlook Report' 79 https://energychamber.org/wp-content/uploads/The-State-of-African-Energy-2025_digital.pdf accessed 11 April 2025.

[61] ibid.

[62] Through the Kenya Off-Grid Solar Access Project (KOSAP) https://www.kosap-fm.or.ke accessed 28 October 2025.

[63] The Nigeria Rural Electrification Agency as Implementing Agency of the Federal Government of Nigeria is tasked with electrification of rural and unserved communities. As part of its Economic Sustainability Plan (ESP), Nigeria launched the Solar Power Naija Programme (SPN) with the aim of deploying five million new solar-based connections in unserved and underserved communities, including off-grid businesses. The programme seeks to expand energy access to 25 million people by providing Solar Home Systems (SHS) or mini-grid solutions, thereby supporting inclusive electrification and stimulating economic activity in remote areas. See https://spn.rea.gov.ng accessed 11 April 2025.

[64] https://www.cif.org/news/rwanda-engaging-private-sector-grid-solar-electrification accessed 11 April 2025 and the full report available at https://www.cif.org/sites/cif_enc/files/knowledge-documents/cdi_rwanda.pdf accessed 11 April 2025.

[65] IEA, 'Financing Clean Energy in Africa' (2023) 53 https://www.iea.org/reports/financing-clean-energy-in-africa accessed 11 April 2025.

[66] IRENA, 'Pay-as-you-go Models: Innovation Landscape Brief' (2020) 10 https://www.irena.org/-/media/Files/IRENA/Agency/Publication/2020/Jul/IRENA_Pay-as-you-go_models_2020.pdf accessed 11 April 2025; African Energy Chamber 'The State of African Energy: 2025 Outlook Report', 78 https://energychamber.org/wp-content/uploads/The-State-of-African-Energy-2025_digital.pdf accessed 11 April 2025.

[67] Success in Rural Electrification Framework Case Studies https://africa-energy-portal.org/sites/default/files/2022-03/Success-in-Rural-Electrification-Case-Study_Sierra-Leone.pdf accessed 11 April 2025.

[68] Global Energy Alliance for People and Planet https://energyalliance.org/zambian-government-partners-unveil-new-financial-mechanism-accelerate-energy-access-minigrids/ accessed 11 April 2025.

[69] Power Moves: South Africa Sets the Pace in Renewable Battery Storage https://www.cif.org/news/power-moves-south-africa-sets-pace-renewable-energy-battery-storage accessed 11 April 2025.

[70] Middle East Business Intelligence https://www.meed.com/morocco-plans-16gw-battery-storage-projects accessed 12 April 2025.

[71] USAID 'Land Tenure & Energy Infrastructure' (2023) 2.

initiatives facing opposition, delays, or even cancellation due to social or environmental conflict.[72]

Recent studies underscore Latin America's increasing commitment to renewable energy, reflected in sustained investments in solar and wind infrastructure and advancements in related technologies. The region's energy transition is driven by the goal of replacing fossil fuels with renewable sources to reduce carbon emissions and lessen environmental impacts.[73] Driven by this objective, wind farms and solar plants have been implemented in Peru, Brazil, and Argentina, contributing to an approximate 25 per cent reduction in the cost of energy. Also, hybrid plants combining wind and solar energy have been developed in Brazil, further enhancing generation efficiency and system flexibility.[74] However, despite clear ambitions to expand renewable energy, the transition is largely being driven by private companies, with minimal involvement from local communities. This top-down approach, marked by large-scale wind projects, closed policy-making processes, and a lack of incentives for community participation, risks undermining the sustainability and inclusiveness of the region's energy transformation.[75] Without integrating rural communities through fair land use, job creation, and participatory planning, Latin America's renewable agenda may fall short of delivering equitable, long-term change.[76] Latin America accounts for 61 per cent of all global human rights abuse allegations linked to renewable energy development, with land and territory rights being the most frequently violated. These abuses include displacement, disregard for communal or ancestral land claims, and infringements on access to water, food, and livelihoods. In South America, 25 per cent of all reported abuses concern land rights, with hydroelectric projects driving most displacement cases, while Central America reports the highest number of individual land-related violations. As renewable energy projects require extensive surface area, their expansion is intensifying land conflicts across the region, especially where land tenure is insecure or contested.[77]

Across Asia, decentralized energy initiatives are gaining traction. In India, the PM-KUSUM scheme supports solar installations on agricultural land by offering financial assistance for farmers to install grid-connected solar pumps and develop solar projects up to 2 MW.[78] However, implementation has faced delays due to issues such as land

[72] For example, the cancellation of the 60.8MW Kinangop Wind Park in Kenya due to hostility from the community: 'Firm pulls the plug on Sh15bn Kinangop wind farm project' https://www.businessdailyafrica.com/Firm-pulls-the-plug-on-Sh15bn-Kinangop-wind-farm-project/-/539552/3088352/-/14l4h5u/-/index.html accessed 11 April 2025, also discussed above. Another example, also from Kenya, is the Lake Turkana Wind Power project, which faced legal challenges from Indigenous communities who claimed that the land acquisition process was irregular and infringed on their rights. The Kenyan Environmental and Land Court found in *Mohamud Iltarakwa Kochale & 5 others v Lake Turkana Wind Power Ltd & 9 others*, ELC Case no 163 (2014) 161, that the land acquisition was 'irregular, unlawful and unconstitutional'. See also P Minadakisa and J Vega-Araújob, 'Rethinking the Social License to Operate? A Theoretical Exploration of its Synergies with Social Acceptance and Energy Justice for a Just Transition' (2024) 113 Energy Research & Social Science 1, 1.

[73] R Seminario-Córdova, 'Latin America Towards Sustainability through Renewable Energies: A Systematic Review' (2023) 16(21) Energies 7422.

[74] ibid.

[75] R Bertam, 'Why community renewable energy projects in Latin America is a pipe dream' (1 April 2021) https://energytransition.org/2021/04/why-community-renewable-energy-projects-in-latin-america-is-a-pipe-dream/ accessed 27 July 2025.

[76] ibid.

[77] Business & Human Rights Resource Centre, 'Renewable Energy (In)justice in Latin America' (August 2021) 11–12 https://media.business-humanrights.org/media/documents/RE_LATAM_final_English.pdf accessed 25 July 2025.

[78] https://pmkusum.mnre.gov.in/#/landing accessed 25 July 2025; A Rahman, S Agrawal, and A Jain, 'Powering Agriculture in India: Strategies to Boost Components A & C Under PM-KUSUM Scheme' (202) (Council on Energy, Environment and Water) 16.

acquisition,[79] community resistance, and delays in statutory approvals.[80] Indonesia launched its US$20 billion Just Energy Transition Partnership (JETP)[81] in November 2022 to support a shift towards renewable energy. However, the country's history of land-intensive development (including the loss of 27.5 million hectares of forest) has raised concerns about the environmental and social impacts of future land acquisition. JETP-related investments in solar, wind, biomass, and geothermal energy risk exacerbating these pressures, particularly in areas with contested or sensitive land rights.[82] Balancing energy transition, deforestation reduction, and Indigenous and community rights with land-intensive investments in extractive sectors demands integrated planning. To meet global climate and human rights commitments, it is essential that environmental protections and land rights are not undermined by commercial interests.[83] Obtaining free, prior and informed consent (FPIC) from Indigenous communities remains particularly challenging, as many affected communities lack formal legal protection for their customary territories.[84]

Despite growing global and regional interest in DERs, the absence of robust, coherent, and justice-oriented legal and policy frameworks continues to hinder their equitable deployment. Across jurisdictions, gaps persist in addressing the intersection of DER implementation with land access, tenure security, environmental protection, and community rights. These legal and policy shortcomings undermine the goal of sustainable decarbonization, where emissions reduction efforts must not come at the expense of environmental justice, Indigenous rights, or equitable land governance. Bridging these legal and policy gaps is essential not only for scaling up DERs but for doing so in a manner that is sustainable, inclusive, and rights-based. The following section explores how these competing interests might be balanced through an energy justice approach.

IV. Balancing Competing Interests through an Energy Justice Approach: Recommendations

As previously mentioned, the implementation of DER projects often gives rise to various human and environmental challenges, as well as conflict with land tenure systems and access, particularly as they are used to decentralize energy systems and expand access. Collectively, these challenges give rise to competing interests in the use of distributed energy resources for decarbonization and the energy transition. To address these challenges, this section argues for the adoption of strategic approaches through four channels namely: (1) mainstreaming human rights and environmental rights into DER projects; (2) ensuring consultation and participatory mechanisms; (3) recognizing land ownership and tenure systems; and (4) developing an effective social licence framework for DER projects. The

[79] Saur News Bureau, 'Waiting To Take Off—The PM-KUSUM Story' (20 April 2023), https://www.saurenergy.com/solar-energy-articles/waiting-to-take-off-the-pm-kusum-story accessed 25 July 2025.

[80] RenewableWatch, 'Solarising Irrigation: PM-KUSUM's promise, progress and challenges' (1 July 2025), https://renewablewatch.in/2025/07/01/solarising-irrigation-pm-kusums-promise-progress-and-challenges accessed 25 July 2025.

[81] Just Energy Transition Partnership https://jetp-id.org accessed 25 July 2025.

[82] International Institute for Environment and Development, 'Bringing Coherence and a Rights-based Approach to Indonesia's Climate and Investment Governance' (August 2023) IIED Briefing 2.

[83] ibid 4.

[84] See in general C Filer, S Mahanty, and L Potter, 'The FPIC Principle Meets Land Struggles in Cambodia, Indonesia and Papua New Guinea' (2020) 9(3) Land 67.

objective of these four approaches is to mitigate the risks associated with DER deployment, thus potentially contributing to the broader objective of achieving sustainable energy transitions. To that end, each strategy is analysed through the lens of the three core tenets of energy justice: distributive justice, procedural justice, and recognition justice.[85]

A. Mainstreaming Human and Environmental Considerations into DER Projects

As mentioned earlier, while DERs are often promoted as clean alternatives to fossil fuels, their deployment is not without human and environmental implications. In the absence of proper legal safeguards and regulatory oversight, DER projects may unintentionally cause negative human and environmental impacts, particularly when deployed in ecologically sensitive or socially complex areas. When deployed in ecologically sensitive areas or near human settlements, DER projects may lead to land degradation, habitat disruption, and water resource strain, risking the livelihoods of local communities.[86]

Despite the rise in DER projects to boost energy security and drive decarbonization, many countries lack comprehensive regulatory frameworks addressing their implications. For example, ESIA requirements may be relaxed or disposed of for smaller decentralized systems.[87] Environmental regulatory frameworks often also do not incorporate DER-specific environmental guidelines.[88] This oversight poses an energy justice issue. From the viewpoint of recognition justice, neglecting traditional uses or land tenure systems, as well as disregarding the connection between impacted Indigenous communities and the environment, leads to the marginalization of vulnerable groups.[89] The deployment of DERs without proper consideration of the environmental and social realities faced by host communities raises concerns about procedural justice. Affected communities often are not

[85] Distributive justice focuses on the fair allocation of benefits and burdens across different social and economic groups, procedural justice emphasizes inclusive and transparent decision-making processes, ensuring that all affected stakeholders have meaningful opportunities to participate, and recognition justice ensures that diverse cultural identities, local knowledge systems, and community values are respected and incorporated into policy and project design. For a more detailed discussion, see L Bhullar, 'Meanings of Energy Justice in the Low-Carbon Transition' in G Bellantuono and others (eds), *Handbook of Energy Law in the Low-Carbon Transition* (Walter de Gruyter 2023) 65; see also MM Akrofi, BC McLellan, and M Okitasari, 'Characterising "Injustices" in Clean Energy Transitions in Africa' (2024) 83 Energy for Sustainable Development 1, 2; see also K Wijsman and M Berbés-Blázquez, 'What Do we Mean by Justice in Sustainability Pathways? Commitments, Dilemmas, and Translations from Theory to Practice in Nature-Based Solutions' (2022) 136 Environmental Science and Policy 377, 379.

[86] Akrofi (n 85) 1; see also T Alex-Oke and others, 'Renewable Energy Market in Africa: Opportunities, Progress, Challenges, and Future Prospects' (2025) 59 Energy Strategy Review 1, 13.

[87] For example, in Zimbabwe and Malawi. See para 9.1 of the National Renewable Energy Policy of Zimbabwe (March 2019) and para 10.4 of the Regulatory Framework for Mini-Grids of Malawi (March 2019); see also C Bhardwaj and S Misra, 'Environmental Impact Assessment and Solar Energy Projects in India: Solutions Creating the same Problems' https://www.nlsir.com/post/environmental-impact-assessment-solar-energy-projects-in-india-solutions-creating-the-same-proble accessed 31 July 2025. Here, the authors faulted the approach to India's energy transition agenda where there is the practice of EIA exemption for renewable energy projects and in the existing laws which provides a rule that does not require solar energy projects to be reviewed through the EIA rules and regulations.

[88] IRENA, 'Policies and Regulations for Renewable Energy Grids' (2018) 20 https://www.researchgate.net/profile/Amalia-Suryani/publication/344931245_Case_Study_Indonesia_within_the_Policies_and_Regulations_for_Renewable_Energy_Mini-grids/links/5f99834d299bf1b53e4bd42b/Case-Study-Indonesia-within-the-Policies-and-Regulations-for-Renewable-Energy-Mini-grids.pdf accessed 31 July 2025.

[89] See Bhullar (n 85) 69; see also Akrofi (n 85) 2.

consulted, lack adequate information, and are not given the opportunity to participate in decisions that impact their land, environment, and livelihoods.[90]

These DER challenges can be tackled through a structured approach that incorporates human and environmental considerations into the planning, execution, and implementation of energy projects. Mainstreaming environmental considerations involves a process where national governments and DER project proponents assess risks and potential violations at the outset and address them proactively. This practice entails integrating any likely human and environmental impacts of DER projects into the workflow to balance competing interests and prevent energy injustice. Consequently, states could adopt a rights-based framework to make the inclusion of human and environmental considerations in DER projects compulsory, which could be accomplished by embedding these considerations into energy laws and policies to ensure compliance by DER project proponents. Such an approach would enable DER project proponents to identify potential human and environmental impacts and demonstrate measures taken to address them. Here, states and project proponents would be able to use the mainstreaming approach to determine the human and environmental rights concerns, the groups likely to be affected, the specific rights that would be affected by the project, and most importantly, can address them before they occur. The aim is to ensure that energy access projects for economic development do not result in negative human rights consequences.[91]

B. Consultation and Participatory Mechanisms in DER Projects

The role of consultation and participatory mechanisms in successfully implementing DERs for a sustainable energy transition cannot be overstated. This process aligns with the DER's role in fostering energy democracy and diversification by engaging citizens to create energy solutions using innovative renewable energy technologies. Moreover, it is a vital key to securing social acceptance of energy projects associated with human and environmental rights issues. Consultation and participatory mechanisms can be used to temper the potential conflicts over land and infringement of community rights, often associated with DER projects. They also play a crucial role in ensuring that the environmental realities experienced by affected communities are adequately understood and incorporated into project planning, thereby reinforcing the integration of human and environmental considerations outlined earlier.

Access to and participation in decision-making regarding DER projects should not merely be a box-ticking exercise.[92] The process should be inclusive in terms of energy decision-making and outcomes.[93] Effective consultation and participatory mechanisms promote public engagement and ensure that appropriate precautions are taken when siting energy projects in local and Indigenous communities. In the development and deployment

[90] Bhullar (n 85) 68.

[91] United Nations Sustainable Development Group (UNSDG), 'The Human Rights Based Approach to Development Cooperation Towards a Common Understanding Among UN Agencies' (September 2004) https://unsdg.un.org/resources/human-rights-based-approach-development-cooperation-towards-common-understanding-among-un accessed 10 August 2025.

[92] ibid 69.

[93] ibid.

of DER projects, it is essential for people to be involved in energy decisions that may impact their rights. Community members should not be excluded from actively participating in the planning and execution of these energy projects.[94]

Jenkins and others identify three mechanisms underpinning inclusiveness to achieve procedural justice. The first is the mobilization of local knowledge. Local knowledge plays a vital role in ensuring fair and sustainable energy decision-making, particularly within the framework of procedural justice. However, local knowledge extends beyond Indigenous contexts and place values on community groups, scientists, and civil society actors who can also contribute valuable insights. The inclusion of diverse stakeholders emphasizes that meaningful participation includes, in addition to presence in decision-making, the integration of local knowledge and perspectives into policy and planning.[95] The second mechanism is the disclosure of information. Procedural justice requires impartial, transparent information disclosure coupled with responsive engagement mechanisms. Merely emphasizing the need for consultation in decision-making is not enough—full information disclosure by public and private actors in energy projects is required.[96] The final mechanism is representation in institutions. Procedural justice requires improved representation of traditionally underrepresented groups, such as women, ethnic minorities, and Indigenous communities in decision-making institutions, including business, local, national, and international governmental bodies, as well as non-state actors.[97]

To address issues concerning the social acceptance of energy projects, it is crucial for states to avoid approving the development of energy projects without clear and adequate citizen participation through a free, fair, direct, and transparent consultative process, as this violates human rights. This underscores the importance of full participation and inclusion in all aspects of society to empower individuals to realize their potential as citizens and contributing members of their communities. When this approach is adopted—ie, integrated into the energy law and policy framework—it can help project proponents collaborate more effectively with host community members to identify, anticipate, and mitigate potential negative human and environmental rights impacts of projects. This approach would reduce the chances of DER projects becoming contentious among host or Indigenous communities, developers, and the government.

C. Recognizing Land Ownership and Tenure Systems

Injustice in energy projects often arises from the location or siting of projects.[98] Effective land-use planning and resource management are essential to ensure that renewable energy development is aligned with both environmental sustainability and the social priorities of affected communities.[99] As mentioned earlier, land tenure systems and ownership

[94] DS Olawuyi, 'Energy (and Human Rights) for All: Addressing Human Rights Risks in Energy Access Projects' in R Salter, CC Gonzalez, and E Kronk Warner (eds), *Energy Justice: US and International Perspectives* (Edward Edgar 2018) 89.

[95] K Jenkins and others, 'Energy Justice: A Conceptual Review' (2016) 11 Energy Research and Social Science 174, 178.

[96] ibid 178.

[97] ibid.

[98] Akrofi (n 85) 5.

[99] Alex-Oke (n 86) 13.

are among the most significant challenges to using DER for sustainable energy transition. The acquisition of land for renewable energy infrastructure can disrupt agricultural livelihoods and lead to the displacement of communities, especially in densely populated rural regions.[100] The failure to recognize land tenure systems and ownership rights may affect the deployment of DERs as a strategy to drive security and transition.

Ensuring the security of land tenure is important to actions aimed at economic and sustainable development. For example, secured access to land is critical for the millions of poor people in rural Africa, especially those who depend on agriculture, livestock, or forests for their livelihoods.[101] Additionally, secured land access for Indigenous communities is vital for reducing their vulnerability to hunger and poverty and enhancing their prospects for better livelihoods, thus contributing to justice, peace, and sustainable development. When proponents of energy projects, including states, fail to recognize land tenure systems and ownership, particularly in the process of allocating land for projects such as hydroelectric dams, wind turbines, and solar panels, this is likely to lead to the displacement of people from their ancestral or spiritual lands.[102] Therefore, it is important that states, as well as project developers, ensure there are clear guidelines anchored on the recognition of relevant land tenure systems and ownership before approving energy projects. Such guidelines would facilitate the preparation of adequate arrangements for relocating affected people and ensure the implementation of a proper plan for resettlement before the displacement occurs.

Beyond issues of tenure systems and land access, project siting also raises significant distributive justice concerns. In many African contexts, energy projects are developed within or near rural communities, yet local households often remain unconnected to the energy produced or underserved.[103] This disconnect underscores a critical flaw in how energy benefits are distributed: energy infrastructure is physically present, but the host or Indigenous communities expected to support it do not benefit. Such outcomes undermine the legitimacy of clean energy transitions and reinforce patterns of exclusion and inequality in energy decisions. A truly just energy transition must, therefore, address not only where projects are sited and whose land is affected but also who benefits from the energy produced. This would require an approach that introduces better laws for the protection of tenure security as well as ensures a just distribution of energy produced from clean energy sources.

D. Developing an Effective Social Licence for DERs

Similar to other carbon-based energy projects that need community approval through social license, this also applies to DER projects, particularly in the areas or host communities where they are located. Due to the perceived acceptance of clean energies as part of transition strategies, social acceptance has been neglected, which is supposed to form part of

[100] ibid 14.
[101] USAID (n 71) 2.
[102] P Oniemola, 'Business Enterprises in Renewable Energy Projects in Africa and the Human Rights Questions Arising from the Duty to Protect' in D Olawuyi and O Abe (eds), *Business and Human Rights Law and Practice in Africa* (Edward Elgar 2022) 197.
[103] Akrofi (n 85) 6.

the procedure for deploying DERs in sustainable energy development and transitions. The consequence of this neglect is the public's rejection of DER technology and projects due to a lack of interaction with projects that may have public ownership and the distribution of benefits at the community level. In practice, social licence helps build stable, constructive, and mutually acceptable relationships between local and Indigenous communities and DER project developers or companies by considering how individuals feel about projects in their communities, and what conditions of approval, legitimacy, and trust are necessary for the smooth execution of energy projects.

Social licence is not a formal legal permit but rather an ongoing level of acceptance and trust by host communities of a company and its operations.[104] Achieving and maintaining this licence requires more than one-off consultations, participation, or information campaigns. It entails early, sustained, and meaningful engagement with stakeholders, recognition of local rights and knowledge, and equitable sharing of the benefits and burdens of DER projects. Social licence invokes and brings together the three main tenets of energy justice and the three strategies mentioned above. Furthermore, the use of an effective social licence framework is likely to promote participation and potentially increase the uptake of DERs. Particularly where states employ a mandatory scheme for DERs, the use of an effective social licence framework would ensure compliance and reduce the costs associated with running the mandatory scheme.

As has been discussed, due to the risks associated with DER projects, their deployment requires acceptance from various stakeholders, including those directly affected by technological developments. This means that many barriers to the successful implementation of DER projects may stem from a lack of public acceptance among different stakeholder groups. Therefore, it is necessary to ensure that some level of social licence accompanies the deployment of DERs—that is, acceptance of decisions related to locating and establishing power plants and renewable energy projects by local residents. When DER project developers adopt an approach focused on securing social licence before planning, executing, and implementing, it is more likely to address issues related to human and environmental rights, lack of consultation and participatory mechanisms, and non-recognition of tenure systems and ownership.

V. Conclusion

The deployment of DER is essential for achieving energy security and sustainable transitions. More so, international experience has shown that DERs can deliver relevant energy and environmental as well as economic and social impacts.[105] With their incorporation into current energy systems, they could help to achieve the goals of decarbonization, energy security, and efficiency.[106] However, they face specific challenges, including issues related to land access, tenure security, human rights, and environmental sustainability. These factors, along with the lack of comprehensive legal and policy frameworks addressing these

[104] Minadakisa (n 72 above) 1.
[105] Gonzalez and Rendon (n 5) 2.
[106] ibid.

challenges, make deploying DER difficult. Addressing these challenges is essential for safe and sustainable energy transitions.

As this chapter has shown, there are tensions and challenges associated with deploying DERs for sustainable energy transitions. These mainly stem from issues related to land access and tenure security, as well as the lack of comprehensive legal and policy frameworks anchored on human and environmental rights consideration in the deployment of DER to support energy access, decarbonization, and efficiency. The chapter has argued through an energy justice perspective that tackling associated risks and challenges in DERs implementation requires actions aimed at integrating human and environmental considerations into DER projects. Integrating these considerations should be accompanied by a process that allows for community participation in energy decisions, recognizing land ownership and tenure systems, and securing social acceptance from Indigenous communities where projects are located. Implementing these approaches within international, regional, and domestic legal and policy frameworks would help ensure that all risks associated with DER deployment are considered and managed, thereby promoting a more just and sustainable energy transition.

5

Sustainability Challenges of Small Modular Reactors in Distributed Energy Systems

Louis de Fontenelle

I. Introduction

This chapter examines sustainability questions and challenges for international, regional, and national law posed by the development of small modular reactors (hereafter SMRs) in distributed energy resource (DER) systems. Key issues related to risk management, safety and security, environmental justice, decommissioning, environmental protection, and stakeholder involvement in the development of SMRs are central to this analysis.

A preliminary observation is that SMRs currently under development present highly diverse technical characteristics, ranging from microreactors of only a few megawatts to larger modular plants designed for grid-scale supply, and including land-based as well as maritime/offshore concepts. This technological heterogeneity complicates their legal apprehension: no single regulatory regime can fully encompass the different uses and deployment contexts of SMRs. Instead, their governance is fragmented across multiple legal spheres—nuclear law, environmental law, and maritime law—and across international conventions with varying scopes and obligations.

The emergence of SMRs is situated at the intersection of two major global trends: the acceleration of the energy transition and the industrial revival of nuclear power. On the one hand, the transition to low-carbon energy systems has created new pressures to decarbonize electricity production while maintaining grid reliability. On the other hand, the energy crises of the 2020s have led to renewed political and industrial interest in nuclear energy, not only as a low-carbon option but also as a sovereign technological sector. SMRs, with their modularity, flexibility, and lower upfront capital cost are increasingly perceived by several states as a technological lever that reconciles decarbonization objectives with industrial competitiveness and energy sovereignty.

The core research question underpinning this chapter is therefore as follows: to what extent does the legal framework governing SMRs—at international, regional, and national levels—enable their development in a manner consistent with the principles of sustainable development? For the purposes of this analysis, sustainable development is understood in its conventional three-pillar structure—environmental, social, and economic—as consolidated by international instruments since the late twentieth century. The detailed implications of this framework will be discussed in Section III. As Del Guayo notes in Chapter 1 of this book, 'sustainability has three dimensions: environmental, economic and social. DERs

Louis de Fontenelle, *Sustainability Challenges of Small Modular Reactors in Distributed Energy Systems* In: *Sustainable Distributed Energy Resources*. Edited by: Íñigo del Guayo, Damilola S. Olawuyi San, Louis de Fontenelle, Milton Fernando Montoya, and José Juan González, Oxford University Press. © International Bar Association 2026. DOI: 10.1093/9780198991946.003.0006

play a significant role in the transition to decarbonization and, thus, to the sustainability of the energy sector.'[1]

In that regard, SMRs are increasingly discussed as a potential complement to renewable energy sources in decentralized microgrid architectures whether in fully autonomous, off-grid systems disconnected from centralized infrastructure, or as part of a broader process of decentralizing the central energy system itself. Their modularity and flexible deployment make them attractive candidates for integration into microgrids powered by variable renewable energy sources. Unlike renewables, SMRs provide dispatchable, controllable power, which enhances grid resilience, ensures energy reliability, and supports system stability in both isolated and decentralized configurations.[2] SMRs are seen as contributing to the transformation of the energy system along four structural axes: decarbonization, decentralization, digitalization, and democratization.[3]

This decentralization, which requires electricity grids to be rethought as bidirectional infrastructures able to accommodate new uses and an unprecedented situation in the field of transmission and distribution, is in fact entirely compatible with the SMR model, which can itself be envisaged in this decentralized logic—depending, of course, on the unit power considered for each reactor.

SMRs could therefore serve a wide range of applications, from supplying electricity and heat to urban areas and energy-intensive industries (eg hydrogen production, steelmaking, or synthetic fuels), to remote or offshore operations, including maritime transport.[4] They are particularly promising for supplying isolated regions—whether rural or islanded—that are not well served by national transmission infrastructure and may be vulnerable to extreme weather or other disruptions.[5] In the United States, for instance, microreactors are being considered for rural and tribal communities facing infrastructural deficits and energy insecurity.[6]

Such prospects suggest that SMRs could contribute to sustainability objectives. However, this alignment remains conditional: their development must be assessed against environmental, social, and economic criteria—including the principles of energy justice, which emphasize equity in access, participation, and intergenerational responsibility.[7]

This chapter adopts a critical legal methodology based on doctrinal analysis of legal instruments and scholarly literature, with particular attention to how legal frameworks at the international level—and within selected national and regional contexts, notably the European Union—address the sustainable deployment of SMRs. The chapter is structured as follows. After this introduction, Section II examines the place of SMRs in the energy transition, focusing on their legal characterization, typology, industrial role, and integration into national and international climate strategies. Section III situates SMRs within the

[1] I Del Guayo, 'Distributed Energy Resources as the New Electricity Paradigm: Legal Context and Principles', in this book.

[2] D Michaelson and J Jiang, 'Review of Integration of Small Modular Reactors in Renewable Energy Microgrids' (2021) 152 Renewable and Sustainable Energy Reviews 111638.

[3] Del Guayo (n 1).

[4] IAEA, 'Small Modular Reactors: Advances in SMR Developments' (Non-serial Publications, IAEA, Vienna, 2024).

[5] IN Kessides and V Kuznetsov, 'Small Modular Reactors for Enhancing Energy Security in Developing Countries' (2012) 4 Sustainability 1806.

[6] HK Fishler and others, 'Challenges to Implementing Microreactor Technologies in Rural and Tribal Communities' (2023) 3 Journal of Critical Infrastructure Policy 61.

[7] BK Sovacool and MH Dworkin, *Global Energy Justice: Problems, Principles, and Practices* (CUP 2014).

broader framework of sustainable development by analysing three dimensions: environmental integrity and intergenerational equity, environmental and energy justice, and economic viability as a condition for sustainable investment. Section IV identifies key legal and policy gaps and develops recommendations for a coherent legal framework for SMRs, emphasizing international harmonization, the integration of energy justice, and democratic governance across scales. Section V concludes by reflecting on the capacity of legal systems to guide the responsible and sustainable deployment of SMRs.

II. Determining the Legal Nature and Regulatory Status of SMRs Within the Energy Transition Framework

A. Definition and Typology of SMRs

There is currently no binding legal definition of SMRs. However, definitions do appear in the grey literature. The most widely accepted is that of the International Atomic Energy Agency (IAEA), which defines SMRs as 'advanced reactors that produce electricity of up to 300 MW(e) per module. These reactors have advanced engineered features, are deployable either as a single or multi-module plant, and are designed to be built in factories and shipped to utilities for installation as demand arises.'[8] SMRs therefore have two defining features: first, their limited electrical capacity—typically capped at 300 megawatts electric (MW(e)) per module; and second, their modular and standardized design intended for serial factory fabrication and on-site assembly.

This definition is echoed by the European Parliament, which, in its resolution of 12 December 2023, stated that 'SMRs can be defined as nuclear reactors in a range of power typically between 10 and 300 MW, designed to be built in factories in a standardised modular form.'[9]

Currently, six main types of SMRs are under development.[10] These include: (i) Light Water Reactors (LWRs) and Heavy Water Reactors (HWRs); (ii) marine-based water-cooled SMRs (floating electrical units mounted on barges); (iii) gas-cooled SMRs, suitable for industrial heat and cogeneration; (iv) liquid metal-cooled, fast-neutron SMRs (using lead, sodium, etc); (v) molten salt SMRs (with molten fuel and/or coolant); and (vi) microreactors (under 30 MW(th)), developed for remote locations and microgrids, using coolants such as light water, helium, molten salt, or liquid metal.[11]

B. Industrial Stakes and Climate Policy

SMRs are being developed globally as part of efforts to combat climate change and as industrial opportunities. They are seen as contributing to the achievement of greenhouse gas

[8] IAEA, 'Small modular reactors' https://www.iaea.org/topics/small-modular-reactors accessed 11 September 2025.

[9] European Parliament Resolution of 12 December 2023 on Small Modular Reactors (2023/2109(INI)).

[10] IAEA, 'Small Modular Reactors: Advances in SMR Developments' (Non-serial Publications, IAEA, Vienna 2024).

[11] ibid.

emission reduction goals under the Paris Agreement of 12 December 2015. Some countries view SMRs as integral to their national decarbonization and industrial strategies.[12]

According to IEA and NEA analyses, innovations in nuclear energy, including SMRs, could help reduce an additional fifteen gigatonnes of CO2 by 2050.[13] SMRs are also perceived as a way to address uranium supply issues. While uranium deposits exist in around fifty countries,[14] and supply disruptions have not historically been used coercively,[15] the true vulnerability lies in the limited number of uranium processing facilities and mastery of fuel-cycle technologies. SMRs may help mitigate this by requiring less fuel and enabling longer operating cycles—between three and seven years, or up to thirty years for some models.[16]

However, nuclear power remains controversial in the energy transition. The IEA notes that nuclear power has avoided 55 GT of CO_2 emissions over the past five decades and that abandoning nuclear would raise the global cost of climate targets by an estimated USD 1.6 trillion.[17] Currently, nuclear provides about 10 per cent of global electricity, with 444 reactors in operation and a combined capacity of 394 GWe.[18]

Several countries have publicly committed to scaling up nuclear capacity by 2050, including via SMRs.[19] At COP28 in Dubai, a Joint Declaration on Tripling Nuclear Energy explicitly supported SMR development: 'Commit to supporting the development and construction of nuclear reactors, such as small modular and other advanced reactors for power generation as well as wider industrial applications for decarbonization, such as for hydrogen or synthetic fuels production.'

C. Applicability of International Legal Frameworks and Development of National Regulations

SMRs actually encompass a variety of technologies with different technical characteristics. Taken separately, each of them will not pose the same difficulties of legal qualification. On the whole, the literature seems unanimous in recognizing the applicability of international conventions on nuclear security[20] and nuclear safety,[21] even if doubts persist about some of them, which are still being developed. Depending on how these SMRs are used, certain conventions will be expressly excluded from application (for example, for maritime or air

[12] OECD, 'The NEA Small Modular Reactor Dashboard: Second Edition' (NEA, No 7671, 2024).
[13] NEA, 'Meeting Climate Change Targets: The Role of Nuclear Energy' (Paris, OECD Publishing 2022).
[14] NEA, 'Uranium 2022: Resources, Production and Demand' (Paris, OECD Publishing 2023).
[15] Teva Meyer, 'L'uranium: un regain d'intérêt pour l'énergie nucléaire' (2023) 1 Questions internationales 27.
[16] EESC, Opinion on the Role of Nuclear Energy in the Stability of EU Energy Prices [2022] OJ C 486.
[17] IEA, 'Nuclear Power in a Clean Energy System' (Paris, IEA, 2019) https://www.iea.org/reports/nuclear-power-in-a-clean-energy-system accessed 11 September 2025.
[18] NEA (n 13).
[19] IAEA, 'Small Modular Reactors: Advances in SMR Developments' (Vienna, Non-serial Publications, IAEA 2024).
[20] BD Jenkins, 'The Efficacy of the Global Nuclear Security Legal Regime and States' Implementation Capacity in Light of the Forthcoming Development of Advanced Nuclear Reactor Technologies' Chapter 8 in International Atomic Energy Agency (ed), *Nuclear Law: The Global Debate* (Asser Press 2022). The author acknowledges the effectiveness of the principles and rules of international law that underpin the nuclear security legal regime for dealing with SMRs (such as the Convention on the Physical Protection of Nuclear Material and Nuclear Facilities concluded on 26 October 1979).
[21] A Van Kalleveen, 'Applicability of the international nuclear legal framework to small modular reactors (SMRs)' (EUR 30992 EN, Luxembourg, Publications Office of the European Union 2022).

transport, and those used for military purposes[22]). It will undoubtedly take a constructive interpretation of certain legal concepts, including in particular that of 'nuclear installation', on the part of the international authorities to determine whether certain SMRs can be included in the scope of application of the international conventions.[23] Clearly, the application of international nuclear law to SMRs will be facilitated to the extent that countries modify their regulatory frameworks to adapt to international nuclear safety and security requirements.[24] But it is certainly in the area of nuclear liability that the problems of legal qualification and applicability of the provisions of international conventions are likely to be the most significant, if certain authors are to be believed.[25]

A number of national frameworks are currently emerging. In the United States, the Congress has actively promoted SMRs through regulatory reforms and legislation to remove market and licensing barriers.[26] In Canada, SMRs are integrated into federal-provincial climate strategies. Federal action plans, increased budget allocations to the Canadian Nuclear Safety Commission (CNSC), and Memoranda of Understanding between four provinces (2019–2020) illustrate strong political support.[27] SMRs are considered strategic for supplying electricity to remote areas, replacing coal, and serving smaller provinces. However, critics point to regulatory opacity, exclusion from impact assessments, and possible conflicts of interest.[28] In India, SMRs are viewed as essential to achieving net-zero objectives. Despite strong public policy interest, regulatory reforms are needed to enable private and foreign investment.[29] SMRs offer benefits over traditional reactors, especially in terms of land use, public acceptance, and localization of energy access. Furthermore, since 2013, the United Kingdom has promoted SMRs through national strategies and major R&D investments.[30] SMRs are also linked to defence policy. By 2050, the UK plans to generate up to 10 GW from 16 SMRs.[31] In France, Law No 2023-491 of 22 June 2023 introduces measures to accelerate nuclear development near existing sites and mandates government planning for SMRs and Gen IV reactors. SMRs are explicitly referenced in Article 6, which foresees strategic national development and technical reporting by public authorities.

While no unified international SMR regime exists, current international nuclear law is broadly adaptable. National strategies are emerging, reflecting different economic, political,

[22] ibid.

[23] VJH Roland, 'Applicability of the Existing Nuclear Liability Conventions to Different Types of Small Modular Reactors Currently Under Development' (2023) 1 Nuclear Law Bulletin OECD 7.

[24] RJ Budnitz, O Heinonen, and A Nilsson, 'Safety, Security, and Safeguards for Small Modular Reactors' (2024) 139 Oxford Energy Forum 29.

[25] J Handrlica and M Novotna, 'The Feast of Insignificance of Small Modular Reactors in International Nuclear Law' (2021) 12 Czech Yearbook of Public and Private International Law 327.

[26] C Merlin, 'Les petits réacteurs modulaires dans le monde: perspectives géopolitiques, technologiques, industrielles et énergétiques' (Etudes de l'Ifri, Ifri 2019).

[27] D Bratt, 'SMRs in Canada: Federal-provincial Cooperation in Pursuing Net-zero Emissions' (2022) 28 Canadian Foreign Policy Journal 305.

[28] MV Ramana and K Blaise, 'Regulation vs Promotion: Small Modular Nuclear Reactors in Canada' (2024) 192 Energy Policy 114228.

[29] H Desai, 'Liberalizing the Nuclear Energy Sector: Small Modular Reactors, Net-Zeros, and Perceptions in India' (2024) 9 International Journal of Nuclear Security Article 12.

[30] P Johnstone, A Stirling, and B Sovacool, 'Policy Mixes for Incumbency: Exploring the Destructive Recreation of Renewable Energy, Shale Gas "Fracking," and Nuclear Power in the United Kingdom' (2017) 33 Energy Research & Social Science 147.

[31] CH Steuer, 'Nuclear Power and the Strategic Role of Small Modular Reactors in the United Kingdom' (2024) 139 Oxford Energy Forum 56.

and legal priorities. These strategies will ultimately shape the sustainability and legal compatibility of SMRs in decentralized energy systems.

III. Sustainable Development and the Governance of SMRs

The legal analysis developed in this section is framed by the three interconnected pillars of sustainable development, understood here not simply as a political aspiration but as a normative framework progressively consolidated in international law. While there is no single legally binding definition of sustainable development, the concept has been consecrated in several international legal instruments. It first appeared in the report *Our Common Future* by the Brundtland Commission in 1987, which defines sustainable development as 'development that meets the needs of the present without compromising the ability of future generations to meet their own needs'.[32]

At the international level, the 1972 Stockholm Declaration marks a foundational milestone. Principle 1 states that 'Man has the fundamental right to freedom, equality and adequate conditions of life, in an environment of a quality that permits a life of dignity and well-being, and he bears a solemn responsibility to protect and improve the environment for present and future generations.' Principle 11 further provides that 'The environmental policies of all States should enhance and not adversely affect the present or future development potential of developing countries [. . .]'. The World Conservation Strategy launched by the International Union for Conservation of Nature (IUCN) in 1980 explicitly referred to sustainable development, and the World Charter for Nature adopted by the UN General Assembly in 1982 reinforced this orientation.

The 1992 Rio Declaration consolidated the international recognition of the concept. Principle 1 affirms that 'Human beings are at the centre of concerns for sustainable development. They are entitled to a healthy and productive life in harmony with nature'. Principle 3 adds that 'The right to development must be fulfilled so as to equitably meet developmental and environmental needs of present and future generations.' The three pillars of sustainable development—environmental, social, and economic—are reflected in Principle 4 ('In order to achieve sustainable development, environmental protection shall constitute an integral part of the development process and cannot be considered in isolation from it'), Principle 8 ('To achieve sustainable development and a higher quality of life for all people, States should reduce and eliminate unsustainable patterns of production and consumption and promote appropriate demographic policies'), and Principle 12 ('States should cooperate to promote a supportive and open international economic system that would lead to economic growth and sustainable development in all countries [. . .]').

These principles have been translated into concrete programmes of action, first with Agenda 21 and subsequently with Agenda 2030 and its seventeen Sustainable Development Goals. They also find expression in specialized legal instruments, such as the 1992 Convention on Biological Diversity, which defines 'sustainable use' as 'the use of components of biological diversity in a way and at a rate that does not lead to the long-term decline of biological diversity, thereby maintaining its potential to meet the needs and aspirations of

[32] World Commission on Environment and Development, 'Our Common Future' (United Nations 1987).

present and future generations'. Similarly, the 1992 United Nations Framework Convention on Climate Change and the 1998 Kyoto Protocol explicitly link climate change mitigation to the promotion of sustainable development.

As Ulrich Beyerlin has observed: 'Notwithstanding the question of whether over the years this concept has gained normativity in the sense of entailing certain political or legally binding rules of State behaviour, it has developed at the international, regional, and national level into a crucial political precept that governs virtually every sphere of activity aimed at embracing and balancing ecology and economy, conservation, and utilization.'[33]

The relevance of this analytical framework for SMRs is twofold. First, it allows for a transversal understanding of the challenges surrounding their development in the context of the energy transition and climate change. Second, it highlights the intergenerational dimension inherent to sustainable development. The doctrine on intergenerational equity emphasizes that sustainable development entails a commitment of fairness towards future generations in the use of natural resources.[34] As Weiss notes, 'The theory of intergenerational equity says that every generation receives a natural and cultural legacy in trust from its ancestors and holds it in trust for its descendants [. . .] These rights and obligations form the corpus of a proposed new doctrine of intergenerational equity.'[35]

According to Weiss's doctrine, the principle of intergenerational equity seeks to address three major risks: resource depletion, environmental degradation, and lack of access to resources. It is based on conserving options (maintaining diversity of resources), conserving quality (preserving environmental integrity), and conserving access (ensuring equitable access to resources across generations).[36] This principle is particularly relevant for environmental issues with long-term effects, such as nuclear waste storage.

Recent international jurisprudence confirms its growing significance. In its advisory opinion of 23 July 2025, the International Court of Justice stated that 'Intergenerational equity is an expression of the idea that present generations are trustees of humanity tasked with preserving dignified living conditions and transmitting them to future generations'.[37] The Court further held that 'Due regard for the interests of future generations and the long-term implications of conduct are equitable considerations that need to be taken into account where States contemplate, decide on and implement policies and measures in fulfilment of their obligations under the relevant treaties and customary international law'.[38] Finally, it emphasized that 'The duty to prevent significant harm to the environment also applies to the climate system, which is an integral and vitally important part of the environment and which must be protected for present and future generations'.[39]

[33] U Beyerlin, 'Sustainable Development' MPEPIL (2013). https://opil.ouplaw.com/display/10.1093/law:epil/9780199231690/law-9780199231690-e1609 accessed 11 September 2025.
[34] E Brown Weiss, 'In Fairness to Our Children: International Law and Intergenerational Equity' (1994) 2 Childhood 22.
[35] ibid.
[36] E Brown Weiss, 'In Fairness To Future Generations and Sustainable Development' (1992) 8 American University International Law Review 19.
[37] *Advisory Opinion on Climate Change* [2025] ICJ para 156.
[38] ibid, para 157.
[39] ibid, para 273.

A. Environmental Integrity and Intergenerational Equity

Nuclear energy, including the development of SMRs, is increasingly viewed as a tool for achieving climate objectives through the decarbonization of electricity systems. However, in order to be considered sustainable, nuclear technologies must meet two fundamental requirements: the efficient use of natural resources and the limitation of adverse environmental impacts throughout their lifecycle.[40] These requirements are particularly critical in the case of SMRs, which introduce new operational modes such as modular construction, mobile applications, and deployment in remote areas. These features challenge existing regulatory frameworks, which were developed primarily for large, stationary nuclear reactors.

Although SMRs differ significantly from conventional nuclear facilities in terms of size, modularity, and potential applications, they remain subject to the principles and rules of international nuclear and environmental law. Yet, the applicability of these legal frameworks raises specific questions due to the diversity of SMR technical designs (including water-cooled, gas-cooled, and molten salt reactors), their intended uses (civilian, industrial, or maritime), and their deployment scenarios (land-based, offshore, or in international waters). At the international level, several legal regimes govern nuclear activities, including those concerning nuclear safety and security, liability, non-proliferation, and safeguards. Many of these instruments incorporate environmental protection considerations. However, as Burns has noted, the case of SMRs highlights structural weaknesses in the existing framework, particularly its limited capacity to adapt to emerging technologies without international harmonization and co-operative interpretation.[41]

The purpose of this section is therefore to assess the ability of international law to respond to the environmental and sustainability challenges posed by the deployment of SMRs, understood here in the broader sense of intergenerational equity. The methodology is based on the identification of legal issues arising from the technical characteristics of SMRs. From this perspective, several themes appear particularly relevant: the principle of environmental impact assessment and risk prevention, including nuclear safety; the right to environmental information and the principle of public participation in environmental decision-making; liability and compensation for environmental damage; site remediation obligations; and the treatment of nuclear waste.

A first structural limitation of international law must be underlined at the outset. Not all States supporting the development of SMRs are parties to the relevant international conventions. In environmental law, certain key treaties, such as the Espoo Convention on environmental impact assessment and the Aarhus Convention on access to information and participation, apply mainly within Europe. Similarly, in nuclear law, not all States operating significant nuclear programmes are parties to the specialized conventions on nuclear safety and liability.[42] This disparity is compounded by particular situations such as Antarctica, where Article 5 of the 1959 Washington Treaty prohibits nuclear explosions and the disposal of nuclear waste, and by regional nuclear-weapon-free zone treaties,[43] which impose

[40] A Al Faruque, *Nuclear Energy Regulation, Risk, and the Environment* (Routledge 2019).

[41] S Burns, 'Milestones in Nuclear Law: A Journey in Nuclear Regulation' in International Atomic Energy Agency (ed), *Nuclear Law. The Global Debate* (Asser Press 2022).

[42] Roland (n 23).

[43] Treaty on the Southeast Asia Nuclear Weapon-Free Zone—Treaty of Bangkok 1995; Treaty Declaring Africa a Nuclear Weapon-Free Zone—Treaty of Pelindaba 1995.

restrictions on radioactive material disposal in specific regions. These limitations explain why national legislation will play a decisive role in framing the sustainable deployment of SMRs. Accordingly, this study highlights legal gaps at the domestic level, drawing on existing literature, without aiming at an exhaustive national analysis.

B. Environmental and Energy Justices

This section adopts the conceptual frameworks of environmental justice and energy justice, with a particular focus on social justice, to provide a critical analysis of the legal framework applicable to SMRs. This approach is useful because it highlights the socio-economic inequalities that could arise from existing legal regimes and public policies, and it enables us to consider how SMR deployment might exacerbate or reduce such disparities. As recent studies have noted, reflecting on the principles of justice is crucial to accompany the development of microreactor and SMR technologies, particularly in vulnerable or marginalized communities.[44]

Environmental justice, which emerged in the United States in the 1970s from social mobilizations led by minority communities disproportionately exposed to industrial hazards, rests on three interdependent pillars: distributive justice, which requires an equitable sharing of environmental benefits and burdens; procedural justice, which entails stakeholder participation in decision-making processes; and recognition justice, which calls for the acknowledgement of cultural and social identities affected by environmental projects.[45] Robert Bullard defines environmental justice as 'the fair treatment and meaningful involvement of all people regardless of race, color, national origin, or income with respect to the development, implementation, and enforcement of environmental laws, regulations, and policies'.[46]

Energy justice constitutes a more recent transposition of these principles to the domain of energy access and governance. Developed mainly in Anglo-Saxon scholarship, it builds upon the same three pillars—distributive, procedural, and recognition justice—while aiming at a global energy system that fairly disseminates both the benefits and the costs of energy services, and one that has representative and impartial decision-making processes.[47]

From the perspective of distributive justice, the development of SMRs raises important questions about the fair distribution of economic benefits. While fiscal rules remain within the competence of national legislations, attention must be paid to the extent to which revenues generated by SMR projects are shared with local communities. Lessons can be drawn from renewable energy value-sharing mechanisms, where social acceptability has been improved by lowering local energy tariffs, creating jobs, granting fiscal advantages, or enabling

[44] Fishler (n 6).

[45] RD Bullard, 'Solid Waste Sites and the Black Houston Community' (1983) 53 Sociological Inquiry 273; RD Bullard and GS Johnson, 'Environmentalism and Public Policy: Environmental Justice: Grassroots Activism and Its Impact on Public Policy Decision Making' (2000) 56 Journal of Social Issues 555; D Schlosberg, *Defining Environmental Justice: Theories, Movements, and Nature* (OUP 2007).

[46] RD Bullard, 'Environmental Justice' in NJ Smelser and PB Baltes (eds), *International Encyclopedia of the Social & Behavioral Sciences* (Pergamon 2001).

[47] BK Sovacool and others, 'New Frontiers and Conceptual Frameworks for Energy Justice' (2017) 105 Energy Policy 677; Sovacool (n 7); BK Sovacool, *Energy & Ethics: Justice and the Global Energy Challenge* (Palgrave Macmillan 2013); BK Sovacool and MH Dworkin, 'Energy Justice: Conceptual Insights and Practical Applications' (2015) 142 Applied Energy 435.

direct financial participation in projects. In the European context, these mechanisms have been critically analysed, showing that distributive justice in renewable energy projects significantly increases local acceptance. Similar mechanisms could be considered for SMRs, especially if they are deployed as decentralized energy resources (DERs).[48]

Procedural justice is equally critical in the governance of SMRs. Several international conventions, such as the Espoo Convention, the Aarhus Convention, and the Escazú Agreement, address issues of access to environmental information and public participation in decision-making. However, their geographical scope is limited, and they do not apply to some of the leading SMR-developing countries, including the United States, Canada, Japan, and China. National legislations thus play a decisive role in preventing procedural injustices. In Canada, for instance, the 2019 Impact Assessment Act appears insufficient to cover SMRs due to their technical features and relatively small power output.[49]

In federal states such as Canada or the United States, procedural justice also intersects with constitutional questions relating to the distribution of competences between federal and subnational authorities. In Canada, nuclear energy falls under federal jurisdiction, while electricity production is the responsibility of the provinces. This division creates complexity that may hinder SMR development. Nevertheless, observers note that co-operation between four provinces and the federal government on SMR deployment is unique in Canada's energy and environmental governance, traditionally marked by federal predominance.[50]

Procedural justice also includes the participation of Indigenous and rural communities, which is particularly relevant in countries such as the United States and Canada. In the US, a recent study emphasized the need for effective involvement of rural and tribal communities in SMR projects, both to ensure respect for sovereignty and cultural values and to address existing service gaps that may take precedence over the deployment of new technologies.[51] Similarly, Canadian initiatives, such as memoranda of understanding with Indigenous groups and the establishment of Indigenous Advisory Councils, have sought to improve Indigenous participation in SMR projects.[52]

Finally, social justice in the context of SMRs requires attention to global inequalities in energy access. A major risk is that SMRs could benefit primarily wealthier countries, further widening the gap between the Global North and South. Extending SMR deployment to developing countries raises legal and policy questions about investment, technology transfer, and financial accessibility. India provides an illustrative case study. Its outdated nuclear regulatory framework has been criticized for preventing private and foreign investment in SMRs, while reform could open the sector to both private actors and international investors, subject to licensing and governmental control mechanisms. Collaboration at the

[48] L de Fontenelle, 'Territorial value-sharing of renewable energies. Critical analysis of the French mechanism introduced by the 2023 Renewable Energy Acceleration Act' in Sébastien Bourdin (ed), *The Social Acceptability of Renewable Energy Projects* (Edward Elgar Publishing, forthcoming 2025).

[49] Ramana (n 28).

[50] Bratt (n 27); R Tscherning and J Dias, 'Emerging Clean Energy Choices in Canada's Net-Zero 2050 Transition: The Role of Nuclear in the Low Carbon and Clean Hydrogen Context' (2024) 16 Yearbook of Antitrust and Regulatory Studies 55.

[51] Fishler (n 6).

[52] A Middleton, 'Mapping the Portrayal of Small Modular Reactors in Canadian Energy Solutions' (2023) 13 The Polar Journal 264.

international level is also identified as a possible pathway to facilitate foreign investment while preserving national sovereignty.[53]

Taken together, environmental justice and energy justice provide a robust analytical framework to interrogate the socio-legal challenges of SMR governance. By integrating distributive, procedural, and recognition-based perspectives, they highlight the risk that SMRs could reinforce existing inequalities unless the legal frameworks that govern them are adapted to ensure fair participation, equitable distribution of benefits, and inclusive recognition of affected communities at both national and international levels.

C. Economic Viability and Sustainable Investment

The economic dimension of SMR deployment raises two central questions. The first concerns the economic rationale of supporting SMRs: to what extent should they be prioritized over large-scale nuclear power in light of the rules on competition law, State aid, and international trade? The second relates to investment strategies and market concentration: could the economic model for SMRs, which relies on industrial scaling and prefabrication, lead to barriers in the market or distortions of competition? These questions are particularly relevant given that, as one recent study has underlined, economic support is a decisive factor for the deployment of SMRs, and the effectiveness of the legal framework must therefore pay special attention to this issue.[54]

There is an ongoing economic debate regarding the attractiveness of developing SMRs as opposed to large-capacity nuclear power. Part of the literature highlights that the rising costs of operating existing plants and building conventional reactors—driven by stricter safety regulations, increased material prices, and recurring construction delays—could make SMRs comparatively more attractive.[55] Other analyses, however, point to potential advantages of SMRs such as lower upfront capital requirements, scalability according to demand, and better compatibility with smaller electricity grids in certain countries.[56]

Despite these arguments, the economic feasibility of SMRs remains contentious. According to estimates by EY, the levelized cost of electricity (LCOE) for the first SMR is expected to be approximately 30 per cent higher than for a large reactor built under normal order conditions.[57] Cost parity might only be achieved if at least ten units are manufactured per year, corresponding to a total capacity of around 5 GWe. Other studies suggest that SMRs have an average investment cost 41 per cent higher per kilowatt than large reactors, estimated at €7031/kW, resulting in an LCOE of approximately €85/MWh. For a 180 MWe

[53] Desai (n 29).

[54] U Kumar Pal and others, 'Comparative Analysis of Regulatory Compliance and Legal Issues in Small Modular Nuclear Reactor Deployment' (2025) 17(3) Journal of Legal Affairs and Dispute Resolution in Engineering and Construction 15.

[55] T Kaberger and LJ Nilsson, 'The Economics of Nuclear Power Revisited', in Oxford Institute for Energy Studies, 'Nuclear Energy in the Global Energy Landscape: Advancing Sustainability and Ensuring Energy Security?' (2024) 139 Oxford Energy Forum 14.

[56] N Van Hee, H Peremans, and P Nimmegeers, 'Economic Potential and Barriers of Small Modular Reactors in Europe' (2024) 203 Renewable and Sustainable Energy Reviews 114743.

[57] EY, 'Can building nuclear power become more cost-effective' (March 2016), at <https://www.researchgate.net/publication/321715136_Small_modular_reactors_Can_building_nuclear_power_become_more_cost-effective> accessed 17 December 2025.

SMR, recovering fixed factory costs of $1 billion would require at least four orders at a unit sale price of $1.5 billion.[58]

The evaluation of SMR economics is further complicated by the fact that most studies focus primarily on investment costs, internal rates of return, and electricity prices, while giving less attention to operation, maintenance, and decommissioning costs. While SMRs are expected to benefit from 'economies of multiples', whereby modular replication reduces costs, such benefits remain hypothetical in the absence of industrial precedent. Some scholars warn of 'diseconomies of scale'[59] highlighting the risk of overestimating the potential of SMRs to accelerate decarbonization, unless accompanied by substantial infrastructure investment and a transition to mass production.

Another important consideration is the dependence of SMRs on government subsidies and public support schemes.[60] The technological immaturity of SMRs increases financial uncertainty, even if their shorter construction times—estimated at three to four years[61] thanks to modular assembly—offer an advantage over conventional nuclear plants, where delays are frequent and costly.

Given these uncertainties, the legal and financial frameworks available to support SMRs deserve particular scrutiny. Within the European Union, the inclusion of nuclear energy in the EU Taxonomy for Sustainable Activities following Delegated Regulation 2022/1214[62] is a significant step. For SMRs, specific reference is made to section 4.26 of Annex I, which concerns 'Pre-commercial stages of advanced technologies to produce energy from nuclear processes with minimal waste from the fuel cycle.' This provides a formal legal basis for qualifying SMR investments as sustainable under EU law. In parallel, the EU's reformed electricity market[63] design includes two-way Contracts for Difference (CfDs) as the preferred mechanism to support investment in new power generation, including nuclear (Art 19d).

However, important gaps remain. The 2022 Guidelines on State aid for climate, environmental protection, and energy exclude nuclear energy, leaving Article 107(3) TFEU as the legal basis for assessing the compatibility of national subsidies to nuclear projects,[64] including SMRs. This creates a grey zone of legal uncertainty, particularly in comparison to renewable energy projects which benefit from clearer and more comprehensive rules on support mechanisms.

In light of these considerations, the economic pillar of sustainability calls for caution. While SMRs could in theory achieve future cost advantages through standardization and modular deployment, their present competitiveness depends heavily on public investment, supportive regulation, and assumptions that have not yet been validated in practice. This situation raises the question of whether SMRs will ultimately represent a genuinely

[58] B Mignacca and G Locatelli, 'Economics and Finance of Small Modular Reactors: A Systematic Review and Research Agenda' (2020) 118 Renewable and Sustainable Energy Reviews 109519.

[59] F Böse and others, 'Questioning Nuclear Scale-up Propositions: Availability and Economic Prospects of Light Water, Small Modular and Advanced Reactor Technologies' (2024) 110 Energy Research & Social Science 103448.

[60] Kaberger (n 55).

[61] EY (n 57).

[62] [2022] OJ L 188.

[63] Regulation (EU) 2024/1747 of the European Parliament and of the Council of 13 June 2024 amending Regulations (EU) 2019/942 and (EU) 2019/943 as regards improving the Union's electricity market design [2024] OJ L 2024/1747.

[64] Case C-594/18 P *Austria v Commission* [2020] CJEU, para 45.

viable and sustainable investment, or whether they risk remaining dependent on exceptional economic support in a way that could distort both energy markets and international competition.

IV. Recommendations

The sustainable deployment of SMRs requires closing critical gaps in international and national governance. The analysis above has revealed several weaknesses that, if left unaddressed, risk undermining the alignment of SMRs with the principles of sustainable development. This section identifies three main areas of reform: harmonization of legal frameworks, embedding energy justice, and ensuring democratic governance across scales.

A. Harmonizing Legal Instruments for SMR Acceleration

One major gap lies in the absence of harmonized legal instruments specifically tailored to SMRs. Although these technologies are governed by existing international nuclear law, the applicability of conventions originally designed for large, stationary reactors remains uncertain. Moreover, many international conventions in nuclear and environmental law delegate significant discretion to States for defining domestic implementing rules. The Convention on Nuclear Safety, for example, leaves States free to establish the modalities of licensing regimes preceding the operation of nuclear installations (Article 7). This flexibility, while respecting sovereignty, risks creating uneven regulatory standards. In practice, operators may be incentivized to develop SMRs in jurisdictions with less stringent environmental or safety requirements, thereby undermining sustainability objectives.

Harmonization is therefore essential not only for safety and environmental protection but also to ensure fair and healthy competition at the international level. Without co-ordinated standards, disparities in domestic implementation could distort market competition, with companies seeking the most permissive legal environments. This raises the broader issue of constructing international market rules for SMRs, a prerequisite for their effective integration into global energy systems.

The problem is compounded by the disparity and ambiguity of international norms themselves, particularly regarding environmental protection and liability. Additional bilateral or regional agreements may therefore be required to fill gaps, especially where States are not parties to core conventions.[65] The Rio Declaration underscores this imperative of co-operation as a condition for sustainable development: Principle 7 affirms that 'States shall cooperate in a spirit of global partnership to conserve, protect and restore the health and integrity of the Earth's ecosystem', while Principle 12 calls on States to 'cooperate to promote a supportive and open international economic system that would lead to economic growth and sustainable development in all countries'.

Although no dedicated international instrument yet exists, soft law standards and technical guidelines issued by the IAEA already provide practical regulatory benchmarks. The

[65] Roland (n 23).

recent revision of the IAEA's *Milestones in the Development of a National Infrastructure for Nuclear Power*[66] reflects growing attention to SMRs and offers updated guidance for States on developing coherent national infrastructures. Similarly, the OECD Nuclear Energy Agency's *SMR Dashboard: Second Edition*[67] demonstrates the diversity of approaches and regulatory frameworks being applied worldwide, underscoring the need for greater transparency and co-ordination.

At the regional level, the EU has taken an important step towards harmonization through the adoption of the Regulation (EU) 2024/1735 of the European Parliament and of the Council of 13 June 2024 on establishing a framework of measures for strengthening Europe's net-zero technology manufacturing ecosystem and amending Regulation (EU) 2018/1724.[68] This regulation explicitly includes nuclear technologies, including SMRs, within its scope. It introduces simplified authorization processes, creates single points of contact in each Member State, and allows for the designation of 'strategic net-zero projects' benefiting from accelerated permitting and priority status.

In sum, the effectiveness of the legal framework for SMRs will depend on the success of international co-operation and harmonization. This requires co-ordinated interpretation of existing conventions, the negotiation of new bilateral or multilateral instruments where necessary, and the careful alignment of national licensing and environmental regimes.

B. Embedding Environmental and Energy Justice in Regulatory Design

A second critical gap concerns the limited integration of energy justice principles into the governance of SMRs. Although the concept of a fundamental 'right to energy' has not yet been formally recognized in international law, Sustainable Development Goal 7 (SDG 7) sets the global objective of ensuring access to affordable, reliable, sustainable, and modern energy for all. While non-binding, this objective underscores the normative importance of equal access to energy services. SMRs could, in theory, contribute to energy justice by expanding quantitative energy access (especially in remote or off-grid areas), providing an alternative low-carbon energy source, and ultimately offering more affordable electricity if economies of multiples materialize. However, these potential benefits must be matched by robust legal guarantees that ensure equitable distribution of costs and benefits, inclusive decision-making, and intergenerational responsibility.

One central challenge lies in the governance of investment flows. At present, there are no clear rules to ensure that subsidies, financial support mechanisms, or international investments in SMRs benefit developing countries rather than being concentrated in wealthier nations. Without reform, the deployment of SMRs risks reinforcing existing global inequalities in access to technology and finance. International financial institutions such as the World Bank and regional development banks should be mobilized to facilitate SMR projects in the Global South, particularly in countries where they could serve isolated regions. National legal reforms are also required to adapt investment regimes, as illustrated by the case of India. Recent scholarship has shown that India's outdated nuclear legislation

66 (IAEA Nuclear Energy Series No NG-G-3.1, Rev 2, Vienna 2024).
67 (Paris, OECD Publishing 2024).
68 [2024] OJ L2024/1735, art 4(i) and (s).

continues to restrict private and foreign participation in SMR development, and that significant legal reform would be needed to open the sector while ensuring state control through licensing systems.[69]

Another key issue concerns the localization of SMR sites, which plays a decisive role in public acceptance. Empirical studies demonstrate that opposition to nuclear projects often arises not from rejection of the technology itself but from the perceived unfairness of siting decisions and the exclusion of communities from early stages of planning.[70] Legal frameworks must therefore embed procedural justice safeguards, including rights of access to information, early and inclusive public participation, and access to judicial review in cases of non-compliance. This is particularly important for indigenous and rural communities, which may face disproportionate impacts from SMR deployment.

Finally, embedding intergenerational justice is crucial to align SMR development with sustainable development. As emphasized in theories of intergenerational equity, each generation acts as a trustee of natural and cultural resources for future generations, with corresponding duties of conservation and equitable access.[71] Legal frameworks should therefore ensure that long-term impacts of SMRs—such as nuclear waste management, decommissioning, and environmental restoration—are adequately regulated and transparently funded, so as not to impose unfair burdens on future generations.

In sum, embedding energy justice in SMR regulation requires rules that guarantee: (i) equitable distribution of financial and energy benefits, including mechanisms for value-sharing at the local level; (ii) meaningful, inclusive, and early participation of affected communities; and (iii) safeguards for intergenerational equity in waste and decommissioning policies. Only by integrating these justice dimensions into both international and national legal frameworks can SMRs contribute meaningfully to a just and sustainable energy transition.

C. Enabling Democratic Governance Across Scales

A final gap concerns the governance of SMRs, which remains fragmented and insufficiently anchored in democratic processes. Ensuring the legitimacy and sustainability of SMR deployment requires governance frameworks that are transparent, inclusive, and accountable across three interdependent levels: international, national, and local.

At the international level, SMRs cannot be treated solely as a matter of national sovereignty or industrial competition. Given the transboundary nature of nuclear risks, the long-term challenges of waste management, and the framing of SMRs as contributors to global decarbonization, nuclear innovation increasingly resembles a matter of global commons. This perspective requires collective deliberation within the international community about the public interest of SMRs, the allocation of international funding, and the equitable distribution of risks and benefits. Theories of global justice highlight the need for institutions capable of addressing inequalities beyond national borders and embedding fairness into

[69] Desai (n 29).

[70] SP Tuler and T Webler, 'The Challenge of Community Acceptance of Small Nuclear Reactors' (2024) 118 Energy Research & Social Science 103831.

[71] Brown Weiss (n 34).

global governance.[72] These approaches remind us that nuclear decision-making cannot remain confined to technocratic elites but should be considered part of a broader 'contract of global society' that reflects the values and interests of humanity as a whole.

At the national level, SMRs revive the fundamental democratic question of the role of nuclear energy within national energy strategies. In many States, renewed nuclear ambitions have been reaffirmed in recent years, often framed by climate imperatives and industrial competitiveness. However, strategic decisions of this magnitude must not bypass broad-based democratic deliberation. Parliaments, civil society organizations, and independent experts should be actively involved in shaping national nuclear policy. This is particularly important to ensure that industrial and environmental objectives are reconciled with social acceptability and long-term stewardship. Democratic debate at the national level must therefore extend beyond purely technical or economic criteria to incorporate principles of energy justice and intergenerational equity.

At the local level, the legitimacy of SMRs depends crucially on community engagement in siting and deployment decisions. Rural and Indigenous communities are often targeted for SMR projects, given their geographic and infrastructural characteristics. Yet without meaningful opportunities for participation, these communities may perceive SMRs as imposed technologies, eroding trust and increasing resistance. Robust public consultation processes must therefore be guaranteed, granting affected populations not only the right to be informed but also the ability to negotiate local benefits and influence governance outcomes. This is not only a matter of procedural fairness but also of distributive justice, ensuring that communities hosting SMRs are compensated for the risks they bear. Studies of renewable energy deployment have shown that mechanisms such as local value-sharing, fiscal benefits, and preferential access to energy can significantly improve public acceptance.[73] Similar approaches could be adapted for SMRs.

Finally, democratic governance must be matched by transparent financial allocation mechanisms. The long-term implications of SMRs—especially decommissioning, nuclear waste management, and site remediation—require robust and clearly funded regulatory frameworks. Without such safeguards, the costs and risks of SMRs may be unfairly transferred to future generations, contravening principles of intergenerational justice.[74] Embedding intergenerational equity into financial governance is thus essential to ensure that SMRs are not only technically and economically viable but also socially legitimate and ethically defensible.

In sum, enabling democratic governance across scales means recognizing SMRs as a technology whose governance extends beyond industrial competitiveness or national energy security. By embedding transparency, participation, and fairness into decision-making at international, national, and local levels, legal frameworks can contribute to building the societal trust and legitimacy required for SMRs to play a meaningful role in the sustainable energy transition.

[72] D Miller, 'National Responsibility and Global Justice' (2008) 11(4) Critical Review of International Social and Political Philosophy 383–99.

[73] de Fontenelle (n 48).

[74] Brown Weiss (n 34).

V. Conclusion

The development of Small Modular Reactors (SMRs) within Distributed Energy Resource systems reflects a profound transformation of the global energy landscape. Positioned at the crossroads of decarbonization, energy security, and industrial innovation, SMRs embody both promise and complexity. Their flexibility, modularity, and potential integration into decentralized infrastructures make them attractive options, particularly for regions underserved by traditional grid systems. Yet, their successful deployment depends not only on technological advances but on the adequacy, coherence, and legitimacy of the legal frameworks that govern them.

This chapter has demonstrated that SMRs raise novel legal questions at the intersection of nuclear law, environmental protection, and energy justice. While existing international instruments remain formally applicable, they were designed for large, stationary reactors and often require adaptive interpretation. Important normative gaps persist, especially concerning mobile reactors, maritime applications, and inter-jurisdictional co-ordination.

From an environmental perspective, SMRs must be regulated by robust standards for safety, waste management, and environmental impact assessment—yet current frameworks remain fragmented. From a social perspective, principles of energy justice and public participation are insufficiently embedded in many national strategies, raising risks for public trust and democratic legitimacy. Economically, the viability of SMRs is still uncertain, heavily dependent on public investment, regulatory flexibility, and long-term industrial scaling.

Three main conclusions emerge from this analysis. First, legal harmonization and procedural flexibility are essential to facilitate SMR deployment without compromising safety or environmental standards, with strategic reforms such as the EU Net-Zero Industry Regulation illustrating how legal innovation can enable responsible nuclear development. Second, environmental and energy justice principles—equitable access, participatory governance, and intergenerational responsibility—must be systematically integrated into SMR governance at international, national, and local levels. Finally, sustainable financing and democratic accountability must underpin public investment in SMRs, since transparent governance mechanisms are indispensable to securing the social licence to operate.

Ultimately, the promise of SMRs does not lie in their technical characteristics alone, but in the ability of legal systems to guide their deployment in ways that are environmentally sound, socially just, and economically prudent. The legal framework must evolve accordingly, combining international co-operation, domestic reforms, and normative innovation. Given the novelty of SMRs, this analysis remains exploratory: further empirical and comparative research—particularly in the fields of lifecycle assessment, cost modelling, and participatory governance—will be critical to evaluate their real-world sustainability as they transition from concept to deployment. SMRs, in the end, will only be as sustainable as the legal frameworks that govern them.

6
Distributed Energy Storage

Elena I Athwal

I. Introduction

As distributed energy resources (DERs) infrastructure increases, legal and policy frameworks are required to ensure the sustainable storage of produced electricity in a manner that enhances sustainability in their use, as well as in their flexible integration with the extant grid to prevent congestion and system collapse. Drawing lessons from the Middle East and North Africa (MENA) region, this chapter examines legal and practical aspects of distributed energy storage, environmental gaps in emerging laws and policies on distributed energy storage, and the barriers to the deployment of storage facilities, especially in areas with insufficient or unstable grid capacity and legal reforms to address those gaps.

DERs—including rooftop solar, wind microgrids, and bioenergy—are central to the creation of decentralized, low-carbon energy systems, yet their success depends on the widespread deployment of electricity storage. Electricity storage is indispensable in the creation of resilient, low-carbon energy systems, serving as a pivotal enabler of the integration of variable renewable energy sources and enhances grid reliability, flexibility, and efficiency.[1] This critical role has been emphasized by recent global guidance, including assessments by the International Energy Agency (IEA)[2] and the International Renewable Energy Agency (IRENA),[3] which identify electricity storage facilities as an essential prerequisite for large-scale renewable energy deployment and energy security. As IRENA underscores:

Electricity storage systems have the potential to be a key technology for the integration of VRE due to their capability to quickly absorb, store and then reinject electricity to the grid.[4]

This recognition of storage's system-wide value provides the foundation for a more precise understanding of what 'electricity storage' entails within the context of this chapter.

For the purposes of this chapter, electricity storage refers specifically to technologies that store electrical energy for later use, including battery storage systems, pumped hydro, and thermal storage—not broader forms such as hydrogen, compressed air, or underground

[1] N Rosetto and V Reif, 'Digitalization of the Electricity Infrastructure: A Key Enabler for the Decarbonization and Decentralization of the Power Sector' in J Montero and M Finger (eds), *A Modern Guide to the Digitalization of Infrastructure* (Edward Elgar Publishing 2021) 217–65. PJ Thimet and G Mavromatidis, 'What-Where-When: Investigating the Role of Storage for the German Electricity System Transition' (2023) 351 Applied Energy 121764.

[2] International Energy Agency, *Grid-Scale Storage* (IEA, 11 July 2023) <https://www.iea.org/energy-system/electricity/grid-scale-storage> accessed 4 May 2025.

[3] International Renewable Energy Agency (IRENA), *Electricity Storage Valuation Framework: Assessing system value and ensuring project viability* (2020) <https://www.irena.org/-/media/Files/IRENA/Agency/Publication/2020/Mar/IRENA_Storage_valuation_2020.pdf accessed 29 October 2025.

[4] Ibid 63.

Elena I Athwal, *Distributed Energy Storage* In: *Sustainable Distributed Energy Resources*. Edited by: Íñigo del Guayo, Damilola S. Olawuyi San, Louis de Fontenelle, Milton Fernando Montoya, and José Juan González, Oxford University Press.
 DOI: 10.1093/9780198991946.003.0007

energy storage.[5] In this regard, electricity storage is not merely a technical solution but a legal linchpin for sustainable DER deployment—necessary to ensure reliability, equity, and regulatory coherence in clean energy systems. However, while the technological promise of storage advances rapidly, regulatory frameworks often fail to keep pace.[6]

This chapter contends that unless regulatory frameworks evolve to recognize and govern storage explicitly, the full potential of DERs will remain legally constrained and economically underleveraged. Despite its growing importance, legal and regulatory frameworks for electricity storage remain fragmented or underdeveloped in numerous regions,[7] particularly in emerging markets. The absence of coherent asset classification, standardized performance criteria, and clear tariff structures for storage has hindered both public and private sector engagement with storage technologies.

Without reliable storage, DERs remain vulnerable to intermittency, curtailment, and limited grid integration, undermining their sustainability and scalability. Yet in many jurisdictions, especially in emerging DER markets such as the MENA region, legal frameworks on electricity storage remain fragmented and inconsistent. Additionally, Behind-the-Meter (BTM) storage systems raise legal questions about consumer rights, data protection, and the enforceability of grid service agreements in deregulated markets. Legal concerns here include ownership rights, interconnection rules, and participation in grid services such as net metering.[8] DER-integrated storage, such as microgrid-linked systems, enhances energy resilience and independence, particularly in decentralized contexts.[9] These systems present regulatory challenges including co-ordination across multiple stakeholders and alignment of incentives.[10] These challenges are further complicated by the lack of cross-sector

[5] C Gu and others, 'Carbon-Oriented Planning of Distributed Generation and Energy Storage Assets in Power Distribution Network with Hydrogen-Based Microgrids' (2022) 14(2) IEEE Transactions on Sustainable Energy 790–802; S Zhang and others, 'Low-Carbon Optimal Operation of Distributed Energy Systems in the Context of Electricity Supply Restriction and Carbon Tax Policy: A Fully Decentralized Energy Dispatch Strategy' (2023) 396 Journal of Cleaner Production 136511; Y Wang and Y Fan, 'Capacity Allocation in Distributed Wind Power Generation Hybrid Energy Storage Systems' (2024) 19 International Journal of Low-Carbon Technologies 2299–2308.

[6] UN Economic and Social Commission for Asia and the Pacific (UNESCAP), *Energy Storage Readiness Assessment Framework* (2022) 4–6.

[7] JJG Márquez and MG Brambila, 'Regulation of Electricity Storage, Intelligent Grids, and Clean Energies in an Open Market in Mexico' in D Zillman and others (eds), *Innovation in Energy Law and Technology: Dynamic Solutions for Energy Transitions* (OUP 2018) 172–92; S Tiwari and others, 'Unit Commitment Problem in Renewable Integrated Environment with Storage: A Review' (2021) 31(10) International Transactions on Electrical Energy Systems e12775; L Hancher and J Rumpf, 'Balancing Power: The Impact of Legal Review on Harmonising the European Electricity Market' (2024) 15(4) European Journal of Risk Regulation 827–45.

[8] A Arena, 'From an Unpaid Electricity Bill to the Primacy of EU Law: Gian Galeazzo Stendardi and the Making of Costa v ENEL' (2019) 30(3) European Journal of International Law 1017–37; R Poudineh and T Jamasb, 'Electricity Market Reform: Review of the EU Electricity Market Design and Lessons for Other Regions' (2019) 8(5) Energy Policy 612–24; A Arena, 'From an Unpaid Electricity Bill to the Primacy of EU Law: Gian Galeazzo Stendardi and the Making of Costa v ENEL' (2019) 30(3) European Journal of International Law 1017; TD Robinson and E Arnould, 'Portable Technology and Multi-Domain Energy Practices' (2020) 20(1) Marketing Theory 3; X Shuai and R Raufer, 'Behind-the-Meter Energy Storage in China: Lessons from California's Approach' (2021) 10(4) WIREs Energy and Environment e394.

[9] A Vosughi and others, 'Cyber–Physical Vulnerability and Resiliency Analysis for DER Integration: A Review, Challenges and Research Needs' (2022) 168 Renewable and Sustainable Energy Reviews 112794; W Rickerson and others, 'The Value of Resilience for Distributed Energy Resources: An Overview of Current Analytical Practices' (2024) https://docs.nrel.gov/docs/fy24osti/90139.pdf accessed 14 October 2025.

[10] IJ Pérez-Arriaga and others, 'A Regulatory Framework for an Evolving Electricity Sector: Highlights of the MIT Utility of the Future Study' (2017) 6(1) Economics of Energy & Environmental Policy 71; K Bell and S Gill, 'Delivering a Highly Distributed Electricity System: Technical, Regulatory and Policy Challenges' (2018) 113 Energy Policy 765; R Bray and others, *Policy and Regulatory Barriers to Local Energy Markets in Great Britain* (University of Exeter 2018); A Sakti and others, 'Review of Wholesale Markets and Regulations for Advanced Energy Storage Services in the United States: Current Status and Path Forward' (2018) 120 Energy Policy 569; L

co-ordination between energy, telecommunications, and land use regulators, especially where storage interfaces with digital infrastructure or shared community assets.

Beyond technical and economic barriers, electricity storage also raises profound human rights and environmental justice concerns—especially in relation to the lifecycle of battery technologies. The mining of cobalt, lithium, and other rare earth minerals—often concentrated in fragile states—has been linked to child labor, displacement, and environmental degradation. As Olawuyi aptly notes, legal frameworks must transcend access-oriented models and embed procedural safeguards, rigorous due diligence standards, and equitable benefit-sharing mechanisms into clean energy governance.[11] In the absence of such safeguards, the energy transition risks replicating historical patterns of resource extraction without local empowerment or legal accountability.[12]

This chapter is organized as follows. First, it provides an overview of electricity storage technologies and their crucial roles in modern energy systems. Next, it explores the regulatory challenges and legal frameworks that impact the deployment of electricity storage. It then conducts a comparative legal analysis between distributors and consumers of stored electricity. This is followed by detailed case studies that illustrate how regulatory decisions influence storage deployment. Finally, it offers recommendations for harmonizing regulations to ensure more efficient integration of storage solutions. Throughout, the chapter emphasizes how effective electricity storage governance is a precondition for achieving the sustainability, scalability, and inclusivity of DER systems.

II. Elements of a Sustainable Legal Framework for Electricity Storage

This section identifies six core elements of a sustainable legal framework for electricity storage in the MENA region: (1) classification and ownership; (2) licensing and grid access; (3) safety and data governance; (4) tariff and compensation regimes; (5) investment and innovation incentives; and (6) consumer rights and participation. Each is addressed in turn below, drawing on MENA examples with selective reference to international practice.

Coherent asset classification is a legal precondition for the sustainability and scalability of storage systems.[13] Without a harmonized legal definition of electricity storage, markets face uncertainty in assigning regulatory responsibilities, enabling fair access, and designing tariffs that reflect multifunctional value.[14] This section also draws on examples from within

Baker and J Phillips, 'Tensions in the Transition: The Politics of Electricity Distribution in South Africa' (2019) 37(1) Environment and Planning C Politics and Space 177; Z Peterson and others, *Overcoming Challenges for DER Interconnection—Spotlight on Energy Storage* (National Renewable Energy Laboratory 2019).

[11] DS Olawuyi, *Environmental Law in Arab States* (OUP 2022) 210–12.

[12] Amnesty International, *Powering Change: Principles for Businesses and Governments in the Battery Value Chain* (2023) <https://www.amnesty.org/en/documents/pol30/6724/2023/en/> accessed 5 May 2025.

[13] IRENA, *Electricity Storage and Renewables: Costs and Markets to 2030* (IRENA 2017) 43–45 <https://www.irena.org/publications/2017/Oct/Electricity-storage-and-renewables-Costs-and-markets> accessed 31 July 2025.

[14] IEA, Grid-Scale Storage (IEA 2024) <https://www.iea.org/energy-system/electricity/grid-scale-storage> accessed 31 July 2025; see also European Commission, Commission Staff Working Document: Guidance on Electricity Storage SWD(2017) 61 final, 5–6 <https://climate.ec.europa.eu/system/files/2017-11/swd_2017_650_p1_en.pdf> accessed 31 July 2025; GC Gissey, PE Dodds, and J Radcliffe, 'Market and Regulatory Barriers to Electrical Energy Storage Innovation' (2018) 82 Renewable and Sustainable Energy Reviews 781–90; CP Agupugo and others, 'Policy and regulatory framework supporting renewable energy microgrids and energy storage systems' (2024) 5(8) Engineering Science & Technology Journal 2589–2615 https://www.fepbl.com/index.php/estj/article/view/1460.

the MENA region—such as Dubai's Law No 27 of 2021, Morocco's Law 13-09, and Oman's regulatory provisions for microgrids—to illustrate how classification, ownership, and market access challenges are currently addressed in diverse national contexts.

A. Classification and Ownership of Electricity

A central legal challenge in electricity storage is the absence of a harmonized classification system. In MENA jurisdictions, storage may be treated alternately as a generation asset,[15] grid infrastructure,[16] or demand-side resource.[17] This tripartite ambiguity creates uncertainty for hybrid systems—such as solar PV paired with battery storage.

The multifunctional character of electricity storage—where a single system may act as a generation, transmission, and demand-side resource—poses profound regulatory challenges. As the IRENA notes, this versatility is a strength from a systems engineering perspective but a liability in legal and policy terms, since most regulatory frameworks are not designed to accommodate hybrid classifications.[18] In Morocco, Law 13-09 does not define storage as a separate asset class;[19] in Egypt, existing electricity laws provide for renewable licensing but lack dedicated provisions for battery systems; in the UAE, Dubai's Law No (27) of 2021[20] incorporates storage within the Dubai Electricity and Water Authority (DEWA)'s regulatory framework, offering greater clarity.[21]

B. Ownership Models and Legal Constraints

The question of who owns and operates storage infrastructure has significant legal and economic implications. In many jurisdictions, Transmission System Operators (TSOs) and Distribution System Operators (DSOs) are restricted from owning or operating electricity storage, except under specific conditions. This aims to prevent cross-subsidization and preserve market neutrality.

[15] GC Gissey and others (n 47); MM Rana and others, 'Applications of Energy Storage Systems in Power Grids with and Without Renewable Energy Integration—A Comprehensive Review' (2023) 68 Journal of Energy Storage 107811.

[16] MM Islam and others, 'Improving Reliability and Stability of the Power Systems: A Comprehensive Review on the Role of Energy Storage Systems to Enhance Flexibility' (2024) IEEE Access.

[17] M Behrangrad, 'A Review of Demand Side Management Business Models in the Electricity Market' (2015) 47 Renewable and Sustainable Energy Reviews 270 <https://www.sciencedirect.com/science/article/abs/pii/S1364032115001860?via%3Dihub> accessed 11 August 2025; B Li and others, 'From Controllable Loads to Generalized Demand-Side Resources: A Review on Developments of Demand-Side Resources' (2016) 53 Renewable and Sustainable Energy Reviews 936–44; MB Anwar and others, 'Harnessing the Flexibility of Demand-Side Resources' (2018) 10(4) IEEE Transactions on Smart Grid 4151; S Panda and others, 'An Insight into the Integration of Distributed Energy Resources and Energy Storage Systems with Smart Distribution Networks Using Demand-Side Management' (2022) 12(17) Applied Sciences 8914.

[18] IEA, *Grid-Scale Storage* (IEA, accessed November 2023) <https://www.iea.org/energy-system/electricity/grid-scale-storage> accessed 31 July 2025.

[19] Loi no 13-09 relative aux énergies renouvelables (Maroc) (2009) Journal Officiel du Royaume du Maroc.

[20] Dubai Law No (27) of 2021 Establishing the Dubai Electricity and Water Authority as a Public Joint Stock Company (12 October 2021).

[21] Dubai Electricity and Water Authority (DEWA), *R&D Centre Overview* (2024) <https://www.dewa.gov.ae/en/about-us/strategic-initiatives/mbr-solar-park/research-and-development-centre> accessed 11 August 2025.

Legal clarity around ownership is especially critical in emerging markets such as in the MENA region, where public utilities may dominate the electricity sector. Without proper legal safeguards, utility ownership can crowd out private investment, distort tariff structures, and reduce transparency. In the UAE, Dubai Law No (27) of 2021 empowers DEWA to own and operate storage assets while also enabling partnerships with private entities, creating a hybrid model. In Oman, the Authority for Public Services Regulation (APSR) regulates microgrid and storage ownership, but most large-scale assets remain under state utility control.[22] Morocco's Law 13-09 allows private renewable generation ownership, yet does not define storage ownership, leaving projects like Noor Midelt reliant on bespoke contractual arrangements. In Egypt, the Electricity Law (Law No 87/2015) does not contain dedicated ownership provisions for storage, resulting in continued utility dominance.[23] Clear and equitable ownership rules are therefore essential for attracting investment, ensuring efficient operation, and aligning storage deployment with broader decarbonization goals.[24]

C. Licensing, Grid Access, Safety, Data, and Interconnection Rule

Electricity storage systems, particularly behind-the-meter and community-based solutions encounter persistent regulatory hurdles in licensing and interconnection. Fragmented, outdated,[25] and inconsistent procedures delay deployment,[26] restrict grid access, and disproportionately burden smaller-scale systems.[27] As storage becomes digitally integrated into smart grids, concerns around data privacy and cybersecurity intensify. As noted by IRENA and recent MENA-focused studies, the absence of clear, harmonized frameworks for data governance, interconnection, and safety not only impedes innovation but also jeopardizes and erodes consumer trust and compromises grid security.[28] Establishing a cohesive, modernized regulatory architecture is therefore imperative to unlocking storage's full value and ensuring resilient, equitable energy transitions.

[22] Authority for Public Services Regulation (APSR), 'Rural Areas Electricity Company License' (2023) <https://apsr.om/pdfs/Licences/RAECLicence.pdf> accessed 4 April 2025.

[23] Climate Change Laws of the World, 'Law n. 87—Electricity Law' Egypt <https://climate-laws.org/document/law-n-87-2015-electricity-law_4853> accessed 11 August 2025.

[24] G Agbaitoro, *Resolving Energy Challenges: Implementing and Evaluating an Energy Justice Framework for Nigeria* (PhD thesis, University of Essex 2022) 209–13 <https://repository.essex.ac.uk/35635/1/Godswill%20Agbaitoro_PhD%20thesis.pdf> accessed 31 July 2025; Directive (EU) 2019/944 of the European Parliament and of the Council of 5 June 2019 on common rules for the internal market for electricity [2019] OJ L158/125, art 36; Directive 2009/72/EC of the European Parliament and of the Council of 13 July 2009 concerning common rules for the internal market in electricity [2009] OJ L211/55, art 7; Council of European Energy Regulators (CEER), *Regulatory Frameworks Report 2024* (CEER 2024) <https://www.ceer.eu/wp-content/uploads/2025/02/GA191_05_2_IRB-Regulatory-Frameworks-Report-2024-Main-report-1-1.pdf> accessed 11 August 2025.

[25] Y Chen and others, 'Investigation on Characteristic Parameters Identification and Evolution of Supercapacitor Energy Storage System from Sparse and Fragmented Monitoring Data' (2023) 11 IEEE Access 56983 <https://ieeexplore.ieee.org/stamp/stamp.jsp?arnumber=10145766> accessed 11 August 2025.

[26] LH Gulbrandsen and others, 'Is Political Steering Gone with the Wind? Administrative Power and Wind Energy Licensing Practices in Norway' (2021) 74 Energy Research & Social Science 101963 <https://www.sciencedirect.com/science/article/pii/S2214629621000566?via%3Dihub> accessed 11 August 2025.

[27] J Shin and others, 'A Review on Mechanistic Understanding of MnO_2 in Aqueous Electrolyte for Electrical Energy Storage Systems' (2020) 65(6) International Materials Reviews 356 <https://journals.sagepub.com/doi/full/10.1080/09506608.2019.1653520> accessed 11 August 2025.

[28] MZ Gunduz and R Das, 'Smart Grid Security: An Effective Hybrid CNN-Based Approach for Detecting Energy Theft Using Consumption Patterns' (2024) 24(4) Sensors 1148 <https://www.mdpi.com/1424-8220/24/4/1148> accessed 11 August 2025.

In the UAE, DEWA's licensing framework for grid-connected storage includes detailed technical approval processes but lacks a dedicated cybersecurity protocol for distributed assets. Morocco's Law 13-09 similarly does not specify safety and data governance standards for hybrid solar-storage projects, leaving such requirements to project-specific contracts.

D. Tariff Regimes and Compensation Structure

The absence of comprehensive, forward-looking tariff regimes and compensation structures continues to undervalue electricity storage, particularly at the consumer and prosumer level, despite its critical role in delivering essential grid services like frequency regulation, peak shaving, and load shifting.[29] In much of the MENA region, tariff design still reflects conventional generation and consumption models, offering no differentiated rates or incentives for dispatchable renewable energy supported by storage. For example, Morocco's blended tariff arrangement under the Noor Midelt project is one of the few in the region to explicitly value both generation and storage functions, while most other jurisdictions provide only generation-based remuneration.[30] As IRENA and the World Bank[31] have noted, cost-reflective tariffs and transparent compensation schemes are essential for scaling storage, attracting private capital, and integrating distributed resources into national energy markets.

E. Investment, Subsidy Schemes, and Innovation Incentives

This market misalignment is further compounded by a lack of dedicated investment and subsidy frameworks tailored to storage technologies. In the MENA region, subsidy schemes for electricity storage remain limited and ad hoc, often embedded in broader renewable energy programs, rather than designed for storage itself. Morocco's Law 13-09 and the Energy Transition Law provide a statutory framework under which storage assets linked to renewable generation may be eligible for investment facilitation, but fiscal incentives remain modest.[32] In the UAE, flagship projects such as the Mohammed bin Rashid Al Maktoum Solar Park have leveraged green bonds and long-term power purchase agreements (PPAs), yet no specific subsidy exists for distributed storage systems.[33]

[29] CJ Meinrenken and A Mehmani, 'Concurrent Optimization of Thermal and Electric Storage in Commercial Buildings to Reduce Operating Cost and Demand Peaks under Time-of-Use Tariffs' (2019) 254 Applied Energy 113630 <https://www.sciencedirect.com/science/article/abs/pii/S0306261919313170?via%3Dihub> accessed 11 August 2025; I Savelli and T Morstyn, 'Electricity Prices and Tariffs to Keep Everyone Happy: A Framework for Fixed and Nodal Prices Coexistence in Distribution Grids with Optimal Tariffs for Investment Cost Recovery' (2021) 103 Omega 102450 <https://www.sciencedirect.com/science/article/abs/pii/S0305048321000591?via%3Dihub> accessed 11 August 2025.

[30] Law No 13-09 Relating to Renewable Energies (Morocco), Official Gazette No 5822 (18 March 2010).

[31] Energy Sector Management Assistance Program (ESMAP), Annual Report 2024 (World Bank 2024) <https://documents1.worldbank.org/curated/en/099552001292511745/pdf/IDU1c4e326371712214d3718f3d199658d68adaf.pdf> accessed 11 August 2025.

[32] Loi no 13-09 (n 30).

[33] ICLG, *Renewable Energy Laws and Regulations—United Arab Emirates 2025* (Global Legal Group, 19 September 2024) 5 <https://iclg.com/practice-areas/renewable-energy-laws-and-regulations/united-arab-emirates> accessed 12 August 2025.

In 2023, the UAE's federal Policy to Regulate the Market for Energy Services Companies introduced a national framework to formalize contractual arrangements between public entities, private sector actors, and energy service providers, offering clearer pathways for investment and risk allocation in energy efficiency and distributed storage projects.[34] As IRENA reports, studies on emerging markets highlight that predictable revenue mechanisms and dedicated innovation funding have been associated with greater private sector participation in storage deployment.[35]

F. Comparative Approaches from Advanced Regulatory Environment

Advanced regulatory environments provide critical lessons for the design of enabling frameworks for electricity storage. While much of the global literature examines EU, UK, and Australian experiences, emerging regulatory approaches within the MENA region—particularly in Morocco, the UAE, and Oman—illustrate early efforts to integrate storage into clean-energy strategies through legal recognition of hybrid systems, gradual tariff reform, and pilot incentive programs. Scholarly analyses, including Olawuyi, also caution that direct transplants of such frameworks can overlook contextual differences in developing regions.[36] These comparative observations are relevant to ongoing MENA discussions on legal definition, ownership, and tariff design; detailed evaluation is addressed in Section III.

III. Distributors vs Consumers—Comparative Legal Analysis

Electricity storage is a cornerstone of modern energy systems, enhancing grid resilience and enabling renewable integration; however, in the MENA region, its legal and regulatory treatment diverges sharply between distributors (utilities, grid operators) and consumers (households, prosumers, and businesses). This often reflect deeper tensions between grid stability and energy independence.

[34] Ministry of Energy and Infrastructure, *Policy to Regulate the Market for Energy Services Companies in the UAE* (UAE Ministry of Energy and Infrastructure, 7 June 2023) <https://www.uaelegislation.gov.ae/en/policy/details/sy-s-tnthym-sok-shrk-t-khdm-t-lt-k-fy-ldol> accessed 12 August 2025.

[35] M Iskandarova and others, 'Who Finances Renewable Energy in Europe? Examining Temporality, Authority and Contestation in Solar and Wind Subsidies in Poland, the Netherlands and the United Kingdom' (2021) 38 Energy Strategy Reviews 100730 <https://www.sciencedirect.com/science/article/pii/S2211467X21001164?via%3Dihub> accessed 11 August 2025; A Masoumzadeh and others, 'Designing Tax and Subsidy Incentives Towards a Green and Reliable Electricity Market' (2020) 195 Energy 117033 <https://www.sciencedirect.com/science/article/abs/pii/S0360544220301407?via%3Dihub> accessed 11 August 2025; CP Agupugo and others, 'Policy and Regulatory Framework Supporting Renewable Energy Microgrids and Energy Storage Systems' (2024) 5 Engineering Science and Technology Journal 5(8) <https://fepbl.com/index.php/estj/article/view/1460> accessed 11 August 2025. IRENA, 'World Energy Transitions Outlook 2024' (IRENA 2024) <https://www.irena.org/-/media/Files/IRENA/Agency/Publication/2024/Nov/IRENA_World_energy_transitions_outlook_2024_Summary.pdf> accessed 11 August 2025.

[36] Olawuyi (n 11 above).

A. Licensing and Regulatory Oversight

In the MENA region, distributors operate under stringent licensing and regulatory frameworks to uphold grid reliability, safety, and market integrity.[37] For example, in the UAE, Dubai Law No (27) of 2021 regulates grid-connected storage through the DEWA licensing regime, while in Oman, the APSR oversees licensing for storage integrated into microgrids and renewable projects.[38] In Morocco, licensing requirements under Law 13-09 cover renewable generation but do not yet contain dedicated provisions for stand-alone storage, resulting in project-specific licensing arrangements.

By contrast, small-scale or behind-the-meter (BTM) systems face fewer licensing obligations,[39] though compliance with local safety, fire, and building codes is mandatory. In Egypt, the Electricity Law (Law No 87/2015) regulates renewable installations but leaves licensing of BTM storage to subsidiary regulations, creating uncertainty for prosumers seeking grid participation.

B. Market Access and Trading Restrictions

Distributors such as utilities and grid operators face regulatory limits on market participation to preserve neutrality and prevent anti-competitive behavior. Though allowed to offer grid services like frequency regulation and peak shaving, their access to wholesale and retail markets is restricted to prioritize infrastructure management over commercial profit.[40]

Consumers and prosumers have broader market access via mechanisms like net metering, feed-in tariffs, and Virtual Power Plants (VPPs), enabling energy arbitrage and localized grid support. However, current incentives often undervalue their contributions.[41] Under Federal Decree-Law No 17 of 2022, producers of distributed renewable energy—including those pairing generation with storage—must obtain mandatory Connection Agreements and adhere to annual capacity limits and interconnection terms overseen by the Competent Authority, thereby formalizing their grid access responsibilities.[42]

In the UAE, DEWA's Shams Dubai program allows prosumers to export surplus energy to the grid under net-metering rules, though stored energy from batteries is not separately remunerated.[43] In Morocco, Law 13-09 enables renewable generators to sell excess power to the grid, yet lacks explicit provisions for market participation by standalone storage,

[37] Economic Regulation Authority, 'Electricity Licensing' <https://www.erawa.com.au/electricity/electricity-licensing> accessed 16 March 2025.

[38] APSR (n 22).

[39] S Bellekom and others, 'Prosumption and the Distribution and Supply of Electricity' (2016) 6 Energy, Sustainability and Society 1.

[40] R Faia and others, 'Local Electricity Markets: A Review on Benefits, Barriers, Current Trends and Future Perspectives' (2024) 190 Renewable and Sustainable Energy Reviews 114006.

[41] B Speetles and others, 'Virtual Power Plants and Energy Justice' (2023) National Renewable Energy Laboratory <https://research-hub.nrel.gov/en/publications/virtual-power-plants-and-energy-justice> accessed 11 August 2025.

[42] Federal Decree-Law No 17 of 2022 Regulating the Connection of Distributed Renewable Energy Production Units to the Electricity Grid (UAE 26 September 2022) art 5 <https://www.uaelegislation.gov.ae/en/legislations/1567> accessed 12 August 2025.

[43] DEWA (n 21).

limiting revenue from arbitrage. Oman's APSR permits microgrid storage to provide ancillary services in isolated systems, but grid-scale market access remains largely reserved for the state utility.[44] In Jordan, the Renewable Energy and Energy Efficiency Law supports net-metering and wheeling for solar-plus-storage projects, but tariff structures still undervalue storage's role in peak shaving and frequency regulation.

C. Grid Integration and Stability vs Energy Independence

Electricity storage deployment exposes a key regulatory tension in the MENA region: grid stability versus energy independence. Across the region, regulators and utilities are testing diverse integration models to address this challenge. In the UAE, the Abu Dhabi Transmission and Despatch Company (TRANSCO), now operating under TAQA Transmission,[45] has implemented grid-scale lithium-ion battery systems at critical substations to provide frequency regulation and peak shaving, in line with Emirate-level grid codes that set stringent technical and operational performance standards for storage integration.[46] Morocco's Office National de l'Électricité et de l'Eau Potable (ONEE) has piloted battery storage integration in rural electrification schemes to stabilize microgrids, with operational protocols designed to maintain voltage and frequency stability during renewable fluctuations.[47] In Oman, hybrid diesel–solar–storage systems deployed in remote governorates are enabled under APSR's self-generation and wheeling policies, allowing generators to store electricity and then dispatch it to microgrids or the national grid under regulated annual caps—an approach that enhances energy independence while safeguarding grid stability.[48] In Jordan, the Renewable Energy and Energy Efficiency Law No 13 of 2012 (as amended in 2014), complemented by the Electricity Law No 64 of 2002 and the Energy and Minerals Regulatory Commission (EMRC) by-laws, provides the statutory basis for the country's net-metering and wheeling regulations. The EMRC has introduced net-metering and wheeling regulations that allow solar-plus-storage systems to operate in both grid-connected and island modes. While this framework promotes consumer flexibility and resilience, it mandates synchronization and protection equipment to preserve system stability.[49]

D. Rights to Discharge and Use Stored Energy

In the UAE, under the Emirates Authority for Standardization and Metrology (ESMA) grid codes, transmission and distribution companies must obtain approvals from the federal

[44] APSR (n 22).

[45] TAQA Transmission <https://taqatransmission.com> accessed 11 August 2025.

[46] ibid.

[47] H El Hafdaoui, A Khallaayoun, and S Al-Majeed, 'Renewable Energies in Morocco: A Comprehensive Review and Analysis of Current Status, Policy Framework, and Prospective Potential' (2025) 26 Energy Conversion and Management: X 100967. https://www.sciencedirect.com/science/article/pii/S2590174525000996?via%3Dihubf> accessed 11 August 2025.

[48] Oman unveils new self-generation and direct sale policy (Oman Observer, 25 January 2025) <https://www.omanobserver.om/article/1165335/business/economy/oman-unveils-new-policy-for-renewable-energy> accessed 11 August 2025.

[49] Energy and Minerals Regulatory Commission (EMRC), 'Instruction Concerning Net-Metering and Wheeling for Renewable Energy Systems' (2023) ss 5–7 <http://www.emrc.gov.jo> accessed 12 August 2025.

regulator before discharging stored energy into the grid. Discharge is authorized only under specific operational conditions to maintain frequency stability and reserve margins.[50] In Morocco, operators connected to the national grid through the Office National de l'Électricité et de l'Eau Potable (ONEE) must comply with dispatch instructions and reserve requirements, ensuring that storage discharge supports system reliability rather than market arbitrage.[51]

By contrast, consumers and prosumers in parts of the MENA region enjoy greater flexibility. In Jordan, residential and commercial solar-plus-storage systems are permitted to discharge primarily for self-consumption, with surplus export allowed under the net-metering framework set by the EMRC, although compensation rates often fall below wholesale market value.[52] In Oman, rural hybrid systems operated under the APSR allow limited discharge to community microgrids, with tariffs and export volumes capped to protect system balance.[53]

The UAE's clean energy regulations further strengthen consumer rights by prohibiting double charging for stored electricity consumed on-site and requiring that prosumer exports to the grid are not subject to disproportionate technical or economic conditions, aligning with international best practice while retaining grid protection measures.

E. Permitting and Regulatory Compliance

Distributors face complex permitting processes, including environmental impact assessments, grid interconnection approvals, and compliance with safety and stability regulations. In Morocco, utility-scale storage projects connected to renewable energy plants must undergo approval by the Moroccan Agency for Sustainable Energy (MASEN) and the ONEE, with procedures covering technical feasibility, environmental safeguards, and grid stability requirements.[54] In the UAE, distribution companies require permits from the federal or emirate-level energy authorities, along with compliance to Emirates Authority for Standardization and Metrology (ESMA) safety codes, before storage assets can be commissioned and discharged into the grid. In Oman, the APSR mandates environmental clearance and operational certification for storage projects exceeding specified capacity thresholds.[55]

In contrast, consumers and prosumers encounter fewer regulatory hurdles. They must obtain local permits for modifications, such as municipal approvals in Jordan for residential solar-plus-storage systems, but interconnection agreements are generally simpler than those for utility-scale projects.

[50] Ministry of Industry and Advanced Technology <https://moiat.gov.ae/en/about-us/about-the-ministry> accessed 11 August 2025.

[51] Loi n° 40-09 relative à l'Office National de l'Électricité et de l'Eau Potable (ONEE), Bulletin Officiel n° 5932 du 5 août 2010 [Moroccan Law No. 40-09 (2010) on the Restructuring of ONEE]. Law No. 40-09 relating to the National Office of Electricity and Drinking Water (ONEE) grants the Office responsibility for planning, operating, and dispatching the national electricity system, ensuring continuity, quality, and security of supply.

[52] EMRC (n 49).

[53] APSR (n 22).

[54] ONEE (n 51).

[55] APSR (n 22).

F. Compensation for Services Provided

Distributors earn compensation primarily through ancillary services like frequency regulation, voltage control, and peak shaving, which support grid stability. For example, in Morocco, compensation for storage-based ancillary services is set through PPA terms negotiated with the ONEE, with tariffs reflecting renewable integration targets rather than open market bidding.[56] In the UAE, distribution companies may recover costs for grid-support services through regulatory tariffs approved by the relevant emirate-level authority, though these do not yet fully reflect the flexibility value of storage. The Abu Dhabi Department of Energy's website outlines how bulk supply, transmission, and distribution tariffs require economic regulation and independent approval, meaning cost recovery mechanisms are clearly governed and applicable to grid-support services.[57] Law No 11 of 2018 authorizes the Abu Dhabi DOE to propose and regulate energy tariffs, including Bulk Supply, TUoS, and DUoS charges—where cost recovery structures exist but don't yet reflect storage flexibility value.[58]

In Jordan, residential and commercial prosumers under the net metering scheme receive credits at the applicable retail tariff, while those exporting under wheeling arrangements are compensated at wholesale-equivalent rates, though both remain below full market value. In Oman, the recently updated Small-Scale Grid-Connected Renewable Energy Regulation allows prosumers to offset consumption with storage-supported generation, but financial credits are capped annually, limiting income potential.

IV. Legal and Policy Barriers to Electricity Storage Regulation in the MENA Region

A. Gaps in Legal Frameworks for Distributed Energy Resources and Storage Definition

A core challenge across the MENA region is the uneven development of legal frameworks that clearly define and classify electricity storage, especially when deployed as part of distributed energy resources (DERs). Key regulations shaping Egypt's storage landscape include the Renewable Energy Law (Decree No 203/2014), Electricity Law (Law No 87/2015), and the Green Hydrogen Projects Incentives Law (Law No 2/2024).[59] However, in practice, there remains a lack of clear regulatory frameworks regarding grid access, tariffs,

[56] ONEE (n 51).

[57] Abu Dhabi Department of Energy, 'Economic Regulations—Tariffs and Charges' (DOE Abu Dhabi) <https://www.doe.gov.ae/en/Legislation-and-Compliance/Economic-Regulations/Tariffs-and-charges> accessed 12 August 2025.

[58] Department of Energy, Abu Dhabi, Law No 11 of 2018 concerning the Establishment of the Department of Energy (Abu Dhabi) 'Economic Regulations—Tariffs and Charges' <https://www.doe.gov.ae/Legislation-and-Compliance/Laws-and-Regulations> accessed 12 August 2025.

[59] IEA, 'Egypt Renewable Energy Law (Decree No 203/2014)' (2016) JOIN IEA/IRENA Policy and Measures Database https://www.iea.org/policies/6104-egypt-renewable-energy-law-decree-no-2032014 accessed 4 April 2025; OECD, 'Green Hydrogen Incentives Law—Government of Egypt' <https://www.oecd.org/content/dam/oecd/en/about/programmes/cefim/green-hydrogen/2024-case-studies/Green-hydrogen-incentives-law-case-study-2024.pdf/_jcr_content/renditions/original./Green-hydrogen-incentives-law-case-study-2024.pdf> accessed 11 August 2025.

and compensation schemes—a gap that risks delaying full renewable integration and limits the role of storage in grid flexibility.

Morocco offers partial progress in integrating storage into renewable policy. Under Morocco's Law 13-09 on renewable energy, the national regulator is authorized to approve hybrid systems for connection to the grid, provided they meet integration standards and grid stability metrics.[60] This framework permits private sector participation, defines conditions for third-party access to the grid, and establishes rights for renewable energy producers to sell electricity both to the national grid and directly to consumers. This recognition of hybrid systems allows for some storage deployment but falls short of establishing a standalone legal category for storage assets.

Oman demonstrates a narrower but targeted approach, focusing on localized renewable integration. Under the Law for the Regulation and Privatisation of the Electricity and Related Water Sector (Royal Decree 78/2004), the APSR regulates microgrid and associated storage projects.

In contrast, the UAE presents one of the region's clearest examples of a defined legal framework for storage integration. Dubai Law No (27) of 2021.[61] reconstituted DEWA as a public joint-stock company and established a clear legal framework for public-private partnerships (PPPs) in clean energy infrastructure. The law enables DEWA to regulate grid-connected infrastructure, oversee clean energy projects, and engage with private sector actors in emerging technologies such as battery storage and electric mobility. This statutory clarity strengthens investor confidence and creates a stable environment for both utility-scale and distributed storage initiatives.

B. Licensing and Procedural Barriers to Storage Deployment

Across the MENA region, electricity storage projects face significant procedural and regulatory hurdles that slow deployment and increase costs. These challenges are particularly acute in jurisdictions where approval processes are fragmented, roles between public and private actors are unclear, and procedural bottlenecks remain unresolved.

In Oman, the integration of microgrids into rural areas presents both benefits and challenges in decentralizing energy via renewable sources.[62] The APSR governs microgrid licensing under the Law for the Regulation and Privatization of the Electricity and Related Water Sector (Royal Decree 78/2004). Microgrid operators must obtain approval from the Oman Electricity Transmission Company SAOC (OETC) for grid integration.[63] Despite the legal framework supporting microgrid development, significant licensing challenges remain, such as bureaucratic delays and the unclear role of private developers versus state utilities. These hurdles often lead to project delays and increased costs.[64]

[60] Law No 13-09 (n 30).
[61] Dubai Law No (27) of 2021 (n 20).
[62] N Hosseinzadeh and others, 'Constructing a Multi-Microgrid with the Inclusion of Renewable Energy in Oman's Rural Power System' (2018) 15(1) The Journal of Engineering Research 61 <https://tjer.squ.edu.om/journal/vol15/iss1/4/> accessed 11 August 2025; AH Al-Badi and others, 'Survey of Smart Grid Concepts and Technological Demonstrations Worldwide Emphasizing on the Oman Perspective' (2020) 3(1) Applied System Innovation 5 <https://doi.org/10.3390/asi3010005> accessed 11 August 2025.
[63] APSR (n 22).
[64] KAS Al-Marboui and E Al-Mahdawi, 'Oman Rural Grid Extension Employing Renewable Energy Technology' (2021) IOP Conference Series: Earth and Environmental Science 779(1) 012035; A Al Abri

Similar licensing and procedural challenges can be seen elsewhere in the region. In Egypt, while the policy framework for renewable energy has advanced, barriers remain in moving from legislation to operational licensing for storage systems.[65] However, a lack of clear regulatory frameworks regarding grid access, tariffs, and compensation schemes risks delaying full renewable integration.[66] Early drafts of the roadmap recommend a three-tier licensing model for (i) consumer-scale residential batteries, (ii) commercial or microgrid applications, and (iii) grid-scale BESS, to ensure scalability and inclusiveness.[67]

Here, the absence of standardized licensing processes creates uncertainty for investors and implementers. Projects can be delayed not due to technological limitations but because of unclear criteria for grid connection, insufficiently defined discharge rights, or inconsistent approval timelines between agencies.

In parallel, the integration of storage into Egypt's Power System Master Plan (2022–2035) remains nascent, underscoring the need for cross-ministerial co-ordination between the Ministry of Electricity and Renewable Energy (MoERE), EgyptERA, and the Ministry of Planning.

Taken together, these examples show that even when national strategies explicitly include energy storage, licensing and procedural barriers can undermine implementation. Legal reforms that clarify institutional roles, standardize technical approval processes, and reduce bureaucratic delays are critical to enabling timely and cost-effective deployment of storage projects across the MENA region.

C. Market Design, Tariff Structures, and Incentives

Across the MENA region, market design and tariff structures remain underdeveloped in their ability to recognize and reward the multi-functional value of electricity storage. In many jurisdictions, storage is still valued only as a generation asset, without clear mechanisms to compensate for grid services such as frequency regulation, peak shaving, or load shifting.

In Morocco, the Noor Midelt project, a hybrid solar and storage initiative, addresses challenges in market design and the integration of storage with renewable energy. Under Morocco's Law 13-09 on renewable energy, the national regulator is authorized to approve hybrid systems for connection to the grid, provided they meet integration standards and grid stability metrics.[68] This has allowed Noor Midelt to benefit from a blended

and P Dharmasa, 'Main-Grid Versus Renewable Micro Energy Supply—Case Study of Isolated Rural Areas in the Sultanate of Oman' (2023) 27th International Conference on Electricity Distribution (CIRED 2023) (IET 2023) 975–79; Al Abri and Dharmasa (n 64) 975–79; APSR (n 22).

[65] IEA (2024), State of Energy Policy 2024, IEA, Paris <https://www.iea.org/reports/state-of-energy-policy-2024> accessed 26 November 2025.

[66] Organisation for Economic Co-operation and Development (OECD), State of Energy Policy 2024 (2024) 34–35 <https://iea.blob.core.windows.net/assets/80794060-b21d-4062-82a6-93dece2c5163/StateofEnergyPolicy2024.pdf> accessed 26 November 2025.

[67] Egyptian Ministry of Electricity and Renewable Energy, 'Integrated Sustainable Energy Strategy to 2035' (ISES 2035) <https://www.egypt-energy.com/content/dam/Informa/egypt-energy/en/pdf/Egypt%20Energy%20Report-16-5%20.pdf> accessed 10 May 2025; African Development Bank (AfDB), 'Financing Change in Africa: AfDB-CIF Annual Report 2024' (AfDB 2024) https://www.afdb.org/sites/default/files/documents/publications/afdb-cif_annual_report_2024.pdf accessed 10 May 2025.

[68] European Investment Bank, 'Environmental and Social Data Sheet: Noor Midelt' (2019) <https://www.eib.org/en/projects/all/20160526> accessed 11 August 2025.

tariff arrangement that compensates both generation and storage functions—pioneering a model for dual-function regulation.[69]

While Morocco provides a rare example of a dual-function tariff model, other jurisdictions have yet to integrate comparable market incentives. Elsewhere in the region, similar gaps persist. In the UAE, tariff structures for grid-support services—such as frequency regulation or flexibility—are still evolving and do not fully reflect the value that storage provides.

Jordan allows solar-plus-storage systems under its net-metering and wheeling frameworks to export electricity at applicable retail or wholesale-equivalent rates. However, these rates remain below full market value and do not differentiate between storage-supported and direct renewable generation, reducing the commercial appeal of integrated systems.

To address regulatory fragmentation,[70] Egypt has recently proposed the creation of a National Energy Storage Roadmap under EgyptERA, with technical assistance from UNECA and funding from the African Development Bank.[71] The roadmap aims to develop harmonized standards for utility and consumer storage, clarify discharge rights, and introduce grid service compensation tariffs.[72] If adopted, this would significantly strengthen Egypt's ability to attract private investment in decentralized renewable integration. However, until these tariffs are implemented and matched with predictable revenue mechanisms, private sector participation in the storage market will remain limited. These examples show that while Morocco's blended tariff model offers a strong regional precedent, most MENA jurisdictions have yet to establish market rules and financial incentives that fully capture storage's value across generation, grid services, and demand management.

D. Institutional Co-ordination and Strategic Alignments

Institutional fragmentation and insufficient co-ordination between regulatory, operational, and policy bodies remain major obstacles to scaling electricity storage across the MENA region. Even where enabling laws and technical standards exist, gaps in inter-agency alignment can slow project execution, dilute accountability, and limit integration of storage into broader energy strategies.

In the UAE, strong legal frameworks have enabled flagship projects. DEWA and Tesla partnered on a large-scale utility storage project at the Mohammed bin Rashid Al Maktoum Solar Park to stabilize the grid by storing surplus solar energy.[73] These institutional

[69] African Energy Chamber, 'The State of African Energy: 2025 Outlook Report' (African Energy Chamber 2024) 64 https://energychamber.org/wp-content/uploads/2024/10/The-State-of-African-Energy-2025_digital-2.pdf accessed 9 May 2025.

[70] SI Salah and others, 'Towards a Sustainable Energy Future for Egypt: A Systematic Review of Renewable Energy Sources, Technologies, Challenges, and Recommendations' (2022) 8 Cleaner Engineering and Technology 100497; A Raihan and others, 'A Review of the Current Situation and Challenges Facing Egyptian Renewable Energy Technology' (2024) 3(3) Journal of Technology Innovations and Energy 29–52.

[71] United Nations Economic Commission for Africa, 'Regulatory Review of the Electricity Market in Egypt: Towards Crowding-in Private Sector Investment' (2023) <https://repository.uneca.org/handle/10855/48484> accessed 11 August 2025.

[72] African Development Bank, 'Morocco—Noor Midelt Solar Central Project—Phase I' (2017) <https://www.afdb.org/en/projects-and-operations/p-ma-ff0-004> accessed 11 August 2025.

[73] A Colthorpe, 'Dubai Electricity and Water Authority Inaugurates Tesla Battery System at World's Largest Solar Farm' (September 2021) <https://www.energy-storage.news/dubai-electricity-and-water-authority-inaugurates-tesla-battery-system-at-worlds-largest-solar-farm/> accessed 4 March 2025.

investments are aligned with the UAE Energy Strategy 2050, which targets 44 per cent clean energy by 2050 and explicitly includes storage in its implementation roadmap.[74] However, even in this advanced context, co-ordination between utility-led initiatives and wider market reforms remains an ongoing challenge.

Oman has made strides in integrating battery storage within rural microgrids, supported by APSR policies for self-generation and direct sale.[75] Yet, implementation requires synchronized action between APSR, grid operators, and market platforms such as the newly launched electricity spot market.[76] Without clear delineation of responsibilities and consistent operational frameworks, integration timelines risk slipping.

Yet, as with other jurisdictions, sector leadership is split between multiple agencies, and the absence of a centralized co-ordination mechanism can hinder the scaling of incentives and market rules beyond flagship projects.

Egypt has recognized the need for co-ordinated governance in its proposed National Energy Storage Roadmap, aiming to harmonize standards and responsibilities across EgyptERA, the Ministry of Electricity and Renewable Energy, and the Ministry of Planning.[77] This shows how fragmented decision-making can undercut even well-designed strategies.

Collectively, these examples underline that legal provisions alone are not sufficient; storage integration also depends on clear institutional mandates, strong inter-agency co-ordination, and alignment with national energy and climate strategies. Without such alignment, storage projects risk remaining isolated pilot efforts rather than forming part of a coherent, scalable infrastructure transformation.

E. Limited Private Sector Participation and Investment Mobilization

Private sector participation in electricity storage across the MENA region remains constrained by a mix of regulatory uncertainty, market dominance of public utilities, and the absence of tailored investment incentives. While large-scale projects often attract international finance, smaller or decentralized initiatives struggle to secure capital without predictable revenue frameworks.

In the UAE, flagship partnerships such as DEWA and Tesla's collaboration at the Mohammed bin Rashid Al Maktoum Solar Park demonstrate how clear licensing authority and targeted reforms can draw in global technology providers.[78] In 2024, DEWA awarded Tesla its first Electric Vehicle (EV) Charge Point Operator license in Dubai, signaling

[74] UAE Ministry of Energy, 'UAE Energy Strategy 2050' <https://u.ae/en/about-the-uae/strategies-initiatives-and-awards/strategies-plans-and-visions/environment-and-energy/uae-energy-strategy-2050> accessed 4 April 2025 DEWA (n 21).

[75] Oman Vision 2040, 'National Priorities: Energy and Natural Resources' (2020) <https://www.oman2040.om/uploads/publication/20231105221146-2023-11-05publication221143_.pdf> accessed 11 August 2025.

[76] APSR (n 22) 42; Ministry of Energy and Minerals, Oman, 'Policy for the Use of Renewable Energy for Self-Generation and Direct Sale' (2024) <https://mem.gov.om/Media-Center/News/ArtMID/608/ArticleID/1422> accessed 1 May 2025.

[77] IRENA, 'Renewable Capacity Statistics 2025' (IRENA, March 2025) 16–17 <https://www.irena.org/-/media/Files/IRENA/Agency/Publication/2025/Mar/IRENA_DAT_RE_Capacity_Statistics_2025.pdf> accessed 9 May 2025.

[78] A Sanfilippo, M Vermeersch, and VB Benito, 'Energy Transition Strategies in the Gulf Cooperation Council Countries' (2024) 55 Energy Strategy Reviews 101512.

regulatory progress and growing confidence in Tesla's technological role in the UAE's energy and transportation sectors.[79]

In support of these goals, Oman has introduced a Renewable Energy Policy for Self-Generation and Direct Sale, which streamlines licensing procedures for small-scale producers and permits direct electricity sales. The policy also outlines provisions for the integration of BESS to enhance grid stability and efficiency. Concurrently, the APSR's 2022 Annual Report highlights the successful launch of the electricity spot market, providing a platform for real-time electricity trading and promoting greater participation from renewable energy producers.[80] However, access to these opportunities remains weighted towards larger, well-capitalized players, with smaller developers facing both licensing and financing hurdles.

In Morocco, hybrid solar-storage projects like Noor Midelt have succeeded in attracting multilaterals such as the EIB and World Bank,[81] supported by a legal framework that permits private sector participation and defines third-party grid access.[82] Yet without dedicated storage incentives, uptake is largely confined to flagship projects, leaving untapped potential for smaller-scale or purely private ventures.

Egypt has sought to address private sector constraints through its proposed National Energy Storage Roadmap, backed by UNECA and the AfDB, which includes tariff reforms and a three-tier licensing model. If implemented alongside predictable revenue guarantees, these reforms could unlock private capital for both grid-scale and distributed storage.

Across the region, the pattern is consistent: without robust, storage-specific incentives and predictable revenue streams, private sector participation will remain the exception rather than the norm. Unlocking this potential requires legal reforms that not only allow but actively encourage private investment through targeted procurement, risk-sharing mechanisms, and stable market rules.

F. Summary

The case studies of the UAE, Oman, Morocco, and Egypt each reflects a different stage in regulatory maturity, market orientation, and institutional co-ordination—yet collectively provide a mosaic of complementary legal strategies to guide regional harmonization and investor engagement.

The UAE's approach emphasizes a top-down innovation ecosystem, with DEWA leveraging strategic partnerships—such as its Tesla 100 MW/300 MWh battery system—to pilot large-scale storage and integrate it into national clean energy strategies.[83] These efforts

[79] Olawuyi (n 11 above) ch 5.

[80] APSR (n 22); Oman Power and Water Procurement Company (OPWP), 'Oman Electricity Market Annual Report 2022' (2023) <https://www.oemo.om/media/sfjbnj2m/market-annual-report-final.pdf> accessed 1 May 2025.

[81] World Bank, 'Morocco Noor Solar Power Project Additional Financing' (2018) <https://documents1.worldbank.org/curated/en/138481528687821561/pdf/Morocco-Noor-AF-project-paper-P164288-May17-clean-05212018.pdf> accessed 4 May 2025.

[82] Ibid; European Investment Bank, 'Environmental and Social Data Sheet' (2019) <https://www.eib.org/attachments/registers/77307870.pdf> accessed 4 April 2025; K Choukri and others, 'Renewable Energy in Emergent Countries: Lessons from Energy Transition in Morocco' (2017) 7 Energy, Sustainability and Society 25 <https://doi.org/10.1186/s13705-017-0131-2> accessed 11 August 2025.

[83] A Sanfilippo and others (n 77).

are grounded in strong regulatory clarity and alignment with the UAE Energy Strategy 2050, which explicitly targets a 44 per cent clean energy share and includes energy storage as a priority technology.[84]

In contrast, Oman's strategy reflects a bottom-up model focused on rural electrification, where microgrid deployment under the authority of APSR (pursuant to Royal Decree 78/2004) offers important lessons in adaptive licensing.[85] However, regulatory bottlenecks, capital access constraints, and overlapping institutional roles still impede rapid deployment—especially in remote communities reliant on diesel generators.

Morocco's Noor Midelt project represents a hybrid legal and market design, enabling dual-function regulation for solar-plus-storage systems. Anchored in Law 13-09 on renewable energy, the project has attracted substantial international finance, such as from the World Bank and EIB, through its blended tariff design and hybrid system classification.[86] These legal innovations are paving the way for greater private sector participation and grid flexibility.

Meanwhile, Egypt's regulatory framework is still in development. Despite enacting supportive laws—including Law No 87/2015 (Electricity), Law No 2/2024 (Hydrogen Incentives), and Decree 203/2014 (Renewables)—key barriers persist.[87] Delays in land acquisition and procurement have hindered the deployment of BESS in strategic locations like Benban. However, a proposed National Energy Storage Roadmap, supported by UNECA and the AfDB, introduces a promising three-tier licensing model (residential, microgrid, grid-scale) that could help crowd in private investment once tariff predictability improves.

Across all four jurisdictions, regulatory certainty and legal reform are essential for scaling up storage. According to the IRENA, the MENA region's installed energy storage capacity remains below 1 GW, but the potential exceeds 20 GW by 2040 if legal, financial, and institutional barriers are addressed.[88]

As storage gains traction in national energy strategies, regulatory alignment with climate objectives and SDG 7 becomes increasingly important. Jurisdictions like Egypt and Oman have incorporated renewable access and equity into strategic visions (ISES 2035 and Vision 2040), yet implementation gaps remain. Future reforms should explicitly link legal instruments governing storage to climate finance access, just transition principles, and international obligations under the Paris Agreement. Without such integration, storage will remain peripheral to national development agendas.

[84] UAE Ministry of Energy, 'UAE Energy Strategy 2050' https://u.ae/en/about-the-uae/strategies-initiatives-and-awards/strategies-plans-and-visions/environment-and-energy/uae-energy-strategy-2050 accessed 4 April 2025.

[85] Al Abri and Dharmasa (n 64) 975–79.

[86] Law No 13-09 (n 30).

[87] IEA, 'Egypt Renewable Energy Law (Decree No 203/2014)' (2016) JOIN IEA/IRENA Policy and Measures Database https://www.iea.org/policies/6104-egypt-renewable-energy-law-decree-no-2032014 accessed 4 April 2025; A Fattah Al-Sisi, 'Law No. 2 of 2024 Regarding Incentives for Green Hydrogen Production Projects and Its Derivatives' (Ecoris, January 2024) https://ecoris.green/en/lawsandregulations/law-no-2-of-2024-regarding-incentives-for-green-hydrogen-production-projects-and-its-derivatives/ accessed 4 April 2025.

[88] IRENA, 'World Energy Transitions Outlook 2024: 1.5°C Pathway' (2024) 73–75 <https://www.irena.org/-/media/Files/IRENA/Agency/Publication/2024/Nov/IRENA_World_energy_transitions_outlook_2024.pdf> accessed 10 May 2025; IEA, 'Grid-Scale Storage' (2023) <https://www.iea.org/energy-system/electricity/grid-scale-storage> accessed 1 May 2025.

V. Policy and Legal Recommendation

The growing centrality of battery storage to energy transitions worldwide requires a multi-dimensional legal response. Drawing from the comparative case studies of the UAE, Oman, Morocco, Jordan, and Egypt, the recommendations below address the four main barrier areas identified in Section IV, namely: (1) gaps in legal frameworks for DERs and storage; (2) licensing and procedural bottlenecks; (3) underdeveloped market design and incentives; and (4) weak institutional co-ordination and limited private sector participation.

First, closing the gaps in legal definitions and classifications for distributed energy resources and storage is a prerequisite for scaling deployment. Legislatures should enact or amend primary energy laws to provide clear statutory definitions covering grid-scale, behind-the-meter, and hybrid systems. This classification must distinguish storage from generation and transmission assets, clarify market roles, and ensure consistent treatment in licensing, incentives, and interconnection procedures.

Second, fragmented permitting frameworks, unclear market participation rights, and inadequate safety standards continue to limit the effective deployment of electricity storage across the MENA region. To address these barriers, regulatory authorities should prioritize the streamlining of licensing and permitting procedures. This involves codifying the eligibility of behind-the-meter and commercial storage operators to provide grid services such as frequency regulation, peak shaving, and demand response, with fair and transparent compensation mechanisms. Adopting comprehensive, enforceable standards—alongside clear restrictions on utility ownership of storage assets would protect consumer rights. Finally, as regional interconnections expand through initiatives such as the GCC Interconnection Authority and North African grid links, regulatory harmonization will be essential.

Third, barriers to consumer participation and private sector investment in energy storage remain significant across the MENA region, reflecting the regulatory gaps and institutional limitations identified in Section 4. In many jurisdictions, consumers lack explicit legal rights to store, self-consume, and trade surplus electricity from distributed systems. National legal frameworks should therefore establish enforceable rights for consumers to own and operate storage, connect to the grid, and participate in market mechanisms such as peer-to-peer trading, Virtual Power Plants (VPPs), and co-operative storage schemes.

Innovation is further constrained by the absence of structured pilot programs and regulatory sandboxes. Creating dedicated legal carve-outs for time-bound experimentation would allow regulators to monitor performance, mitigate risks, and gather data to inform future market rules. Private sector engagement is also hindered by uncertainty over ownership structures, risk allocation, and revenue guarantees. Establishing transparent, storage-specific provisions within public–private partnership (PPP) frameworks—covering procurement processes, performance benchmarks, and dispute resolution mechanisms—would help crowd in private capital.

Fifth, differences in system classification, safety standards, tariff methodologies, and licensing requirements create legal uncertainty for investors and prevent the seamless integration of storage assets into regional interconnection projects such as the Gulf Cooperation Council Interconnection Authority (GCCIA) and North African power links.

Harmonizing these frameworks is essential for unlocking the full value of storage in enhancing cross-border system flexibility, reliability, and climate resilience. Governments and regulatory bodies should work through existing intergovernmental platforms to develop mutually recognized standards for technical compliance, grid codes, and market participation. Legal provisions for cross-border storage operations must be embedded in electricity and energy market laws, ensuring that operators have predictable rights to connect, trade, and receive remuneration for storage services across jurisdictions.

VI. Conclusion

This chapter has traced the evolution of regulatory frameworks governing distributed electricity storage across utility-scale, behind-the-meter, and hybrid DER systems, revealing persistent legal ambiguities, procedural bottlenecks, market and tariff misalignments, and institutional fragmentation that hinder widespread deployment. Through in-depth case studies of the UAE, Oman, Morocco, and Egypt, the analysis has emphasized the critical role of adaptive legal systems in enabling secure, inclusive, and climate-aligned electricity storage ecosystems.

The case studies illustrate varying levels of legal innovation and maturity: the UAE's centralized yet investor-oriented approach; Oman's microgrid-focused licensing adaptations; Morocco's dual-function classification regime fostering tariff flexibility; and Egypt's roadmap-led reforms underpinned by AfDB and UNECA partnerships.

Yet despite these developments, regional deployment remains significantly below potential. As IRENA highlights, MENA's current installed storage capacity is under 1 GW, while its technical potential could exceed 20 GW by 2040. Meanwhile, electricity demand across MENA is projected to increase by over 70 per cent by 2040. Without legal clarity on asset classification, streamlined and adaptive licensing, robust market access and tariff frameworks, and co-ordinated institutional mandates, energy storage will remain stranded: technically feasible, yet legally constrained.

To avoid this outcome, legal frameworks must recognize energy storage as core infrastructure—on par with generation and transmission—not a regulatory afterthought.

PART II
SUSTAINABLE DECENTRALIZATION AND DIGITALIZATION OF ENERGY SYSTEMS

7
Cross-border Distributed Energy in the EU

Martha M Roggenkamp

I. Introduction

This chapter will examine the developments of 'distributed energy sources' (DER) in the EU and specifically focus on cross-border DER. The concept of DER generally refers to the establishment of small-scale energy resources and innovative techniques (eg electricity storage) close to consumption sites and connected to the distribution grid.[1] In Europe this concept is not new. During the energy crisis of the 1970s several European governments allowed regional energy distribution companies to install and operate wind turbines in their supply areas. Although a breach of the electricity producers monopoly position, this development was considered necessary for securing energy supply. Currently the reasons for developing DER in Europe are more diverse. Apart from security of supply, the main driver is the decarbonization process and the subsequent need to phase out the use of fossil fuels. The energy market liberalization process of the last three decades has also facilitated this process as it enables consumers and small producers to invest in distributed energy.

The concept of 'distributed generation' was introduced in the 2003 Electricity Directive and defined as 'generation plants connected to the distribution system'.[2] Although the Gas Directives do not contain a similar provision, references can be found to 'renewable gases' and 'distributed production'.[3] Apart from this, it seems that the terminology used does not exclude the broader concept of DER. Furthermore, the concept of distributed generation and DER changes the traditional top-down approach to the energy system where (major) producers are connected to the transmission system, which in its turn is connected to various regional distribution systems who ultimately are supplying end users. DER, however, is based on a bottom-up approach where small producers (including active consumers) are connected to the distribution system, which no longer can be considered as a passive 'pass-on' facility. Given its position in the energy chain, the EU and EU member states (MS) have considered DER as a matter of national interest.

This chapter will take a different approach and assess the possibility of cross-border DER within the EU. The reason for this approach is both conceptual and practical. First, the aim of the EU since the 1990s is to develop one internal energy market, ie a market where energy can be freely produced and traded across borders. Such an approach should necessarily also include DER and thus the possibility to connect DER to a distribution system in another

[1] See Chapter 1 of this book. See also International Energy Agency (IEA), *Unlocking the Potential of Distributed Energy Sources* (IEA 2022) 14.

[2] See art 2(31) Directives 2003/54/EC, 2009/72/EU, and 2019/944/EU. The wording in these articles is almost similar.

[3] See art 40(d) Directive 2009/73/EU and arts 2(2), 77(d) Directive 2024/1788.

Martha M Roggenkamp, *Cross-border Distributed Energy in the EU* In: *Sustainable Distributed Energy Resources*. Edited by: Íñigo del Guayo, Damilola S. Olawuyi San, Louis de Fontenelle, Milton Fernando Montoya, and José Juan González, Oxford University Press. DOI: 10.1093/9780198991946.003.0009

MS and/or to a cross-border distribution system. From a more practical perspective, cross-border DER might offer a solution for problems involving congestion and high network costs. Currently, network congestion is an issue in several MS. Smaller and larger producers and consumers are increasingly not allowed a grid connection due to insufficient capacity on the transmission and/or distribution system. If DER is being developed close to national borders, a cross-border connection to a neighbouring distribution system (with sufficient capacity) might be a solution. In other words, cross-border DER may solve congestion issues and lower costs, but only if the law so allows.

Below section II will first discuss some relevant concepts of EU market liberalization for developing DER and how this might impact cross-border DER. Thereafter section III will discuss two examples of cross-border energy co-operation on distribution level and illustrate how EU law impacts such co-operation. Finally, section IV will provide a conclusion and some recommendations.

II. Relevant Concepts for Developing (Cross-border) DER

This section presents the main principles governing energy markets in the EU. Since the early 1990s the EU has been aiming at developing one internal energy market. This concept involves two—potential opposite—principles. It entails the need for market liberalization as well as market integration. Market liberalization basically requires an abolishment of exclusive rights and as such a freedom of production and consumption. In case of network-bound energy sectors, this freedom is accompanied by strict rules governing access to and use of these networks in order to avoid market abuse by the owners and/or operators of the transmission and distribution systems. The EU legislator issued a range of EU laws (directives and regulations) to ensure that all market parties have non-discriminatory access to the networks. Market integration entails the establishment of one internal (liberalized) EU market and thus requires the free movement of energy across borders and thus removing obstacles for cross-border trade and an increase in the number and the capacity of cross-border energy systems. Notably EU policy makers and legislators have primarily focussed on and regulated the network-bound electricity and gas markets. Hardly any attention was paid to heat supply, which is also a network-bound energy sector but to date primarily local and/or regional in scope. However, it is becoming of increasing interest when developing cross-border energy co-operation as will be illustrated in section III below.

A. The Energy System and Potential Cross-border Solutions

Traditionally the energy market depends on the proper functioning of the transmission and distribution systems. Notably, EU law does not provide a clear definition of both terms. Basically, transmission involves the transport of electricity or gas via the extra or high-voltage or high-pressure networks, respectively, and distribution is defined as electricity or gas transport via medium/low voltage or local networks. There is no general rule as regards classifying a cable or pipeline as high, medium, or low voltage. Hence, MS have the possibility of classifying networks as being part of the transmission or the distribution system. Difference in qualification may hamper cross-border connections. Anyway, whichever classification is used, these systems need to be operated by an independent system operator and subject to a regime of non-discriminatory access and several unbundling options.

1. Independent system operators

A key element of the EU regime is to achieve that transmission system operators (TSOs) and distribution system operators (DSOs) can operate independently. Independence from market parties is guaranteed by a regime of regulated non-discriminatory access and unbundling. Consequently, all market parties are entitled to have regulated non-discriminatory access to the system, which means that tariffs and conditions need to be non-discriminatory and transparent.[4] In other words, all users are being charged a non-discriminatory tariff for using the system and neither their size, their relationship with the suppliers and/or network operators can affect the tariffs and transport conditions. Moreover, these tariffs, or the methodologies underlying their calculation, need to be approved by the national regulatory authority (NRA) prior to their entry into force.[5]

Despite the requirement of non-discriminatory access, a risk of abuse continues to exist if suppliers and network operators are part of one vertically integrated company (VIC). Therefore the EU aims to unbundle network operations from commercial activities in order to ensure a level-playing field for all system users. The level of unbundling has gradually increased since the 1990s. The initial requirement of administrative unbundling involving that VICs should keep separate accounts for their commercial (production and supply) and network activities[6] was soon replaced by provisions on legal and functional unbundling[7] as a result of which VICs had to establish a legal entity for the transmission and/or distribution network activities separate from the production and supply activities, although the network company could belong to the same group or holding of companies. Since 2009, even stricter unbundling rules apply for transmission systems (see further below).

Despite these far-reaching unbundling rules, MS are free to decide whether the energy systems will be publicly or privately owned as long as they meet the general requirements of EU law. In practice, national laws of some MS restrict the ownership of energy systems to national, provincial, and/or municipal governments (eg the Netherlands), others allow for private parties to own and operate these systems (eg Germany).

2. National and cross-border transmission systems

Transmission systems are generally national in scope. These systems transport energy from large producers (eg gas production facilities and/or electricity generators) via several distribution systems to end users. The proper functioning of these transmission systems (eg the national energy highways) is vital to market liberalization. In order to enhance the independence of these TSOs, three new types of unbundling apply since 2009: (i) 'ownership unbundling', which means that producers and suppliers cannot own the operating transmission operators and vice versa;[8] (ii) 'independent system operator' (ISO), which requires owners of the transmission system to appoint an ISO to operate the system;[9] and

[4] See art 6 Directive 2019/944/EU and art 31 Directive (EU) 2024/1788.
[5] See for these tasks art 59 Directive 2019/944 and art 78 Directive (EU) 2024/1788.
[6] See Directive 1996/92/EC and Directive 1998/30/EC.
[7] Directive 2003/54/EU and Directive 2003/55/EU.
[8] Art 9 Directive 2009/72/EC.
[9] ibid arts 13–14.

(iii) 'independent transmission operator' (ITO), which means that the network company remains legally unbundled, but independence is achieved by adding several strict rules to prevent the mother company from interfering in the decision-making process of the network company.[10] Once opted for a specific unbundling regime a MS can opt for a higher degree of unbundling, but once a regime of ownership unbundling has been introduced there is no longer the possibility to return to an ITO or ISO. Generally, MS have either opted for ownership unbundling or an ITO.[11]

A separate category of transmission systems are the so-called cross-border transmission systems or interconnectors. These cross-border interconnectors emerged after World War II for several reasons. Where cross-border gas pipelines were constructed for trading and supply purposes, national electricity systems were primarily interconnected for reasons of supply security[12] and thus limited in size and not equipped to trade large volumes of electricity in a liberalized and integrated market. As such it is no surprise that since the 1990s EU energy law finds cross-border interconnectors of increasing importance to develop one internal energy market. Currently, EU law defines an interconnector generally as a 'transmission line which crosses or spans a border between MS and which connects the national transmission systems of the member states'.[13] In principle only (unbundled) TSOs have the right to construct and use such interconnectors. However, if necessary for developing an internal market and to promote competition a (temporary) exemption can be granted to other (private) parties to invest in 'merchant' interconnectors.[14] Since new interconnectors often are being developed because existing cross-border connections lack available capacity, the EU legal regime focuses to a large extent on (cross-zonal) capacity allocation, tariffication, and congestion management.[15] Clearly these interconnectors are not of direct relevance for DER, but a lack of available interconnector capacity may indirectly trigger the interest in other cross-border solutions.

3. Distribution systems and cross-border connections

Where transmission systems are national in scope, distribution systems operate regionally. National laws govern the designation of the regional energy supply areas. In the Netherlands, for example, the Code on Area Division establishes the respective areas of operation for DSOs.[16] Usually, the designated supply areas involve one of more regions, provinces, and/or

[10] ibid arts 17–23.

[11] CEER, *Status Review TSO/DSO Unbundling*, 20 March 2024 <https://www.ceer.eu/wp-content/uploads/2024/06/CEER-Status-review-TSO-DSO-unbundling.pdf> accessed 27 July 2025.

[12] The Union for the Coordination of Production and Transmission of Electricity (UCPTE) was founded in 1951 in Paris and aimed to ensure the optimum operation of electric power plants and to balance production and to prevent a loss of surplus production. In line with market liberalization, the activities of the Union focussed on transmission and was renamed to UCTE in 2001. See UCPTE/UCTE, *The 50 Year Success Story—Evolution of a European Interconnected Grid*. <https://eepublicdownloads.entsoe.eu/clean-documents/pre2015/publications/ce/110422_UCPTE-UCTE_The50yearSuccessStory.pdf>.

[13] Art 2(1) Electricity Regulation 2019/943/EU. Art 2(39) Gas Directive 2024/1788 holds an almost similar definition: 'transmission line which crosses or spans a border between Member States for the sole purpose of connecting the national transmission systems of those Member States . . .'

[14] See also T van der Vijver, 'Third Party Access Exemption Policy in the EU Gas and Electricity Sectors: Finding the Right Balance between Competition and Investments' in MM Roggenkamp and others (eds), *Energy Networks and the Law* (OUP 2012) and A Gautier, 'Merchant Interconnectors in Europe: Merits and Value Drivers' (European University Institute, Policy Brief no 2020/05, 2020).

[15] See for more details Electricity Regulation 2019/943/EU

[16] The regional codes (*Gebiedsindelingscode elektriciteit*, BWBR0037943 and *Gebiedsindelingscode gas*, BWBR0037927) are drafted on the basis of input of the DSOs. See www.acm.nl accessed 30 October 2025.

municipalities.[17]

By contrast to transmission systems, DSOs are subject to one minimum unbundling requirement, which involves legal and functional unbundling.[18] The DSOs are also responsible for connecting the (small) generation plants to the distribution system.[19] In order to facilitate DER, MS need to establish specific authorization procedures, which take into account their limited size and potential impact.[20] In addition, specific rules apply for connecting these facilities to the system. Apart from the general requirements that conditions and tariffs need to be objective, transparent, and non-discriminatory,[21] the connection tariffs for DER also need to take account of the avoided network costs from distributed generation.[22] In addition, DSOs need to take into account DER when developing their network development plans and the possibility to procure flexibility services, including congestion management services, from providers of DER.[23]

A provider of DER will need a physical connection to the distribution system.[24] Although EU law increasingly provides minimum standards and grid connections rules,[25] MS are basically free to govern the connection of system users to the grid as long as the process is based on objective and non-discriminatory criteria. Hence, national laws govern individual connection requests. Usually such a request depends on the size of the connection: large producers need a larger (industrial size) connection than household consumers. In order to create transparency and increase free trade within the EU energy market, each connection is identifiable on the basis of a European Article Number (EAN code).[26] A request for a connection can only be denied if there is no capacity available.[27] Notably, increasingly DSOs are unable to grant connection requests due to a lack of capacity.[28]

Such lack of capacity might be a reason to request a connection to a distribution system in another MS, at least if that system is located close to the DER facility. Moreover, if the distance of a cross-border cable is shorter, such a connection might even be more cost-effective. EU law does not hold an explicit barrier to such an approach as it defines distributed generation generally as 'generation plants connected to the distribution system'. This does not exclude possible cross-border connections. However, there will be some national legal obstacles as such a connection will be situated outside the dedicated regional supply area and requires a decision on the 'nationality' of the EAN-code. If electricity production involves RES, another potential obstacle is the use of national financial support schemes. Often national laws provide that producers are only eligible to financial support if the

[17] Art 5 Division Code Electricity and art 4 Division Code Gas. See for other MM Roggenkamp and others (eds), *Energy Law in Europe—National, EU and International Regulation* (3rd edn, OUP 2016).

[18] Art 26 Directive 2009/72/EC.

[19] Art 2(31) Directive 2003/54/EC.

[20] Art 5 para 3 Directive 2003/54/EC.

[21] And the NRA needs to ensure this pursuant to art 23 para 1(f) Directive 2003/54/EC.

[22] Recital 81 Directive (EU) 2019/944.

[23] Art 32 Directive (EU) 2019/944 and MM Roggenkamp, 'How Net Zero Goals Impact *Electricity* Systems and Supply Security—An Analysis of the EU Legal Framework' in DS Olawuyi and others (eds), *Net Zero and Natural Resources Law: Sovereignty, Security, and Solidarity in the Clean Energy Transition* (OUP 2024) chapter 15.

[24] The ECJ distinguishes between the terms 'connection' and 'access'. See C-239/07 *Julius Sabatauskas and Others* [2007] ECR II-7523, para 42.

[25] See, for example, Regulation (EU) 2016/631 of 14 April 2016 establishing a network code on requirements for grid connection of generators.

[26] The EAN codes of eighteen digits.

[27] Art 6(2) Directive (EU) 2019/944.

[28] Eurelectric, *From Backlog to Breakthrough: Managing Connection Queues in Distribution Networks*, April 2025.

production installation and grid connection are located on the territory of the same MS.[29] In such a case, the RES-producer and thus also the potential developer of the cross-border connection would be deprived of financial support.

B. Exemptions and Alternative Regimes

In addition to the above systems, EU law also governs other networks such as small isolated or connected systems and closed distribution systems as well as infrastructure, which is not qualified as a system such as direct lines. Different regulatory regimes apply to these networks. This is usually endorsed by the need to avoid too high administrative costs.

1. Small isolated and/or connected systems

Since the 1990s a separate regime applies to small isolated or connected systems.[30] Whereas a small connected system serves less than 100,000 connected customers, a small isolated system is a system with a consumption of less than 3,000 GWh in the year 1996 and where less than 5 per cent of annual consumption is obtained through a connection with other systems. The difference between both small systems is the extent to which the connection to the main grid can be used. Whereas the connection between a small isolated system and the main grid can only be used to a very limited extent, it is unclear whether the limit of 5 per cent of the annual electricity consumption refers to the year 1996 as a fixed reference point in time, or whether the 5 per cent of imports rule refers directly to 'any system', which would allow for a yearly assessment of the exports/imports balance via the existing connection. Anyway, small systems operate mostly independently as production within the system usually satisfies (all) consumers connected to the system. If a system qualifies as a small system, MS may apply for an exemption from the unbundling requirements[31] as well as further derogations if it can be proven that there are 'substantial problems with the operation of small isolated or connected systems'.[32] As these exemptions generally deviate from the main market liberalization principles, they can only be applied for a limited period of time and subject to specific conditions that also ensure that transition towards RES.[33] In general these small systems may be relevant for developing DER but presumably only if integrated in existing systems.

2. Closed distribution systems

Another and more recent example of a small energy system is a so-called '*closed distribution system*' (CDS). The concept evolved from the ruling in the case of *Citiworks* where the ECJ examined whether a small industrial network at the airport of Halle Leipzig (Germany) could be considered as an isolated system and be granted an exemption from the principle of non-discriminatory network access.[34] Accordingly, the 2009 Electricity and Gas

[29] Spatial Foresight, Eurac, and Eureconsult, *Handbook on Cross-border Energy Communities—Final Version, Contract: 2022CE160AT275*, January 2025 (hereafter Handbook), 16 and 21.
[30] Art 35(4) Directive (EU) 2019/944.
[31] ibid art 35(4). CDS can also be exempted from the obligation to draft network development plans pursuant to art 32(5).
[32] ibid art 66(1).
[33] ibid art 66(2).
[34] C-439/06 *Citiworks AG* [2008] ECR I-03913.

Directives introduced the concept of a CDS: a system which distributes electricity within a geographically confined industrial, commercial, or shared services site[35] and which in essence does not supply household customers.[36]

Although a CDS is considered as a distribution system[37] and thus subject to the general provisions governing these systems, operators of a CDS can be exempted from some specific requirements such as the requirements regarding (i) the prior approval of tariffs,[38] and (ii) the procurement of energy to cover energy losses and non-frequency ancillary services.[39] The latter allows operators of a CDS to design their own system in locally and clearly defined (industrial) areas. So, where small connected and/or isolated systems primarily serve household customers, a CDS involves an industrial site that may be used by (industrial) owners of the system or require specific operational standards to ensure the optimal efficiency of the integrated supply. Hence, a CDS will be less relevant for developing DER by active (household) consumers.

3. Direct lines

Another concept is a so-called 'direct line', which has been included in EU energy law since the 1990s. Currently, the Electricity Directive defines a 'direct line' as '[. . .] either an electricity line linking an isolated generation site with an isolated customer or an electricity line, linking a producer and an electricity supply undertaking to supply directly their own premises, subsidiaries and customers.'[40] By contrast, the Gas Directive defines a 'direct line' as a natural gas pipeline complementary to the interconnected system so that natural gas undertakings established within the territory of a MS can supply customers through a direct line and vice versa.[41] Although the Electricity Directive does not explicitly state that the direct line has to be complementary to the interconnected system, this can be assumed as otherwise a customer connected to the direct line would be deprived of its basic right to switch to another supplier.[42]

Direct lines are a distinct category of cables/pipelines and not qualified as a system; they are complementary to the general electricity and gas system and operate on a very small scale. Therefore, the rules on unbundling and non-discriminatory access do not apply. However, as a general rule, access needs to be negotiated so that operators of direct lines cannot abuse their (monopoly) position. MS also need to establish authorization criteria for developing direct lines, which are objective and non-discriminatory.[43] As these direct lines are complementary to the system, it can be argued that they can play a role in developing DER, for example, if these energy sources cannot supply customers directly via the distribution system.

In practice and due to differences in implementation, the provisions in national law will differ. Dutch law, for example, applies a more limited approach towards the term 'isolated'

[35] Recital 66 Directive 2019/944/EU refers to, inter alia, train stations, airports, hospitals and large camping sites, and chemical industry sites.
[36] ibid art 38. Art 38(2) refers to two specific situations where a CDS can supply households.
[37] ibid art 32(2).
[38] ibid art 6(1).
[39] ibid art 31(5) and (7).
[40] ibid art 7.
[41] Art 2(41) and art 34 Directive (EU) 2024/1788.
[42] cf art 6 Directive (EU) 2019/944.
[43] ibid art 7 and art 34 para 2 Directive (EU) 2024/1788.

than German law.[44] Also national laws can hold different administrative and registration procedures regarding 'direct lines'. Additionally, some MS find that production installations connected to a 'direct line' are eligible to getting guarantees of origins and financial support for RES. Finally, the approach towards the isolated consumers may differ. Some national laws regard consumers connected to a 'direct line' as 'customers' in accordance with EU law (and thus subject to consumer protection) and others do not. Despite these differences, the construction and use of 'direct line' offers opportunities for private developers, including the option to develop cross-border direct lines for DER or other purposes.[45]

C. Cross-border Energy Projects

The above sections have addressed the EU legal framework governing the transmission and distribution systems as well as some alternative regimes that can be applied for reasons of functionalities or to limit the administrative burden. These regimes offer very limited possibilities to develop cross-border DER, except for possible cross-border connections to distribution systems and direct lines. However, meanwhile the EU has also introduced new concepts involving cross-border energy projects that might be relevant for DER. These will be discussed below.

1. Energy communities

The concept of energy communities may be relevant for developing DER. Energy communities involve a group of collective active customers co-operating under certain conditions. EU law defines active customers as 'final customers, or a group of jointly acting final customers, who consume or store electricity they have generated within their premises or who sell self-generated electricity [. . .], provided that these activities are not their primary commercial or professional activity'.[46] When and how these active customers participate also depends on the type of community, ie a citizen energy community or a renewable energy community.[47] These two concepts reflect the different scopes of the two directives involved. Whereas the E-Directive provides market rules that aim at creating a level playing field for all participants in the electricity market, RED provide measures for reaching the pre-determined shares of RES.[48] A citizen energy community is thus broader in scope then a renewable energy community. In fact, a renewable energy community is generally also a citizen energy community. This overlap can also be found in the key requirements for the development of these communities.

Both energy communities require the establishment of a legal entity, which can be a partnership, a limited liability company, a foundation, an association, or a co-operative.

[44] The word 'isolation' has been translated to 'einzeln' (single) in German. See L Diestelmeier and MM Roggenkamp, 'Current Legal Framework for Cross-Border Local Energy Markets—National Legal Frameworks', WP4 Interreg Project Smart Energy Region Emmen-Haren (SEREH) (2021) 33 <www.sereh.eu/sereh> accessed 27 July 2025.

[45] A typical example of a cross-border direct line is the cable linking an offshore wind park Riffgat in the German EEZ with a gas production facility N05-A of ONE-Dyas in Dutch part of the North Sea. See https://onedyas.com/ accessed 11 September 2025.

[46] Art 2(8) Directive (EU) 2019/944. Art 2(14) Directive (EU) 2018/2001 refers to 'renewables self-consumers'. Despite the different terminology, the definition is otherwise similar.

[47] See art 16 Directive (EU) 2019/944 and art 2(16) Directive (EU) 2018/2001.

[48] M Jasiak, 'Energy Communities in the EU: Challenges for the Implementation of the EU Legal Framework' in M Roggenkamp and C Banet (eds), *European Energy Law Report XIV* (Intersentia 2021) 200.

MS and/or communities are free to choose the legal form they find suitable as long as it is based on open and voluntary participation and effectively controlled by shareholders or its members. In case of a renewable energy community some additional requirements apply. First, these communities need to involve the use of RES and members of the community need to be located 'in the proximity of the renewable energy projects of that entity'. Secondly, the legal entity has to be 'autonomous' in order to avoid abuse and ensure broad participation.[49] Thirdly, participants in renewable energy communities are limited to natural persons, SMEs, or local authorities, including municipalities. Also it has as its primary aim to provide environmental, economic, or social community benefits to its shareholders or members or the local areas where it operates, rather than financial profits. In order to promote such renewable energy communities, MS need to carry out an assessment of the existing barriers and the potential of developing these communities in their territories in order to remove any 'unjustified regulatory and administrative barriers to renewable energy communities'. The RES directive seems to provide a proactive approach in favour of renewable energy communities.

Notably, the Electricity Directive explicitly provides that citizen energy communities may involve cross-border participation and be entitled to own, establish, purchase, or lease distribution networks and to autonomously manage them.[50] Hence, the community should be able to operate the system in their area and thus also to conclude an agreement with the DSO or TSO to which the network is connected and to avoid any discrimination or harm vis-à-vis those customers who remain connected to the distribution system. Since the system is considered to be some sort of distribution system, a citizen energy community may also be granted an exemption like a CDS. Because renewable energy communities are considered a specific category of citizen energy communities, it can be argued that the provisions on system operations also apply to renewable energy communities.[51] However, as a CDS primarily has to consist of industrial customers and the energy communities mainly of citizens (households/small consumers), it remains to be seen how this particular exemption could impact the developments of energy communities. Similarly it is unclear how any cross-border participation of citizen energy communities will be developed, what this will mean in practice, and how it might impact the development of cross-border energy systems and DER.

2. Joint projects

The Renewable Energy Directive provides for several cross-border co-operation mechanisms that serve the ultimate goal to meet the EU (and national) renewable goals.[52] One of these mechanisms includes the possibility to develop joint projects with either other MS or third countries. It entails that 'two or more Member States [or a third country] may co-operate on all types of joint projects relating to the production of electricity, heating or cooling from renewable energy sources. That cooperation may involve private operators'.[53] It has been acknowledged that joint projects can be suitable to, inter alia, jointly develop

[49] Recitals 16, 71 Directive (EU) 2018/2001.
[50] Article 16(2) Directive (EU) 2019/944.
[51] Jasiak, 214.
[52] Arts 8, 9, and 13 Directive (EU) 2018/2001 provide for three mechanisms: statistical transfers, joint projects, and joint support schemes.
[53] Art 9 Directive (EU) 2018/2001.

technologies and save costs of meeting RES targets.[54] In order to promote the development of joint projects the Directive was amended in 2023 and now obliges MS to establish before 2026 a framework for co-operation with one or more other MS to develop joint RES projects. More particularly, they shall endeavour to establish at least two joint projects by 31 December 2030 and a third joint project by 31 December 2033 with an annual electricity consumption of more than 100 TWh.[55] These co-operation agreements need to ensure a fair distribution of all costs and benefits of a joint project and to take into account existing EU and national legal frameworks.

Although the EU aims to promote joint projects, the concept is still quite broad. It could involve a project in one MS developed jointly by two or more MS but could potentially also involve a cross-border joint project. As it involves the use of renewable energy sources, it could also be qualified as a joint DER project if it is connected to a distribution system or a comparable system. Potentially, such a joint project could also be qualified as a (cross-border) renewable energy community. Given these ambiguities there might be a need for more legal certainty to develop such projects.

D. Gaps in Existing EU and National Laws and Policies

The above has shown that the EU has initiated new concepts—energy communities and joint projects—that may be relevant for developing DER and that some of these may even be cross-border in scope. However, so far EU law provides very little guidance on how this could be materialized and how it may impact the use of energy systems. Moreover, some MS have not taken into account the possibility of cross-border co-operation and a cross-border distribution system when implementing the concept of energy communities into national law.

When zooming in on the options for cross-border infrastructure, it is clear that one needs to distinguish energy systems and direct lines. Where transmission and distribution systems are heavily regulated and their development is often restricted to a specific group of entities, fewer rules apply to smaller and/or exempted distribution systems and direct lines. So far Regulation (EU) 2019/943 seems to limit cross-border connections to transmission systems only. This restriction clearly has an historical background as well as the existing provisions in national law limiting the operation of distribution systems to designated national supply areas. However, examples can be found of small cross-border medium and/or low voltage electricity systems. In Norway, for example, it has been possible since 1969 to develop small cross-border electricity systems for socio-economic reasons. In other words, in sparsely populated border areas it could be more economically viable to supply consumers directly from the neighbouring country (primarily Sweden) via a cross-border distribution line.[56] It can thus be argued that the provision in EU law limiting interconnectors to cross-border transmission systems does not

[54] C Klessmann and others, 'Cooperation between EU Member States under the RES Directive Cooperation between EU Member States under the RES Directive', Task 1 Report (Ecofys 2014) 16.

[55] See art 1(4) Directive (EU) 2023/2413. The Commission needs to be informed of these co-operation agreements, including the date on which the joint projects are expected to become operational.

[56] Art 4-2 para 3 of the Energy Act 1991 replaces the earlier provision in the Industry Concession Act as amended in 1969. See also U Hammer, *Tilrettelegging av Kraftmarkedet* (Cappelen 1999) 107.

necessarily have to restrict MS in developing cross-border distribution systems. After all, MS may deviate from EU law as long as this serves the general public interest and/or to achieve an internal energy market.[57] Such reasoning can also apply to the need for cross-border distribution systems, as illustrated by the example in Norway. By doing so, the possibility of cross-border DER could go beyond projects involving a cross-border direct line or a cross-border connection, ie connecting a DER-project to a distribution system in another MS. Even so, these options may still encounter major obstacles and gaps in national law. Some examples will be discussed below.

III. National Approaches Involving Cross-border Local Energy Supply: Some Examples

The above section has discussed the EU legal framework involving potential cross-border DER. It shows that some possibilities exist but these might face some limitations in EU as well as national law. This section will present two examples involving cross-border local energy supply. First, it will discuss local energy supply in enclaves, which are pieces of land belonging to one country (or region etc) but surrounded by another country (or region). Thereafter, this section will examine some recent attempts to develop cross-border energy communities, including some 'alternative' cross-border connections. Both examples illustrate the limitations arising from the EU rules governing the liberalization of the electricity and gas markets.

A. Energy Supply in Enclaves

Several enclaves have emerged in Europe due to historic border disputes and/or border adjustments. Examples can be found at the borders of Belgium-the Netherlands, Belgium-Germany, Germany-Switzerland, Switzerland-Italy, France-Spain, and Slovenia-Croatia.[58] The size of these enclaves can vary as well as their use (eg industrial, agriculture, etc). A relevant example can be found at the Belgium-Dutch border. It involves the town Baarle, which consists of Baarle Nassau in the Netherlands and Baarle-Hertog in Belgium. However, this division is not straightforward as Baarle Hertog has twenty-two enclaves in Baarle Nassau and Baarle Nassau has eight in Belgium (Baarle Hertog). The reason for this complex situation is historic and dates back to the fourteenth century and was finally settled in a border agreement in 1973.[59] However, the border agreement did not change the position of the enclaves. The situation remains complicated, which is also reflected in the way in which energy supply has been organized in Castelré.

[57] See (i) Joined Cases C-105/12–C-107/12 *Essent (Unbundling Act)* ECLI:EU:C:2013:677 as a result of which the ECJ agreed on ownership unbundling of DSOs and (ii) the reasons for exempting merchant interconnectors from unbundling/TPA requirements as discussed by H Vedder and others, 'EU Energy Law' in M Roggenkamp and others (eds), *Energy Law in Europe—National, EU and International Regulation* (OUP 2016) chapter 4.

[58] https://en.wikipedia.org/wiki/List_of_enclaves_and_exclaves accessed 31 July 2025.

[59] Agreement between the Kingdom of the Netherlands and the Kingdom of Belgium, Tractatenblad, 1974, no 188A.

The Dutch village of Castelré (approximately 130 inhabitants) in Baarle Nassau is an enclave almost completely located on Belgium territory.[60] When developing the electricity supply system in the early 1950s, it was decided to connect Castelré to the Belgium electricity grid of IVEKA (*Intercommunale Vereniging voor de Elektriciteitsdistributie in de Kempen en het Antwerpse*) as a connection to the Dutch grid was not economically viable.[61] Hence, IVEKA was at that time in essence acting as a cross-border vertically integrated electricity distribution company. However, due to EU market liberalization at the turn of the century the situation had to be reassessed.[62] First, pursuant to article 12 of the Dutch Electricity Act 1998 all DSOs operating in the Netherlands need a prior approval by the Dutch Minister of Economic Affairs. It was not clear whether the legal successor of IVEKA (EANDIS, currently Fluvius) would have been granted such approval. Secondly, different technical standards apply in the Netherlands and Belgium.[63] Thirdly, it was questioned whether Dutch customers in Castelré could be required to pay energy taxation in the Netherlands as well as in Belgium as the grid connection was identifiable via a Belgian EAN-code.[64]

In order to solve the situation, it was considered whether the electricity consumers in Castelré could be directly connected to the Dutch distribution system of Enexis.[65] However, this was not considered feasible given the distance to the Enexis system (6 km to the closest medium voltage substation) and thus the high investment costs and subsequent high connection costs per household. Instead EANDIS and Enexis concluded an agreement whereby Enexis bought from EANDIS the part of the low voltage system on Dutch territory[66] and that EANDIS (now Fluvius) continues to operate the low voltage system on the basis of a separate service agreement as a subcontractor for Enexis. Moreover, all customers in Castelré (seventy connections) are offered a Dutch EAN-code and a smart meter. In other words, since 2018 the low voltage electricity grid located in Castelré is owned by the Dutch DSO Enexis but all day-to-day operations are still executed by a Belgian DSO (now Fluvius).

The above example presents an interesting example of how market liberalization can impact cross-border energy supply systems. It is striking that a functioning cross-border distribution system had to be abolished due to market liberalization, which in essence aims to create one internal market and provide consumers with more choice.

B. Cross-border Energy Communities

Section II.C.1 above discussed the concept of energy communities. It also highlighted the possibility to develop cross-border energy communities, which is particularly relevant for

[60] This enclave is not completely surrounded by Belgium territory as it still has direct access to the Netherlands via a strip of land of 200 meters.

[61] The situation is thus comparable to the above-mentioned scenario (section II D) for developing small cross-border systems crossing the Norwegian-Swedish border.

[62] See for some background information a ruling by the Gerechtshof (Court of Appeal) 's-Hertogenbosch, 29 May 2009, LJN: BJ1415.

[63] Belgium law holds different medium voltage requirements. Also the Belgium low voltage grid is partly situated above ground.

[64] See fn 62 for the ruling of the Court of 's Hertogenbosch.

[65] Baarle Nassau is situated in the supply area of Enexis. As Enexis is also operating the gas distribution system in the Baarle area, consumers in the centre of Baarle-Hertog are connected to the system of Enexis.

[66] The medium voltage system remains the ownership of EANDIS.

cross-border DER. Notably, the interest in developing cross-border energy communities rapidly increased since the concept was launched by the EU. In 2025 some twenty-seven initiatives were identified in seventeen EU MS and at fourteen different borders.[67] However, strictly speaking some of these initiatives seem to go beyond the scope of the relevant directives as they include district heating networks, e-bike sharing, and green hydrogen projects.

Two successful projects involve solar PV projects: (i) 'EfiDuero' at the Spanish-Portuguese border operated by European Cooperative Society; and (ii) 'Zusamme Solar! Colmar' at the French-German border operated by Altora-PV Sarl (a limited company under French law). Both projects aim to promote the use of solar PV. Efi-Duero, for example, aims to assist municipalities on both side of the border in developing PV (and thus DER) and thus enhance energy independency.[68] The company Altora-PV Salt that operates the 'Zusamme Solar Colmar' was already established in 2014 and based on a collaboration of existing energy co-operations: the German co-operative 'Fesa Energie Geno e.G.' (50 per cent); the French national co-operative structure 'Energie Partagée Investissement' (40 per cent); and the French regional co-operative 'Energies Partagées en Alsace' (10 per cent). The company aims to install large volumes of solar panels on (unused) commercial buildings. Currently a total of 2,600 m2 of solar panels on the roofs of former textile factories produce about 400 kWh per day. Although 'Zusamme Solar Colmar' is considered as a cross-border energy community, it is only generating and distributing electricity on the French side of the border.[69]

Apart from these two cases, there are examples of attempts to develop cross-border energy communities based on the use of renewable energy sources as well as attempts to involve some sort of cross-border energy infrastructure. These will be discussed below.

1. Attempts to develop cross-border (local) energy systems

Two potential cross-border projects are situated on the border of the Netherlands and Germany, ie the initiative to develop the Smart Energy Region Emmen-Haren (SEREH) and the Energy Park Kerkrade-Herzogenrath.

The SEREH project was an ambitious project initiated by the municipalities of Emmen (the Netherlands) and Haren (Germany) and supported by the EU as an Interreg project.[70] The project involved the development of wind parks in Germany (Fehndorf-Lindloh) and the Netherlands (Zwartenbergerweg). However, instead of connecting each of these wind parks to the national distributions system, SEREH tried to go a step further and aimed to develop a novel energy system by means of connecting the regional energy systems and exchanging energy across the border. It also assumed that on each side of the border more wind (and solar) energy would be produced than actually consumed, and that the surplus could be stored in batteries or converted into hydrogen and potentially be used by industrial consumers on each side of the border. The physical exchange of energy between the Netherlands and Germany would require (i) a connection between both distribution systems at the medium voltage level and/or connections of the wind parks to each other's electricity systems, and (ii) the conversion locally of renewable energy to green hydrogen

[67] Handbook, 76.
[68] Handbook, 28.
[69] Handbook, 29.
[70] Art 174 TFEU aims to reduce regional disparities and forms the basis for Interreg, which is a programme for European Territorial Co-operation and has funded several strategic cross-border projects since 1990.

and the transport (and/or storage) of hydrogen via an existing cross-border pipeline (and/or underground storage facilities) operated by the Dutch gas producer NAM. The ambitions of SEREH were manifold: cross-border connections on medium voltage level would lower system costs and ensure the development of a (true) cross-border energy community but at the same time also allow for cross-border system integration and accelerate the process of energy transition whilst keeping the (financial) revenues of the project in the region.[71] Despite all efforts, the project has not come about due to financial considerations, a lack of support by the national government (see below), as well as several legal and regulatory obstacles involving the qualification and development of the necessary cross-border infrastructure.[72]

Another German-Dutch energy project involves co-operation between the German town Herzogenrath and the Dutch town Kerkrade. These towns are situated on each side of the southern border of the province of Limburg. Via the public entity 'Eurode' both towns already co-operate in improving the living conditions of their inhabitants and alleviating daily problems of, amongst other, commuters. Although this co-operation does not involve any energy co-operation, both towns launched the initiative of Energy Park Kerkrade-Herzogenrath. It aimed to host a large solar park (14 MW) in Herzogenrath and supply waste heat from a glass producer in the town of Kerkrade. The intended cross-border co-operation involved the need to share the energy produced and to develop cross-border infrastructure. Just like SEREH, this project encountered several obstacles regarding the cross-border supply of electricity. First, German law does not allow neighbouring municipalities to regulate the supply and demand of electricity amongst themselves. Secondly, there is no agreement in place between Germany and the Netherlands on how to share the financial support of RES-projects. Finally, there was uncertainty about the possibilities of physically connecting the Dutch and German electricity systems.[73]

Both projects illustrate the intention to develop (true) cross-border energy communities and/or projects, which also involve a physical cross-border connection and the use of a (low/medium voltage) cross-border electricity cable. Given this apparent need, some Dutch municipalities[74] requested the Minister of Economic Affairs and Climate in 2020 to assess the possibility of cross-border energy co-operation, including the development of cross-border infrastructure. The minister's answer was disappointing and not very supportive.[75] First, he stated that cross-border exchange of electricity does not lead to social benefits and does not contribute to increasing the available grid capacity. This is astonishing as grid congestion was well known at the time.[76] Secondly, the Minister took the formal approach that EU law does not allow for other cross-border connections than interconnectors operated by TSOs. Again, this argument can be questioned as well (see above). It is therefore not

[71] https://sereh.eu/ accessed 31 July 2025.

[72] This relates amongst others to the qualification of the 110 kV grid involved. Whereas the Dutch TSO operates the 380/220 kV and 110 kV grids, the German TSO operates only the 380/220 kV grid and the DSO the 110 kV grid.

[73] Handbook, 36. Please note that the Handbook by mistake refers to the German Electricity Act 1996. The electricity sector in Germany is governed by the Industry Energy Act.

[74] The request was sent by Emmen (province of Drenthe), Kerkrade (province of Limburg), and the town Losser (province of Overijssel).

[75] Tweede Kamer 2020–2021, 30 196 and 29 023, nr 732.

[76] See R Hakvoort and others, 'A System for Congestion Management in the Netherlands' (D-Cision/The Brattle Group 2009); Energyproof, 'Analyse van congestive op Nederlandse energienetten' (2020) https://energyproof.nl/analyse-van-congestie-op-nederlandse-energienetten/ accessed 11 September 2025.

surprising that municipalities are focussing on developing other types of cross-border energy networks.

2. Cross-border heat supply

By contrast to the complex and slow procedures to develop cross-border low/medium voltage electricity connections, it is striking to see the larger and increasing numbers of cross-border heat supply systems. In the EU almost ten such systems exist.[77] Some have already been developed in the early 1990s but others are of recent date. Although these examples involve cross-border energy projects, they cannot be considered as citizen or renewable energy community.[78] Moreover, several projects involve waste heat from fossil-based industries (eg Calorie Kehl-Strasbourg and Energy Park Herzogenrath-Kerkrade) albeit there is a clear trend towards replacing industrial waste heat for heat from renewable sources.

As mentioned earlier in section I, EU law has hardly addressed the development of district heating. Initial provisions referring to the need to consider combined heat and power when tendering new generation capacity,[79] have been replaced with provisions stimulating heat supply from RES.[80] These provisions relate, inter alia, to the role of renewable heat in energy transition and system integration and call upon MS to integrate heat systems in their network development plans and to encourage third party access to the heat systems on the basis of non-discriminatory criteria set by the competent authority.[81] Although rules regarding third-party access emerge, they are still limited in scope and do not include unbundling requirements nor affect the governance structure of the sector. Therefore no legal obstacles on an EU level exist (yet) for developing and using cross-border heat supply infrastructure. It is thus not surprising that communities consider such projects as an alternative for cross-border electricity systems. A clear example is the above-mentioned co-operation between Kerkrade and Herzogenrath. Following a feasibility study and finding common solutions regarding the initial investments and some tax issues, they signed a declaration of intent to transport waste heat from the glass producer Saint-Gobain in Herzogenrath across the border to the city Kerkrade.

C. Assessment

The two examples discussed illustrate how Internal Energy Market law impacts cross-border electricity and gas distribution systems. Whereas prior to market liberalization such systems have developed due to socio-economic reasons, the governance structure had to be amended in order to be in line with EU law. Similarly new cross-border local electricity and

[77] Such projects have been developed at the following borders: DE-DK, DE-PL, SE-FI, AT-IT, AT-DE, Fr-DE, DE-NL. See for an overview Handbook, Annex 1.

[78] Although heat like electricity is a secondary energy source, they are not comparable and heat is not to be treated as electricity. Heat is neither per definition a renewable energy source nor is it covered by the definition of art 2 Directive (EU) 2018/2001.

[79] See eg recital 43 Directive (EU) 2019/944.

[80] Notably Directive (EU) 2018/2001 and the amending Directive (EU) 2023/2413 several provisions governing heat and district heating.

[81] Directive (EU) 2023/2413 amends amongst others art 24 Directive (EU) 2018/2001 to ensure that operators of district heating systems above 25 MWth capacity will provide third parties access.

gas systems are difficult to achieve due to existing legislation. It is therefore not surprising that MS and developers turn to less restrictive options, such as cross-border heat supply systems. Although such systems may play a role in the energy transition process, strictly speaking they cannot be considered as DER.

IV. Conclusion

This chapter has analysed the needs, obstacles, and some examples of cross-border distributed energy in the EU. In the EU distributed energy entails the connection of (small) energy production (and storage) facilities and/or active customers to distribution systems. So far this development has been national in scope. However, this chapter argues that there is a need for facilitating cross-border distributed energy and considering other regional/local networks.

The need for cross-border DER is evident, especially given the challenges in meeting the EU and national sustainability goals. MS are increasingly confronted with DER that cannot be connected to the system due to a lack of capacity. Traditionally congestion existed at cross-border transmission systems, but congestion also appears more and more at national and regional level due to the rising amounts of DER. It thus seems obvious to consider cross-border regional connections in so far as capacity is available in the neighbouring system. This may speed up the integration of DER and reduce the need for local grid development and reinforcement costs. Another, more conceptual argument is that regional cross-border connections make sense from an internal market perspective and the need to enhance free movement of energy in Europe. From an internal energy market perspective, cross-border connections at distribution level are thus a prerequisite.

The main obstacle for cross-border DER seems to be the provision in EU law that limits cross-border interconnectors to transmission systems. This focus is historically inspired and should be reassessed. Apart from this, several obstacles exist on national level. These vary from restrictions involving dedicated regional supply areas, the impact and structure of national support schemes, and differences in technical standards on each side of the border. It is striking that increased energy market regulation has worsened the situation (eg Castelré case) albeit the aim of market liberalization is to enhance free movement.

On the other hand, the EU acknowledges at the same time the need for more cross-border energy co-operation. Recent legislation provides for the non-mandatory development of joint projects and energy communities. Although MS have shown interested in developing energy communities, the examples show that developing cross-border low/medium voltage electricity systems are extremely challenging. So far, such joint energy projects and cross-border energy communities (including joint infrastructure) have not been developed. Currently, the possibilities are limited to the construction of cross-border connections at medium/low voltage level and/or the development of cross-border direct lines as EU law offers MS considerable regulatory freedom in this regard. The absence of EU law restrictions also explains why local communities tend to prefer to develop cross-border heat supply systems.

How to proceed? A first step forward would be to amend the definition in EU law of cross-border interconnectors to a: '[. . .] line which crosses or spans a border between Member States and which connects the national systems of the Member States.' By removing the reference to transmission systems, use can be made of alternative cross-border connections and provide more flexibility as well as regulatory certainty. Alternatively, EU law could be expanded to include an exemption regime. Similarly to the regime that enables private parties to develop and use cross-border transmission systems ('merchant interconnectors'), an exemption could be included to allow for small cross-border systems on a distribution level for reasons of socio-economic interests (see Norwegian Energy Act) or to meet sustainability goals. Moreover, it could be envisaged to stimulate MS to include a specific cross-border DER project in the proposal for joint RES projects they need to establish before 2030. Clearly this will not be sufficient given the differences in national laws with regard to the designation of dedicated regional supply areas, technical requirements on each side of the border, and restrictions involving national support schemes. Some of the technical obstacles may be solved via the EU DSO Entity, which, amongst others, is charged with facilitating the integration of DER into the distribution systems, promoting operation and planning of these systems, and participating in the development of network codes relevant to the operation and planning of distribution systems.[82] Such network codes could be instrumental for increased regional co-operation, which would be in line with the aim of art 174 TFEU to increase territorial cohesion. Restrictions relating to national support schemes are currently considered to be a major obstacle as the electricity producer with a connection to a 'foreign' system is not entitled to financial support. However, it can be questioned whether these support schemes will remain in force if most electricity is generated on the basis of renewable energy sources. Also the matter of financial support for cross-border DER could be solved via an exemption regime and/or bilateral agreements. Generally, there will be a need for bilateral agreements to facilitate and develop cross-border DER. In the absence of applicable EU and/or national law, such agreements will enable MS to present a local legal framework governing technical, financial, and/or regulatory issues relevant for developing cross-border DER.

This chapter has discussed the development of cross-border DER. In order to meet EU and national sustainability goals, there is a need for cross-border DER. For socio-economic reasons and due to increasing congestion levels in national energy systems, cross-border connections for DER seem an obvious solution. The situation in the EU is complex as on the one hand such connections could facilitate the establishment of an internal energy market, but on the other hand these market rules also limit the development of cross-border DER. Nevertheless cross-border DER will also be relevant in other parts of the world as it serves common sustainability interests.

[82] The DSO Entity was established in June 2021 on the basis of arts 52–55 Regulation 2019/943/EU, which governs the establishment and tasks of this association of DSOs.

8

Decentralization with Distributed Energy Resources in New Zealand

Barry Barton

I. Introduction

Straight away, let us tackle the question why should we include New Zealand in a book study of this kind when there is so little provision for distributed energy resources there? The short answer is that it is worth examining why that is so. It is certainly true that distributed energy resources (DER) and decentralization have not advanced far in New Zealand, and that legal and regulatory action to enable it has only begun recently.[1] Since the sweeping reforms of the 1990s, the New Zealand energy system (the electricity system in particular), has stood out for its reliance on market forces and minimal regulation. One would think that neoliberal characteristics would favour DER and decentralization; once unleashed, the theory would say, market competition, neutral as to technology, will allow new entrants with innovative ideas and new technology to take on the incumbents. Why did that not happen? This chapter argues that when the market forces were too weak then minimal regulation meant that little happened, but when a concerted regulatory effort began, with strong government backing, change started to happen, and that because of the special characteristics of DER the change is very detailed and granular.

DER offers significant benefits in the New Zealand context. The decentralization that it represents offers better resilience and reliability in the face of natural disasters such as earthquake and flooding. It offers better management of peak demand, increased competition, and lower energy prices—all of which are serious current concerns. It offers sustainability benefits in reducing greenhouse gas emissions and reducing impacts on land, water, and other resources. Although New Zealand's power system has advantages in hydro, wind, and geothermal resources, about 12 per cent of its generation uses natural gas and coal,[2] and replacing that remaining fraction with renewables is proving difficult. (The New Zealand Emissions Trading Scheme puts only modest price pressure on fossil fuels.) There is limited hydro storage to deal with seasonal fluctuations, and a 'dry winter' has become a persistent risk to energy security. Electricity demand is growing rapidly as households, transport, and industry move away from hydrocarbons for their energy needs. At the same time, high electricity prices, at least since 2018, and volatility in the wholesale electricity market are putting pressure on energy-intensive industries and low-income households. In

[1] I use the definition of distributed energy resources (DER) in Chapter 1. I thank Daniela Aguilar for comments on a draft but the views I offer are my own.

[2] Ministry of Business, Innovation, and Employment, 'Energy in New Zealand 2024' (2024).

Barry Barton, *Decentralization with Distributed Energy Resources in New Zealand* In: *Sustainable Distributed Energy Resources*. Edited by: Íñigo del Guayo, Damilola S. Olawuyi San, Louis de Fontenelle, Milton Fernando Montoya, and José Juan González, Oxford University Press. © International Bar Association 2026. DOI: 10.1093/9780198991946.003.0010

this context, the potential of DERs has been recognized for some time, especially when it is in the form of demand-side flexibility (DSF).[3] The main contribution on the subject in the legal literature is the work of Daniela Aguilar, who produced an excellent conceptual and comparative analysis of the emerging active role of electricity consumers.[4] Other contributions have considered microgrids[5] and the economics of DER.[6]

II. The Unfulfilled Promise of Market Forces

Reform of the New Zealand electricity system in the 1990s was animated by trust in market competition, removing legal monopolies and restructuring all aspects of the industry. Neoliberal political thought guided the reform and led to the expectation that market competition would break out, without any need for a regulator, price control, or central planning.[7] Before 1988 the whole electricity sector was in public ownership; from 1988 to 2003, energy companies were deregulated and restructured, and industry self-regulation set up the wholesale market and other co-ordination institutions.[8] There are four dominant generator-retailers, and the transmission company (Transpower) and local distribution network companies are separate from them.[9] It was not until the period 2003 to 2010 that the industry was put under statutory regulation and price control was imposed on transmission and distribution services. From 2010 a period of regulatory consolidation began, but there have been no major reforms since then that would mark the beginning of a new period.

For DER, what is particularly noticeable is the promise that market forces would lead to innovation; that competitive pressure would encourage experimentation and reward the companies that could introduce new technology and business models that attracted customers away from the old ways of doing business.[10] Customers would be able to respond to price signals by cutting consumption.[11] Regulation, where necessary, would be technology

[3] The importance of DSF or demand-side management is also emphasized in Chapter 13 of this book.

[4] D Aguilar Abaunza, *The Law for Energy Prosumers: The Case of the Netherlands, New Zealand and Colombia* (Springer 2022). Also D Aguilar Abaunza, 'Getting Ready for the Future' [2018] New Zealand Law Journal 292. The concept of the active consumer is analyzed in detail in the chapter by Huhta in Chapter 15 of this book.

[5] S Mohseni and A Brent, 'Unlocking the Transformative Potential of Community Microgrids in Aotearoa New Zealand' (2024) Journal of the Royal Society of New Zealand, doi 10.1080/03036758.2024.2391347.

[6] D Reeve, T Stevenson, and C Comendant, 'Cost-Benefit Analysis of Distributed Energy Resources in New Zealand' (Sapere 2021).

[7] The writer has chronicled these reforms in earlier books in this series and elsewhere: B Barton, 'Reaching the Limits of what the Market will Provide: Energy Security in New Zealand' p 373 in B Barton and others (eds), *Energy Security: Managing Risk in a Dynamic Legal and Regulatory Environment* (OUP 2004); B Barton, 'Electricity Regulation in New Zealand: the Early Stages of a New Regime' (2008) 26(2) Journal of Energy and Natural Resources Law 207; B Barton, 'Law and Regulation for Energy Networks in New Zealand' p 274 in M Roggenkamp and others (eds), *Energy Networks and the Law: Innovative Solutions in Changing Markets* (OUP 2012).

[8] Barton, Networks 2012 (n 7) 290.

[9] In this chapter, these companies will generally be called distribution companies or distributors; they can also be called lines companies, local network companies, or electricity distribution businesses. The high-voltage transmission grid has not been a point of controversy: contrast Banet in this volume.

[10] J Culy, 'Electricity Restructuring: Towards a Competitive Wholesale Market' (1992, discussion paper, NZ Institute of Economic Research) 22; D Kidd, 'To the NZ Wholesale Electricity Market Congress' (7 June 1996, press release, Minister of Energy).

[11] D Kidd, 'The Lights Will Stay On!' (1 October 1996, press release, Minister of Energy).

neutral. Ostensibly, these features should make it easy for the energy system to adopt DER technology and business models.

Nonetheless, New Zealand has lagged in DER, whether it be in distributed generation, rooftop solar, demand-side flexibility, energy storage, or integration of EV charging and smart buildings. In 2023 the International Energy Agency observed that less than 2 per cent of New Zealand's installed capacity is in DER, mainly in the form of rooftop solar installations, and that the deployment of DER is lower than in other IEA countries, including neighbouring Australia.[12] It ascribed this slower rate of uptake to the lack of incentives (such as net metering), and the lack of urgency for the electricity market to develop flexibility options such as DER; and it warned that some distribution companies lack the ability to respond well to DER.

At present, most retailers are offering domestic consumers plans with solar buy-back, time-of-use pricing, and water heating control incentives, but they are limited and the buy-back rates are usually much lower than the sale price. A few independent retailers are pushing the boundaries, but they encounter numerous obstacles. They call for policy measures to promote stronger competition and remove barriers to competition and growth.[13] One says that the market dominance of the large generator-retailers makes it difficult for smaller retailers to continue.[14] It is argued that independents cannot access hedge products at competitive rates, that there is high market concentration and persistently high prices, so that the retail market has stagnated, with fewer participants, less innovation, and less decentralization, not more.[15] Two solar companies have failed, including solarZero that had strong financial backing to offer 'energy as a service'.[16] Some of the dominant retailers and distribution companies are experimenting with retail DER, especially demand-side flexibility, but it is basically exploratory.[17] There is more success developing demand-side flexibility at an industrial and commercial scale, to deal with winter shortages and high prices; one company is working with food processing and cold storage operators who can reduce power use for periods.[18] In 2024 an aluminium smelter owned by Rio Tinto entered into a power supply contract that provided unprecedented levels of flexibility. But care is needed before saluting such arrangements; it is not a company's core business to refrain from manufacturing product.

[12] International Energy Agency (IEA), 'New Zealand 2023 Energy Policy Review' (OECD / IEA 2023) 120. Similarly Electricity Authority, 'Issues Paper: Updating the Regulatory Settings for Distribution Networks' (December 2022) 12.

[13] 2degrees and others, 'Level Playing Field Measure Critical for Promotion of Effective Competition' (submission to Energy Competition Task Force, 5 November 2024). <https://www.google.com/url?sa=t&source=web&rct=j&opi=89978449&url=https://www.ea.govt.nz/documents/6168/Independent_retailers_FJeyQOf.pdf&ved=2ahUKEwj417WvxsCRAxUP3jgGHdoeOx4QFnoECB4QAQ&usg=AOvVaw0qipaTFukQX0vmhontCKyf>.

[14] Radio NZ, 'Independent Retail Does "Not Make Financial Sense"—Power Company' (3 December 2024). This was Octopus Energy, a well-funded international player. <https://www.rnz.co.nz/news/business/535498/independent-retail-does-not-make-financial-sense-power-company>.

[15] Consumer NZ, 'Submission on Level Playing Field Measures' (7 May 2025). <https://www.ea.govt.nz/projects/all/energy-competition-task-force/consultation/level-playing-field-measures/>.

[16] Grant Thornton, 'Liquidators' Initial Report on the State of Affairs of solarZero Ltd' 3 December 2024 <https://www.grantthornton.co.nz/solarzero-ltd-in-liquidation2/>; C Smith, 'Solar Group in Liquidation' NZ Herald, 15 July 2025. <https://www.nzherald.co.nz/business/solar-group-in-liquidation-kiwibank-among-creditors-owed-more-than-3-million/LQO5EZZ7UZCLXP377MANOXQNYE/>.

[17] F Wolfe, 'Orion and Ecotricity trial 60-cent buy-back offer' Energy News, 9 July 2025. <https://www.energynews.co.nz/news/electricity-supply/822445/orion-and-ecotricity-trial-60-cent-buy-back-offer>.

[18] Simply Energy, Submission to Electricity Authority on Future Security and Resilience (12 April 2022). <https://www.ea.govt.nz/projects/all/future-security-and-resilience/consultation/phase-2-draft-roadmap/>.

Many of the problems are sector-wide and not DER-specific. Many observers believe that there is not enough competitive pressure on companies in either the wholesale or retail market, resulting in too little decentralization.[19] Wholesale and retail prices have risen sharply and energy affordability is a problem for households and industry alike, but for many years there were few new entrants into generation, and the incumbents were slow to build new capacity. Although collaborative exercises for change do occur, the industry indulges in a lot of squabbling and finger-pointing.

The electricity distribution companies, which run the local low-voltage distribution networks, have a central role in DER, but it is one that is new to them and they are having to adjust.[20] It is unlikely that the smaller companies have the capacity to press forward with the new technologies, smart grids, and DER generally, and some of them are very small by international standards.[21] Often their ownership structure produces weak governance and poor capital management.[22] Distribution has traditionally been a relatively passive activity, channelling predictable, unilateral flows of power from substations to local consumers, with only a few relationships to manage. However, radically new uses of networks are being called for, with volatile power flows, and thousands of connected parties to be managed. DER companies say that the distributors are difficult to deal with, slow to respond, and miserly with consumer and network data.[23]

New Zealand moved to modern metering faster than many countries, but in ways that hinder DER. The retail companies, and not the distributors, rolled them out in the period 2006-2008; by 2022, 92 per cent of meters were modern.[24] Rules appeared after the main roll-out period, but they did not dictate a standard meter type or insist on the high-specification requirements that they could have, for communication with home area networks or in-home displays.[25] Arguably they are not 'smart' meters.[26] Metering equipment providers were made responsible for handling meter data, allowing customers to have access to it, and keeping it secure, but there are problems with access for legitimate use, even by the distributor companies.[27] Metering is inconsistent and complex; every retailer has its own system, and changing retailer can mean changing meters. The metering equipment providers have a de facto monopoly once their meters are deployed.[28] Others, such as EV

[19] IEA 2023 (n 12); Consumer NZ, 'Submission on Level Playing Field Measures' (7 May 2025).

[20] Barton, Networks 2012 (n 7). They must operate distribution separately from generation and retailing: Electricity Industry Act 2010, ss 6A, 72, and 73(1) and the Electricity Industry Participation Code Part 6A, and this sometimes affects DER: IEA 2023 (n 12) 143.

[21] IEA 2023 (n 12) 106 and 121.

[22] International Energy Agency, 'New Zealand 2017 Energy Policy Review' (OECD / IEA 2017) 148.

[23] For example, Drive Electric, 'Submission on Electricity Authority Issue Paper—Targeted Reform of Distribution Pricing' submission 15 August 2023; Ecotricity, submission letter to Electricity Authority on Issues Paper on Targeted Reform of Distribution Pricing, 2 August 2023.

[24] J Gumz and others, 'A Mixed-Method Analysis of New Zealand's Smart Meter Rollout Experience' (2024) 90 Utilities Policy 101817.

[25] Code Part 10. It only required separate recording of imports and exports in 2021, in clause 10.13A.

[26] Parliamentary Commissioner for the Environment, 'Smart Electricity Meters: How Households and the Environment Can Benefit' (2009) and Parliamentary Commissioner for the Environment, 'Update Report on the Original Investigation' (2013).

[27] Especially Code clause 10.15 and Schedule 10.6. The Customer and Product Data Act 2025 now authorizes regulations requiring retailers to make usage data more accessible to consumers, which may make it easier for companies to package DER services. Electricity Authority, 'Enabling Consumer Mobility by Improving Access to Electricity Product Data' (consultation, 15 July 2025). On data privacy and digitalization, see the chapters in this book by Godden and Paddock.

[28] Ara Ake, 'Ara Ake EDB Challenge: Learnings and Insights Report' (2024) https://cdn.prod.website-files.com/670669f4e6068197aafd0771/6736b99218ce332e1b0ffd4a_EDB%20Challenge_Learnings%20and%20insights%20report_Final.pdf accessed 18 October 2025.

charger operators and DER operators, also face delay, trouble, and expense in obtaining anonymized data. Leaving it to the market has meant that the government missed a chance to steer technical development and facilitate DER, with a result that is cumbersome, expensive, and only weakly competitive.

III. The Absence of Political and Regulatory Pressure

If market competition was not a strong driver for DER, what pressure came from the state? It is convenient in asking the question to distinguish between government action, in the sense of legislation and political leadership by ministers, and regulatory action of a more detailed kind, in DER mainly carried out by the Electricity Authority. The overall picture is that many years went by without ministers or the legislature taking any real action at all on DER, or on the electricity system as a whole. The industry structure has not been modified since 1998, and the Electricity Industry Act 2010 has not been amended much since its enactment. A formal Government Policy Statement was issued in 2006 but the first update was in 2024, eighteen years later. Possibly the Government felt that it could not interfere with the degree of independence conferred on the Electricity Authority in 2010, but that is not much of an excuse for so little action.[29]

As for DER specifically, the first mention of distributed energy resources in policy documents may be one in 2001, but it was the first of several that mentioned it only in a general way as a possibility for the future.[30] Climate plans did not focus on DER either.[31] As the IEA noted in 2023, there are no targets for DER, and no subsidies for it.[32] It is not surprising that progress was sluggish.

Legislation in 2001 authorized regulations providing for terms and conditions for purchase by electricity retailers of surplus electricity generated by small units owned by consumers, subject to the retailer not incurring ongoing financial losses as a result.[33] The Government Policy Statement in 2006 directed the regulator to develop non-regulatory arrangements for this purpose, or regulations as a backup.[34] The Statement also said that there should be no unnecessary barriers to the development of distributed generation,[35] so in 2007 regulations were made for a procedure for distributed generators to apply to local distribution companies for a connection (and requiring approval to be granted if the application complied with the rules), subject to regulated terms, a dispute resolution process,

[29] Electricity Industry Act 2010, s 17 requires the Authority to have regard to a statement of government policy; generally see Barton, Early Stages (n 7) and Barton, Networks (n 7).

[30] Ministry of Economic Development, 'National Energy Efficiency and Conservation Strategy 2001' 17; 'National Energy Efficiency and Conservation Strategy 2011' 26; Ministry of Economic Development, 'New Zealand Energy Strategy to 2050' (2007) 67, 77; Ministry of Economic Development, 'Developing Our Energy Potential: New Zealand Energy Strategy 2011-2021 and New Zealand Energy Efficiency and Conservation Strategy 2011-2016' (2011) 26. Plans for another energy strategy in 2022 came to naught.

[31] Ministry for the Environment, 'Te Hau Mārohi ki Anamata: First Emissions Reduction Plan' (2022) and 'Our Journey towards Net Zero: Second Emissions Reduction Plan' (2024).

[32] IEA 2023 (n 12) 120. There are a few isolated programs for financial support such as co-funding of certain energy-efficiency projects by the Energy Efficiency and Conservation Authority; but the situation is very different from Australia and California where the subsidies have been massive: see Godden and Paddock in this volume.

[33] Electricity Act 1992, s 172D(23), inserted by the Electricity Amendment Act 2001.

[34] Government Policy Statement, 30 October 2006, paras 111–13.

[35] ibid para 109. The capacity limit was 40,000 kWh per annum.

pricing principles, prescribed maximum fees, and assurances of non-discrimination.[36] Connecting wind farms was undoubtedly the main purpose of the regulations, but they were a step towards DER. They are now in Part 6 of the Electricity Industry Participation Code (the Code).

IV. Earlier Regulatory Action Affecting Distributed Energy Resources

Regulatory attention in the Electricity Authority turned in the 2010s to matters that relate to DER, even if not directly so. The Authority's main statutory objective is 'to promote competition in, reliable supply by, and the efficient operation of the electricity industry for the long-term benefit of consumers'.[37] While the Authority may not act inconsistently with this objective, it has a broadly expressed power as an expert body, and may promote one or more of the three elements of competition, reliable supply, and efficient operation as it sees fit, consistently with the overall goal of long-term benefit for consumers.[38] It oversees the operation of the wholesale electricity market, and it regulates Transpower as the system operator. Most of its regulatory activity is in the form of amendments to the Electricity Industry Participation Code,[39] a detailed and technical set of rules, about 1,500 pages long, which controls the activities of all market participants. Although its regulatory mandate was a broad one, it is reasonably clear that for some time DER, and indeed distribution as a whole, did not get much of the Authority's attention; its main concerns in the 2010s and early 2020s were transmission pricing and the functioning of the wholesale electricity market.[40] But it carried out three sets of reform that are important in making DER progress.

A. Distribution Agreements

The first of these reforms was for distribution agreements, also called use of system agreements. They mainly concern the conventional industry operations of retailers using local distribution companies' networks to convey electricity to consumers. The arrangements for making them, under Part 12A of the Code, had been criticized for being too complex, forming a barrier to entry for new retailers, who if they want a national presence must

[36] Electricity Governance (Connection of Distributed Generation) Regulations 2007. The pricing principles emphasized the avoided cost of transmission, which may have resulted in uneconomic distributed generation being installed: IEA 2017 (n 22) 88 and 139; but it was removed from the Code in 2022: Electricity Authority, 'Avoided Cost of Transmission (ACOT)—TPM-related Amendments' (decision paper, 20 December 2022).

[37] Electricity Industry Act 2010 s 15(1).

[38] *Manawa Energy Ltd v Electricity Authority* [2022] NZHC 1444 [60]-[69]; *Nova Energy Ltd v Electricity Authority* [2023] NZCA 275.

[39] Electricity Industry Act 2010, ss 38-39. At times the Authority also produces guidance that sits outside the Code, such as the distribution pricing principles discussed below.

[40] 'Our strategy; we have been successful in promoting strong competitive markets. Now we're thinking more broadly to ensure our regulation responds to a changing world.' Electricity Authority, 'Statement of Intent 1 July 2020–30 June 2024' 11. An advisory group scoped out the issues in creating equal access to networks in 2019: https://www.ea.govt.nz/about-us/our-people/our-advisory-and-technical-groups/ipag/ accessed 26 October 2025. A cross-industry initiative in 2022 identified the benefits of DER (demand side flexibility in particular), and regulatory changes that would allow it to play a greater role in the system. Flexforum, 'A Flexibility Plan 1.0' (31 August 2022).

negotiate with each of the twenty-nine distributor companies, many of which are disposed to protect their own operations and avoid having costs imposed on them.[41]

The Authority sought a degree of standardization, and in 2012 proposed voluntary model agreements, but the distributors resisted them, so in 2014 it proposed Code rules for a default set of terms. The distributors resisted again and obtained a declaration that the Act did not go as far as to authorize a complete standardization.[42] A year later the distributors were less successful arguing that the Authority could not regulate anything that the Commerce Commission could regulate on quality standards. But the Electricity Industry Amendment Act 2022 clarified and strengthened the Authority's jurisdiction, and it can now set requirements for distribution quality, information, and pricing methodology, in the same way as it can for transmission. It can impose a distribution agreement for connection to, and use of, the distributor's network, with prescribed default terms and conditions.[43]

In consequence, Part 12A of the Code now requires distributors to publish their default distributor agreements containing default terms (based on a template it provides) for retailers wanting to trade on the networks. The core terms must be consistent across all agreements, while operational terms may vary. The Authority used the 2022 Act to reduce retailers' costs and transaction costs, improve consumer consumption data procedures, and prevent charges accruing during outages.[44] The new rules should make it easier for innovative retailers to provide DER offerings to their customers on the network.

B. Real Time Pricing

Although DER is characterized by energy resources connected to a local distribution network, or 'behind the meter' on the consumer's premises, the regulatory framework of the broader electricity system is also relevant. In New Zealand, one key element has been real-time pricing in the wholesale market, which was introduced in 2022, and is important because it allows DER supply and demand to respond to spot market signals. Before it was introduced, after each thirty-minute trading period, the only price information available was an indicative price; the final price only appeared two days later, and the delay prevented consumers, especially large industrial users, from modifying their production or consumption in real time in order to manage volatility and avoid high prices. They now have price certainty (with few exceptions) at the end of each trading period.[45] Real-time pricing was thought to be the biggest change in the operation of the wholesale market since it began in 1996.

A second phase of real-time pricing, in April 2023, allowed 'dispatch notification purchasers' or aggregators of small-scale DER to bid and offer resources into the wholesale market, subject to the approval of the system operator. Through such aggregators, small participants such as the owners of EV chargers, solar and battery installations, and

[41] Also D Caygill, S Wakefield, and S Kelly, *Inquiry into the Electricity Industry* (Ministry of Economic Development 2000) para 224.

[42] *Vector Ltd v Electricity Authority* [2018] NZCA 543.

[43] Electricity Industry Act 2010, ss 32 and 44A.

[44] Electricity Authority, 'Changes to the Default Distributor Agreement Template, Consumption Data Template, and Related Part 12A Clauses' (decision paper, 22 October 2024).

[45] Electricity Authority, 'Implementing Spot Market Settlement on Real-Time Pricing: Decision' (28 June 2019).

commercial or residential buildings can participate in the wholesale market. Soon after, aggregators were allowed to nominate one grid exit point for their wholesale market transactions, a simplification that should not affect the system's nodal pricing principles while there is so little aggregator activity.[46] Aggregators and virtual power plants (VPPs) have not yet emerged on a commercial scale in New Zealand,[47] but it is evident that for VPPs other demand-side flexibility and DER to progress, the wholesale market, with all its technicalities, is an indispensable part of the puzzle.

C. Market Development Advisory Group

At much the same time, the Authority established an industry-wide Market Development Advisory Group (MDAG) to improve practical insight and stakeholder involvement. In 2021 MDAG began work on how the wholesale market should operate with the foreseeable massive expansion of renewable energy, especially with higher levels of intermittent supply. Its recommendations in late 2023 delivered a blueprint for a redesign of the market that would be underpinned by flexibility in supply and demand to balance the system.[48] Its primary targets were in the wholesale market, to produce accurate pricing, risk management, competition, and confidence, but its secondary targets were never out of sight in distributed resources and demand-side flexibility. It foresaw that the system would soon have far more diversity in technology and non-traditional players, many more active participants, and many more devices connected; and that distribution network activity would no longer be separate from the wholesale market.[49] Its proposals would activate demand-side flexibility and allow dynamic pricing for responsive supply and demand. Distribution pricing would need to change, and there would need to be better use of the data gathered by modern meters. It recommended price-driven methods of dispatching DER and integrating it into the wholesale market. The Electricity Authority accepted MDAG's proposals in early 2024 and the Government Policy Statement of October 2024 confirmed that they are to be implemented in full.[50]

D. Innovation and Non-Traditional Solutions Allowance

One specific problem in driving this innovation forward is the price control that the Commerce Commission imposes on electricity distribution businesses, because its price-quality path regime gives regulated companies an incentive to increase efficiency, rewarding them with higher profits if they can control their expenditure; but this discourages

[46] Electricity Authority, 'Dispatch Notification Enhancement and Clarifications' (26 January 2024).

[47] There has been a successful pilot: M Daalder, 'Solar "virtual power plant" helps get NZ through cold snaps' Newsroom, 27 February 2024. On VPPs in more detail, see Robinson in this volume.

[48] Market Development Advisory Group, 'Price Discovery in a Renewables-based Electricity System' (final recommendations paper, Electricity Authority, 11 December 2023).

[49] ibid 12, 21, 27, 35, and 116.

[50] Minister for Energy, 'Statement of Government Policy to the Electricity Authority under section 17 of the Electricity Industry Act 2010: New Zealand Electricity Industry' (October 2024) https://www.beehive.govt.nz/sites/default/files/2024-10/Government%20Policy%20Statement%20on%20Electricity%20-%20October%202024.pdf accessed 31 October 2025.

spending on innovation.[51] For example, if a distribution company makes improvements to its consumption and connection information, or explores DER as an alternative to network expansion, it cannot charge network users for the work.[52] The companies tend to default to well-tested traditional solutions if there are not enough incentives to embark on innovation.[53] The companies think that DER entrepreneurs want them to drop everything and rebuild their businesses but are not keen on paying for it. To address this problem, and to respond to statutory direction,[54] the Commission in 2020 introduced an innovation allowance that companies could apply for, but at 0.1 per cent of allowed revenue it was too small to attract attention. For the next regulatory period, the Commission was persuaded to go higher, with an innovation and non-traditional solutions allowance of 0.8 per cent of allowed revenue, a quarter of which is ringfenced for projects where two or more companies collaborate.[55] Companies must apply for the allowance for projects such as trialling batteries or managed EV charging where the benefits are sufficiently uncertain that they might not otherwise be undertaken, but with potential benefits for customers. But the allowance is still a meagre one in relation to overall distributor revenues.

V. The Point of Change in 2024

The New Zealand story reached a turning point in 2024 when the industry, regulators, and government realized that they needed to do much more, faster, and that DER was a major part of it. The main reason was serious problems of energy security and affordability.[56] Wholesale electricity prices had risen in 2018 and had stayed high. Few observers other than the big companies and the regulator believed that there was enough competition in the wholesale or futures electricity markets to keep prices down or drive real technical and commercial change.[57] Companies seemed to have little interest in building new generation capacity far enough ahead to maintain adequate levels of security of supply; certainly they had to be prudent managers of capital, but undoubtedly an oversupply of electricity would bring prices down. New generators and retailers found it difficult to enter the market.

In each of the last four or five winters, there had been apprehension that New Zealand had insufficient fuel and capacity to meet the demand for electricity. It is a regular pattern that demand is higher in winter, and for the hydro storage in the South Island to decline until spring; but in addition in recent years natural gas supplies were constrained because of production problems. The risk of scarcity produced sharp increases in wholesale market prices. In 2024, demand-side flexibility was actively deployed, but at a cost. Several large industrial sites reduced their demand for electricity (in steelmaking, forestry, dairy products,

[51] IEA 2017 (n 22) 135, 154.

[52] Flexforum, 'A Flexibility Plan 1.0' (31 August 2022) 28.

[53] Commerce Commission, 'Default Price-quality Paths for Electricity Distribution Businesses from 1 April 2025' (issues paper, 2023) 46–47. Thirteen of the smaller EDBs are exempt from price-quality regulation.

[54] The Commerce Act 1986, s 54Q requires the Commission to promote incentives, and to avoid disincentives, for distribution companies to invest in energy efficiency and demand side management, and to reduce energy losses.

[55] The Innovation and Non-Traditional Solutions Allowance (INTSA): Commerce Commission, 'Default Price-Quality Paths for Electricity Distribution Businesses from 1 April 2025: Final Decision' (20 November 2024) chapter 3.

[56] L Hanly, 'New Zealand has an "energy security crisis"—Christopher Luxon' Radio NZ, 26 August 2024.

[57] E Ricketts, 'What the energy shortage shows about future power supply' Radio NZ, 12 September 2024.

and in particular aluminium smelting) and a methanol plant allowed large amounts of natural gas to be diverted to power generation, but the reductions in production inflicted material damage on the country's economic performance.[58] Two wood pulp mills closed, apparently because of their unhedged exposure to high electricity spot prices, causing significant regional unemployment. Some independent electricity retailers found themselves in financial difficulties. An unreadiness to use DER had come under a harsh spotlight in an incident in August 2021 when the system operator mistakenly believed that it had to shed load urgently and imposed blackouts across wide areas. It turned out that it had not known that it could have avoided the blackouts by using the substantial discretionary demand management that distributor companies could provide through ripple control of domestic hot water heating.[59]

In 2024, political action was inevitable. The government issued a formal Government Policy Statement was issued, giving direction to the Electricity Authority.[60] It noted that distributed resources and demand-side response were becoming more significant. Competitive markets continued to be at the centre of policy, and the government would not insulate companies from supply or price risks, but the Authority was to ensure that competition is facilitated and remained technology-neutral. The recommendations of the MDAG report, which the Authority accepted in early 2024, were all to be implemented. The demand was for faster regulatory action, not radical change. However, deeper change may emerge from a different review that the government has commissioned from international consultants on industry structure, ownership, market rules, and institutions, and dry year supply.[61]

In August 2024, the Electricity Authority and the Commerce Commission set up an Energy Competition Task Force to accelerate work already under way and oversee rapid implementation of the MDAG recommendations.[62] Most of its work, discussed below, is under the existing legal settings, but some of it could be far-reaching, such as non-discrimination rules as a backstop to provide a level playing field between independent retailers and the incumbent generator-retailers.[63]

This political impulsion—a real shove—came at a time when change was already in the air. The Authority had turned its attention away from transmission pricing and the wholesale market (it completed its work on real-time pricing in 2022) and towards the distribution sector. Its efforts to improve distribution agreements went back to 2012 (although it was actually in 2024 that they were completed), and MDAG's major reform project had started in 2021. But the political urgency galvanized those efforts to produce rapid results, with major implications for DER.

[58] A Hallett, 'Economy taking a hit from "long electricity winter": Enerlytica' Energy News, 23 September 2024.

[59] Ministry of Business, Innovation, and Employment, *Investigation into Electricity Supply Interruptions of 9 August 2021* (Peter Hodgson 2021). Ripple control is an early form of DER where distributed equipment can be controlled by superimposing a current on the power supply, most often to turn domestic hot-water cylinders off for a few minutes during peak load periods. TW Ross and RMA Smith, 'Centralized Ripple Control on High-Voltage Networks' (1948) 95 Journal of the Institution of Electrical Engineers 470.

[60] Statement (n 50).

[61] S Brown and S Jones, 'Government Launches Review to Ensure Electricity Market is Fit-for-Purpose' (ministers' press release, 15 November 2024) https://www.beehive.govt.nz/release/government-launches-review-ensure-electricity-market-fit-purpose accessed 27 October 2025.

[62] Electricity Authority, 'Energy Competition Task Force Set Up to Improve Electricity Market Performance' press release, 28 August 2024.

[63] Electricity Authority, 'Level Playing Field Measures' (options paper, 27 February 2025).

VI. Recent Regulatory Action for Distributed Energy Resources

On DER specifically, the Electricity Authority's work has a background in analysis that had been going on for several years,[64] even though 2024 and 2025 have seen an extraordinary flood of regulatory decisions. In 2022 it outlined a programme of work on the matter, with five themes for regulatory settings: access to data, market settings, capability and capacity in the industry, operating agreements for flexibility services, and standards.[65] In relation to market settings, it identified the problems of distributor companies favouring network solutions when non-network solutions could be more efficient, favouring in-house investment or subsidiary firms, and using their monopoly position in distribution to secure an advantage. It reformulated its work programme in 2023.[66]

What we see in the regulatory actions of 2024 and 2025 is, firstly, that political and regulatory effort can bring about DER change when companies are unwilling or unable to produce it themselves, and secondly, as we examine these actions, we see that DER reform requires a bewildering number of detailed individual changes.

A. Distribution Pricing

The Authority seeks to improve the ways that distributor companies price their services.[67]

(Most end-users of electricity of course have these prices paid for by their retailer.) The overall revenue of a non-exempt distributor company is regulated by the Commerce Commission, but the Electricity Authority has jurisdiction over how it is allocated or paid for.[68] The Authority is interested in the basic question of how the burden should be allocated to different pricing areas and different consumer groups, with minimal cross-subsidy; but at this point its work on the matter is only at an early stage. Traditional tariffs charge customers for their use of the network based on the volume of electricity they use, but the investment that a network requires is to carry the peak load, on the coldest evening of the year, and it is more logical to charge based on the customer's peak load.

The Authority is encouraging distributors to increase the contrast between their peak and off-peak prices, especially by reducing prices off peak when the cost of network usage is effectively zero, and it is investigating the extent to which the retailers pass these price signals on to their customers in their tariffs. It would also like to see the benefits of time-of-use meters reflected in tariffs. However, it has not produced rules for an elaborate distribution pricing methodology, and instead it has made statements of pricing principles which call, in general terms, for prices that signal the economic cost of service provision.[69] So far,

[64] Electricity Authority, 'Enabling Mass Participation: Response and Next Steps' (decision paper, 4 October 2017).

[65] Electricity Authority, 'Issues Paper: Updating the Regulatory Settings for Distribution Networks' (December 2022).

[66] Electricity Authority, 'Delivering Key Distribution Sector Reform: Work Programme' (16 October 2023). This did not include the Authority's work on distribution pricing reform.

[67] Electricity Authority, 'Distribution Pricing Reform: Next Steps' (7 May 2024) and Electricity Authority, 'Open Letter to Distributors' (20 May 2024).

[68] Electricity Industry Act 2010, s 32(4).

[69] Electricity Authority, 'Distribution Pricing: Practice Note Second Edition v2.2' (2022) p 4.

prescriptive regulation has been introduced for some specific matters, detailed below, but more is foreseeable, and is likely to stimulate DER operations.

B. Connecting to a Distribution Network

Earlier we encountered the rules for distribution companies to connect distributed generation, made in 2007 and now in Part 6 of the Code. The Authority has identified several problems with them, both for connections of load (energy use) and distributed generation. The distribution companies have all being using their own practices and procedures for applications to connect to their networks, and for setting the cost. Companies with load projects such as EV charging stations could not get information about the existing state of the network, and faced uncertain, inconsistent, and often protracted approvals processes from the monopoly network company[70] and distributed generation faced similar obstacles. The Authority therefore decided to regulate both procedures and prices.[71] As for procedures, applications for larger-capacity load will be brought under Part 6 to be regulated there along with distributed generation. (For example, Part 6 will cover an application for an EV charging station as well as one for a solar installation.) Application processes will be tightened up. Distributors must publish information about their network capacity and the pipeline of likely network connection proposals, and the regulated terms for distributed generation are being updated. But the Authority has signalled that this will not be all; further change will follow.[72]

As for the prices that the monopoly distribution companies can impose for connections, the Authority is dealing with load and generation separately. For load, prices must now be set with reference to a 'minimum scheme' with any enhancement being paid by whichever party requests it, and with a power for the customer to request that a cheaper scheme be considered.[73] A pioneer applicant will be entitled to a rebate from subsequent users of the network extension it paid for. Distributors will have to disclose more information about how their charges are made up, and they will have to disclose rates for charging upstream capacity costs, which must be allocated as the capacity is consumed, not as it is built. For smaller customers, companies can continue to use posted prices for low-cost connection types.

For connection prices for distributed generation, the Authority has not yet made final decisions. It believes that there are disadvantages in the existing 2007 pricing principles which say that charges are to be no more than the incremental cost of providing the connection.[74] Those principles were intended to ensure a level playing field, but they prevent distributor companies from recovering a number of the less immediate costs in running the network as a whole, fewer in fact than would be paid by a generator connected to the

[70] F Wolfe, 'Network Connection Reform Lacks Ambition, Needs Rewrite—Sapere' Energy News, 24 January 2025.

[71] Electricity Authority, 'Network Connections Project (Stage One)' (decision paper, 18 July 2025). Larger capacity was decided as being greater than 69 kVA.

[72] Network Connections Project (n 71) 10.

[73] Electricity Authority, 'Distribution Connection Pricing Code Amendment' (decision paper, 18 July 2025).

[74] Electricity Authority, 'Distributed Generation Pricing Principles: Issues Paper' (12 February 2025). The incremental cost rule may unduly influence DER locational decisions, with a similarity to behaviour in California: see Paddock in this volume.

national grid. The principles also open up questions of a 'first mover disadvantage' where the first party pays for the upgrade and latecomers use it for less; or, the first mover might take up the available capacity, throwing the expense of an upgrade onto the next mover.[75] The answer may be rules similar to those developed for load customers, with pricing principles outside the Code.

C. Consumer-Related Distribution Changes

In July 2025 the Authority decided to require distributor companies to pay a rebate when a residential or small-business consumer injects electricity at peak times, which is strong support for DER and distributed generation in particular.[76] The rebate is to be calculated on the basis of the long-range marginal cost of peak demand that can be avoided by the injection, and the peak times are described as the times that on average are likely to drive future network investment. These are broad signals and the Authority will need to provide guidance on how they are to be applied.

Also in July 2025 the Authority made rules for retailers to offer their customers time-of-use plans, which must provide 'a financial benefit to each customer for consumption and injection patterns that reduce pressure on system costs'.[77] Again, this is general language, but it strongly favours DER and treats consumption and injection alike. The rules also require distributors to charge retailers by the time-of-use behaviour of the retailers' customers, which ensures that retailers get the benefit of work with their customers to reduce peak-period load. Again, the Authority signalled that more reform will be required in the longer term.

The Authority also proposes that consumers be allowed to use different retailers for their electricity consumption and their generation.[78]

D. Decentralization

The Authority has begun to think specifically about energy decentralization, and to engage with the broader community about it.[79] While its work on this phenomenon is at a very preliminary stage, at least it is addressing it directly. It suggests that decentralization will make the governance of the sector even more complex than at present, especially at the local level. Distribution service operators may also be required as well as the national transmission-level system operator.[80] Decentralization may be an important strand in the regulator's activity in the future.

[75] IEA 2023 (n 12) 106.

[76] Electricity Authority, 'Requiring Distributors to Pay a Rebate when Consumers Supply Electricity at Peak Times' (decision paper, 16 July 2025). It will be in the Code clause 12A.7.

[77] Electricity Authority, 'Improving Pricing Plan Options for Consumers: Time-Varying Retail Pricing for Electricity Consumption and Supply' (decision paper, 16 July 2025), amending Code Subpart 11A.1 and 12A.4 and 12A.5.

[78] Electricity Authority, 'Evolving Multiple Retailing and Switching' (consultation paper, 3 June 2025).

[79] Electricity Authority, 'Working Together to Ensure our Electricity System Meets the Needs of All New Zealanders' (green paper, 30 April 2025).

[80] Electricity Authority, 'The Future Operation of New Zealand's Power System' (consultation paper, 24 June 2025).

E. Multiple Technical Adjustments to Allow Distributed Energy Resource Activity

Many small, detailed, technical changes are needed in the electricity system to allow DER. Going back several years, one finds batteries as an example; the Electricity Authority made Code changes in 2018 to allow direct-current producers (batteries and solar farms) to offer energy into the wholesale market.[81] Another change in 2022 allowed batteries to take part in the instantaneous reserve market, for which they are ideal.[82] Until then the rules assumed that reserves would only be thermal or hydro generation. In 2024 another adjustment was made so that energy storage systems need not pay double transmission charges if they could both take and inject power.[83] Other adjustments to technology-specific rules have been made to include wind, solar, storage, and embedded generation more fully in the system's quality standards.[84] Another proposal is to broaden the permitted voltage range for supply installations from 6 per cent to 10 per cent, to accord with international developments; the change will allow more rooftop solar and EV chargers to be employed.[85] DER also needs voltage-related changes in the common quality requirements.[86] Construction rules are being modified to make rooftop solar installation easier.[87] We see that DER regulation is granular in character, and that sweeping statements of principle are not enough.

VII. Enabling Sustainable Distributed Energy Resources

At a broader level, there seem to be three main insights to be had from this New Zealand experience with DER. The first is the effect of ideas about market competition. It is no surprise that market forces alone do not bring about important kinds of change; there is no need to make another attack on neoliberalism. We see problems of imperfect competition in the wholesale market and of course none at all in the distribution sector. Leaving companies alone produced very qualified results in metering. The more interesting point is why there was so little state pressure for change even when it became clear that innovations like DER could substantially alleviate real problems like scarcity in dry winters. The failure to use ripple control DER in 2021 was telling. DER was on the sidelines for many years. Ministers did not seem engaged. How did the regulator formulate its priorities, and why did ministers do so little when households and businesses were under price and supply pressure? Why is there still so little active planning for the sector?[88] However, the recent spate of

[81] Electricity Authority, 'Enabling Energy Storage Systems to Offer Instantaneous Reserve' (final decision 29 March 2022) 3, amending Code Part 13, Trading Arrangements.

[82] Enabling Energy Storage Systems (n 81).

[83] Electricity Authority, 'Transmission Pricing Methodology Amendments: A Level Playing Field for Emerging Technologies' (decision paper, 26 November 2024). On storage generally, see the chapter by [Athwal] in this volume.

[84] Electricity Authority, 'Part 8 Code Terminology and Network Information' (decision paper, 1 April 2025).

[85] The rule is under the Electricity Act 1992 in reg 28 of the Electricity (Safety) Regulations 2010.

[86] Electricity Authority, 'Promoting Reliable Electricity Supply—a Voltage-related Code Amendment Proposal' (consultation paper, 3 June 2025).

[87] S Watts and C Penk, 'Supercharging Residential Solar Power Generation' (ministers' press release, 13 June 2025) https://www.beehive.govt.nz/release/supercharging-residential-solar-power-generation accessed 31 October 2025.

[88] Compare Australia's Integrated System Plan, 2024, discussed by Godden in this volume.

activity shows what a determined regulator, with government backing, can do. It also shows the benefits that a collaborative policy approach like MDAG can bring. Whether it will be enough for serious progress in demand-side flexibility and other forms of DER is still unclear; there is much less in the nature of targets, planning, and incentives than we find in many of the other jurisdictions discussed in this volume.

The second main insight, and one that may be a general feature internationally, is that decentralization is complicated; it is technically complex, it involves many parties, many more than wholesale market reform or transmission pricing, for example, and all of them defending their vested interests. Why change sometimes struggles to emerge from niche innovation into the broader socio-technical mainstream is a challenging question,[89] one that is only partly susceptible to legal analysis. Change is made especially complicated in New Zealand by the separate ownership of meters and the smallness of many distribution companies, but complexity is also visible in the number of regulatory changes that are needed; even for distribution companies alone regulation is required for connection procedures and prices for both load and generation, large versus small load and generation, network information, network use, time-of-use of the network, time-of-use in retail plans. We see that distribution networks are vital to DER and decentralization, and until they attracted significant regulatory attention DER was not advancing. And at the wholesale and transmission levels there need to be rules for aggregators, rewarding demand flexibility, and hedging products. Rules need to be changed for voltage tolerances, direct-current specifications, instantaneous reserve, and batteries. These rules must allow a complex electricity system to function safely, with supply matching demand minute by minute and second by second. The regulation is therefore granular in character, and although it needs legal and political weight behind it, with broad statements of principle, it also needs a myriad of small but essential regulatory changes.

The third insight concerns sustainability. Although sustainability and decentralization are not the same thing, there are significant sustainability advantages in the decentralization that distributed energy resources represent. One is that the growth of the energy system can encompass rapid electrification, while (in the New Zealand case at least) relying entirely on renewable energy sources, thereby reducing or avoiding greenhouse gas emissions, other forms of pollution, and environmental degradation in the extraction of fossil fuels; and that, of course, is sustainability in an improvement in the ability of future generations to meet their own needs. Renewables can be promoted with policy tools like government subsidies, but even if such policies are little used renewables can also be promoted by firm government support for the many granular regulatory changes needed for DER. The other sustainability benefit that we can look for in advances in DER is in meeting the needs of the present, specifically in improving access to energy at affordable prices. DER can decentralize, increase flexibility, and weaken monopoly, putting downward pressure on prices and allowing customers to take an active role in providing for their energy needs.

[89] These concepts are rooted in the multilevel perspective of analysis of socio-technical transitions: FW Geels and others, 'The Socio-Technical Dynamics of Low-Carbon Transitions' (2017) 1 Joule 463–79.

9

Decentralizing China's Energy Future

*Hao Zhang**

I. Introduction

China's power sector is widely recognized for its carbon intensity, and there is a strong consensus among policymakers and commentators that decarbonizing this sector is essential for achieving the country's mid- and long-term climate goals, namely reaching peak carbon emissions by 2030 and achieving carbon neutrality by 2060.[1] Supported by robust policy measures, such as the enactment of the Renewable Energy Law at the national level and various supporting mechanisms at the local level, China has seen remarkable growth in renewable energy deployment over the past decade. By the end of 2023, renewable energy accounted for over 50 per cent of the country's total power generation capacity, marking a historic milestone as it surpassed thermal power in installed capacity.[2] This reflects China's significant progress in transitioning towards cleaner and more sustainable energy sources.

However, the majority of renewable energy capacity additions have come from centralized generation, particularly large-scale wind and solar installations in northern and northwestern China. These projects are often located far from load centres and points of consumption, necessitating long-distance electricity transmission.[3] This not only increases grid construction costs but also poses challenges to the security and reliability of the power system. Furthermore, China's slow progress in power market reform, coupled with a decentralized electricity governance structure where provinces operate as individual units, has exacerbated the difficulties associated with long-distance and cross-provincial renewable energy transmission.[4]

In contrast, distributed energy resources (DERs) have received less attention in the central government's policy agenda and thus represent a smaller share of China's overall

* The work described in this chapter was fully supported by a grant from the Research Grants Council of the Hong Kong SAR (24602421).

[1] S Mallapaty, 'How China could be Carbon Neutral by Mid-century' (2020) 586 Nature 482; M Meidan, 'Unpacking China's 2060 Carbon Neutrality Pledge' (The Oxford Institute of Energy Studies, December 2020) <https://www.oxfordenergy.org/wpcms/wp-content/uploads/2020/12/Unpacking-Chinas-carbon-neutrality-pledge.pdf> accessed 30 July 2025; National Development and Reform Commission (NDRC), '14th Five-year Plan for Renewable Energy Development' (June 2022) <https://www.ndrc.gov.cn/xwdt/tzgg/202206/P020220602315650388122.pdf> accessed 30 July 2025 (in Chinese).

[2] National Energy Administration (NEA), 'The Share of Total Installed Capacity for Renewable Energy Power Generation in China has Exceeded 50%' (22 December 2023) <https://www.gov.cn/lianbo/bumen/202312/content_6921782.htm> accessed 30 July 2025 (in Chinese).

[3] Energy Foundation, 'Breaking the Bottleneck of Consumption: The Path to Sustainable Development of Distributed Photovoltaic Power in China' (July 2024) <https://www.efchina.org/Blog-zh/blog-20240718-zh> accessed 30 July 2025 (in Chinese).

[4] S Zhang and others, 'To What extent will China's Ongoing Electricity Market Reforms Assist the Integration of Renewable Energy?' (2018) 114 Energy Policy 165; H Zhang, 'Energy Market Reform in China in the Context of Carbon Neutrality: Regulation of Network-dependent Energy Industries' (2023) 21(2) Oil, Gas & Energy Law 1.

Hao Zhang, *Decentralizing China's Energy Future* In: *Sustainable Distributed Energy Resources*. Edited by: Íñigo del Guayo, Damilola S. Olawuyi San, Louis de Fontenelle, Milton Fernando Montoya, and José Juan González, Oxford University Press.
 DOI: 10.1093/9780198991946.003.0011

generation mix.[5] Since 2013, however, policy support for renewable DERs has begun to take shape, notably through the promotion of distributed solar photovoltaic (PV) during the initial phase.[6] Among distributed renewable technologies, distributed solar PV projects are preferred by policymakers and stakeholders due to their flexibility in site selection, relatively low investment requirements, and short construction cycles.[7] To promote DER development, the Chinese central government has established a nascent regulatory framework and implemented pilot programs to facilitate project proliferation. With vast development potential and strong production capacity, China's distributed PV industry has experienced rapid growth in recent years. By the end of 2023, PV installations accounted for 21 per cent of the country's total power generation capacity and contributed over 8 per cent of total electricity consumption.[8] Within this, distributed PV systems, particularly those with capacities at the megawatt level or below, have seen especially rapid growth. In 2023, distributed PV accounted for 42 per cent of total PV installed capacity and 9 per cent of total power generation capacity. In 2024, China added 278,000 megawatts of new PV capacity, a year-on-year increase of 28 per cent, with distributed PV making up approximately 42.5 per cent of this growth.[9]

Geographically, distributed PV development is concentrated in central and eastern provinces, where electricity demand is high. The presence of well-developed industrial and commercial sectors, coupled with abundant rooftop resources, has driven rapid growth in these regions. Provinces such as Shandong, Henan, Zhejiang, Jiangsu, Hebei, and Anhui account for over 90 per cent of the national total installed capacity, with distributed PV representing more than 50 per cent of total PV capacity in these areas.[10] This has effectively addressed local power supply shortages. Alongside the rapid deployment of DERs in the PV sector, the central government has initiated further reforms to establish a regulatory framework for DERs and launched pilot programs, such as the Solar Energy for Poverty Alleviation Program (SEPAP),[11] and experiments with direct transactions between distributed generators and nearby users since 2017 (commonly known as 'selling electricity over the fence'),[12] to promote DERs.

Despite the explosive growth of distributed PV, integrating DERs into the distribution network has presented significant social justice and network regulation challenges. These include insufficient distribution grid capacity, which has led to restrictions on new distributed PV connections in an increasing number of areas, and a lack of market mechanisms for

[5] Energy Foundation (n 3).

[6] NEA, 'Interim Measures for the Management of Distributed Photovoltaic Power Generation Projects' (18 November 2013) (in Chinese).

[7] M Li and J Li, 'Development Status and Prospects of Distributed Photovoltaic Power Generation in China Under the "Dual Carbon" Goals' (2023) 5 Solar Energy 5 (in Chinese).

[8] Energy Foundation (n 3).

[9] NEA, '2024 Renewable Energy Grid Integration Operation Status' (China Energy News Network, administered by NEA, January 2025) <https://www.cpnn.com.cn/news/nytt/202501/t20250127_1769671.html> accessed 30 July 2025 (in Chinese).

[10] Z Pei and others, 'Analysis and Recommendations on Distributed Photovoltaic Grid Integration Issues' (2018) 51(10) Electric Power 80 (in Chinese); J Xu and Q Cao, 'Current Status and Application Research of Distributed Photovoltaic Grid-Connected Power Generation in China' (2020) 6 Energy and Energy Conservation 45 (in Chinese).

[11] NEA, 'Implementation Opinions on Accelerating Energy Development and Construction in Poverty-Stricken Areas to Promote Poverty Alleviation' (24 December 2015) (in Chinese).

[12] NDRC and NEA, 'Notice on Launching Pilot Programs for Market-Based Trading of Distributed Generation' (31 October 2017) (in Chinese).

local DER consumption. This chapter examines the key legal arrangements under China's nascent DERs regulatory framework, providing insights into the challenges hindering their mass adoption. Section II reviews literature on the rise of DERs and needed regulatory shifts, contrasting China's DER advocacy with its traditional centralized power focus. Section III examines China's power sector reforms and emerging DER regulations, analyzing gaps through case studies like SEPAP and 'selling electricity over the fence'. Section IV proposes legal and policy recommendations, drawing on China's experiments and the relevant regulatory development in the EU. Section V concludes with future directions and research gaps.

II. Network Regulation and Sustainable DERs

The global energy transition is driving a fundamental shift from centralized to decentralized energy systems, a transformation that is particularly relevant in the context of energy regulatory governance.[13] Centralized systems, characterized by large-scale power plants and extensive transmission networks, have long dominated electricity generation and distribution. However, the rise of DERs and the growing emphasis on sustainability, energy security, and democratization are compelling energy regulatory frameworks to evolve, particularly for jurisdictions and nations, such as China, that are extensively dependent on energy imports.[14] As discussed in Chapter 1 of this book, this shift is encapsulated in the four key vectors of transformation, namely decarbonization, decentralization, digitalization, and democratization (the 'four Ds'), which collectively underscore the need for regulatory systems to adapt to the complexities and opportunities of decentralized energy systems.[15]

In decentralized systems, energy generation occurs closer to consumption points, often through small-scale, renewable technologies such as solar panels, wind turbines, and microgrids. This shift not only reduces reliance on fossil fuels and minimizes transmission losses but also empowers consumers to become active participants in the energy market as 'prosumers' that actively participate in the energy market by consuming, generating, and often selling excess electricity back to the grid, blurring the traditional lines between producers and consumers.[16] However, this transition requires regulatory frameworks to move beyond traditional top-down approaches and embrace more adaptive, inclusive, and innovative governance models.[17] The 'four Ds' provide a conceptual framework for

[13] MA Heldeweg and S Saintier, 'Renewable Energy Communities as "Socio-Legal Institutions": A Normative Frame for Energy Decentralization?' (2020) 119 Renewable and Sustainable Energy Reviews 109518; M Markantoni, 'Low Carbon Governance: Mobilizing Community Energy through Top-down Support?' (2016) 26(3) Environmental Policy & Governance 155.

[14] R Leal-Arcas and others, 'Prosumers: New Actors in EU Energy Security' in F Amtenbrink, D Prévost, and RA Wessel (eds), *Netherlands Yearbook of International Law 2017* (Springer 2018) 139–72; K Chmutina and CI Goodier, 'Alternative Future Energy Pathways: Assessment of the Potential of Innovative Decentralized Energy Systems in the UK' (2014) 66 Energy Policy 62.

[15] Í del Guayo Castiella, 'Introduction: Distributed Energy Resources as the New Electricity Paradigm: Legal Context and Principles', Chapter 1 of this book.

[16] D Aguilar Abaunza, *The Law for Energy Prosumers. The Case of the Netherlands, New Zealand and Colombia* (Springer 2022).

[17] ibid; Leal-Arcas (n 14); del Guayo Castiella (n 15); SB Jacobs, 'The Energy Prosumer' (2017) 43 Ecology Law Quarterly 519; Y Parag and BK Sovacool, 'Electricity Market Design for the Prosumer Era' (2016) 1 Nature Energy 1.

understanding the multifaceted nature of this transition, highlighting the interconnected challenges and opportunities that DERs systems present.

In China, the transition to DERs is driven by two key factors: market reforms aimed at introducing competition to certain segments of the electricity system for better system flexibility and sustainability and the rapid expansion of renewable energy technologies for decarbonization purposes.[18] These developments offer significant opportunities to address energy poverty, reduce environmental impact and emissions, and enhance energy security. All these aspects have been emphasized by the Chinese central government's energy planning and socioeconomic policies, particularly those concerning DERs.[19] China's power market reforms and technological advancements align with the principles of the 'four Ds' in several ways. The large-scale deployment of renewable energy, particularly solar and wind power, along with regulatory measures to improve energy efficiency and reduce emissions, underscores efforts towards decarbonization.[20] Decentralization is evident in policies such as SEPAP and 'selling electricity over the fence', which promote DERs and community-driven initiatives.[21] Digitalization is advancing through smart grid technologies, AI-driven grid management, and blockchain-enabled peer-to-peer energy trading, improving efficiency and integration.[22] Meanwhile, democratization is emerging as households and businesses become prosumers, both generating and consuming energy, while public participation in energy policy consultations grows.[23] Despite these advancements, challenges such as regulatory barriers and grid integration issues persist, underscoring the need for continued reforms to fully realize the 'four Ds' in China's energy transition.

However, the progress towards DERs in China has been slow and fragmented, reflecting the challenges of aligning regulatory frameworks with the principles of the 'four Ds'. China's centralized energy governance model, historically focused on large-scale infrastructure and state-controlled utilities, has struggled to accommodate the decentralized nature of DERs, as commentators have observed in the implementation of the SEPAP initiative.[24] Market reforms launched in 2015 sought to diversify the market by allowing distribution system operators other than the two major grid companies—State Grid and China Southern Power Grid and introducing competition in retail markets.[25] While these reforms represent a step towards market liberalization, which is conducive to accommodate renewable energy integration and DERs development, their impact has been limited by regulatory barriers and the dominance of state-owned enterprises that have historically owned transmission

[18] Pei (n 10); Xu and Cao (n 10); H Zhang, 'The Essential Elements of the Regulatory Frameworks Governing Renewable Energy Communities: A Comparative Analysis' in MM Sokołowski and A Visvizi (eds), *Routledge Handbook of Energy Communities and Smart Cities* (Routledge 2023) 172–84.

[19] Eg NDRC and other agencies, 'Opinions on the Implementation of PV-based Power Generation for Poverty Relief' (23 March 2016) (in Chinese); NEA, 'Interim Measures for the Management of Distributed Solar PV Power Generation Projects' (18 November 2013) (in Chinese).

[20] Mallapaty (n 1).

[21] S Geall and others, 'Solar Energy for Poverty Alleviation in China: State Ambitions, Bureaucratic Interests, and Local Realities' (2018) 41 Energy Research & Social Science 238; H You, 'Economic Analysis of Distributed Photovoltaic Projects Considering the "Selling Electricity Over the Fence" Model and Carbon Emission Trading' (2022) 38(4) Energy Research and Information 203 (in Chinese).

[22] Yuquan Liu, 'Empowering the Development of Digital Power Grids with AI' (People's Daily, February 2025) <http://paper.people.com.cn/rmrbhwb/pc/content/202502/10/content_30055608.html> accessed 30 July 2025 (in Chinese).

[23] Geall (n 21); You (n 21).

[24] Geall (n 21); Zhang (n 18).

[25] H Guo and others, 'Power Market Reform in China: Motivations, Progress, and Recommendations' (2020) 145 Energy Policy 111717.

and distribution network assets.[26] For instance, the lack of effective market mechanisms to support high levels of renewable energy integration into the transmission system led to significant renewable energy curtailment in the mid-2010s.[27] Similarly, key central government initiatives to promote DERs, such as the SEPAP and 'selling electricity over the fence', have advanced decentralized renewable energy deployment mainly on the generation side, but remain constrained by inadequate regulatory support and grid integration challenges, which will be discussed in detail in section III below.

The Chinese context illustrates the critical role of regulatory governance in enabling the transition to include DERs in the power sector. However, the existing legal scholarship has not studied DERs development in China through the legal and regulatory perspectives.[28] This chapter aims to fill this gap by examining the regulatory framework governing DERs in China. It argues that achieving the full potential of DERs in the country requires addressing the regulatory gaps identified in this chapter, notably distribution grid system upgrade and expansion, system flexibility and integration, fair and non-discriminatory access, and the lack of supportive frameworks for bottom-up energy initiatives. By embracing these principles, China can unlock the transformative potential of DERs, contributing to the national policy imperative to achieve the carbon peaking and carbon neutrality objectives.

III. China's Regulatory Framework Governing DERs and Experiments

Over the past decade, China's renewable energy sector has experienced unprecedented growth, with the country investing more in solar and wind generation capacity than any other nation globally.[29] This rapid expansion has been largely driven by the development of large-scale solar and wind farms, which have attracted significant investment due to their ability to leverage financial incentives under China's Renewable Energy Law, ensuring more stable long-term returns.[30] However, despite the Chinese government's policy support for DERs, its development has consistently fallen short of policy targets and expectations. For instance, in 2014, the National Energy Administration (NEA) set a target of eight GW for solar DERs, but actual installations reached only 2.25 GW, achieving just 28 per cent of the goal.[31] This trend persisted in 2015, with reports indicating that approximately 90 per cent of newly installed capacity came from large-scale, centralized projects.[32] Recognizing that the existing regulatory framework disproportionately favoured centralized generation, energy regulators began revising both substantive and procedural aspects of the framework

[26] ibid.

[27] H Zhang, 'Prioritizing Access of Renewable Energy to the Grid in China: Regulatory Mechanisms and Challenges for Implementation' (2019) 3(2) Chinese Journal of Environmental Law 167–202

[28] The existing scholarship mainly focuses on market development and reform in the context of DERs development in China. Eg, L Pingkuo and T Zhongfu, 'How to Develop Distributed Generation in China: In the Context of the Reformation of Electric Power System' (2016) 66 Renewable and Sustainable Energy Reviews 10; Y Li and others, 'A Review of Photovoltaic Poverty Alleviation Projects in China: Current Status, Challenge and Policy Recommendations' (2018) 94 Renewable and Sustainable Energy Reviews 214; Geall (n 21).

[29] Zhang (n 27).

[30] ibid.

[31] NDRC Energy Research Institute (ERI) and Renewable Energy Development Center (REDC), 'Study on the Policy Roadmap of Distributed PV Power Generation in China' (The Energy Foundation Research Report 2015) <https://www.efchina.org/Reports-zh/report-cre-20150601-2-zh> accessed 30 July 2025 (in Chinese).

[32] NEA, 'Opinions on Accelerating the Construction of Energy Projects in Poverty Areas to Support the Battle of Eradicating Poverty' (24 December 2015) (in Chinese).

in 2013 to create a more enabling environment for DERs. In this context, renewable energy DERs, especially solar PV generation, have been given particular policy focus, offering a unique opportunity to align the goals of energy decarbonization, sustainability requirements, and social goals, notably poverty alleviation. By promoting DERs, particularly through solar PV projects, Chinese local residents can become 'prosumers', thereby supporting strategic energy goals while delivering economic benefits to communities.[33] This approach not only advances renewable energy development but also contributes to broader social and economic objectives. That said, the development of DERs and the regulatory barriers hindering its further expansion will need to be understood in the context of China's power market structure and regulation.

A. Power Market Structure and Regulation

China's power market has gone through several key stages of reform to liberalize the market for better system efficiency and flexibility. The 2002 reforms dismantled China's vertically integrated power sector, breaking up the State Power Corporation to foster competition in power generation.[34] This round of reform diversified investors and decentralized regulatory governance, granting provincial governments greater authority over sector development, including project approvals and system operations.[35] Between 2002 and 2015, China's power sector, particularly coal power, expanded significantly. However, economic slowdowns led to declining operating hours for coal plants, exacerbating overcapacity issues. Despite this, post-Covid recovery plans have driven renewed coal power investments, creating tensions with China's renewable energy targets.[36] China remains a global leader in renewable energy investment, particularly in solar and wind. However, capacity growth has slowed compared to previous years as policy priorities have shifted from installation to integration and consumption.[37] The 2002 reforms also fell short in introducing key competitive mechanisms, such as dispatch based on economic considerations rather than administrative orders, and the separation of distribution from transmission. China's major grid companies, ie State Grid and China Southern Power Grid, continue to control both transmission and distribution, with electricity prices regulated under a benchmark system. While this framework provides investor stability, it lacks the flexibility needed for high renewable integration. This has created a 'renewable paradox', where renewables' zero marginal costs depress wholesale prices while simultaneously increasing overall system costs.[38]

[33] Geall (n 21); Li (n 21).

[34] Hongye Guo (n 25).

[35] ibid.

[36] Greenpeace, 'China has already Approved more New Coal in 2023 than it did in all of 2021' (24 April 2023) <https://www.greenpeace.org/eastasia/press/7939/china-has-already-approved-more-new-coal-in-2023-than-it-did-in-all-of-2021-greenpeace/> accessed 30 July 2025.

[37] C Song and others, 'Unveiling Energy Transition Strategy: A Deep Dive into China's Ambitious Renewable Energy Policy and its Impact on Carbon Emission Dynamics' (2024) 475 Journal of Cleaner Production 143684; H Zhang, 'Regulatory Pathways of Renewable Energy Integration in the Guangdong-Hong Kong-Macao Greater Bay Area' (2024) 24(2) The China Review 97.

[38] S Feng, 'As China's Energy Mix Shifts to Renewables, How can Power Markets Evolve?' (China Dialogue, April 2021) <https://chinadialogue.net/en/energy/power-markets-chinas-energy-mix-shifts-to-renewables/> accessed 30 July 2025.

Regulated retail prices further discourage grid companies from integrating more renewable energy.[39]

In 2015, China launched new reforms aimed at enhancing system flexibility, focusing on deregulating electricity prices for industrial users and liberalizing the retail market. Industrial consumers can now negotiate power purchase agreements directly with generators or through retailers, bypassing benchmark pricing.[40] Grid companies are shifting from revenue collectors to service providers, with transmission and distribution tariffs regulated under a 'cost-plus-reasonable-profit' model.[41] This hybrid system blends regulated market with market liberalization, as benchmark prices still apply to non-participating industrial users, while residential, agricultural, and public sectors remain under regulated pricing.[42] The reform seeks to balance market efficiency with affordability and system stability.

B. Policy and Regulatory Framework Governing DERs

China's policy and regulatory framework for DERs places particular emphasis on solar DERs, given the country's abundant solar resources, advanced technology, and favorable application conditions (eg large rooftop spaces in industrial and commercial areas). The regulatory framework for solar DERs has evolved through four distinct stages: (1) an initial phase focused on securing investment returns and streamlining project approvals; (2) a second stage emphasizing technical standardization; (3) a market-oriented phase aimed at phasing out subsidies; and (4) the current stage, which seeks to integrate solar DERs with other system flexibility mechanisms such as energy storage. Together, these policy measures have established a regulatory environment conducive to the growth of solar DERs in China, despite the challenges discussed in section III(C). This section provides an overview of key policy instruments before examining experimental approaches to solar DER deployment.

Solar DERs are regarded as a key strategy to address weak international demand for solar PV and China's overcapacity in solar PV manufacturing.[43] The State Council's overarching policy on solar PV development outlines several measures, with particular emphasis on solar DERs under the guiding principle of 'optimal layout, local grid integration, and on-site consumption'.[44] This principle highlights the need for efficient and strategic placement of solar DERs, integration with local distribution grids, and localized use of generated solar energy to minimize reliance on long-distance transmission. A significant policy shift introduced by the *Opinions* is the exemption of solar DERs from power generation business licensing, granting priority support for both large-scale projects in industrial and commercial areas and small-scale projects in residential, rural, and remote areas.[45]

Following the formal endorsement of solar DERs as a complement to centralized solar PV generation, the NEA issued the *Interim Measures for the Management of Distributed Solar PV Power Generation Projects*.[46] The *Interim Measures* established fundamental

39 ibid.
40 Hongye Guo (n 25).
41 Zhang (n 4).
42 Zhang (n 4); Hongye Guo (n 25).
43 State Council's Several Opinions on Promoting the Healthy Development of the Photovoltaic Industry, 4 July 2013 (in Chinese).
44 ibid part 3(2).
45 ibid part 3(1).
46 NEA (n 19).

principles that have since guided subsequent implementation policies. Core elements of the fundamental principles include streamlining project filing procedures,[47] clarifying grid enterprises' responsibilities in connecting distributed solar PV projects,[48] providing economic incentives,[49] and ensuring operational safety through further technical standardization.[50] These principles collectively form a comprehensive policy and regulatory framework supporting solar DER development and laying the groundwork for future policy evolution.

A key policy instrument during the initial stage was the introduction of subsidies aligning economic incentives for solar DERs with funding sources for centralized PV projects. The NDRC later specified a general subsidy level of 0.42 CNY/kWh for electricity generated by solar DERs, funded by the Renewable Energy Development Fund, for up to twenty years.[51] For surplus electricity fed into the grid, purchase prices were benchmarked against coal power tariffs at the locality.[52] Estimates suggest that this level of support enhanced the economic viability of both small-scale and large-scale solar DER projects in over twenty-five and fifteen provinces, respectively, provided that at least 80 per cent of the electricity generated was used for self-consumption.[53] However, these subsidies were insufficient to drive widespread adoption in residential, public, and agricultural sectors, which explained the wider adoption of solar DERs in industrial and commercial sectors than others.[54] Beyond subsidies, the NDRC also restructured the payment mechanism, requiring grid enterprises to pay subsidies and purchase prices directly to solar DER operators. This ensured timely payments, attracting greater private investment into the sector.[55]

The second stage of policymaking emphasizing technical standardization is driven collaboratively by the NEA, the Ministry of Industry and Information Technology (MIIT), and the Certification and Accreditation Administration of China. In order to promote the more advanced technology products and to accelerate the phase-out of outdated ones, the three central government agencies have initiated some key regulatory measures to promote the application of advanced PV technology products and facilitate industrial upgrading. The key measures include: (1) raising market access standard for PV products to prioritize products that meet the more advanced technical standards; (2) leveraging fiscal funds and government procurement to support technological progress; (3) strengthening testing, certification, and quality management to ensure the reliability and quality of PV products and project engineering; and (4) establishing technology monitoring and supervision mechanisms to guarantee product quality through robust oversight. The technical standardization is explicitly linked to economic incentives, which requires that grid-connected PV power generation projects and off-grid independent PV power generation projects receiving government subsidies must use PV products that meet the relevant standards and certified by approved certification bodies.[56] The standardization process has laid the foundation to

[47] ibid ch 3.
[48] ibid ch 5.
[49] ibid arts 25–27.
[50] ibid arts 17 and 18.
[51] NDRC, 'Notice on Utilizing Price Leverage to Promote the Healthy Development of the Photovoltaic Industry', 26 August 2013, parts 2(1) and 4(2) (in Chinese).
[52] NDRC Notice (n 51).
[53] NDRC ERI and REDC (n 31).
[54] ibid.
[55] Ministry of Finance, 'Notice on the Implementation of Electricity-Based Subsidy Policies for Distributed Solar PV Power Generation and Related Issues' (24 July 2013) part 2 (in Chinese).
[56] NEA and other agencies, 'Opinions on Promoting the Application of Advanced Photovoltaic Technology Products and Industrial Upgrading' (1 June 2015) (in Chinese).

enable the proliferation of solar DER projects through the SEPAP initiative, allowing multiple policy objectives, including poverty alleviation, access to electricity and development of DERs, to be achieved concurrently.[57]

The market-oriented phase of policymaking aimed to phase out subsidies, primarily because solar PV systems in many regions of China had reached grid parity, making them competitive with coal power. In 2017, the NDRC and the NEA jointly issued two regulatory documents outlining the framework and guidelines for implementing a mechanism called 'selling electricity over the fence' through distributed generation, particularly solar PV, via pilot projects.[58] The NDRC and NEA define this mechanism as enabling DER units, including individuals, to trade electricity with nearby users within the distribution network. Grid enterprises, including those investing in incremental distribution networks with social capital, are responsible for transmitting electricity from distributed generation sources. They must co-operate with relevant power trading institutions to facilitate market-based trading and charge a 'grid access fee' according to government-approved standards.[59] Given the regional differences in the progress of electricity market-based trading, regulators allow the adoption of one or more of the three following models.

The first model is the direct electricity trading between DER projects and electricity users, with payment of the 'grid access fee' to the grid enterprise. The scope of trading is limited, generally confined to the supply range of the upstream transformer at the connection point. The second model allows the DER units to entrust grid enterprises to sell electricity on their behalf. The grid enterprise sells electricity at the comprehensive electricity selling price. After deducting the 'grid access fee' (which includes network loss electricity), the remaining revenue from the sale is transferred to the DER units. The third model enables the grid enterprises to purchase electricity at the national benchmark on-grid prices for various types of DER generation. However, the national subsidy per kilowatt-hour for electricity to grid enterprises is reduced by the corresponding transmission and distribution costs for users at the highest voltage level within the distribution network area.[60]

The NDRC and NEA required each province to select one or two localities to experiment with the 'selling electricity over the fence' model. Provincial energy regulators, in collaboration with provincial pricing management authorities, were tasked with establishing and determining trading rules, tariffs, and prices (particularly 'grid access fees'), and settlement methods.[61] Local grid enterprises are responsible for connecting DERs to local distribution networks and integrating them based on grid capacity. To incentivize DER development, energy regulators explicitly required grid enterprises to provide services for the integration and trading of DER electricity, allowing them to charge only 'grid access fees'.[62] These experiments in market-based trading for DERs open the door for decentralized, digitalized, and democratized system management, offering better system flexibility and greater renewable energy deployment. This is considered a significant milestone in breaking down the country's centralized electricity management system.[63]

[57] Zhang (n 18); Geall (n 21).

[58] NDRC, 'Notice on Launching the Pilot Program for Market-Oriented Trading of Distributed Generation' (31 October 2017) (in Chinese); NDRC, 'Supplementary Notice on Launching the Pilot Program for Market-Oriented Trading of Distributed Generation' (28 December 2017) (in Chinese).

[59] NDRC Notice, 31 October 2017 (n 58) part 2.

[60] ibid.

[61] NDRC Notice, 28 December 2017 (n 58) part 1.

[62] ibid.

[63] Xu and Cao (n 10); You (n 21).

Building on the experiences and lessons learned from the initial market-oriented phase, the current stage of policymaking aims to further liberalize market trading by fully removing pricing controls on renewables, including DERs, and reinvigorating the 'selling electricity over the fence' initiative.[64] This new phase of market liberalization takes a more comprehensive approach, requiring improvements to grid infrastructure, especially energy storage and digitalization (eg smart grids), to enable a higher level of integration of electricity from DERs.[65] Recognizing that China's power market is still in its early stages and that the high volatility of spot market prices could significantly impact the commercial viability of solar DERs, the NDRC and NEA encourage local governments to adopt transitional measures, such as using price floors for solar DERs in rural areas to facilitate their growth.[66] For commercial and industrial solar DER projects, fluctuating electricity prices, driven by supply and demand, expose these projects to higher financial risks. In response, local governments are encouraged to explore various financial tools to help commercial and industrial solar DER projects hedge against these risks.[67]

This more holistic approach to market-oriented reforms and infrastructure improvements signals a shift in China's solar DER strategy from being installation-driven to integration-oriented. The incentives for capacity expansion have also evolved from relying solely on subsidies to harnessing greater market forces. Given that the current policy and regulatory framework is largely built on the previous stage of policy experimentation, the following section will discuss and examine the solar DER experiments through the lens of SEPAP and 'selling electricity over the fence', identifying gaps in existing laws and policies. This will serve as the foundation for exploring legal approaches to address these challenges in section IV.

C. Experiments with Solar DERs and Gaps in Existing Laws/Policies

The experiments with solar DERs through the initiatives of SEPAP and 'selling electricity over the fence' provide insights as to how the market structure and regulatory arrangements governing the distribution market have constrained the proliferation of solar DERs. To recall, the SEPAP experiment, initially initiated in 2014, aimed to systematically deploy distributed solar PV in rural areas to eradicate poverty, including energy poverty. The SEPAP projects are structured in a decentralized approached and managed by local communities, with institutional and financial support provided by central, provincial, and local governments. The initial experiments were conducted in thirty counties across six provinces where local income is far below national average and the solar resources are abundant.[68] In 2015, the SEPAP experiments received the highest political endorsement and policy support in China after President Xi Jinping announced the central government's commitment

[64] NDRC and NEA, 'Notice on Deepening Market-Oriented Reforms for New Energy On-Grid Electricity Prices to Promote High-Quality Development of New Energy' (27 January 2025) (in Chinese).
[65] NDRC and NEA, 'Guiding Opinions on Promoting the Integration of Power Source-Grid-Load-Storage and the Development of Multi-Energy Complementarity' (25 February 2021) (in Chinese).
[66] ibid part 3.
[67] ibid.
[68] Geall (n 21); C Liao, 'Targeted Poverty Alleviation through Photovoltaic-based Intervention: Rhetoric and Reality in Qinghai, China' (2021) 137 World Development 105117.

to completely eradicating poverty by 2020.[69] The high political endorsement has been translated into financial support for SEPAP projects. It is reported that around 70 per cent of projects were supported by government subsidies, while other projects were sponsored by policy banks and corporate investors.[70]

The experiment of 'selling electricity over the fence' represents a critical step in advancing China's power market reform. This initiative aims to reduce transaction costs and foster the growth of the renewable energy sector, particularly the solar PV industry, which benefits from China's extensive manufacturing capacity for solar PV panels and related components.[71] By enabling direct electricity transactions between wind and solar DERs and nearby users, this experiment seeks to decentralize power supply and enhance grid efficiency. Initially piloted in twenty-six regions, the program has faced significant challenges, with very few of pilots achieving success.[72] The experiment was designed to address the inefficiencies of traditional distribution grids, which are built and operated on a one-way power supply model. In this conventional setup, electricity flows from higher-level transmission grids to end-users, with no power generation occurring within the distribution grid itself. However, the integration of solar DERs into the distribution grid fundamentally disrupts this model, necessitating a two-way flow of electricity. This shift requires substantial upgrades to grid infrastructure and operational practices, leading to increased costs for investments and maintenance.[73] As a result, the transition to a more flexible and decentralized grid system has proven to be a complex and costly endeavour.

While the 'selling electricity over the fence' initiative focuses on improving system flexibility and diversifying power supply, it shares a common dependency with the SEPAP initiative: both rely heavily on the regulatory framework governing China's distribution markets. To recall, despite incremental reforms aimed at liberalizing the power sector, the transmission and distribution segments remain largely monopolized by state-owned grid enterprises. Since the 2015 power market reforms, progress in separating transmission and distribution has been slow, leaving the distribution segment underdeveloped and lacking the regulatory support needed to enhance system flexibility. A key issue is the absence of clear trading rules and sufficient grid capacity to accommodate decentralized energy resources. This regulatory gap has created significant barriers to integrating solar DERs into the distribution network. In practice, solar DERs are effectively barred from connecting to the grid, undermining the potential of initiatives like 'selling electricity over the fence' and SEPAP.

One of the most pressing gaps in China's regulatory framework supporting DERs is the lack of integration between sustainability goals and system flexibility. While large-scale renewable energy projects receive considerable policy attention, smaller-scale, decentralized systems remain underprioritized. For instance, the 2015 electricity market reforms opened distribution markets to new stakeholders but failed to address the technical and operational challenges of decentralized grids. Issues such as grid stability,

[69] Decision of the Central Committee of the Communist Party of China and the State Council on Winning the Battle Against Poverty, 29 November 2015 (in Chinese).
[70] Liao (n 68).
[71] NDRC Notice, 31 October 2017 (n 58).
[72] NDRC and NEA, 'Notice on Announcing the First Batch of Grid-Parity Wind Power and Photovoltaic Power Projects in 2019' (20 May 2019) (in Chinese); Energy Foundation (n 3).
[73] Energy Foundation (n 3).

distribution tariffs that incentivize DER integration, and local market participation remain unresolved. Moreover, initiatives like SEPAP and 'selling electricity over the fence' have not been supported by broader regulatory reforms. This has left grid enterprises with disproportionate decision-making power, as they lack incentives to invest in infrastructure upgrades or develop transaction mechanisms for DERs. The integration of solar DERs requires significant infrastructure expansion to address the temporal and spatial imbalances in power supply. For example, the mismatch between solar generation peaks (midday) and demand peaks (evening), as well as the disparity between abundant rooftop solar generation in rural areas and limited local demand, necessitates a more flexible and capacious grid.[74]

However, the current 'grid access fees' charged by grid enterprises for DER integration do not adequately address the cost allocation for infrastructure and operational upgrades. In practice, grid companies are reluctant to invest in the necessary upgrades, as integrating solar DERs would incur additional costs and reduce their electricity sales revenue. Under the existing integrated transmission, distribution, and retail management system, any move towards self-consumption or local transactions of solar DERs directly impacts grid companies' profitability. Grid enterprises' concerns over revenue generation further complicate the development of market trading mechanisms for DERs. While these companies are legally and politically obligated to support renewable energy integration, their primary responsibility remains ensuring safe and reliable power supply.[75] This obligation often takes precedence over renewable energy goals, particularly given the intermittent nature of solar and wind generation.

Local governments also face a delicate balancing act. On one hand, they aim to promote large-scale DER development and reduce electricity costs to attract new industries and stimulate economic growth. On the other hand, they must ensure energy supply security, prevent power shortages during extreme conditions, and maintain social stability. These competing priorities create a complex regulatory environment, where energy regulators must navigate the conflicting interests of grid enterprises, local governments, and DER project developers.

The success of initiatives like SEPAP and 'selling electricity over the fence' hinges on comprehensive regulatory reforms that address the technical, operational, and financial challenges of integrating solar DERs into China's distribution grids. Without clear policies to incentivize infrastructure investment, establish fair cost-sharing mechanisms, and enhance grid flexibility, the potential of decentralized renewable energy will remain untapped. As China continues its transition towards a more sustainable energy system, bridging these regulatory gaps will be essential to achieving its renewable energy targets and ensuring a reliable, flexible, and inclusive power market. The below section discusses the legal approaches for addressing the challenges by focusing on the lessons learned from the two experiments and other jurisdictions.

[74] Pei (n 10).
[75] Zhang (n 4).

IV. Legal Approaches for Addressing the Challenges: Lessons Learned

To effectively address the regulatory gaps identified earlier, China must adopt a comprehensive policy approach that integrates lessons from both domestic pilot programs and international experiences, especially the regulatory development in the European Union (EU). Regulatory approaches from more liberalized power markets demonstrate that streamlined permitting processes, simplified licensing requirements, fair grid access rules, and inclusive participation mechanisms are fundamental to fostering decentralized energy systems.[76] By synthesizing insights from these models with findings from China's own experiments, policymakers can develop a regulatory framework that enhances system flexibility, encourages local stakeholder engagement, and ensures the reliable integration of DERs. This section examines key legal and regulatory lessons, proposing policy recommendations based on China's own circumstances and international experiences to better support DER development in China.

A. The EU's Renewable Energy Directive (RED-II): A Model for Decentralized Energy Integration

The EU's Directive 2018/2001 (RED-II) provides a valuable blueprint for designing regulatory frameworks that promote energy justice and community participation.[77] By requiring member states to establish an 'enabling framework' for renewable energy communities (RECs), the directive exemplifies how policy can facilitate a just and inclusive energy transition.[78] However, its implementation also reveals challenges that Chinese policymakers must consider. Key regulatory lessons from RED-II include four essential aspects. The first aspect focuses on balancing regulatory flexibility with clarity. RED-II adopts a principle-based approach, allowing member states to tailor rules to local conditions while safeguarding core rights for RECs. This flexibility is essential given the diversity of energy systems, but inconsistent transposition into national laws has sometimes created uncertainty.[79] China can draw from this lesson by establishing clear minimum standards for DER integration while permitting regional adaptations. The country's own experiments demonstrate that overly rigid frameworks lead to fragmentation, whereas adaptable yet well-defined regulations encourage broader participation without compromising consistency.

The second aspect highlights the need to reduce administrative and market barriers. The directive emphasizes simplifying permitting, licensing, and grid access for RECs while ensuring their non-discriminatory participation in energy markets.[80] Bureaucratic inefficiencies and unfair market access disproportionately hinder small-scale DER producers. China's pilot programs similarly highlight the importance of streamlined administrative

[76] I Capellán-Péreza and others, 'Renewable Energy Cooperatives as an Instrument towards the Energy Transition in Spain' (2018) 123 Energy Policy 215; Chmutina and Goodier (n 14).

[77] Directive (EU) 2018/2001 of the European Parliament and of the Council of 11 December 2018 on the Promotion of the Use of Energy from Renewable Sources [2018] OJ L 328 (RED-II).

[78] ibid art 22(4).

[79] MM Sokołowski, 'Renewable and Citizen Energy Communities in the European Union: How (not) to Regulate Community Energy in National Laws and Policies' (2020) 38(3) Journal of Energy & Natural Resources Law 289.

[80] RED-II, art 22(2).

procedures. However, even with simplified permitting, DER growth remains constrained if distribution networks lack the capacity or operational protocols to accommodate higher renewable penetration. Thus, regulatory reforms must go beyond paperwork reduction to address physical and market access limitations.

The third aspect emphasizes promoting equitable participation through financial and technical support. RED-II mandates inclusive access for all consumers, including low-income and vulnerable groups, to prevent exclusion from the economic and environmental benefits of decentralized energy. It supports this goal through financing mechanisms, grants, and capacity-building programs that help RECs overcome initial investment hurdles.[81] Two critical insights emerge based on this observation, including dedicated funding and technical assistance; and proactive policies, such as targeted subsidies, awareness campaigns, and skill development initiatives to prevent energy democracy from becoming an exclusive privilege. China's SEPAP initiative illustrates the transformative potential of state-backed financial support in addressing energy poverty, particularly in rural and remote areas. However, SEPAP also reveals limitations: while effective in driving initial installations, it has struggled to ensure the long-term sustainability and economic efficiency of DERs, which depend heavily on fair access rules and seamless integration with the broader energy system.

The fourth aspect aims to ensure seamless integration with existing energy systems. RED-II requires active collaboration between RECs and distribution system operators to harmonize decentralized and centralized grid operations. This underscores the need for regulatory frameworks that facilitate interoperability between traditional and modern energy systems.[82] China faces significant challenges in this regard. As evidenced by its pilot programs, the dominance of grid enterprises, sluggish progress in unbundling transmission and distribution since 2015, and the absence of transparent trading mechanisms for DERs continue to obstruct local energy transactions. Without structural reforms, DERs will remain sidelined in China's energy transition.

B. Policy Recommendations for China's DER Development

While RED-II offers a pioneering model, its effectiveness depends on robust enforcement and continuous adaptation to emerging challenges in a national/jurisdictional context. Commentators argue that reforming the Member States' domestic legal systems to make it compatible with DERs is a daunting task, given the various projects in terms of legal structures, scale, and complexity.[83] For China, the following measures are critical to creating a DER-friendly regulatory environment: first, it is important to establish non-discriminatory grid access rules by setting out clear, minimum technical and operational standards for DER integration. These standards should permit regional variation to reflect local grid conditions, thereby achieving a balance between national regulatory coherence and the flexibility

[81] CE Hoicka and others, 'Implementing a Just Renewable Energy Transition: Policy Advice for Transposing the New European Rules for Renewable Energy Communities' (2021) 156 Energy Policy 112435.

[82] Heldeweg and Saintier (n 13); E Minghui Gui and I MacGill, 'Typology of Future Clean Energy Communities: An Exploratory Structure, Opportunities, and Challenges' (2018) 35 Energy Research & Social Science 94.

[83] Sokołowski (n 79); Gui and MacGill (n 82).

required for implementation across diverse settings. Second, grid infrastructure must be modernized, and incentive structures must be aligned accordingly. Grid enterprises, such as State Grid and China Southern Power Grid, should be legally obligated to upgrade infrastructure to accommodate bidirectional electricity flows. Performance-based incentives, such as those promoting investment in smart grid technologies and energy storage, should be introduced to lower integration costs and improve system flexibility. Third, it is necessary to reform grid access fees by introducing cost-reflective pricing mechanisms. These mechanisms should equitably allocate the costs associated with grid upgrades among DER owners, grid operators, and end-users, thereby avoiding disproportionate financial burdens on any one stakeholder group. Fourth, there should be a shift away from broad capacity-based subsidies in favour of more targeted financial support mechanisms. Funding should be directed towards addressing infrastructure deficiencies in regions with high levels of DER penetration, with financial incentives, such as those linked to grid access fee structures, tied to actual integration outcomes rather than mere increases in installed capacity. Finally, the dual mandates of grid operators must be clarified and compliance strictly enforced. Renewable energy targets should be expanded to explicitly include DERs, with sanctions imposed for failure to meet these targets. Moreover, the dual responsibilities of grid operators, ie to facilitate renewable energy integration while maintaining system reliability, should be clearly articulated to reduce institutional bias against DER deployment.

On a positive note, China's decentralized energy transition is gaining momentum recently through innovative regulatory developments at the provincial level. Recent initiatives to establish zero-carbon industrial parks exemplify how localized policies are translating theoretical recommendations into actionable frameworks, with DERs playing a central role. These pilots demonstrate a growing recognition of DERs as critical tools for meeting both decarbonization goals and corporate demand for green electricity in supply chains. For instance, the *Work Plan for Zero-Carbon Industrial Parks Pilot Program*, adopted by Sichuan Province, embodies several of the key reforms discussed earlier.[84] Firstly, the *Plan* prioritizes on-site and nearby DERs, including solar, wind, and small hydropower, to supply industrial parks with localized clean energy. By enabling direct power purchase agreements between DERs and industrial consumers, the model reduces reliance on centralized grids and lowers transaction costs. Secondly, financial support for infrastructure modernization is explicitly mentioned. Provincial authorities will subsidize grid infrastructure upgrades, but only when investments are directly tied to DER integration (eg enabling bidirectional power flows or energy storage deployment). This performance-linked funding mechanism aligns with the recommendation to shift from blanket subsidies to incentives that reward DER consumption. Finally, the *Plan* focuses on policy synergies with green supply chain demands by leveraging corporate sustainability mandates (eg multinationals' net-zero commitments) to create a market-driven demand for DERs. It is expected that through the pilot programs, industrial tenants in these parks gain preferential access to low-cost, traceable renewable power, helping them comply with both domestic carbon policies and international ESG standards.

[84] Sichuan Provincial Department of Economy and Information Technology and Department of Ecology and Environment, 'Notice on Jointly Issuing the Work Plan for Zero-Carbon Industrial Parks Pilot Program of Sichuan Province' (3 March 2025) (in Chinese).

V. Conclusion

China has made significant strides in decentralizing its electricity sector, but the existing regulatory framework is not yet sufficient to fully support renewable DERs. To realize the full potential of DERs, China must adopt more integrated regulatory frameworks that ensure grid flexibility, reliability, and sustainability. By learning from the EU's regulatory model and refining its own reforms, China can create an enabling environment for DERs that contributes to both environmental and social goals. The policy recommendations discussed in section IV(B) shed some light on the regulatory pathway for scaling up DERs in China through improved regulatory governance focusing on non-discriminatory access rules, regulatory incentives on modernizing grid infrastructure, grid access fees that are reflective of changing market conditions for DERs, refined funding support that are more consumption-driven, and institutional support to reduce systematic bias against DERs.

The shift to include renewable energy DERs represents a profound transformation in energy regulatory governance, driven by the imperatives of 'four Ds'—decarbonization, decentralization, digitalization, and democratization. This transition is particularly relevant in the Chinese context, where market reforms and renewable energy expansion offer significant opportunities but are hindered by regulatory and institutional challenges. As energy systems evolve, regulatory frameworks must adapt to support the complexities of DERs, ensuring that the benefits of this transition are realized equitably and sustainably. The 'four Ds' and the specific regulatory approaches in national/jurisdictional context provide a valuable lens for understanding and navigating this transformation, offering a roadmap for aligning energy law and policy with the demands of a rapidly changing energy landscape. As China's renewable DERs is gaining momentum through local and innovative policy and regulatory tools, further research is needed to examine the localized initiatives and practices to better understand how DERs can play a more positive role in China's energy decarbonization process.

10
Energy Decentralization and Human Rights in Mexico

José Juan González Márquez, Ivett Montelongo Buenavista, Ana Laura Silva López, and Juan Manuel Zamora Velázquez

I. Introduction

This chapter examines a topic that has not been thoroughly studied in Mexico: the relationship between energy decentralization and the protection of human rights, as well as its contribution to achieving the universal goal of inheriting a sustainable energy sector for future generations. It attempts to answer the question of how energy decentralization can contribute to realizing the human right to energy. Only a few studies have examined energy decentralization in Mexico from a human rights lens. Therefore, this chapter aims to contribute to filling this gap.[1]

The Mexican energy legal framework does not explicitly regulate Decentralized Energy Resources (DER), unlike European law.[2] However, at different times, Mexican legislation has referenced mechanisms intended to regulate energy decentralization projects. The reform of the Electricity Public Service Law (EPSL), enacted in 1992, introduced the concepts of self-supply, cogeneration, and small-scale energy production,[3] which permitted private companies to generate energy for their own consumption or delivery to related companies. However, this system was abolished by the Electricity Industry Law (EIL) in 2014. In 2025, the Federal Congress passed a new Energy Sector Law (ESL), which repealed the EIL. The ESL does not explicitly regulate Distributed Energy Resources (DERs), but it includes a few regulations related to DERs activities. The ESL addresses distributed generation and self-consumption in its third title, and in Chapter 2 of Title Six, it rules small generation systems and briefly refers to electricity storage.

[1] Among the limited studies available on this topic, the following are noteworthy: LF Cortés and JM Ortega, 'Perspectivas del desarrollo de sistemas de generación distribuida en México' in Ciemat (ed), *Prespectivas de Sustentabilidad* (Centro de investigaciones energéticas, medioambientales y tecnológicas 2017); M Cruz Arellano, 'Generación de energía eléctrica mediante sistemas de generación distribuida fotovoltaica' (2017) 4(7) Revista Iberoamericana de Producción Académica y Gestión Educativa; E Ramos, 'La generación distribuida: El camino hacia la producción descentralizada de electricidad y pautas para su reglamentación' (2020) 8(11) Revista de derecho; NJ Camacho, 'Derecho a la electricidad' (2020) 37(232) El Cotidiano; LG Samaniego Santamaría, *El derecho de acceso a la energía eléctrica: un derecho humano del Siglo XXI* (Tirant Lo Blanch 2024).

[2] In that legal system, DERs 'include not only generation, but also other resources used on-site, such as storage'. See chapter 1 of this book.

[3] Within the scope of its powers, between September 1994 and May 2019, the CRE granted a total of 468 permits for self-supply electricity generation. See J Carrillo, D Díaz, and O Ocampo, 'El autoabasto eléctrico en datos' (*Instituto Mexicano para la Competitividad AC*, 5 May 2022) <https://imco.org.mx/wp-content/uploads/2022/05/Autoabasto-ele%CC%81ctrico-en-datos_IMCO_05052022.pdf> accessed 24 July 2025.

José Juan González Márquez, Ivett Montelongo Buenavista, Ana Laura Silva López, and Juan Manuel Zamora Velázquez, *Energy Decentralization and Human Rights in Mexico* In: *Sustainable Distributed Energy Resources*. Edited by: Íñigo del Guayo, Damilola S. Olawuyi San, Louis de Fontenelle, Milton Fernando Montoya, and José Juan González, Oxford University Press. © International Bar Association 2026. DOI: 10.1093/9780198991946.003.0012

Despite the above, the National Development Plan for 2025–2030[4] outlines a Mexican energy policy rooted in the principles of sustainable electricity generation from renewable sources, energy justice, and providing access to electricity for the population, particularly in remote communities. Additionally, the Plan sets as one of its environmental policy goals the achievement of the Sustainable Development Goals (SDGs) and compliance with Mexico`s climate obligations. In the same vein, the Climate Change Law[5] and the Energy Transition Law[6] committed Mexico to ensuring that electricity generation from clean energy sources would account for at least 35 per cent of the energy mix by 2024. However, while various laws refer to the subject analysed in this chapter, the legal framework for energy decentralization remains underdeveloped and fragmented. Although the Mexican constitutional and secondary legislation have adopted the principle of sustainability, neither the Political Constitution of the United Mexican States (Federal Constitution) nor its secondary legislation explicitly addresses energy decentralization projects.

Furthermore, the legal system defines the framework of a centralized and monopolistic electricity sector in some cases. For example, article 25 of the Federal Constitution recognizes the principle of sustainability. It states that corporations in the economy's social and private sectors should be supported and promoted based on social equity, productivity, and sustainability criteria;[7] however, the sustainability of the energy sector has often been overlooked because Mexican energy law is founded on three key principles that may conflict with the idea of sustainable development: the importance of the energy sector as a lever of national economic development, its role in representing national sovereignty, and the consideration of energy generation, transmission, distribution, and supply activities as a public service. Conversely, in some instances, the new ESL appears to promote energy sustainability while acknowledging the essential role of the state in the electricity sector. This is evident in the law's emphasis on the state's obligation to ensure the continuity, accessibility, security, and reliability of the public electricity service and the National Electric System. Thus, the ESL is inconsistent.

Regarding human rights, the Mexican Federal Congress has evaluated a constitutional reform project to recognize access to energy as a human right. However, approval for this project is still pending. The new Electricity Sector Law (ESL), passed in 2025, tackles a significant issue by acknowledging two conflicting principles. Firstly, it asserts that basic electricity supply is a strategic activity essential for national development, aimed at delivering electricity to the people at the lowest possible cost.[8] Conversely, it also mandates that the Mexican government must provide the public service of electricity to anyone who requests it, provided it is technically feasible to do so.

As discussed in this chapter, a constitutional reform is urgently needed to promote decentralization, sustainability, and human rights protection. This reform should be

[4] Presidencia de la República, 'Plan Nacional de Desarrollo 2025–2030' (*Official Gazette of the Federation*, 15 April 2025) <https://www.dof.gob.mx/nota_detalle.php?codigo=5755162&fecha=15/04/2025> accessed 24 July 2025.

[5] Climate Change Law [2012]. <https://www.diputados.gob.mx/LeyesBiblio/pdf/LGCC.pdf> accessed 13 August 2025.

[6] Energy Transition Law [2015]. <https://www.diputados.gob.mx/LeyesBiblio/pdf/LTE.pdf> accessed 13 August 2025.

[7] Political Constitution of the United Mexican States [1917]. <https://www.diputados.gob.mx/LeyesBiblio/pdf/CPEUM.pdf> accessed 7 July 2025.

[8] Electricity Sector Law [2025], art 2, para 3.

complemented by the implementation of sustainable and clean energy policies outlined in the National Development Plan for 2025–2030, following the regulations established under the new energy legal framework published and approved in 2025. In line with Mexico's commitments under trade and environmental treaties, there is a need for increased reliance on renewable energy and a transition towards more sustainable energy practices. The constitutional reform should recognize the human right to energy and lay the groundwork for an energy policy that aligns with Goals 1 and 7 of the 2030 Agenda for Sustainable Development. Achieving energy decentralization is essential to meet this objective.

In that context, this chapter explores sustainability and human rights challenges that arise with electricity decentralization. Drawing lessons from Mexico, it evaluates how legal and regulatory systems can better integrate human rights norms in decentralization programs to guarantee energy justice. This chapter is organized into six sections. After the introduction, the second section analyses Mexican public policies and the legal framework for decentralization. The third section explores distributed energy and the right to electricity. The fourth section presents a case study of an energy decentralization project in a fishing community in Yucatan, Mexico. The fifth section addresses the legal challenges and offers recommendations for energy decentralization. Finally, the sixth section provides conclusions and insights based on the analysis presented in this chapter. These insights underscore the potential of energy decentralization to empower local communities and highlight the importance of recognizing electricity as a human right within the Mexican context.

II. Mexican Public Policies and Legal Framework for Decentralization

A. Background

In Mexico, 3.4 million people lack access to electricity. Sixty per cent of them belong to Indigenous communities located far from distribution grids.[9] These communities face other challenges such as a lack of basic services and infrastructure, including roads, water, telecommunications, schools, and hospitals. This lack of access to energy primarily affects women because women are traditionally mainly responsible for household chores and family care.[10]

The future in this regard is not promising. Due to the trend towards population growth, the number of people without access to electricity could increase by 20 per cent over the next ten years.[11]

[9] World Bank Group, 'Impulsando el acceso a la electricidad en comunidades remotas de México' (*World Bank Group*, 1 November 2017) <https://www.worldbank.org/en/results/2017/11/01/switching-on-remote-communities-through-electricity-access-in-mexico#:~:text=In%20the.%20first%20decade%20of,%2C%20proved%20particularly%20challenging%2C%20however> accessed 1 August 2025.

[10] A Campos, 'En México, hasta 4.9% de los hogares carecen del servicio de energía eléctrica' (*El Sol de Tampico*, 13 January 2025) <https://oem.com.mx/elsoldetampico/mexico/4-9-de-hogares-sin-electricidad-en-mexico-urgen-soluciones-en-el-plan-nacional-21116655> accessed 1 August 2025.

[11] Grupo Banco Mundial, 'Comunidades rurales alejadas de la red eléctrica logran acceso a la electricidad en México' (*Grupo Banco Mundial*, 1 November 2017) <https://www.bancomundial.org/es/results/2017/11/01/switching-on-remote-communities-through-electricity-access-in-mexico> accessed 18 July 2025.

Moreover, the excessive demand for electrical energy in industrial settings can lead to interruptions or fluctuations, resulting in substantial economic losses, equipment damage, decreased productivity, and a noticeable lack of energy security in Mexico.[12] One reason suggested for the failure to provide public electricity services to remote communities is the technical challenge of extending distribution networks to these areas.[13]

Among the strategies that could resolve these problems of electricity access, energy decentralization projects have been highlighted. Similarly, it has been suggested that recognizing the right to electricity as a human right would contribute to consolidating the strategy of decentralizing the electricity sector. However, as will be analysed in the next sections, there are several gaps and inconsistencies in the legislation that hinder the achievement of this goal.[14]

B. Energy Decentralization

In Mexico, the roots of decentralized generation can be traced back to the country's early days of electricity power generation. Initially, the electricity industry focused on generating power near the consumption sites.[15] However, with demographic growth and increasing demand for assets and services, the industry evolved towards a centralized generation model.[16]

Power plants were strategically located at the centre of consumption, with consumers gradually developing around them.[17] The technological limitations of direct current electricity generators, specifically the low-voltage capability, restricted power transport to only 30 to 57 kilometres.[18] As time progressed, the structure of electricity generation became what we know today, relying on alternating electrical currents and transformers.[19] This development enabled electricity to be transmitted far from generation centres. Consequently,

[12] PRODENSA, 'Addressing Mexico's Energy Crisis with Prodensa Energy: Role and Solutions' (*PRODENSA*, 21 May 2024) <https://prodensahr.com/mexicos-energy-crisis/> accessed 24 July 2025.

[13] Promateriales. 'Estrategias de transición energética en zonas rurales: Análisis de impacto social y económico' (*Prososstenible*, 17 July 2025) <https://prosostenible.es/estrategias-de-transicion-energetica-en-zonas-rurales-analisis-de-impacto-social-yeconomico/#:~:text=Desaf%C3%ADos%20y%20oportunidades%20en%20zonas%20rurales%20La,numerosas%20comunidades%20sin%20acceso%20a%20energ%C3%ADa%20confiable> accessed 1 August 2025.

[14] M del Carmen Aurora Carmona Lara, 'Hacia un enfoque integrado de justicia energética en México: agua y energía los retos del futuro. Aspectos jurídicos' in M Anglés Hernández and M Palomino Guerrero (coord), *Aproximaciones comparadas sobre el sector eléctrico en Iberoamérica* (UNAM/IIJ 2022).

[15] For example, it was the case of the first-generation plant built in 1879 in León, Guanajuato, that provided electricity for the textile factory The American. See CFE, 'Nuestra Empresa, Historia de la CFE' (*CFE*) <https://www.cfe.gob.mx/nuestraempresa/pages/historia.aspx> accessed 7 July 2025.

[16] L de Jesús Ramos-Gutiérrez and M Montenegro-Fragoso, 'Las centrales hidroeléctricas en México: pasado, presente y futuro' (2012) 3(2) Tecnología y ciencias del agua 106

[17] By the early twentieth century, industries began to locate near power plants, either buying electricity directly from the plants installed for that purpose or acquiring surplus electricity from plants installed by other industries. See A Parra, 'Los orígenes de la industria eléctrica en México: las compañías británicas de electricidad (1900-1929)' (1988) 19 Historias <https://revistas.inah.gob.mx/index.php/historias/article/view/14943> accessed 24 July 2025.

[18] Comisión Nacional para el Uso Eficiente de la Energía, '¿Qué es la generación distribuida? Grandes Usuarios de la Energía' (*Gob.mx*, 6 June 2014) <https://www.gob.mx/conuee/acciones-y-programas/que-es-la-generacion-distribuida-estados-y-municipios#:~:text=Sin%20embargo%2C%20exist%C3%ADan%20restricciones%20tecnol%C3%B3gicas%20de%20los,que%20era%20de%2030%20a%2057%20kil%C3%B3metros> accessed 7 July 2025.

[19] Parra (n 17).

decentralized generation became less relevant, as large power plants were built far from consumption areas but near fuel and water supply sources.[20]

In the 1970s, several factors—including the oil crisis, environmental concerns such as climate change, and a rapidly growing electricity demand—highlighted the need for technological alternatives.[21] These alternatives aimed to ensure a reliable and high-quality electrical energy supply, as well as the efficient use of natural resources. One such alternative is generating electricity as close as possible to the point of consumption. This approach draws inspiration from the methods employed in the early days of the Mexican electricity industry while incorporating the benefits of modern technology and the support of the electricity grid.[22]

Although energy decentralization projects do not play a significant role in Mexico's electricity industry, according to official data, Mexico's distributed generation market expanded by 32 per cent in 2024 compared to the previous year, reaching a total installed capacity of 4.4 GW.[23] This growing distributed generation supports the energy transition and creates new opportunities for electricity decentralization in Mexico. Despite that, as this chapter analyses, Mexico faces significant challenges in achieving energy decentralization.

In terms of public policies, the Mexican Energy Sector Program 2020–2024[24] had already highlighted the strategic importance of energy decentralization. This document stated that decentralization is crucial for achieving a more sustainable and resilient energy system. The program held, 'The generation of distributed or interconnected renewable energy is promoted to achieve the energy transition.'[25]

In the same direction, the new president of Mexico, who assumed power in 2024, appears committed to advancing energy decentralization. The 2025–2030 National Development Plan identifies sustainability as one of the key pillars for the country's development. It emphasizes the importance of energy sustainability, which aims to provide affordable electricity to the population, increase the share of renewable energy sources in the energy mix, and accelerate the transition to cleaner energy. This approach seeks to reduce polluting emissions while ensuring energy coverage and access, particularly in marginalized

[20] ibid.

[21] Centro de Estudios de las Finanzas Públicas, 'Evolución y Perspectiva del Sector Energético en México, 1970–2000' (Cámara de Diputados 2001) <https://www.cefp.gob.mx/intr/edocumentos/pdf/cefp/cefp0512001.pdf> accessed 24 July 2025.

[22] This type of electricity generation is referred to as *in-situ*, dispersed, or distributed. Distributed generation refers to producing energy using multiple small sources, primarily from photovoltaic systems and, to a lesser extent, from cogeneration processes in manufacturing. Its primary goal is for a producer to also act as a consumer, often referred to as a 'prosumer'. This generation occurs at or near the consumption sites, allowing for a direct connection to electrical distribution networks. See A Jonker and A Gomstyn, '¿Qué es la generación distribuida?' (*IBM*, 9 June 2024) <https://www.ibm.com/mx-es/think/topics/distributed-generation> accessed 7 July 2025.

[23] Bnamericas, 'Mercado de generación distribuida crece 32% en México' (*Bnamericas,* 4 February 2025) <https://www.bnamericas.com/es/noticias/mercado-de-generacion-distribuida-crece-32-en-mexico#:~:text=El%20mercado%20de%20generaci%C3%B3n%20distribuida,seg%C3%BAn%20cifras%20de%20la%20secretar%C3%ADa> accessed 21 February 2025.

[24] Published 08 July 2020 on the Official Gazette of the Federation.

[25] The National Strategy for the Energy Sector, presented at the President's conference on 6 November 2024, is based on four guiding principles that seek to strengthen national electricity sector planning, promote energy justice, guarantee a robust, reliable, and secure electricity system, and establish clear rules to increase private investment. See Presidencia de la República, 'Conferencia encabezada por la presidenta de los Estados Unidos Mexicanos, Claudia Sheinbaum Pardo, desde Palacio Nacional' (*Gob.mx*, 6 November 2024) <https://www.gob.mx/presidencia/articulos/version-estenografica-conferencia-de-prensa-de-la-presidenta-claudia-sheinbaum-pardo-del-6-de-noviembre-de-2024> accessed 25 June 2025.

communities. The plan also strives to guarantee that all regions of the country benefit from clean and sustainable energy sources.[26]

In the same avenue, the inclusion of various electricity generation models in the new ESL supports the move towards decentralization. Similarly, both the National Electricity Sector Strategy[27] and the Plan to Strengthen and Expand the National Electricity System 2025–2030[28] have as their fundamental axes the generation of affordable, reliable, and clean energy, in accordance with SDG 7, and the achievement of energy justice, which, in general terms, means achieving SDG 1 and thereby also fulfilling the reduction of climate gas emissions committed to when ratifying the Paris Agreement. All of this is aimed at ensuring the human right to access electricity.[29] Finally, the new laws and regulatory measures may play a significant role in facilitating its implementation.[30] However, eliminating the centralized nature of the Mexican electricity sector could be complicated in the near future.

C. The Centralized Nature of the Mexican Electricity

Electricity generation in Mexico is largely centralized. Various factors contribute to this issue. Generation is centralized due to its subordination to state leadership, its integration into the planning and management of the national electricity system, and its significant reliance on public services for the transmission and distribution of electricity.[31]

The Mexican centralized legal framework has driven the development of power plants for decades, allowing them to be installed and interconnected to the national electricity system. This setup enables large quantities of energy to be delivered to areas far from production centres, serving most of the population. However, this centralized model has notable drawbacks, including discrimination against users located near generating plants, social impacts, and significant contributions to carbon emissions.[32]

On the other hand, decentralized energy generation is only minimally regulated by Mexican law under the categories of distributed generation, self-consumption, small-scale electricity systems, and energy storage. This type of generation allows electricity to be produced at the point of consumption or nearby, addressing the immediate needs of the population. It encourages smaller-scale investments, lessens the burden on general transmission and distribution networks, and promotes the use of clean energy sources, among other

[26] Presidencia de la República, 'Plan Nacional de Desarrollo 2025–2030' (*Official Gazette of the Federation*, 15 April 2025) <https://www.dof.gob.mx/nota_detalle.php?codigo=5755162&fecha=15/04/2025> accessed 24 July 2025.

[27] This strategy was presented on 6 November 2024, during the President's press conference.

[28] This plan was presented on 5 February 2025, during the President's press conference.

[29] Strategy 4.3.3 of Objective 4.4: Guarantee equitable access to energy for populations and regions with supply problems, promoting their inclusion and integration into the sustainable energy development of the National Development Plan 2025–2030. See Presidencia de la República, 'Plan Nacional de Desarrollo 2025–2030' (*Official Gazette of the Federation*, 15 April 2025) <https://www.dof.gob.mx/nota_detalle.php?codigo=5755162&fecha=15/04/2025> accessed 24 July 2025.

[30] Gobierno de México and CFE, 'Plan de Fortalecimiento y Expansión del Sistema Eléctrico Nacional 2025–2030' <https://www.proyectosmexico.gob.mx/wp-content/uploads/2025/02/Plan-Fortalecimiento-y-Expansion-Sistema-Electrico-Nacional.pdf> accessed 24 July 2025.

[31] Energy Sector Law [2025] art 2 and fifth paragraph of art 25.

[32] JC Anchundia and others, 'Impactos de la generación distribuida en redes de distribución' (2023) 2(1) Cadrnos Latino-Americanos de Engenharia, Tecnologia e Ciencias Aplicadas <https://tech.sapienzaeditorial.com/index.php/CLADETEC/article/view/632/466> accessed 13 August 2025

advantages. Nevertheless, decentralized generation has its challenges, including its intermittent nature, the lack of training for operators, limited funding for development, and a lack of interest from government agencies in transforming it into a national model that democratizes access to this essential service.

Choosing between these two models is unnecessary; they can coexist and complement each other, mitigating their disadvantages while maximizing the benefits of each.

III. Distributed Energy and the Right to Electricity

A. Background

In the 1990s, because of the commitments assumed by Mexico in the North American Free Trade Agreement (NAFTA), the Public Electricity Service Law was amended[33] to introduce the concepts of self-supply, cogeneration, and small-scale energy production, allowing private companies to generate electricity on a smaller scale than large power plants for the first time in decades. They could produce electricity near consumption centres and through *in-situ* generation for their use. In all cases, there was an obligation to make any excess energy available to the state-owned corporation. In fact, energy decentralization resurfaced in Mexico under the protection of the regulatory model before the constitutional reform of 2013.[34] However, as analysed in the next section, the legal framework applicable to decentralized energy projects is fragmented and still insufficient.

B. The New Electricity Sector Legal Framework

During the last three decades, there have been three major reforms to the electricity legal framework in Mexico. The first occurred in 1992, when, due to the impossibility of modifying the constitution to open the sector to private investment, the Federal Congress passed a reform to the Public Service of Electricity Energy Act (1975), thereby opening opportunities for private investment in electricity generation not intended for public service.[35] This reform did not have the objective of promoting energy transition but of complying with the commitments assumed by Mexico in the NAFTA negotiations. Considering that reform, private parties were allowed to generate electricity through the modalities of self-supply, co-generation, and small production.[36]

The second major energy reform took place in 2013 when the Federal Congress, along with the thirty-two local congresses, amended the Federal Constitution.[37] This amendment was designed to distinguish between strategic and non-strategic activities in the electricity

[33] Public Electricity Service Law [1975] (amended 23 December 1992) <https://www.dof.gob.mx/nota_to_imagen_fs.php?codnota=4705440&fecha=23/12/1992&cod_diario=202787> accessed 7 July 2025.

[34] In 2013, the federal government amended the Constitution and related laws to provide clarity and assurance to investors, as opening the energy sector to private investment had previously contradicted the Constitution. These amendments explicitly allowed private investment in specific areas of the energy sector. For further details, see JJ González, *Nuevo Derecho Energético Mexicano* (Universidad Autónoma Metropolitana 2017).

[35] González (n 34).

[36] By Decree published in the Official Gazette of the Federation, arts 101 and 103 of the regulations are modified.

[37] González (n 34).

industry. Planning, control, transmission, and distribution were considered strategic activities reserved exclusively for the state monopoly. In contrast, generation and supply, as non-strategic activities, were opened to private investment.[38]

As a result of the new constitutional basis, in 2014, the Federal Congress passed nine new laws and amended another eleven.[39] The main objective of this reform was to open the energy sector to private investment.

The third major electricity sector reform took place in 2024, when the Federal Congress and the thirty-two local congresses passed a new constitutional amendment to reverse the constitutional and legal reform of 2013–14.[40]

This reform eliminated the separation of the electricity industry's value chain activities into strategic and non-strategic. Instead, Article 27 of the Federal Constitution states that the planning and control of the electricity sector and the activities of electricity transmission and distribution are considered public services and therefore no private contracts or concessions may be granted in these areas, thereby modifying the structure of the electricity industry in Mexico. In addition to the above, the reform under discussion does not grant the character of public services to electricity generation and supply activities, nor does it leave them entirely open for private investment.

The new constitutional text provides that 'The laws shall determine how private individuals may participate in other activities of the electricity industry, which in no case shall have supremacy over the public company of the State.'[41] In this regard, Article 12, Section VI of The ESL states, 'The State must maintain at least fifty-four per cent of the average energy injected into the grid in a calendar year.' Thus, electricity generation is a semi-state monopoly because the participation of the private sector in these activities may never exceed that of the public sector.[42]

[38] Presidencia de la República 'DECRETO por el que se reforman y adicionan diversas disposiciones de la Constitución Política de los Estados Unidos Mexicanos, en Materia de Energía' (Official Gazette of the Federation, 20 December 2013) <https://www.dof.gob.mx/nota_detalle.php?codigo=5327463&fecha=20/12/2013#gsc.tab=0> accessed 18 July 2025.

[39] The secondary energy reform comprised the following laws: the Hydrocarbons Law, the Electricity Industry Law, the Geothermal Energy Law, the Petróleos Mexicanos Act, the Federal Electricity Commission Act, the National Agency for Industrial Safety and Environmental Protection in the Hydrocarbons Sector Act, and the Coordinated Energy Regulatory Bodies Act. The following laws were also reformed and amended: the Foreign Investment Law, the Mining Law, the National Water Law, the Public-Private Partnerships Law, the Organic Law of the Federal Public Administration, the Federal Law on Parastatal Entities, the Law on Public Sector Procurement, Leasing and Services, and the Law on Public Works and Related Services. See J Cárdenas Gracia, 'La nueva legislación secundaria en materia energética de 2014' (2015) 48(143) Boletín mexicano de derecho comparado <http://www.scielo.org.mx/scielo.php?script=sci_arttext&pid=S0041-86332015000200003&lng=es&tlng=es> accessed 18 July 2025.

[40] Presidencia de la República, 'DECRETO por el que se reforman y adicionan diversas disposiciones de la Constitución Política de los Estados Unidos Mexicanos, en Materia de Energía' (*Official Gazette of the Federation*, 20 December 2013) <https://www.diputados.gob.mx/LeyesBiblio/ref/dof/CPEUM_ref_212_20dic13.pdf> accessed 25 July 2025.

[41] Sixth paragraph of art 27 of the Political Constitution of the United Mexican States. Presidencia de la República, 'Decreto por el que se reforman, adicionan y derogan diversas disposiciones de la Constitución Política de los Estados Unidos Mexicanos, en materia de simplificación orgánica' (*Official Gazette of the Federation*, 20 December 2024) <https://www.diputados.gob.mx/LeyesBiblio/ref/dof/CPEUM_ref_268_20dic24.pdf> accessed 18 July 2025.

[42] Presidencia de la República, 'DECRETO por el que se reforman el párrafo quinto del artículo 25, los párrafos sexto y séptimo del artículo 27 y el párrafo cuarto del artículo 28 de la Constitución Política de los Estados Unidos Mexicanos, en materia de áreas y empresas estratégicas' (*Official Gazette of the Federation*, 31 October 2024) <https://www.diputados.gob.mx/LeyesBiblio/ref/dof/CPEUM_ref_262_31oct24.pdf> accessed 18 July 2025.

Based on the new Constitutional reform, in 2025, the Federal Congress passed eight new laws[43] and amended two others.[44] Regarding energy decentralization, the reform included a new Electricity Sector Law and a Planning and Energy Transition Law (PETL).

The ESL has a section on electricity generation that covers not only generation for the wholesale market but also energy decentralization projects. It addresses self-supply (Articles 30 to 34), small electricity systems (Articles 79 to 81), energy storage (Articles 82 to 85), and distributed generation (Articles 25 to 29). In addition, the PETL establishes the Fund for Energy Transition and Sustainable Energy Use, a financial mechanism designed to enhance financing for various purposes, including clean distributed generation.[45]

The analysis of the new energy laws from a human rights perspective indicates that the Mexican energy legislation offers a solid foundation for integrating energy policies that consider human rights concerns.

Among the changes introduced by these laws to promote decentralized generation projects, the following points can be highlighted. The ESL has a Title Eight that includes a chapter on energy justice, a chapter on social impact and sustainability, and a chapter on clean energies, energy transition, and decarbonization of the energy sector. Although these are programmatic precepts, their inclusion in the text of the law attests to the importance that the energy transition is beginning to take on in Mexico. To make the right to access electricity a reality, these provisions deserve to be improved.

Besides, the PETL designates the National Energy Commission (NEC) as the authority responsible for issuing interconnection contract templates and compensation for clean distributed generation.[46] Under this framework, the NEC holds the authority to grant, modify, terminate, and supervise electricity generation and supply permits and is also empowered to issue relevant authorizations and sector-related acts.[47] Furthermore, according to Article 8, Section VII of the law, the NEC must promote the orderly, continuous, and safe development of electricity generation activities. Article 129 of the ESL establishes the Universal Energy Service Fund to support Energy Justice initiatives. This fund prioritizes electrification for rural communities and marginalized urban areas, providing electricity to individuals and communities in vulnerable situations or experiencing energy poverty. The following section will explore these issues further.

[43] On 18 March the Executive published into de Official Gazette of the Federation the following laws: Law of the Public Corporation Federal Commission of Electricity; Law of the Public Corporation PEMEX; Law of the Energy Sector; Law of Hydrocarbons; Law of Planning and Energy Transition; Law of Bioenergetics; Law of Geothermic; and Laws of the National Commission of Energy.

[44] On 18 March, the Executive Branch published the reforms to the Mexican Petroleum Fund for Stabilization and Development Law and the Organic Law of the Federal Public Administration in the Official Gazette of the Federation.

[45] Energy Planning and Transition Law [2025] art 43.

[46] Energy Planning and Transition Law [2025] art 14, s II. Under the reforms approved in 2025 by the Congress of the Union on energy matters, the NEC replaced the Energy Regulatory Commission.

[47] Energy Planning and Transition Law [2025] art 8, s IV.

C. The Law for Energy Decentralization

1. Legal framework for distributed generation

Unlike the EIL, the new ESL clarifies the differences between clean, distributed, and small-scale electricity generation. The new law classifies the different types of generation and establishes a legal framework with greater scope and precision for decentralized generation; without being a finished scheme, Article 25 redefines distributed generation as a modality of electricity generation carried out by exempt generators with a capacity of less than 0.7 MW interconnected in a distribution circuit that contains a high concentration of load centres in terms of the Market Rules,[48] with the possibility of consuming it or selling it to the market, including its associated products.

Under ESL, the Federal Electricity Commission is the only provider for all sales of energy or its associated products.[49] This is a significant change from the previous law, where the sale of energy could also be made through qualified suppliers, and exempt generators could participate in the whole sales market through a qualified supplier.

Furthermore, for the deregulation and promotion of distributed generation, the ESL exempts distributed generators from the requirement to conduct studies to determine the specific characteristics of the required infrastructure when submitting their interconnection applications, provided they comply with the general technical specifications and provisions for interconnecting new distributed generation plants.

Besides, the PETL empowers the Ministry of Energy to plan electrical infrastructure that supports the development of clean distributed generation (CDG) when it is economically viable and promotes social development. In Article 43, this law also establishes the Fund for Energy Transition and Sustainable Energy Use, a financial mechanism aimed at enhancing financing for various purposes, including CDG.

Similarly, the PETL promotes distributed generation in various ways. For example, it empowers the Ministry to plan the construction of infrastructure works in the Energy Sector that facilitate the development of projects for the Energy Transition and Clean Distributed Generation in conditions of economic viability and that promote social development (Article 8, Section VIII). However, many of the articles mentioned are written in a very general manner, so it is necessary to enhance their scope through regulatory provisions that have not yet been issued.

The regulation of distributed generation remains uncertain under the new ESL. However, Articles 26 to 29 provide several benefits for distributed generators. These benefits include access to general distribution grids when technically feasible, access to markets, and the opportunity to sell electricity and associated products through basic service providers.

In summary, regarding distributed generation, the new legislation introduces positive changes; however, it also presents several drawbacks. For instance, while the limit on installed capacity has increased from 0.5 MW to 0.7 MW, the 2025 legal framework has removed specific mechanisms that previously made this generation model attractive. These

[48] An exempt generator does not require or have a permit to generate electrical energy. According to art 19 of EIL, no permit is required for the generation of electricity when power plants carry it out with a capacity equal to or greater than 0.5 MW.

[49] Electricity Sector Law [2025] art 6.

include open access to general distribution networks and the ability to sell energy to qualified suppliers.

2. Legal framework of small-scale generation

According to Article 79 of ESL, small electrical systems are those used to supply electrical energy to the public that are not permanently connected to the National Transmission Grid or General Distribution Networks and that supply a demand of no more than 100 MW. These systems include an interconnected set of devices, equipment, and components designed to generate, transmit, distribute, and consume electrical energy at reduced levels. Unlike distributed generation projects, small-scale systems require a permit to operate since they are not exempt generators.

As in the case of EIL, the ESL includes only a few provisions addressing small electricity systems. In addition to the definition given in Article 79, the only regulations for this mode of generation are found in Articles 80 and 81.

Article 80 refers to the collaboration between participants in the electricity sector and small electricity systems to provide efficient electricity supply services. In contrast, Article 81 allows the RM to regulate small electricity systems more specifically.

3. Legal framework of self-consumption

The third path for energy decentralization in Mexico is self-consumption.[50] The new ESL refers to self-consumption as power plants with a capacity equal to or greater than 0.7 MW. This modality of generation requires a permit issued by the Ministry, and it aims to meet their own needs *in-situ*. Self-consumption can assume two modalities: isolated or interconnected.[51]

Article 31 establishes that isolated self-consumption is the production of the Power Plant that is exclusively intended for *in-situ* consumption. In contrast, Article 32 establishes that interconnected self-consumption is considered the production of a Power Plant intended for *in-situ* consumption and interconnected to the National Transmission Network or General Distribution Networks.

Self-consumption allocates its entire production exclusively for *in-situ* consumption. In contrast, interconnected self-consumption, in addition to allocating its production for on-site consumption, is interconnected to the National Transmission Grid or General Distribution Networks. This modality allows the injection of surpluses into the National Electric System without compensation or, if applicable, sold to the State-owned Enterprise.[52]

4. Legal framework of electricity storage

Electricity storage is crucial for decentralization due to the intermittent nature of renewable energy sources. Electricity storage is mentioned in several provisions of the ESL, with specific regulations outlined in Articles 82 to 85. However, these provisions are more programmatic than regulatory, leaving the development of the necessary legal framework to

[50] This type of electricity generation was included in the Electricity Industry Law of 2014 as *Isolated Supply*.

[51] F Zendejas Reyes, 'Generación distribuida y abasto aislado: usuarios complementando al sistema eléctrico nacional' in M Anglés and M Palomino (eds), *Aproximaciones comparado sobre el sector eléctrico en Iberoamérica* (Universidad Nacional Autónoma de México 2019).

[52] See arts 30 al 34 of the ESL on this matter.

lower-level regulations. One important aspect introduced by the ESL is the recognition of electricity storage as an independent stage in the electricity industry's value chain. This marks a departure from the previous view under the LIE, which considered storage to be a form of generation.

D. Energy Decentralization from the Human Rights Perspective

Some scholars argue that 'sustainable Distributed Energy Resources (DERs) are founded on principles and concepts from international energy law, international environmental law, international climate change law, and international human rights law'.[53] However, this does not imply that access to electricity has been universally recognized as a human right in all legal systems across the globe.

There are diverse opinions regarding the convenience of recognizing the right to access electricity as a human right. Some argue that it is not a right in itself, but rather a contractual right that depends on the way electricity is supplied to users, which, in principle, would exclude those without access from this right.[54] On the other hand, some scholars consider access to electricity a universal human right to which every human being is entitled, given the importance of electricity in their lives and its role in the enjoyment of other human rights.[55] Access to electricity is implicit in different human rights.[56]

In Mexico, while access to energy is not explicitly recognized as a human right in the Federal Constitution, it is implicitly included among other human rights that are established in the constitutional text, such as the right to healthy environment, which is affected when, for example, due to the absence or shortage of electricity, people have limited access to essential services, including medical care, personal well-being, economic development, quality of life, and access to education and food.[57] The next section analyses the state of the art regarding the constitutional recognition of access to energy as a human right in Mexico.

E. First Attempts to Recognize the Human Right to Electricity

The human right to access electricity is closely tied to the right to an adequate standard of living, as outlined in Article 25, paragraph 1, of the Universal Declaration of Human Rights.[58] These rights guarantee everyone a basic level of security that promotes their health and overall well-being, covering personal, family, and social aspects. This right also encompasses essential needs such as food, clothing, housing, medical care, and access to water.

[53] See Chapter 2 of this book.

[54] E Paola Carmona Díaz de León, 'El acceso a la energía eléctrica como derecho humano' (2023) 47(47) Revista de Investigaciones Jurídicas.

[55] L Löfquist, 'Is There a Universal Human Right to Electricity?' (2019) 24(6) The International Journal of Human Rights 711–23.

[56] D Olawuyi, 'Energy (and human rights) for all: Addressing Human Rights Risks in Energy Access Projects' in R Salter, CG González, and EA Kronk Warner (eds), *Energy Justice: US and International Perspectives* (Edward Elgar Publishing Ltd 2018) pp 73–104.

[57] N Peniche Sala, 'Universal Access to Electricity as a Strategy to Close Socioeconomic Gaps in Mexico' in M Anglés Hernández and M Palomino Guerrero (coord), *Lessons in Energy Law in Mexico* (UNAM 2023).

[58] Universal Declaration of Human Rights, 1948.

The connection between the right to access electricity and an adequate standard of living is evident, especially since many of the rights mentioned in the Universal Declaration of Human Rights involve electricity use. For instance, the United Nations has stated that nearly 6 billion people use the Internet; this would be impossible without electricity.[59] On the other hand, access to drinking water is closely linked to electricity generation.[60] Similarly, water extraction, distribution, transportation, and treatment require electricity.[61] That is why, according to CEPAL data, access to water is limited in regions of Latin America without electricity.[62] Finally, access to health is impossible without affordable and accessible electricity. According to the WHO, approximately 8 per cent of health facilities in Latin America and the Caribbean lack access to electricity, rendering it impossible to ensure that essential medical devices function correctly and potentially jeopardizing lifesaving efforts.[63]

In 2012, the Inter-American Court of Human Rights ruled that 'access to electricity is essential for the guarantee of other human rights' in the case of the *Río Negro Massacres v Guatemala*.[64] This ruling is relevant because the State was ordered to restore the rights of the survivors and the affected community, including security, and this could only be achieved if other human rights, such as access to electricity, were also guaranteed.[65]

This ruling acknowledges that in the twenty-first century, electricity is an essential resource. It is vital for the functioning of daily activities for people in both urban and rural areas, regardless of their social status. Electricity supports numerous tasks, including operating servers, computers, and intelligent systems in buildings, as well as maintaining hospital functions and providing access to water, among other essential activities.[66] Precisely

[59] Information taken from the statistics page of the International Telecommunication Union (ITU). ITU is the UN specialized agency for ICTs. ITU, 'Statistics' (ITU 2025) <https://www.itu.int/en/ITU-D/Statistics/Pages/stat/default.aspx#:~:text=You%20are%20here%20ITU%20%3E%20Home,Unexpected%20Error> accessed 25 June 2025.

[60] C Cleveland, 'What is the relationship between energy use and access to safe water?' (*Visualising Energy*, 30 January 2023) <https://visualizingenergy.org/what-is-the-relationship-between-energy-use-and-access-to-safe-water/> accessed 18 July 2025.

[61] A Perea Moreno, QH Escobedo, and MJ Aguilera Ureña, 'Seguridad en el suministro del agua y energía limpia: una propuesta de proyecto para los regantes del río Torrox' (2017) 8(3) Tecnología y Ciencias del Agua.

[62] CEPAL, 'Los servicios básicos de agua potable y electricidad como sectores clave para la recuperación transformadora en América Latina y el Caribe' (*CEPAL*, 7 September 2022) <https://www.cepal.org/es/enfoques/servicios-basicos-agua-potable-electricidad-como-sectores-clave-la-recuperacion> accessed 16 February 2025.

[63] World Health Organization and others, 'Energising health: accelerating electricity access in health-care facilities' (World Health Organization 2023) <https://www.who.int/publications/i/item/9789240066960> accessed 16 February 2025.

[64] See *Masacres de Río Negro v Guatemala* (Inter-American Court of Human Rights, 4 September 2012). The case is related to the destruction of the Mayan community of Río Negro, through a series of massacres carried out by the Guatemalan Army and members of the Civil Self-Defense Patrols in 1980 and 1982; the persecution and elimination of its members, as well as various violations against the survivors of that attack that involved violations of their rights recognized in the American Convention on Human Rights (Pact of San José, 1978), such as the right to life, to personal integrity, the protection of the family, the rights of children among others, which the Guatemalan State should guaranteed as a member of this Convention. This case is essential because the Inter-American Court of Human Rights resolved in its ruling that to guarantee all the human rights violated, the State should provide, among other things, electrical energy to that community. Up until now, the relationship between access to electrical energy and the guarantee of internationally recognized human rights has evolved. Access to electricity is increasingly being recognized not just as a means to support human rights, but as a human right in itself.

[65] The judgment, which covers preliminary objections, merits, reparations, and costs, was issued on 4 September 2012. The State was notified of the judgment on 19 October 2012.

[66] OECD, *A New Rural Development Paradigm for the 21st Century: A Toolkit for Developing Countries* (OECD Publishing 2016) <https://www.oecd.org/en/publications/a-new-rural-development-paradigm-for-the-21st-century_9789264252271-en.html> accessed 24 July 2025.

for this reason, this human right demonstrates its impact on people's lives and, therefore, its universality.

The judiciary has played a crucial role in recognizing the human right to access electricity in Mexico. In 2018, the Mexican Judicial Power issued a jurisprudence thesis[67] titled 'Access to Electricity: It Must Be Recognised as a Human Right', which addresses the relationship between the human right to access electricity and other fundamental rights. The thesis states that access to electricity is an essential prerequisite for exercising multiple fundamental rights. This ruling, which underscores the necessity of electricity for daily activities and the enjoyment of fundamental rights, is a positive step forward.[68]

Access to electricity is crucial for ensuring and safeguarding other human rights. However, the Mexican Federal legal system does not yet fully recognize it as a human right. Congress reviewed several initiatives for constitutional reform to establish this recognition in 2017, 2022, and 2024, but none were approved.

F. The Initiatives for a Constitutional Amendment

The 2017 initiative, presented by the National Assembly of Electric Energy Users and the Mexican Union of Energy Sector Workers, proposed adding a paragraph to Article 4 of the Constitution that states, 'Every person has the right to electricity.'[69]

This initiative aimed to support the 2013 energy reform, acknowledging that electricity is necessary for fostering social, economic, cultural, and political development.

In May 2022, Mexico's City Congress approved a proposal to amend Article 4 of the Constitution,[70] elevating the 'human right to electric energy' to constitutional status.[71] This proposal was submitted to the House of Representatives for further consideration. One of the main arguments supporting this initiative was the belief that electricity should not be considered a commodity but instead recognized as a human right essential for fulfilling societal needs.

In 2023, the Congresses of two states submitted proposals for reforms to Article 4 of the Federal Constitution to include the human right to access electricity. In September 2023, the Congress of Yucatán,[72] and in October 2023, the Congress of Baja California Sur, approved

[67] Jurisprudence is the doctrine established repeatedly by the Supreme Court of Justice of the Nation and by the Collegiate Circuit Courts when interpreting the laws, to establish the correct meaning and scope of the legal norms.

[68] This jurisprudence thesis was issued by the Third Collegiate Court of the First Circuit in Civil Matters. It was published on Friday, 7 December 2018 at 10:19 am in the Judicial Weekly of the Federation.

[69] Asamblea Nacional de Usuarios de Energía Eléctrica & Sindicato Mexicano de Electricistas, 'Iniciativa con proyecto de Decreto por el que se adiciona un párrafo al artículo 4o. De la Constitución Política de los Estados Unidos Mexicanos' (*Gaceta Parlamentaria*, 30 November 2017) <https://gaceta.diputados.gob.mx/PDF/63/2017/nov/20171130-I.pdf> accessed 18 July 2025.

[70] Articles 71, ss III, 72, and 122 of the Political Constitution of the United Mexican States and art 29, para D, sub-s (c) of the Political Constitution of Mexico City establish the powers of Mexico City's Congress to send constitutional reform initiatives to the Chamber of Deputies.

[71] Official Gazette of Mexico City number No 183, 17 May 2022. <https://www.congresocdmx.gob.mx/media/documentos/5d7dd6b58cbaaf841882617b1b905111b9982719.pdf> accessed 25 July 2025.

[72] This proposal states: 'Every person has the right to access to electricity. The State shall guarantee a sustainable, affordable, and socially oriented supply following the terms established by this Constitution and the relevant laws.' See F Loeza Novelo, 'Iniciativa por la que se reforma la Ley de Víctimas del Estado de Yucatán, en materia de protección inmediata en caso de incumplimiento de obligaciones de asistencia familiar' (*Congreso de Yucatán*, 8 November 2023) https://congresoyucatan.gob.mx/index.php/gaceta/iniciativas> accessed 21 July 2025.

and sent to the Federal Congress the proposal to recognize the human right of every individual to access and utilize electricity, considering the various climate phenomena affecting our country.[73]

In 2024, the Senate reached an agreement to finalize the analysis of pending initiatives presented by senators. It was decided that any initiatives submitted by senators before 30 April 2024, which had not yet been considered by the relevant committees, would be deemed concluded in their legislative process. These initiatives would then be sent to the Historical Archive and Legislative Memory. Furthermore, it was noted that initiatives presented after this date would follow the appropriate procedures. Following that agreement, two initiatives, including the right to access electricity, were submitted: one to the House of Representatives and the other to the Senate. The *Movimiento Ciudadano Party* promoted the first one.[74] The initiative stated that 'every person has the right to access electricity for personal, domestic, and educational consumption in an affordable, reliable, sustainable, renewable, modern, safe and quality manner'.

In August 2024, the Official Party, MORENA, promoted a new initiative that considers the human right to access electricity and aims to reduce the tariff for individuals in extreme poverty and those over sixty-five years old. This initiative recognizes the right of every person to an electricity supply, including the obligation of the state to grant discounts on electricity supply to people over sixty-five years old who are in poverty. Both initiatives are now being discussed in the specialized commissions to follow their process in each House.

All proposed initiatives emphasize a crucial point: access to electricity is a fundamental human right. Without electricity, life and the ability to exercise rights are unimaginable for many people worldwide. Furthermore, acknowledging this right would strengthen the State's obligation to ensure all individuals can access this vital resource, guaranteeing other human rights, such as health and food security. Unfortunately, both initiatives are still pending approval.

Despite the above, while the constitution does not explicitly recognize a human right to energy, unlike the right to a suitable environment or access to water, it does indirectly acknowledge this goal in Article 28. This article states that providing electricity to the people at the lowest possible price, while avoiding profit, is a strategic area aimed at ensuring national security and sovereignty.

G. Sustainability of the Energy Sector

Access to electricity should grant individuals the right to receive energy, ensuring it is available at a low cost and located close to their homes or workplaces. This approach will

[73] The proposed text to include in art 4 of the Federal Constitution states: 'Every person has the right to access and use sufficient, acceptable, affordable, and quality electricity for personal and domestic use and to ensure family subsistence. The State shall guarantee access to energy by establishing general, payable rates. In the event of climatic phenomena such as heat waves, hurricanes, natural disasters, pandemics, and economic crises, the State may decree a general basic service rate for the entire population or, where appropriate, suspend payments and even grant waivers or exemptions, as well as grant reasonable payment extensions without interruption of supply.' See O Mayor, 'Acta de la sesión pública ordinaria. Jueves 5 de octubre de 2023' (*Poder Legislativo del Estado de Baja California Sur*, 5 October 2023) <https://www.cbcs.gob.mx/index.php/actas-de-sesiones/lista-de-actas/433-acta-sesion-xvi-legislatura/6213-actas-sesiones-xvi-legislatura-cbcs> accessed 18 July 2025.

[74] Submitted by Deputy Braulio López Ochoa Mijares, Co-ordinator of the Parliamentary Group of the Citizen Movement on 21 August 2024.

enhance efficiency and help reduce expenses. Achieving this goal is crucial for attaining SDG 7, which aims to provide people with affordable energy and contribute to efforts to combat climate change.

Furthermore, when people have access to affordable energy, they also experience an improvement in their quality of life. As a synergetic consequence, they contribute to ending poverty, as SDG 1 establishes, because governments and companies invest in delivering energy infrastructure, including clean energy.

Finally, the new package of energy legislation published in March 2025 did not include any reference to the human right to access to electricity and its connection with sustainability. However, Article 133 of the ESL emphasizes that infrastructure projects must comply with sustainability principles and respect the human rights and natural resources of Indigenous and Afro-descendant peoples and communities. Similarly, the PETL establishes the objective of energy planning as promoting energy justice through efficient and sustainable mechanisms to reduce energy poverty and provide the population with the necessary infrastructure to meet its basic energy needs, including water heating, cooking, refrigeration, lighting, and other essential requirements.[75]

Neither the Federal Constitution nor energy legislation provides mechanisms or instruments that directly offer the population located far from electricity distribution points the possibility of enjoying this service. Thus, to address energy poverty and ensure energy justice for all, it is essential to incorporate the human right to electricity into the Constitution.

For this reason, in recent years, efforts have been made to develop mechanisms that address the issue of energy poverty. These mechanisms, in terms of their characteristics and objectives, are like those of energy decentralization projects. They involve not only the generation of electricity at or near the point of consumption but also other activities such as storage, collective self-consumption, and citizen energy communities, among others.[76]

Sustainable Rural Energy Systems aim to provide electricity to remote and often marginalized communities that lack access to the electricity grid. As a result of this lack of access, these communities experience energy poverty and face various human rights challenges.[77] These systems aim to provide energy to approximately 75 per cent of rural residents who lack access to it through eco-technologies that thereby meet domestic and productive energy needs efficiently and affordably.[78] Additionally, there are other mechanisms for bringing electricity to isolated communities, such as energy co-operatives[79] and solar ejidos.[80] Both mechanisms not only seek to make communities self-sufficient but also

[75] Energy Planning and Transition Law [2025] art 2, s VII.

[76] See Chapter 1 of this book.

[77] C Octaviano Villasana and others, *Electrificación rural para comunidades fuera de la red utilizando generación de energía renovable con sistemas híbridos* (SEMARNAT, INECC, Clean Energy Investment Accelerator 2021).

[78] M Torres Wong and O Masera, 'Sistemas Energéticos Rurales: Justicia y Sustentabilidad' (2023) 7 Ciencias y Humanidades, Transición Energética y Cambio Climático <https://www.researchgate.net/publication/371965810_Sistemas_Energeticos_Rurales_Justicia_y_Sustentabilidad> accessed 20 June 2025.

[79] GIZ and DGRV, *Guía: cooperativas de energía sustentable en México. Cooperativas de consumo de energía solar (prosumidoras)* (GIZ 2020).

[80] The Mexican Climate Initiative developed this model, which aims to promote community development in rural Mexico through distributed solar photovoltaic energy generation models on communal lands owned by ejidos. This model requires the collaboration of both state governments and the private sector. Iniciativa Climática de México, 'Ejido Solar' (*Iniciativa Climática de México*) <https://www.iniciativaclimatica.org/ejidosolar/> accessed 20 June 2025.

become a viable business model by considering the possibility of selling electricity under the terms of Article 27 of the ESL.[81]

To achieve the Sustainable Development Goals (SDGs), the world must focus on two key objectives. First, we need to ensure that most of the population has access to electricity, as this access is essential for fulfilling fundamental rights. Second, it is crucial that this electricity is generated sustainably to minimize pollution and combat climate change. In this context, access to electricity should be recognized as a human right, encompassing both accessibility and affordability.

H. The Acknowledgement of Access to Energy as a Human Right Through Secondary Legislation

Secondary legislation since 1983 recognized the right to access electricity, but not as a human right. The Law of Public Service of Electricity of 1975, as amended in 1983,[82] stated in article 25, 'The Federal Electricity Commission must supply electricity to all who request it, unless there is a technical impediment or economic reasons to do so, without establishing any preference within each tariff classification.'[83] This provision was also integrated into the EIL which in its article 4 section II said 'The following are considered public and universal service obligations: II. Offer and provide the Electricity Supply to anyone who requests it, when technically feasible, under conditions of efficiency, quality, reliability, continuity, safety and sustainability.'[84] Finally, the ESL reproduces the same provision.[85] However, complete recognition of access to energy as a human right demands a constitutional amendment.

As mentioned, the Federal Constitution has not yet explicitly recognized the human right to access energy. According to ESL, the Mexican Government is obliged to provide the public service of electricity to each person who requests it. However, this right is contingent upon technical feasibility. This requirement reduces the scope of the right to energy postulated by the ESL.

Energy decentralization is a way of democratizing access to electricity. It enables remote or hard-to-reach areas to generate energy, reducing dependence on the national distribution grid. It also contributes to energy resilience by significantly reducing the number of blackouts or failures in regions prone to them, allowing users to improve the quality and timeliness of services that depend on electricity, such as heating, cooling, and lighting. In this sense, small-scale, distributed generation, and self-supply are straightforward means of guaranteeing the human right to access electricity and a healthy environment, as they promote the sustainable use of renewable energy sources for electricity generation.[86]

[81] Article 27 of the Energy Sector Law establishes: Distributed Generation can sell electricity and associated products through the Basic Services Provider.

[82] Presidencia de la República, 'DECRETO por el que se reforma la Ley del Servicio Público de Energía Eléctrica' (*Official Gazette of the Federation*, 27 December 1983) <https://dof.gob.mx/nota_to_imagen_fs.php?codnota=4841459&fecha=27/12/1983&cod_diario=208318> accessed 28 July 2025.

[83] Law of Public Service of Electricity [1975] (amended in 1983) art 25 <https://www.diputados.gob.mx/LeyesBiblio/abro/lspee/LSPEE_abro.pdf> accessed 24 July 2025.

[84] Electricity Industry Law [2014] art 4 s II.

[85] ibid.

[86] M Tzika, 'Decentralising and Democratising Energy with Neel Tamhane, Solar Strategy Lead from SPACE10' (2021) 4(1) The Urban Transcripts Journal <https://journal.urbantranscripts.org/article/decentralising-and-democratising-energy-guest-editor-maria-tzika-talks-to-neel-tamhane-solar-strategy-lead-from-spac

Recognizing the human right to access electricity cannot rely solely on the government. The State often lacks the resources necessary to provide energy to every corner of its territory, particularly in Mexico, due to insufficient investment in transmission lines, poor maintenance of existing infrastructure, and a lack of development for new projects.[87] Furthermore, according to the PETL, the Energy Transition and Sustainable Energy Use Plan (PLATEASE) will establish goals for reducing energy poverty, which could improve the development of distributed generation. This Plan, along with the National Electricity Sector Strategy, serves as a fundamental basis for ensuring access to electricity in areas where it is currently lacking, through on-site generation projects, in some cases for social and community purposes.

In that sense, promoting small-scale, distributed generation, and self-supply projects by the Mexican Government in collaboration with other stakeholders can contribute to achieving the goals of SDG 7, particularly to ensuring universal access to affordable, reliable, and modern energy services and guarantee the human right to access energy, as well as other human rights already acknowledged by the United Nations.

It is crucial for the legal system to create mechanisms that ensure access to electricity for everyone. If this is accomplished, access to electricity will become a reality, even if it is not considered a right.

IV. Case Study: Energy Decentralization Project in the Fishing Community of Javier Rojo Gómez, Commonly Known as Punta Allen, Located Within the Sian Ka'an Biosphere Reserve in Yucatán, Mexico

In Mexico, several energy decentralization projects have been implemented through small-scale, distributed generation, and self-supply energy projects.[88] These initiatives aim to provide access to electricity in areas far from the national grid and to support the energy transition. According to an official report,[89] various efforts have been made since 2015 to employ energy decentralization projects in different communities. These efforts have involved participation from various government bodies, civil society organizations, and support from international agencies.[90]

Among the various cases documented in Mexico, the authors specifically selected the 'Punta Allen' project, located in the fishing community of Javier Rojo Gómez, Tulum,

e10/#:~:text=La%20estrategia%20de%20Rutas%20de,sobre%20SPACE10%20y%20su%20visi%C3%B3n?> accessed 25 July 2025.

[87] R Fuentes, 'El problema energético de México' (*EGADE Business School. Tecnológico de Monterrey*, 27 August 2024) <https://egade.tec.mx/es/egade-ideas/investigacion/el-problema-energetico-de-mexico> accessed 25 July 2025.

[88] Deutsche Gesellschaft für Internationale Zusammenarbeit, 'Hoja de Ruta sobre modelos de negocio comunitarios y cooperativos de energía sutentable' (Ministerio Federal de Cooperación Económica y Desarrollo (BMZ) Germany 2022), <https://www.gob.mx/cms/uploads/attachment/file/721173/Hoja_de_Ruta_sobre_modelos_de_negocio_comunitarios_y_cooperativos_de_energ_a_sustentable.pdf> accessed 2 August 2025.

[89] SEMARNAT, INECC, and Clean Energy Investment Accelerator, 'Electrificación rural para comunidades fuera de la red utilizando generación de energía renovable con sistemas hibridos' (SEMARNAT, INECC, and Clean Energy Investment Accelerator 2021) <https://www.gob.mx/cms/uploads/attachment/file/690391/INFORME_Electrificacion_integrado18112021.pdf> accessed 18 July 2025.

[90] See Chapter 19 of this book.

Quintana Roo. The outcomes of these projects provide valuable lessons on effective implementation strategies.[91]

A. The Problem: Lack of Access to Energy in the Community of Punta Allen

'Punta Allen', whose official name is Javier Rojo Gómez, is a fishing community located within Tulum, Quintana Roo, in the Sian Ka'an Biosphere Reserve.[92] The community has a population of approximately 400 residents; its primary economic activities are fishing and tourism.[93] This issue arises from the community's geographic isolation, as the nearest electrical grid is located 55 kilometres away in the municipal capital, Tulum. This considerable distance made it expensive for the federal government to supply electricity to the area. Given that women are traditionally primarily responsible for household chores and family care, the lack of electricity significantly increases the time and effort required for these tasks, further limiting their available time for educational, economic, or leisure activities.[94]

In addition, this energy deficit hinders access to essential health information, negatively affecting the nutrition and safety of women, girls, boys, older people, and individuals with disabilities.[95] In summary, the lack of electricity significantly hinders equitable development opportunities and personal growth for vulnerable groups in the community.

B. The Unsuccessful Solution

To address the lack of energy access, the community was provided electricity via a diesel generator, which offered eleven hours of power daily, using 1,000 litres of diesel fuel. This system was costly[96] and highly polluting. Additionally, it was impractical to transport diesel from Tulum.

[91] ibid.

[92] According to the General Law of Ecological Balance and Environmental Protection [1988], biosphere reserves are biogeographic areas that represent one or more ecosystems not significantly altered by human action, which is why they require preservation. These protected natural areas are home to representative species of national biodiversity, including those considered endemic, threatened, or in danger of extinction. Due to its fragility, the activities permitted within a protected natural area are limited, excluding infrastructure construction for the electricity industry.

[93] Its population is composed of people from other communities who have migrated in search of economic growth, so its population includes Indigenous people. See J Guadalupe Cabrera Borraz, L Arroyo Arcos, and A Imbach, 'Pesca y turismo en una comunidad costera, a partir del aprovechamiento del patrimonio de la reserva de la biosfera de Sian Ka'an, Quintana Roo' (2018) 4(3) International Journal of Scientific Management and Tourism 55 <https://dialnet.unirioja.es/servlet/articulo?codigo=6773956> accessed 23 July 2025. According to the 2020 census conducted by INEGI, Punta Allen is inhabited by 196 women and 197 men. See INEGI, 'Censo de población y vivienda 2020' (*INEGI*) <https://www.inegi.org.mx/programas/ccpv/2020/#microdatos> accessed 24 July 2025.

[94] D Medina, 'El impacto de la pobreza energética en las mujeres mexicanas' (*Expansión*, 12 March 2024) <https://mujeres.expansion.mx/opinion/2024/03/12/el-impacto-de-la-pobreza-energetica-en-las-mujeres-mexicanas> accessed 24 July 2025; E Jaime, *Vivir a oscuras: la pobreza energética en México* (México evalúa 2021).

[95] A Driver, '5 maneras en que la falta de acceso a la energía afecta a las mujeres' (*Bid, Mejorando vidas*, 1 February 2025) <https://blogs.iadb.org/sostenibilidad/es/5-maneras-en-que-la-falta-de-acceso-la-energia-afecta-las-mujeres/> accessed 24 July 2025.

[96] Inhabitants were required to pay $550 MXN per house, which is a significant expense considering their income levels. YM García Beltrán, 'Generación de energía a baja escala en México: obstáculos y alternativas' (2024) 21(55) Andamios <https://doi.org/10.29092/uacm.v21i55.1107> accessed 21 February 2025.

The most suitable option identified was a hybrid wind and photovoltaic system designed to supply electricity to the community. However, this option proved too expensive for the residents, so they sought assistance from the Universal Service Fund,[97] which provided solar panels and batteries for residences in 2018.[98] Unfortunately, these batteries had a limited capacity of just 1 kWh, which was inadequate for the Punta Allen community, known for its vibrant economic activity and high electricity demands stemming from its tourism sector.

In October 2018, the non-governmental organization (NGO) 'Friends of Sian Ka'an' initiated the process of installing a renewable energy plant powered by photovoltaic technology, funded by the Fund for Energy Transition and the Use of Renewable Energy (FOTEASE).[99] The objective was to create a photovoltaic park capable of generating substantial energy to provide round-the-clock electricity for residents and the nearly 100,000 tourists visiting this fishing community each year. Additionally, the project would benefit nearby hospitals and schools.[100]

The photovoltaic park took five years to complete due to frequent policy changes and personnel turnover. In August 2022, a centralized photovoltaic generation system with lithium battery storage was installed. The project features a microgrid with a capacity of 167 kWp, consisting of 308 photovoltaic panels, each rated at 545 Watts.

Additionally, it incorporates a 500 kW conversion system and storage provided by 95 lithium batteries with a total capacity of 540 kWh. To ensure reliability, there are also two backup diesel generators, each with a capacity of 232 kVA. An Energy Management System (EMS) manages all components.[101]

The project successfully powered 202 homes, two health centres, three schools, seven co-operatives, thirteen hotels, fifteen restaurants, three small businesses, one meat processing plant, one auditorium, one library, and two religious centres.[102] It brings several benefits to the community. Due to this project's transformative impact, diesel consumption has plummeted by an impressive 50 per cent. It enabled the community to enjoy a significantly increased electricity supply, providing consistent daily access to energy. Residents have widely recognized this enhancement, as hundreds of essential establishments have benefited from the change. The reduction in fuel use has a socially and environmentally positive impact, and the territorial occupation of the

[97] The Universal Service Fund was a program aimed at achieving 99 per cent national electricity coverage by 2018, through support via financing for the installation of isolated systems consisting of photovoltaic cells.

[98] R Mora, 'En Punta Allen buscan soluciones ante ausencia de electricidad' (*La Jornada Maya*, 15 May 2018) <https://www.lajornadamaya.mx/quintanaroo/102096/en-punta-allen-buscan-soluciones-ante-ausencia-de-electricidad> accessed 20 February 2025.

[99] The Fund for Energy Transition and the Use of Renewable Energy is a public policy initiative established by the Mexican government to promote the development and implementation of renewable energy projects. It was created in 2008 under art 48 of the Energy Transition Law.

[100] Secretaría de Energía, 'Fondo para la Transición Energética y el Aprovechamiento Sustentable de la Energía' (*Gob.mx*, 6 December 2024) <https://www.gob.mx/sener/en/articulos/el-fondo-para-la-transicion-energetica-y-el-aprovechamiento-sustentable-de-la-energia-es-un-instrumento-de-politica-publica-de-la-secretaria> accessed 21 February 2025.

[101] J Zarco, 'Sian Ka'an' (*PV Magazin*, 8 August 2022) <https://www.pv-magazine-mexico.com/2022/08/08/siank-kaan/> accessed 21 February 2025.

[102] Secretaría de Desarrollo Social, 'Ficha técnica del proyecto Implementación de un sistema eléctrico renovable y sustentable en Punta Allen, Reserva de la Biosfera de Sian Ka`an, Quintana Roo' (Chetumal: Secretaría de Desarrollo Social 2019).

photovoltaic park did not affect the community's inhabitants. They only became aware of the plant's existence when they noticed the increased availability of electricity for more extended periods.[103]

Despite the above, since April 2023, the plant has stopped operations without any official explanation. The only available information indicates that 'technical details in its planning' have affected the functionality of the solar panels.[104]

C. Factors that Explain the Failure

The project's failure can be attributed to several factors. First, the community's energy needs were not adequately met because tourism demands significantly exceeded the electricity required for personal use. The project was designed to generate less than 0.20 MW, which classified it as an isolated supply under Article 22 of the EIL of 2014. This classification meant that it did not require a power generation permit and would not be connected to the national power grid or any interconnected load centre. However, according to the legal provision, the project had the potential to generate a maximum of 0.5 MW without needing a permit. This oversight resulted in the community not receiving the energy it needed.[105] This oversight likely occurred because social impact assessments were not conducted prior to the project's implementation, resulting in an inaccurate evaluation of energy requirements.[106]

Second, there was a lack of consistency between the project and the community's energy needs. The plant was originally designed to supply electricity exclusively to homes. However, private businesses, which account for 60 per cent of Punta Allen's energy consumption, were connected to the system.[107] This misalignment raises serious questions about the project's planning, execution, and ability to meet the community's needs.[108]

Third, there was inefficient communication between the committee, which included five residents, and the local population. This communication was crucial for residents to understand the importance of transitioning from a diesel-powered electricity generator to a clean energy production system through the photovoltaic park.

Finally, the unclear legal framework distinguishing distributed generation from isolated supply created uncertainty within the community.

As a result, even though clean energy generation offers significant benefits through increased energy availability, the project was seen as ineffective. This perception arose

[103] García Beltrán (n 96).

[104] A Nava, 'QRoo: fracaso de paneles solares en Punta Allen obliga a usar el diésel' (*Reportur*, 10 June 2024) <https://www.reportur.com/agencias/2024/06/10/qroo-fracaso-de-paneles-solares-en-punta-allen-obliga-a-usar-diesel/> accessed 21 February 2025.
MILENIO, 'Comunidad de Punta Allen sin electricidad por parque fotovoltaico fuera de servicio' (*MILENIO*, 19 December 2023) <https://www.milenio.com/videos/estados/comunidad-punta-allen-electricidad-parque-fotovoltaico-servicio> accessed 21 February 2025.

[105] Additionally, under the ESL, the new limit for power generation is set at 0.7 MW.

[106] In terms of Article 137 of the ESL, it is not mandatory to submit a social impact statement, as power plants with a capacity lower than 7 MW are exempt from this requirement.

[107] García Beltrán (n 96).

[108] See Chapter 9 of this book.

because the community had to pay two separate fees: one for the photovoltaic plant and another for the diesel needed for the combustion generator. The issue was compounded by the fact that the batteries did not have enough storage capacity to ensure a continuous 24-hour energy supply for the community.

Notwithstanding the above, this case demonstrates that distributed energy resources could serve as a mechanism to ensure the human right to access energy by providing electricity to communities isolated from the electrical grid. Additionally, this case illustrates that distributed energy resources can help ensure the human right to energy access by providing electricity to communities that are isolated from the electricity grid. What is more, these projects play a vital role in the transition to a zero-emissions economy. However, it is crucial to analyse the needs that must be addressed during the planning phase to ensure the project's suitability. Furthermore, social participation mechanisms should involve the beneficiaries and raise awareness about the project's significance regarding energy availability, its positive environmental impact, and the long-term benefits it will offer.

D. Lessons Learned

The analysis of the fishing community's case in southern Mexico, considering constitutional, legal, and regulatory provisions, reveals several important insights.

The legal mechanisms established by the law, such as the social impact assessment, were not properly conducted. Furthermore, the benefits established by the legislation to operate decentralized generation projects without prior permission up to a limit of 0.5 MW under the 2014 EIL (0.7 in the current legislation) were not taken advantage of.

The legal framework concerning energy prior to 2015 lacked suitable financial mechanisms to implement the right to energy effectively. This highlights the need for constitutional reforms to provide the necessary support for the energy transition. The new ESL has made significant progress by establishing the Universal Energy Service Fund; however, more specific regulations regarding the fund are necessary.

Although the new ESL contains mechanisms aimed at the energy transition, they are not sufficient. It is a reform that attempts to re-establish the public monopoly over the electricity sector and, at the same time, recognizes that without self-sufficiency, small-scale production, and distributed generation, the transition will not be achieved, nor will the human right to energy be realized. The case analysis also shows that without social participation, these decentralized energy projects cannot succeed.

Besides, the case analysed also reveals that for projects to be successful, it is necessary to combine accessible technology, governmental support, and community participation, as reflected in the growth of energy decentralization and its potential to democratize access to clean energy. Additionally, financial constraints remain a significant barrier to expanding decentralized energy project initiatives, which hinders energy justice for communities like the one analysed.

Thus, the case examined suggests that to make a reality of the right to energy, a more systematic and robust legal framework is essential. The energy reform approved in 2025 could be the missing link.

V. Legal Challenges and Recommendations

A. Legal Obstacles and Opportunities to Energy Decentralization

The analysis of the legal framework of the Mexican energy sector allows for the identification of several legal obstacles to energy decentralization projects. A significant impediment to energy decentralization in Mexico is the absence of a constitutional basis for it. The Federal Constitution does not mention energy decentralization in any of its sections. In contrast, the Mexican Constitutional model for the electricity industry relies on two main principles: centralized energy generation and the state's dominant role in various activities within the electricity industry's value chain, which are recognized as public services. These principles do not favour energy decentralization.

Considering the constitutional text as amended in 2024, planning, control, transmission, and distribution of electricity are considered public services and are managed by the government corporation, the Federal Commission of Electricity (CFE). Generation and supply, while not explicitly defined as public services, are not strictly reserved for a state monopoly, allowing for private investment. However, the constitutional text, together with the provisions of ESL, establishes a state semi-monopoly on electricity generation, which reduces the possibility of decentralization.

Another major obstacle is weak, unclear, inefficient, and fragmented legislation. The Mexican energy laws have undergone significant changes due to the constitutional reform of 2013. Following this amendment, the Federal Congress passed a new Law of the Electricity Industry in 2014, aimed at opening the sector to private investment. However, when a new government took office in 2018, it prioritized reversing these constitutional and legal reforms. In 2025, the Federal Congress repealed the 2014 law with a new law governing the electricity sector. As discussed in this chapter, the 2014 law was ambiguous in distinguishing between distributed generation and small-scale projects.

A third major obstacle is the centralist character of the regulatory framework of the electricity sector. As discussed in the previous sections of this chapter, the regulatory framework for the electricity sector in Mexico is designed to support a centralized electricity industry, where the State plays a dominant role through its public company, CFE. This structure limits private participation in the sector, as well as the growth of decentralized electricity, including distributed generation. The new law governing the electricity sector in Mexico presents a contradiction. On one hand, it grants the state a semi-monopoly over electricity generation. On the other hand, it aims to ensure that electricity is accessible to all who need it, thereby strengthening the right to electricity. However, this right is conditioned on the existence of technical feasibility. This circumstance raises concerns that the very lack of technical feasibility is what justifies the involvement of private enterprises in electricity generation. Consequently, the state's semi-monopoly may hinder the realization of this right.

However, it is also possible to identify a few elements that could support and encourage the development of energy decentralization projects in Mexico. Among such elements are, first, the push from countries with advanced electricity technology, social movements, and international commitments, particularly those established in the 2015 Paris Agreement;

second, the Mexican Constitution adopted the principle of sustainability in 1999, and a constitutional reform was eventually approved, recognizing the right to electricity as a human right.

In addition, the commitments adopted by Mexico in the 2030 Agenda and the approval of a new Energy Sector Law in 2025 are also key factors that could encourage the development of energy decentralized projects in Mexico.

Isolated self-consumption energy projects, which do not require permits, provide a significant opportunity to ensure access to energy for communities that still lack this service. In these projects, social participation is crucial. Therefore, it is important to develop strategies that encourage and ensure community involvement. It is encouraging that the ESL not only regulates the three-generation models but also includes provisions for financing, which could support these decentralization projects.

All of this, along with the primary goal of decentralizing energy generation, has the potential to reduce energy poverty and promote gender equality, inclusion, and sustainable development. The new law clarifies this distinction but only includes a few provisions addressing decentralized energy projects.

B. Legal Proposals

Energy decentralization in Mexico requires several changes in legislation. First, considering the principle of national treatment postulated by the USMCA, Mexico faces a critical obligation to repeal the rule established by the ESL that gives priority to electricity generation for the state-owned corporation. This rule clearly breaches the principle of national treatment enshrined in the trade agreement. By eliminating this rule, Mexico would unlock the electricity generation sector to private investment, significantly enhancing the decentralization of the electricity market and fostering a more competitive and innovative energy landscape. Second, the Federal Constitution requires reform to recognize the right to electricity as a human right without any conditions.

Unlike the situation with the EIL, the new ESL contains a specific section addressing various forms of decentralized generation, including distributed generation. However, these references are primarily formal and come with reservations due to the inherent limitations of the law. The regulations under ESL could complement these legal frameworks.

The rule of state supremacy on electricity generation should be repealed to incentivize energy decentralization. Additionally, incentives should be established, such as training programs for the population, users, generators, and investors. This would promote investment through funds, financing options, tax reductions, and simplified contracts for the purchase and sale of energy. Furthermore, systems should be created with the necessary equipment and technical characteristics to facilitate access to the grid. This includes the installation of transformers, substations, or storage systems needed to maintain the voltage required for small generators to sell electricity.

Despite the prioritization of electricity generation for the state-owned corporation in Mexico's energy policy, there is still an opportunity for participation from the private and social sectors in decentralized energy projects, whether interconnected or isolated.

VI. Conclusions

In México, the right to access electricity remains an aspiration. In this country, 3.4 million people lack access to electricity. Due to their distance from the transmission and distribution grids, it is not technically possible to provide them with such a public service. Technical obstacles are not the primary drivers explaining this failure of energy policy; the legal framework has been inefficient in supporting an appropriate strategy in this regard. Frequent changes in energy legislation and inconsistencies between legal provisions have created uncertainty and sometimes discouraged the development of decentralized generation projects. Fragmentation of the legal framework is also a factor that discourages decentralization.

Before 2024, legal issues have discouraged the implementation of decentralized energy projects. First, the EIL of 2014 did not establish economic incentives that make it easy for remote communities to support decentralized energy projects. Second, the Mexican energy legal framework undergoes frequent changes, generating uncertainty for those interested in conducting a decentralized energy project. Third, the principal aim of changes in energy law has not been energy transition, but the idea of strengthening the importance of the energy sector as a lever of national economic development, its role in representing national sovereignty, and the consideration of energy generation, transmission, distribution, and supply activities as a public service.

At the end of 2024, a constitutional reform led to a comprehensive overhaul of the legal framework governing the electricity industry. The new Energy Sector Law passed in 2025 includes provisions that rule decentralized generation projects, but administrative authorities must issue more specific regulations.

But the energy reform of 2024–2025 is not enough; additional modifications are required to stimulate decentralized energy projects in Mexico.

The Federal Constitution and the ESL do not explicitly recognize the right to access energy as a human right. However, various constitutional and legal provisions imply this right by stating that the Mexican government must provide electricity as a public service to anyone who requests it. Nonetheless, this obligation is considerably limited, as these same laws specify that the government will only fulfil this duty when it is technically feasible to do so.

Although decentralized energy projects appear to be the best avenue to guarantee access to energy, the analysis of the case of the fishers' community in southern Mexico shows that such strategies are not always successful. Furthermore, according to available data, decentralized energy projects do not play a significant role in energy generation in Mexico.

Fortunately, not all is negative; several factors could support and encourage the development of decentralized energy projects in Mexico. First, Mexico is committed to fulfilling the pledges made in the Paris Agreement of 2015 and the 2030 Agenda for Sustainable Development. Second, the current government's public policies recognize distributed generation as a key mechanism for achieving energy transition. Third, despite its intention of reestablishing the state monopoly on the energy sector, the new legal framework passed in 2025 includes new mechanisms aimed at promoting decentralized energy projects.

Decentralized generation projects in Mexico face obstacles that seem insurmountable: first is the idea that the state must always generate at least 54 per cent of the electricity, which establishes a state semi-monopoly. Second, the state is obliged to provide electricity to anyone who requests it, because it is a public service. Third, it seems that decentralized generation is only justified when it is technically impossible to extend the national transmission grid to hard-to-reach areas.

Finally, the primary barrier to decentralization in Mexico, is the nationalist and centralist nature of energy policy, which has prevailed except for brief intervals and appears unlikely to change in the foreseeable future.

11

Behind-the-Meter Distributed Generation

Lessons from the United States

LeRoy Paddock

I. Introduction

A sustainable approach to DER at the Federal level has proven elusive in the United States. Over the last two decades dramatic differences among the major political parties and their Presidential candidates has resulted in major shifts in policy. The Obama Administration supported efforts to quantify GHG releases and increase reliance on renewable energy, and emphasized energy efficiency.[1] These policies were rejected by the pro-fossil fuel first Trump Administration[2] and then revived by a Biden Administration for whom climate change and environmental justice were top priorities.[3] The second Trump Administration is even more focused on fossil fuel[4] and has terminated environmental and energy justice initiatives. Further, the current Republican-controlled Congress has repealed many DER-related financial incentives adopted during the Biden Administration.[5] Many states, however, have consistently supported policies that advance DER.[6]

In the United States most regulatory authority over electric utilities and the operation of the retail energy grid resides at the state level. The Federal government through the Federal Energy Regulatory Commission (FERC) has jurisdiction over the wholesale sale of energy working with voluntary organizations in most of the country known as Regional Transmission Operators (RTOs) that manage wholesale energy transactions. Both FERC through its policies on methods for meeting energy needs such as demand response programs and RTOs through their use of demand response programs and openness to incorporating aggregated distributed energy resources into their markets can encourage or constrain expansion of BTM DER.

[1] The White House, 'FACT SHEET: President Obama Announces New Actions to Bring Renewable Energy and Energy Efficiency to Households across the Country' (24 August 2015) <https://obamawhitehouse.archives.gov/the-press-office/2015/08/24/fact-sheet-president-obama-announces-new-actions-bring-renewable-energy> accessed 23 July 2025.

[2] See Executive Order 13783.

[3] See Executive Order 14008.

[4] See Executive Order 11991.

[5] One Big Beautiful Act (HR 1 2025).

[6] See eg J St John, '5 States Blazing the Trail for Integration Distributing Energy Resources' (*gtm* 9 September 2019) <https://www.greentechmedia.com/articles/read/the-top-states-for-distributed-energy-integration> accessed 23 July 2025; National Governors Association, 'State Approaches to Equitable Distributed Energy Resource Deployment' (2 March 2023).

LeRoy Paddock, *Behind-the-Meter Distributed Generation* In: *Sustainable Distributed Energy Resources*. Edited by: Íñigo del Guayo, Damilola S. Olawuyi San, Louis de Fontenelle, Milton Fernando Montoya, and José Juan González, Oxford University Press.
 DOI: 10.1093/9780198991946.003.0013

Perhaps the most powerful way the Federal government could support DERs would be adoption of national legislation that drives significant reduction of GHG emissions. Although there have been attempts to do so for more than twenty-five years, the United States still does not have a national law restricting GHG emissions. The Federal government could also encourage growth in DERs by adopting, as many states have, a renewable energy portfolio standard (RPS) requiring that a significant portion of the energy within each electric utility's service area either be derived from the utilities' own renewable energy facilities, through power purchase agreements, or by counting customers' DER in meeting their RPS. While legislation has been introduced to establish a national RPS, the legislation has not succeeded. Congress, however, has long provided financial incentives in the form of investment tax credits and tax rebates for renewable energy facilities such as solar panels and energy saving equipment such as heat pumps that are essential elements of DER.[7] The Biden Administration promoted legislation that dramatically expanded financial incentives for renewable energy facilities, DER-supportive electrical grid upgrades, and DER equipment.[8]

At least through these financial incentives, DER support at the Federal Level had appeared sustainable, although the level of financial support has ebbed and flowed. That support has recently changed with the Republican-led Congress, supported by the Trump Administration, passing the 'One Big Beautiful Act' that rescinds most renewable energy-related incentives.[9] The Trump Administration's pointed effort to focus on development and use of fossil fuels indicates that at the Federal level there is no longer a sustainable policy supporting DER. Instead, the Federal government may be antagonistic to DER and is attempting to limit some of the tools used to support DER by seeking to limit state laws that regulate GHG emissions.[10]

Many states have long been active supporters of DER-related policies. The policies in several states demonstrate that a sustainable approach to supporting DER is possible.[11] One state in particular, California, has staked out a world leadership position in developing a comprehensive approach to DER that addresses energy costs, energy reliability, and climate change. California's experience provides a good platform for examining the critical themes of decarbonization, digitalization, decentralization, and democratization that are the focus of this book.

7 See eg B Lips, 'The Past, Present and Future of Federal Tax Credits for Renewable Energy' (19 November 2024) <https://nccleantech.ncsu.edu/2024/11/19/the-past-present-and-future-of-federal-tax-credits-for-renewable-energy/> accessed 23 July 2025.

8 See Fiscal Responsibility Act of 2023 (HR 3746).

9 One Big Beautiful Act (n 5).

10 See White House, 'Executive Order Protecting American Energy from State Overreach' s 2 (8 April 2025) that provides: 'The Attorney General, in consultation with the heads of appropriate executive departments and agencies, shall identify all State and local laws, regulations, causes of action, policies, and practices (collectively, State laws) burdening the identification, development, siting, production, or use of domestic energy resources that are or may be unconstitutional, preempted by Federal law, or otherwise unenforceable. The Attorney General shall prioritize the identification of any such State laws purporting to address "climate change" or involving "environmental, social, and governance" initiatives, "environmental justice," carbon or "greenhouse gas" emissions, and funds to collect carbon penalties or carbon taxes.'

11 Hawaii, New York, Massachusetts, and Arizona have been recognized as state leaders in DER integration. St John (n 6).

II. Distributed Energy Resources in the United States

Small scale distributed energy resources have played an increasingly important role in the US energy portfolio over the last decade. The growth in DER has been driven by several factors including Federal tax incentives and FERC policies; state GHG goals, renewable energy portfolio standards, net metering laws; decreasing costs for solar energy and batteries; state tax incentives and changes in state utility regulations that facilitate the creation of microgrids; state laws that support demand response aggregation and virtual power plants (VPPs); and growing concerns about energy access/reliability issues due in part to increasing climate-related disruptions including floods and forest fires.[12] For the purposes of this chapter, DER includes behind-the-meter generation and storage, electric vehicles, VPPs, and flexible time-of-use smart electric equipment including appliances, heat pumps, and programmable thermostats.[13]

DER account for less than 5 percent of the energy supply in the US but estimates indicate US DER capacity will increase by 216 gigawatts (GW) through 2028.[14] In 2023, small-scale solar had a capacity of 48 GW, more than six times the amount in 2014.[15] BTM solar installations in commercial building have also increased dramatically over the past decade,[16] as have the deployment of microgrids.[17]

Battery storage, especially battery storage linked to new solar installation is another aspect of DER that has rapidly grown. In 2020 the US had about 3,000 MW of battery storage, almost one third of which was behind the meter.[18] The BTM storage capacity is expected to expand to over 7,500 MW by 2025.[19] One of the factors driving this expansion is state incentives for BTM storage in connection with new solar installations.[20] Similarly, the expansion of electric vehicles has created a significant new DER resource. One study indicated that 'A car with a 30 kWh battery stores as much electricity as the average U.S. residence consumes in a day.'[21] In 2022 there were fewer than five million electric vehicles in the US. The

[12] Deloitte, 'Households transforming the grid: Distributed energy resources are key to affordable clean power' figure 2 (12 June 2024) <https://www2.deloitte.com/us/en/insights/industry/power-and-utilities/der-grid-modernization.html> accessed 29 July 2025.

[13] Not all behind-the-meter resources rely on renewable energy. For example, a large data centre may have dozens of these backup generators that remain idle much of the time. These generators are increasingly seen as DER that can be called on when needed by a grid operator. See eg D Kaeding, 'Microsoft to use diesel-fired generators as backup power for data centers' (Wisconsin Public Radio, 21 November 2024) <https://www.wpr.org/news/microsoft-to-use-diesel-fired-generators-as-backup-power-for-data-centers> accessed 29 July 2025.

[14] M Quinlan and L Ford, 'Distributed Energy Resources: Technology for affordable, resilient power' (Pew 24 March 2025) <https://www.pewtrusts.org/en/research-and-analysis/articles/2025/03/24/distributed-energy-resources-technology-for-affordable-resilient-power> accessed 29 July 2025.

[15] Climate Central, 'A Decade of U.S. Solar Growth' (3 April 2024) <https://www.climatecentral.org/climate-matters/a-decade-of-us-solar-growth-2024> accessed 29 July 2025.

[16] Facilities Management Advisor, 'Distributed Energy Resources "Rapidly Expanding" in Commercial Buildings' (19 October 2021) < https://facilitiesmanagementadvisor.com/green-building/distributed-energy-resources-rapidly-expanding-in-commercial-buildings/> accessed 23 July 2025.

[17] H Nilsson, 'US microgrid market to grow 19% annually through 2027' (8 February 2023) <https://www.utilitydive.com/news/us-microgrid-market-wood-mackenzie/642341/> accessed 23 July 2023.

[18] JJ Cook and others, 'Check the Storage Stack: Comparing Behind-the-Meter Energy Storage State Policy Stacks in the United States' (National Renewable Energy Laboratory 2022) 1 <https://www.nrel.gov/docs/fy22osti/83045.pdf> accessed 28 July 2025.

[19] ibid.

[20] A Colthorpe, 'California, Hawaii drive US to busiest ever quarter for behind-the-meter energy storage' (*Energy Storage News* 7 September 2017) <https://www.energy-storage.news/california-hawaii-drive-us-to-busiest-ever-quarter-for-behind-the-meter-energy-storage/> accessed 28 July 2025.

[21] C Nelder, J Newcomb, and G Fitzgerald, 'Electric Vehicles as Distributed Energy Resources' (Rocky Mountain Institute 2016) 6 <http://www.rmi.org/pdf_evs_as_DERs> accessed 28 July 2025.

Department of Energy projects even under a slow adoption rate scenario that the number will increase to over 30 million by 2030.[22] Finally, other flexible time-of-use equipment such as smart thermostats and smart appliances that are supported by smart electric meters[23] have dramatically expanded.[24] One estimate shows that flexible time-of-use resources including EVs and smart appliances could grow from 9 GW in 2023 to 107 GW in 2035,[25] a potentially valuable resource in meeting peak load needs.[26] A significant element of smart appliance transformation is the growing shift to heat pumps with air source heat pump furnace deliveries now exceeding those of gas furnaces.[27]

Wood Mackenzie's latest analysis shows that 262 gigawatts (GW) of new DER and demand flexible capacity will be installed from 2023 to 2027, close to matching the 272 GW of utility-scale resource installations also expected during that period. The study found that grid insecurity concerns 'of homeowners and business owners represents the strongest demand pull, which will cause the distributed fuel-based generation market to grow 240% from 2022 to 2027, and the distributed storage market to soar 460%, with storage reaching nearly US $10 billion per year'.[28]

III. National Policy Related to DER

Legislation that requires significant reduction in GHG emissions is an important incentive to use DERs that can reduce fossil fuel electricity generation. The US House of Representatives enacted cap-and-trade climate legislation in 2009, but the bill did not pass in the US Senate.[29] National legislation regulating greenhouse emissions remains out of reach in the country. Instead, Congress has relied primarily on financial incentives to promote renewable energy generation and DERs including investment tax credits.[30]

22 Department of Energy, 'Impact of Electric Vehicles on the Grid' (2024) 5 <https://www.energy.gov/sites/default/files/2024-10/Congressional%20Report%20EV%20Grid%20Impacts.pdf> accessed 28 July 2025.

23 'In 2022, U.S. electric utilities had about 119 million advanced (*smart*) metering infrastructure (AMI) installations, equal to about 72% of total electric meters installations. Residential customers accounted for about 88% of total AMI installations, and about 73% of total residential electric meters were AMI meters.' US Energy Information Agency, 'How many smart meters are installed in the United States, and who has them? (undated) <https://www.eia.gov/tools/faqs/faq.php?id=108&t=3> accessed 9 August 2025.

24 Statistica, 'Smart Appliances—United States' (undated) <https://www.statista.com/outlook/cmo/smart-home/smart-appliances/united-states> accessed 29 July 2025.

25 Deloitte (n 12).

26 ibid.

27 M Malhotra and others, 'Heat Pumps in the United States: Market Potentials, Challenges and Opportunities' (undated) <https://heatpumpingtechnologies.org/national-report-heat-pumps-in-the-united-states-market-potentials-challenges-and-opportunities/> accessed 28 July 2025.

28 W McKenzie, 'US Distributed Energy Resource market to almost double by 2027' (20 June 2023) <https://www.woodmac.com/press-releases/us-distributed-energy-resource-market-to-almost-double-by-2027/> accessed 29 July 2025.

29 See American Clean Energy and Security Act of 2009 (HR 2454).

30 Carbon dioxide has been recognized by the US Supreme Court as an air pollutant under the Clean Air Act's (CAA) definition of pollutant in *Massachusetts v EPA*, 49 US 497 (2007). Because carbon dioxide is a pollutant under the CAA, EPA was compelled to decide whether carbon emissions 'endangered' public health. The agency did find carbon emissions endangered public health and based on that finding established vehicle emissions standards designed to address carbon emissions. The Trump Administration's EPA Administrator has announced that the Agency intends to adopt a new rule revoking the current EPA finding that carbon emissions 'endanger' public health. US Environmental Protection Agency, 'EPA Releases Proposal to Rescind Obama-Era Endangerment Finding, Regulations that Paved the Way for Electric Vehicle Mandate' (29 July 2025) <https://www.epa.gov/newsreleases/epa-releases-proposal-rescind-obama-era-endangerment-finding-regulations-paved-way> accessed 30 July 2025.

Reducing greenhouse gases was a top priority of the Biden Administration. One of President Biden's first Executive Orders focused on climate change providing that the Federal government would 'deploy the full capacity of its agencies to combat the climate crisis'.[31] Among the areas of focus for the Administration was research on digitally enabled technology that supports DER including work on batteries and battery management systems, inverter controls and power conversion equipment, distributed control systems, building management systems, electric vehicles, and electrical vehicle supply systems.[32] This initiative is emblematic of strong support during the Biden Administration for DER as a means of reducing GHG emissions.

The Administration's signature step in supporting DER was the 2022 Inflation Reduction Act. The Biden Administration encouraged Congress to adopt legislation that would provide new financial incentives to encourage rapid adoption of renewable energy facilities and DER including battery storage, electric vehicles including charging infrastructure, and smart appliances such as heat pumps and thermostats. The Inflation Reduction Act of 2022 invested $369 trillion in decarbonization.[33] The majority of these tax credits supported larger scale renewable energy generation facilities. However, a significant portion of the tax incentives supported BTM DER. The Act provided a range of tax credits including up to 30 per cent income tax credits for energy efficiency projects including installation of heat pumps and for renewable energy generation.[34] The Act also provided tax credits of up to $7,500 for certain electric vehicles and up to $1,000 for electric charging stations. Several other tax credits and deductions were provided by the Inflation Reduction Act including credits for energy storage.[35] While tax credits for renewable energy projects date back to the 1970s, the scope of the Biden Administration credits dwarfed earlier efforts.

The Biden Administration approach has proven to be unsustainable, however. On day one of the Trump Administration Executive Order 14154 titled 'Unleashing American Energy' announced a national policy of encouraging energy production on federal lands,[36] eliminating the electric vehicle mandate,[37] revisiting energy efficiency standards,[38] revoking all of President Biden's climate and energy related Executive Orders,[39] and ordering a government-wide review of policies that might hinder development of fossil-fuel resources.[40] This Order and subsequent actions by the Trump Administration indicate that there will be little support for DER.

Further, the Republican-led Congress in the 2025 budget bill, referred to as the 'One Big, Beautiful Act' by the Trump Administration, phases out production and investment tax

[31] Executive Order 14008 (27 January 2021).

[32] White House, 'Fact Sheet: Biden-Harris Administration Announces Priorities for Enhancing the Digital Ecosystem to Support a Secure Energy Future' (9 August 2024) <https://bidenwhitehouse.archives.gov/oncd/briefing-room/2024/08/09/fact-sheet-biden-harris-administration-announces-priorities-for-enhancing-the-digital-ecosystem-to-support-a-secure-energy-future/> accessed 29 July 2025.

[33] A Aznar, 'The Inflation Reduction Act, Distributed Resources, & Energy Efficiency' (18 April 2023) <https://virtual-peaker.com/blog/the-inflation-reduction-act-distributed-energy-resources-energy-efficiency/> accessed 9 August 2025.

[34] ibid.

[35] Internal Revenue Service, 'Credits and deductions under the Inflation Recovery Act of 2022' <https://www.irs.gov/credits-and-deductions-under-the-inflation-reduction-act-of-2022> accessed 28 July 2025.

[36] Executive Order 14154 s 2(a).

[37] ibid s 2(e).

[38] ibid s 2(f).

[39] ibid s 4.

[40] ibid s 3.

credits for renewable projects, eliminated tax credits for purchasing electric vehicles at the end of September 2025, and ends tax credits for a variety of household DER such as heat pumps and battery storage at the end of 2025.[41]

The FERC has had a significant impact on DER in the US through the issuance of Order 2222 in 2020. Since FERC oversees the wholesale sale of electricity in the United States, it has the authority to set the rules for the Regional Transmission Organizations. RTOs typically deal with large sources of generation so their rules for participating in the energy market for their region often did not provide a way for smaller sources to participate in the wholesale market. Order 2222 directed all RTOs to develop procedures that would make it possible for DER aggregators (entities that work with several DER sources to accumulate a significant quantity of generation or curtailment resources) to participate in the regional energy markets. The Order could have a significant impact in opening new markets for DER although its actual impact is still unclear since the process of modifying RTO procedures has been slow.[42] However, there are indications that the Trump appointees to FERC may not support Order 2222.[43]

Federal government support of DER has been important in expanding the use of DER but that support has dramatically diminished and has proven unsustainable. However, states have long had a much greater impact on DERs. No state has had a bigger influence on DER than California. The remainder of the chapter focuses on the financial, public policy, and legal issues in California that have made it one of the world leaders in DER but also have exposed several significant issues that arise as DER becomes a much bigger part of the energy picture.

IV. DER in California—A High Implementation Case Study

Because states have historically been responsible for electric utility regulation and are responsible for retail rate regulation they have had a larger impact on DER implementation. No state has played a more significant role than California. California has the world's fifth largest economy, is a major energy user, has set aggressive GHG reduction goals, and as a coastal state and a state susceptible to major forest fires is vulnerable to the impacts of climate change. The state's climate goals include reducing greenhouse gases by 85 percent by 2045[44] including 100 percent of electrical sales to end-use customers from renewable resources.[45] This goal envisions 12,000 MW of distributed generation to support the states zero carbon goal.[46] The state's zero carbon initiative also

[41] See One Big Beautiful Act (n 5) ch 5.

[42] R Walton, 'FERC Order 2222 hurdles require new options for deploying aggregated DERs: Guidehouse' (11 January 2024) <https://www.utilitydive.com/news/ferc-order-2222-implementation-challenges-Guidehouse-alternative-methods-DER-participation/704250/> accessed 23 July 2025.

[43] See JL Key, 'Wither Order No. 2222' (20 May 2025) <https://www.steptoe.com/en/news-publications/purpa-and-distributed-energy-resources-blog/wither-order-no-2222.html> accessed 22 July 2025.

[44] Governor G Newsom, 'California Releases World's First Plan to Achieve Net Zero Carbon Pollution' (undated) <https://www.gov.ca.gov/2022/11/16/california-releases-worlds-first-plan-to-achieve-net-zero-carbon-pollution/> accessed 23 July 2025.

[45] 100 per cent Clean Energy Act of 2018 (SB 100).

[46] Power, 'Distributed Generation: California's Future' (1 January 2013) 4 <https://www.powermag.com/distributed-generation-californias-future/> accessed 23 July 2025. In 2023 California had 3,000 MW of distributed generation in place with another 6,000 MW under development or authorized.

provides for 100 per cent zero emissions vehicles by 2035,[47] creating a significant DER.

A variety of converging circumstances early in the century led California to become the leading state in distributed generation. Among those circumstances were dramatically spiking energy costs and blackouts resulting from deregulation in 2000.[48] Prices for wholesale energy that had averaged between $20 to $50 per megawatt hour (MW/H) soared to over $400 per MWH causing financial turmoil including bankruptcy of one of the state's three primary investor-owned utilities (IOUs).[49] The disruptions occurred following deregulation of the electricity generation sector in 1996[50] and led to further restructuring of the state's electricity regulatory system.[51] At about the same time concerns were increasing in the state about climate change. California's first climate assessment report was issued in 2006.[52] The Assessment helped support the adoption of the country's only economy-wide, state-based greenhouse gas cap and trade system with the passage of the Global Warming Solutions Act (AB32) in 2006. These circumstances led California to pursue a wide range of policies that could help stabilize costs while relying very heavily on renewable energy and DER to meet the state's energy needs.

California is a world leader in addressing climate change. Its DER-related legislation, policies, and initiatives have been a major part of the effort. They provide an interesting menu of approaches for promoting DER. Experience in implementing DER programs in California has also identified a number of problems that can arise when incorporating DER in the energy transition including equity concerns as DER reduce the number of customers paying for the operation of the legacy grid, significant transmission constraints that have resulted in curtailment of DER resources, the availability of public data needed to understand where DER would be best located, excess solar energy production during day time hours that do not match demand peaks in early evening hours creating the California 'duck curve' (showing that energy demands grow rapidly in the early evening hours just a solar generation diminishes), difficulties in utilities' ability to 'see' what DER resources are available to the grid, high energy costs, blackouts or brownouts resulting from severe weather events or high winds that could cause forest fires, and constraints on two-way energy flows. In 2022, the California Independent System Operator (CALISO) curtailed 2.4 million MW hours of wind and solar generation either because of limited transmission capacity or excess generation.[53] The curtailment was slightly less than five percent of the wind and solar energy generated in 2023.[54] In addition, California has had to curtail power at times when there is a high potential for forest fires because of

[47] California Air Resources Board, '2020 Mobile Source Strategy' (28 October 2021) <https://ww2.arb.ca.gov/sites/default/files/2021-12/2020_Mobile_Source_Strategy.pdf> accessed 23 July 2025.

[48] Pacific Northwest National Laboratory, 'Energy Policy Case Study—California: Renewable and Distributed Energy Resources' (September 2016) 2–3.

[49] C Weare, 'The California Electricity Crisis: Causes and Policy Options' (Public Policy Institute of California 2003) 1-2 <https://www.ppic.org/wp-content/uploads/content/pubs/report/R_103CWR.pdf > accessed 30 July 2025.

[50] ibid v.

[51] ibid vii.

[52] California's Fourth Climate Change Assessment, 'About the Assessment' (undated) <https://www.climateassessment.ca.gov/about/> accessed 29 July 2025.

[53] P Maloney, 'Wind and Solar Curtailments are Rising in California Because of Congestion: EIA' (American Public Power 14 November 2023) <https://www.publicpower.org/periodical/article/wind-and-solar-curtailments-are-rising-california-because-congestion-eia> accessed 23 July 2025.

[54] NRDC, 'Feasting on the Glut: Putting Renewable Curtailment in Context' (4 December 2023) <https://www.nrdc.org/bio/mohit-chhabra/feasting-glut-putting-renewable-curtailment-context> accessed 30 July 2025.

the history of fires caused by sparking transmission lines resulting in much more frequent blackouts.[55] IOUs now shut down electric service when high winds are expected in an area to minimize the possibility of sparking fires.

The remainder of the chapter addresses several of the approaches California has used to promote clean energy and DER as well as how the state has attempted to address the problems discussed above.

A. Self-Generation Incentive Program (SGIP)

The SGIP was enacted in 2001 through AB970 in response to the 2000-2001 energy crisis to encourage distributed energy generation that could be called upon to help reduce peak demand.[56] Incentives for BTM solar generation were removed from the SGIP in 2006 with the advent of the California Solar Incentive program.[57] Most of the SGIP today supports battery storage, in significant part to respond to increasing power outages associated with forest fires.[58] Anticipating the fire season, the California Public Utilities Commission appropriated one billion dollars for battery storage installation rebates during 2024.[59] Rebates for an average size 10kwh battery is $1,500 but can be much higher for low-income residents, residents in certain fire-prone areas, areas that have experienced multiple blackouts, and residents with serious medical conditions.[60] SGIP covers approximately 25 percent of the cost of typically residential storage battery.[61]

The current version of SGIP addresses two critical problems that have arisen as the focus on DER increased. First, it helps address the duck curve issue by being able to store excess energy during off-peak hours and discharge energy during the early evening surge in demand in California. Second, SGIP helps address equity concerns by providing higher rebates to communities that either have a lower ability to pay the cost for battery storage or that are at higher risk.

[55] California's inverse condemnation regime makes electric utilities strictly liable for fire damage caused by malfunctions of energy transmission equipment. See G Petek, 'Allocating Utility Wildfire Costs: Options and Issues for Consideration' (Legislative Analyst's Office 19 June 2019) <https://lao.ca.gov/reports/2019/4079/allocating-wildfire-costs-062119.pdf> accessed 30 July 2025.

[56] San Diego Gas & Electric, 'Self-Generation Incentive Program: Background and History' (undated) <https://sgipsd.org/background> accessed 29 July 2025.

[57] A Fero and PJ Ferguson, 'Evolution of the Self-Generation Incentive Program' (17 June 2020) <https://www.dwt.com/blogs/energy--environmental-law-blog/2020/06/california-sgip-wildfire-power-shutoffs> accessed 23 July 2025.

[58] ibid. California utilities increasingly shut down power lines in areas where fire danger is high to avoid sparking fires. These instances are called Public Safety Power Shutoffs. About 10 per cent of the forest fires in California are caused by powerlines. See California Public Utilities Commission, 'Public Safety Power Shutoffs (PSPS)' (undated) <https://www.cpuc.ca.gov/psps/> accessed 23 July 2025. In a two-year period between 2017 and 2019 there were 2,374 PSPSs. Bloomberg Energy, 'California Power Outage Map' (undated) <https://www.bloomenergy.com/bloom-energy-outage-map/> accessed 23 July 2025.

[59] California Public Utilities Commission, 'Participation in the Self-Generation Incentive Program (SGIP)' (undated) <https://www.cpuc.ca.gov/industries-and-topics/electrical-energy/demand-side-management/self-generation-incentive-program/participating-in-self-generation-incentive-program-sgip> accessed 23 July 2025.

[60] S Wigness, 'California SGIP Battery Rebate: How to Lower Your Battery Costs' (25 October 2024) <https://www.solar.com/learn/sgip-info-battery-storage-rebate-up-to-100-percent/> accessed 9 August 2025.

[61] California Public Utilities Commission (n 59).

B. Renewable Portfolio Standard

California was one of the first states to adopt a renewable portfolio standard. In 2002 Senate Bill 1078 required the state's electric load serving entities to procure 20 per cent of their energy from renewable sources by 2017. Today, the ambitious California RPS created by Senate Bill 100 requires load serving entities to procure 60 per cent of their energy from renewable sources by 2030.[62] All of the state's electricity generation must come from renewable sources by 2045.[63] In 2022, 52 per cent of the energy delivered by the three largest investor-owned utilities was generated using renewable energy sources.[64] While customer-generated renewable energy does not directly count towards the state's RPS goal, BTM resources do reduce the amount on energy that is served by the load serving entities decreasing the total amount of renewable energy the utilities must generate using renewable energy sources.[65] In 2019 California had 8.0 GW of customer-generated solar energy, about two-thirds of solar generation that year. Customer-generated solar energy is expected to grow to over 12 GW in 2030 and over 28 GW by 2045, about 40 percent of all California solar generation in 2045.[66] These figures indicate that DER will make a significant contribution to meeting the state's RPS.

C. Net Metering

Today fourty-four states and the District of Columbia have net metering laws. California was a relatively early adopter of net metering in 1995.[67] California's initial net energy metering program (NEM-1) allowed residential solar customers to sell up to 10kw of energy they did not need back into the grid at the full retail rate of the energy.[68] NEM-1 remained in place until the California IOUs reached 5 percent of their power demand from residential solar installations in 2016.[69] NEM-2 made a few changes to the net metering approach including the price paid for electricity fed into the grid but largely continued the program spurring continuing growth in BTM solar energy. Payments under NEM-2 were based on time of use rates that ranged from about 35 cents off-peak to 45 cents for peak use periods.[70] NEM-1

[62] California Public Utilities Commission, 'RPS Program Overview' (undated) <https://www.cpuc.ca.gov/industries-and-topics/electrical-energy/electric-power-procurement/rps/rps-program-overview> accessed 23 July 2025.

[63] California Public Utilities Commission, 'Renewable Portfolio Standard (RPS) Program' (undated) <https://www.cpuc.ca.gov/rps/> accessed 23 July 2025.

[64] ibid.

[65] See J St John, 'Rooftop solar as nothing but lost load: The limits of the RPS construct for valuing DER' (*gtm* 31 August 2018) <https://www.greentechmedia.com/squared/dispatches-from-the-grid-edge/the-grid-edge-implications-for-californias-100-by-2045-clean-energy-plan#:~:text=The%20first%2C%20and%20most%20obvious%2C%20point%20about,in%20meeting%20the%20state's%20renewable%20energy%20targets> accessed 23 July 2025.

[66] SB 100 Joint Agency Report Summary, 'Achieving 100 Clean Electricity in California: An Initial Assessment' (3 September 2021) < https://www.energy.ca.gov/publications/2021/2021-sb-100-joint-agency-report-achieving-100-percent-clean-electricity> accessed 9 August 2025.

[67] SB 656 (1995).

[68] Sunova, 'The Solar Gold Rush: Why Californians are Racing to Go Solar' (6 March 2023) <https://www.sunnova.com/watts-up/california-net-metering-changes> accessed 23 July 2025.

[69] ibid.

[70] J Miller, 'NEM-3 Explained: What Every California Homeowner Needs to Know' (3 September 2024), <https://www.stackrackbattery.com/blogs/stackrack-residential-battery-systems-blog/nem-3-explained-what-every-california-homeowner-needs-to-know> accessed 23 July 2025.

and NEM-2 spurred a dramatic increase in household solar generation but also resulted in two significant concerns. The first was an equity concern that a disproportionate number of the NEM payments were going to higher income households resulting in a higher proportion of the costs of the legacy grid having to be paid by the remaining customers who were increasingly lower income customers.[71] A second concern was that the payments under NEM spurred too rapid growth of BTM solar for the grid to accommodate. These and other reasons led to a significant revision of the net metering requirement in NEM-3.

NEM-3 dramatically changed the compensation model moving to rates closer to wholesale electric prices rather than retail rates, utilizing an avoided cost model rather than a time of use model for rates. Avoided cost reflects the costs that utilities do not have to incur because of the existence of the distributed energy resource.[72] The NEM-3 rates are about 75 percent lower than NEM-2 rates. CPUC Commissioner Clifford Rechtschaffen noted: 'The decision strikes the right balance between many competing priorities and advances our overarching goals of ensuring California meets its climate and clean energy goals equitably.' To address the time of generation issue, NEM-3 provides additional compensation for expansion of BTM battery storage.[73] NEM-1 and NEM-2 had helped support the installation of 1.5 million BTM solar installations that produce 12,000 MW of energy.[74]

The NEM-3 rates have resulted in a slowdown in BTM solar installations in California. Industry estimates indicate 17,000 solar jobs were lost by the end of 2023 and rooftop solar sales were down by more than 66 percent.[75] While BTM solar generation is expected to continue it is likely to do so at a significantly slower rate than under NEM-2. The net metering experience in California demonstrates the importance of understanding the implications of significant subsidizes for DER on the entire electricity system, the importance of adjusting to changes produced by subsidies, and the impact those adjustments can have on the industries involved.

D. Rooftop Solar Initiative

In 2005 California had 20,000 household solar installations. As part of the response to the 2000-2001 energy crisis, the Million Rooftops Initiative was enacted in 2006 providing $3 billion in declining rebates for household solar installations with a goal of adding 3,000 MW of power. The legislation contributed to a rapid growth in household solar generation that by 2019 included 8,700 MW of installed household solar energy.[76] The legislation

71 See S Borenstein, 'California's Solar Equity Challenge' (Kleinman Center for Energy Policy 7 May 2024) <https://kleinmanenergy.upenn.edu/commentary/podcast/californias-solar-equity-challenge/> accessed 23 July 2025.

72 ibid.

73 CPUC, 'CPUC Modernizes Solar Tariff to Support Reliability and Decarbonization' (15 December 2022) <https://www.cpuc.ca.gov/news-and-updates/all-news/cpuc-modernizes-solar-tariff-to-support-reliability-and-decarbonization> accessed 23 July 2025.

74 ibid.

75 K Balaraman, 'California rooftop solar had a bad year following NEM-3. Can the industry bounce back' (2 January 2024) <https://www.utilitydive.com/news/california-rooftop-solar-nem-30-outlook/702498/> accessed 23 July 2024.

76 D Jacobson, 'Three Lessons from California's Million Solar Roofs Milestone' (1 January 2020) <https://environmentamerica.org/california/articles/three-lessons-from-californias-million-solar-roofs-milestone/> accessed 9 August 2025.

was amended in 2019 to require a solar installation that meets the energy demands of the building on all new low-rise residential buildings beginning in 2020.[77] The size of the solar installation can be reduced if it includes battery storage.[78] While the financial incentives for adding solar facilities has been reduced under NAM-3, the new residence solar mandate will continue to promote additional DER that includes both solar and battery storage.

E. Financial Incentives

In 2018, the California legislator[79] created an exclusion from sale and use tax for an active solar energy system that is used 'to provide for the collection, storage, or distribution of solar energy'.[80]

F. Electric Vehicles

California began focusing on zero emissions vehicles in 1990 to address air pollution which in the Los Angeles Valley was among the worst in the country.[81] Because of the severe air quality problems in California, Congress in the Clean Air Act[82] allowed California to apply for waivers that would allow the state's vehicle emission standards to be more stringent than national vehicle emissions standards.[83] California has been granted over 100 waivers over more than thirty years.[84] In 2022 California adopted new vehicle emissions standards for passenger vehicles and light trucks requiring that all such vehicles sold in 2035 achieve zero emissions.[85] California applied for and was granted a Clean Air Act waiver for the standard.[86]

About 23 percent of all new cars sold in California were zero emissions[87] during the first three months of 2025. Californians own 2.2 million zero-emission cars.[88] California had

[77] California Energy Commission, '2022 Single-Family Solar PV' (undated) <https://www.energy.ca.gov/programs-and-topics/programs/building-energy-efficiency-standards/energy-code-support-center/2022-0> accessed 23 July 2025.

[78] ibid.

[79] AB 1817 (2018).

[80] California State Board of Equalization, 'Active Solar Energy System Exclusion' (undated) <https://www.boe.ca.gov/proptaxes/active-solar-energy-system?> accessed 3 December 2025.

[81] California Air Resources Board, 'Zero Emissions Vehicle Program: About' (undated) <https://ww2.arb.ca.gov/our-work/programs/zero-emission-vehicle-program/about> accessed 23 July 2025.

[82] Clean Air Act, s 209(b).

[83] California Air Resources Board (n 82).

[84] Congress.gov, 'California and the Clean Air Act (CAA) Waiver: Frequently Asked Questions' (undated) <https://www.boe.ca.gov/proptaxes/active-solar-energy-system.htm#Description> accessed 23 July 2025.

[85] 13 California Code of Regulations s 1962.4.

[86] Congress recently approved legislation that was signed by President Trump revoking the waiver using a tool that is designed to allow Congress to rescind recent Federal regulations. California is challenging the revocation. D Quinley and KE Shea, 'California Attempts to Catch a "Waive": State Leads 10 Others Suing Trump Administration Over Revocation of Vehicle Emissions Waivers' (17 June 2025) <https://www.dwt.com/blogs/energy--environmental-law-blog/2025/06/california-epa-clean-air-emission-waivers-lawsuit> accessed 30 July 2025.

[87] For details on the numerous California laws related to EVs, incentives for EVs and EV charging stations, and the use of EV batteries as a DER see US Department of Energy, 'Electricity Laws and Incentives in California' (undated) <https://afdc.energy.gov/fuels/laws/ELEC?state=CA> accessed 23 July 2025.

[88] A Lazo and A Reyes-Valarde, 'US Senate blocks California's electric car mandate in historic vote' (22 May 2025) < https://calmatters.org/environment/2025/05/california-electric-car-mandate-senate-revoke-waiver/> accessed 30 July 2025.

provided rebates of up to $7,500 for electric vehicle purchases but that program was discontinued in November 2003.[89]

G. Solar Energy Assistance Program

As part of the California 'Go Solar Program', the California Solar Initiative provided $3 billion to support for a variety of DER programs, The Initiative closed on 31 December 2016. The Initiative included single-family and multifamily affordable solar programs, a solar hotwater heating program, and a low-income solar thermal program.

H. Virtual Power Plants

While demand response programs have had limited impact in California, a new approach called 'virtual power plants' (VPPs) is being tested in the state. The Rocky Mountain Institute describes VPPs as 'grid-integrated aggregations of distributed energy resources such as batteries, electric vehicles, smart thermostats, and other connected devices [that] can help balance electrical loads and provide several grid and community benefits, such as supplying critical grid services (including capacity, energy, ancillary services, and resilience), alleviating stressed transmission and distribution systems, and integrating more renewables for a cleaner energy supply'.[90] A 2024 report observed that increasingly extreme weather events, the growth in dependence on intermittent energy sources, reduced flexibility with hydro-electric assets, and reduced availability of energy imports have created serious problems and have resulted in the doubling of resource adequacy (RA) costs since 2017 in California.[91] The study looked at the 'market potential' of VPPs in California if all cost-effective VPP capacity can be developed at achievable, voluntary participation rates by 2035. The study looked at smart thermostats for air conditioning, BTM batteries, residential EV chargers, grid-interactive water heating, and automated demand response for commercial and industrial facilities.[92] The study found that VPPs could provide over 7,500 MW of peak demand energy by 2035 as compared to 1600 MW in 2024. The 2035 figure could provide over 15 percent of peak load needs.[93] The study estimated that $750 in utility costs could be avoided each year per customer in the VPP.[94]

Deployment of VPPs in California has been slow in part because VPPs have not been widely recognized as a reliable source of energy. A few pilot projects have been approved which have proven successful in regularly drawing on DER batteries to supply energy in the

[89] California Clean Vehicle Rebate Project, 'CVRP Overview' (undated) <https://cleanvehiclerebate.org/en/cvrp-info> accessed 23 July 2025.

[90] K Brehm and M Tobin, 'Virtual Power Plant: Flipbook' (undated) <https://rmi.org/wp-content/uploads/dlm_uploads/2024/06/VP3_flipbook_v1.2.pdf> accessed 23 July 2025.

[91] Brattle Group, 'California's Virtual Power Potential: How Five Consumer Technologies Could Improve the State's Energy Affordability' (2024) 4 <https://www.brattle.com/wp-content/uploads/2024/04/Californias-Virtual-Power-Potential-How-Five-Consumer-Technologies-Could-Improve-the-States-Energy-Affordability.pdf> accessed 23 July 2025.

[92] ibid.

[93] ibid 5.

[94] ibid 6.

early evening during hot months to meet increased demand.[95] However, the absence of a clear state policy supporting VPPs has retarded investment in VPPs. Legislation has been introduced in California that would direct state officials to develop a VPP program for the state.[96] Statewide build out of VPPs would likely result in significant additional growth in DER and could alleviate some of the problems with meeting peak energy demands as well as providing important ancillary services.

I. Limited Generation Profiles and Integrated Capacity Analysis

To address the problem of power line congestion from the rapid expansion of DER, California has created an approach referred to as a 'limited generation profile' (LGP). A March 2024 CPUC order authorized the use of LGPs and resolved technical and safety issues that allowed more rapid approvals of DER interconnections. LGPs limit the timing and amount of DER energy that can be fed into the grid enhancing the ability to respond to grid fluctuations,[97] thereby reducing the need for infrastructure investments. The idea of LGP would not have been possible without the availability of hosting capacity data referred to as 'Integrated Capacity Analysis' in California.[98] Following extensive negotiations a 2017 CPUC decision required investor-owned electric utilities to publish their integrated capacity data creating the ability to build LCPs.[99] As DER expand, the data available through integrated capacity analyses likely will be critical in ensuring that DER are sited in locations that can be accommodated by the existing grid.

J. Battery Storage

Because of the high level of DER, overproduction of solar energy at certain hours of the day, the increasingly frequent blackouts, and more frequent energy interruptions caused by forest fires and floods, California has strongly supported battery storage as part of the state's DER program. California is the leading state for battery storage with more than twice the capacity of the next highest state, Texas.[100] The state has developed a suite of incentives and legal requirements designed to promote battery storage as an integral element of its 2045 climate goal. Battery storage increased 66 percent in 2024 and has grown from 2 GW in 2020 to 26 GW in 2024.[101]

[95] J St John, 'California's patchwork push to scale up virtual power plants' (*Canary Media* 22 February 2024) < https://www.canarymedia.com/articles/grid-edge/californias-patchwork-push-to-scale-up-virtual-power-plants > accessed 30 July 2025.

[96] L Jenifer, 'Power Play: California's Virtual Power Plant Revolution' (Carbon Credits 22 April 2024) <https://carboncredits.com/power-play-californias-virtual-power-plant-revolution/> accessed 23 July 2025.

[97] IREC, 'California Regulators Open the Door for DERs to Avoid Interconnection Upgrades and Unlock Flexibility through Export Scheduling' (21 March 2024) <https://irecusa.org/blog/regulatory-engagement/california-regulators-open-the-door-for-ders-to-avoid-interconnection-upgrades/> accessed 23 July 2025.

[98] ibid.

[99] ibid.

[100] Environment America, 'Which States are Poised to Lead on Battery Storage' (3 November 2024) <https://environmentamerica.org/center/articles/which-states-are-poised-to-lead-on-battery-storage/> accessed 23 July 2025.

[101] Energy Information Agency, 'Today in Energy' (12 March 2025) <https://www.eia.gov/todayinenergy/detail.php?id=64705> accessed 23 July 2025.

California's 2022 building code requires that all commercial and high-rise residential buildings beginning in 2023 install solar panels that can support at least 60 percent of the building's energy demand paired with battery storage, becoming the first state to mandate battery storage for new buildings. In addition, all new residential buildings must be constructed battery storage ready.[102] The number of new solar installations with storage has been growing even without the mandate with half of the new solar installations in 2024 having battery storage at least in part due to the increasing number of power interruptions that are occurring in California.[103] BTM storage can also participate in California's 'Demand-Side Grid Support' program that provides financial incentives as part of 'a suite of programs to alleviate tight energy supplies on the grid caused by heatwaves, wildfires, and other ongoing impacts of climate change'.[104]

K. 2025 California Building Code

The 2025 California building code, one of the strictest in the US, includes several provisions that advance DER. For residential buildings it continues the requirement that all new single family residential building be built battery storage-ready. The code also requires new single-family buildings to use heat pumps, prohibits use of natural gas or propane for heating water, and places restriction on the power sources that can be used to heat swimming pools.[105] The prohibition on the use of gas and propane are likely to expand the use of flexible demand appliances in the state.

L. Flexible Demand Appliance Standards

In 2019 California became the first state in the US to authorize the creation of flexible demand appliance standards.[106] The standards are to be designed to 'promote technologies to schedule, shift, and curtail appliances operations to support grid reliability, benefit customers, and reduce greenhouse gas emissions associated with electricity generation'.[107] The rule is expected to deliver nearly 400 MW of shift flexible capacity by

[102] California Energy Code title 24 (2022). See California Energy Commission, 'Solar PV, Solar Ready, Battery, and Electric Ready' (undated) <https://www.energy.ca.gov/programs-and-topics/programs/building-energy-efficiency-standards/energy-code-support-center/solar> accessed 23 July 20225.

[103] Energy Information Agency, 'California residents are increasingly pairing battery storage with solar installations' (18 July 2024) <https://www.eia.gov/todayinenergy/detail.php?id=62524> accessed 23 July 2025.

[104] California Energy Commission, 'Demand Side Grid Support (DSGS)' (undated) <https://dsgs.olivineinc.com> accessed 23 July 2025.

[105] See California Energy Commission, '2025 Building Energy Efficiency Standards for Residential and Nonresidential Buildings: Title 24, Part 6, and Associated Administrative Regulations in Part 1' (undated) <https://www.energy.ca.gov/publications/2025/2025-building-energy-efficiency-standards-residential-and-nonresidential> accessed 23 July 2025.

[106] Cal Stat 2019, ch 697.

[107] California Energy Commission, 'Flexible Demand Energy Standards' (undated) <https://connectedcommunities.lbl.gov/sites/default/files/2022-02/CEC%20SB49%20Introduction%20to%20Flexible%20Demand%20Appliance%20Standards.pdf> accessed 23 July 2025; California Energy Commission, 'SB 49 on Appliance Standards (24 June 2021) (undated) <https://www.energycodes.gov/sites/default/files/2021-07/Model_Policies_CA.pdf> accessed 23 July 2025.

2030 and 560 MW by 2033.[108] In 2023 California adopted a flexible demand standard for swimming pools.[109]

V. Conclusions: Lessons Learned from the Federal and California Experience with DER

Several lessons related to decarbonization, digitalization, decentralization and democratization can be gleaned from the experience with DER policy in the United States.

A. Decarbonization

Decarbonization requires a clear and consistent focus on carbon reduction. In the US Congress has failed to adopt clear national carbon reduction goals. While Congress has supported various tax incentives, both Congress itself and the Executive Branch of government have vacillated in their support of carbon reduction policies with the contrast between the Biden and Trump administration demonstrating dramatically different policies. This inconsistency has made it difficult for the renewable energy industry and backers of DER to plan and execute new projects. In contrast, California has for more than twenty years focused on GHG reduction establishing a first-in-the-nation economy-wide cap and trade program, setting a clear goal for carbon-free energy generation and for EVs, and for the use of DER. These long-standing and clearly articulated policies and tax incentives have allowed the DER industry to flourish in California.

B. Digitalization

The Federal government has played a role in digitalization through funding under the Infrastructure and Jobs Act of 2022 that supported data analytics processes and system software that better allow integration of DER into the electric grid.[110] The Department of Energy also has supported work to streamline DER interconnections through better sharing of data needed to determine whether capacity is available for new DER-supplied energy.[111] The Trump Administration may support additional digitalization efforts that aid in the Administration's priority of supporting centralized, dispatchable power generation but not because digitalization would enhance the role of DER.[112]

108 California Energy Commission, 'California Energy Commission Adopts Flexible Demand Energy Standards for Pool Controls' (18 October 2023) <https://www.energy.ca.gov/news/2023-10/california-energy-commission-adopts-flexible-demand-appliance-standards-pool> accessed 23 July 2023.

109 ibid.

110 Infrastructure and Jobs Act of 2022 (HR 3684) s 40107.

111 See US Department of Energy, 'Interconnection Innovation e-Exchange' <https://www.energy.gov/eere/i2x/interconnection-innovation-e-xchange> accessed 27 July 2025.

112 B Martucci, 'As Trump targets clean energy, will utilities embrace DERs and VPPs? (Utility Dive 21 April 2025) <https://www.utilitydive.com/news/trump-clean-energy-utilities-distributed-der-vpp/745797/> accessed 10 August 2025.

California has viewed information availability as a key aspect of the State's clean energy strategy and as central to a sustainable approach to DER. The state's 2017 decision to require electric utilities to release capacity data made it much easier to identify locations that could accommodate DER and two-way transmission of energy resources. This allowed the state to use limited generation profiles to expedite approval of DER installations. The California experience indicates that broad access to information related to transmission, congestion, and the nature and location of DER is key to sustainable DER. The state's experimentation with VPPs may also advance digitalization to accommodate new aggregated DER including EVs and flexible demand appliances.

C. Decentralization

At the Federal level, FERC Rule 2222 supporting aggregation of DER has been expected to significantly increase the role of DER in wholesale markets. In addition, the Biden Administration through research at the Department of Energy, support for energy efficiency standards, and its Executive Orders clearly supported DER. The Administration also was the driving force behind the tax provisions of the Infrastructure and Jobs Act and the Inflation Reduction Act that provided incentives for installation of heat pumps and other DER equipment, tax incentives for EVs and EV charging stations, and tax credits for renewable energy facilities. The Trump Administration's policies, however, have reversed course and explicitly favor centralized power generation using fossil fuels.

California, in contrast, has maintained a long-standing approach to lowering carbon emissions that relies in significant part on an explicit goal for energy supplied through DER. This goal has been supported by a wide range of policies for more than twenty years. This multifaceted approach has been a key to sustaining its DER effort. The tools have included a strong renewable portfolio standard, net metering, subsidies for rooftop solar installation, battery storage mandates and subsidies, an EV mandate, a building code that provides strong support for DER, and flexible demand appliance standards. The California experience indicates that a sustainable DER program benefits from the use of a diverse set of tools that are part of a sustainable DER strategy.

D. Democratization

The Biden Administration focused on two aspects of energy democratization—enabling consumer participation in energy markets and energy justice. The Administration supported expansion of DER through research, tax credits, and other programs. And, from the outset the Biden Administration emphasized both environmental justice and energy justice. The Justice 40[113] initiative directed that at least 40 per cent of the benefits from climate and clean energy programs should inure to disadvantaged communities. This approach is based on the idea that these communities had been historically subject to excess emissions and disproportionate costs related to energy production and use. One example

[113] The White House, 'Justice 40' (undated) <https://bidenwhitehouse.archives.gov/environmentaljustice/justice40/> accessed 10 August 2025.

of the focus on justice and democratization is the Clean Energy Low-Income Bonus tax credit created in the Inflation Reduction Act.[114] The Trump Administration's focus, as has been noted, has been on centralized generation and it has required that environmental and energy justice programs be terminated. Energy democratization appears to be no longer a national priority in the US.

California, however, has taken justice and consumer generation issues into account for nearly twenty years. Environmental justice groups in California had raised concerns that subsidizing rooftop solar would favor higher income households leaving lower income households with the burden of paying a larger share of the cost of the legacy grid.[115] Several of California's DER programs including its Solar Assist, battery storage, and EV programs now take justice considerations into account in the way that they are structured. The stability of the DER program over two decades and the ability to learn from implementing the programs and adjusting has helped identify democratization and justice issues and adjust the program to address the issues.

114 See Internal Revenue Service, 'Clean Energy Low-Income Bonus Credit Amount' (undated) <https://www.irs.gov/credits-deductions/clean-electricity-low-income-communities-bonus-credit-amount-program> accessed 28 July 2025.

115 BR Lukinov and EM Krieger, 'Distributed solar and environmental justice: Exploring the demographic and socio-economic trends of PV adoption in California' (2019) 134 Energy Policy 110935 <https://www.sciencedirect.com/science/article/abs/pii/S0301421519305221> accessed 28 July 2025.

12

Financing Distributed Energy Infrastructure

The Role of Islamic Finance

Tasniem Ahmed Elyass Hussain

I. Introduction

This chapter evaluates the potential role of Islamic finance as a tool to bridge the gap in the current financing of Distributed Energy Resources (DER) infrastructure. Drawing lessons from Africa, it examines the legal and institutional challenges to the financing of DER projects through Islamic financing and offers recommendations on addressing them.

Over 600 million people in Africa, over 40 per cent of the population, lack access to reliable electricity, with the Democratic Republic of the Congo (DRC), Ethiopia, Nigeria, Tanzania, and Uganda accounting for over half of those without access to electricity.[1] The majority of those without access to electricity are concentrated in rural areas, disproportionately affecting rural and low-income communities.[2] Almost one billion people are then forced to rely on environmentally polluting fuel sources for cooking.[3]

In South Sudan, where only 7.7 per cent of the population has access to electricity, most depend on kerosene, firewood, sunlight, and personal diesel generators.[4] Traditional cooking indoors using these methods causes poor air quality, which is one of the leading causes of premature death among women and children in Africa.[5] To meet Sustainable Development Goal (SDG) 7, Africa needs to connect 90 million Africans and shift 130 million people from using dirty cooking fuels annually.[6]

There is a clear connection between poverty and energy access. Access to electricity promotes social, economic, and environmental development, which are the three foundational elements of sustainable development.[7] Ensuring universal electricity access for all is crucial for achieving other SDG targets; replacing wood fuel with electricity for cooking can

[1] International Energy Agency and African Development Bank Group, 'Financing Clean Energy in Africa' (IEA/AfDB 2023) 43 https://www.iea.org/reports/financing-clean-energy-in-africa accessed 4 September 2024.
[2] ibid 43.
[3] ibid 7.
[4] S Thiak and A Hira, 'Strategic Options for Building a New Electricity Grid in South Sudan: The Challenges of a New Post-Conflict Nation' (2024) 109 Energy Research & Social Science 103417 https://doi.org/10.1016/j.erss.2024.103417 accessed 6 March 2025.
[5] International Energy Agency, 'A Vision for Clean Cooking Access for All' (IEA 2023) https://www.iea.org/reports/a-vision-for-clean-cooking-access-for-all accessed 6 January 2025.
[6] IEA and AfDB (n 1) 7.
[7] K Nyarko, J Whale, and T Urmee, 'Drivers and Challenges of Off-Grid Renewable Energy-Based Projects in West Africa: A Review' (2023) 9 Heliyon e16710 https://doi.org/10.1016/j.heliyon.2023.e16710 accessed 6 March 2025.

Tasniem Ahmed Elyass Hussain, *Financing Distributed Energy Infrastructure* In: *Sustainable Distributed Energy Resources*.
Edited by: Íñigo del Guayo, Damilola S. Olawuyi San, Louis de Fontenelle, Milton Fernando Montoya, and José Juan González,
Oxford University Press. © International Bar Association 2026. DOI: 10.1093/9780198991946.003.0014

alleviate human health issues (SDG 3) and mitigate deforestation (SDG 13). Access to electricity enhances educational quality (SDG 4) by providing light for nighttime studying and enabling internet access for research.[8]

As discussed in Chapter 1 of this book, DERs such as small-scale microgrids and standalone renewable systems offer a viable solution in addressing energy poverty and advancing sustainable development in Africa. Unlike traditional grid extensions, DERs have the potential to be more affordable, scalable, and well-suited for integrating renewable energy. DERs also create opportunities for economic advancement in remote regions where traditional grid connections are uneconomical and impractical. In Ghana, for example, decentralized mini-grids have become the most cost-effective method for extending electricity access to remote communities.[9] However, their implementation faces several challenges, including high upfront costs, energy storage systems, and fragmented regulations, which make it difficult to secure financial resources and investments.[10]

One of the key challenges of energy decentralization through DERs is financing. Africa receives only 0.6 per cent of global renewable energy investments despite having some of the world's most significant renewable energy resources.[11] Sub-Saharan Africa has primarily benefited from DER investments made by private investors, governments, and development partners, with committed funding covering 377 initiatives at USD 3.1 billion.[12] Concessional funding from major donors, including the World Bank and the African Development Bank, has committed more than $9 billion to the sector. However, according to the World Bank, to provide electricity to 380 million people in Africa by 2030, over 160,000 mini-grids will be needed, requiring $91 billion.[13] Despite an increase in funding for the mini-grid sector over the last fifteen years, a notable financing gap remains. Africa needs to bridge this financing gap by tapping into alternative financing options, and Islamic finance has already begun to yield results in this field.

The engagement of Islamic finance in the promotion of renewable energy offers substantial advantages in advancing global sustainability initiatives. The Islamic financing sector is still a niche industry, accounting for only 0.6 per cent of global wealth.[14] However, the sector is projected to experience continued expansion, with assets anticipated to reach USD 6.7 trillion by 2027.[15] This upward trajectory is fuelled by a growing global recognition of the essential nature of ethical financing. The green sukuk market has demonstrated

[8] D Bukari and others, 'Towards Accelerating the Deployment of Decentralised Renewable Energy Mini-Grids in Ghana: Review and Analysis of Barriers' (2021) 135 Renewable and Sustainable Energy Reviews 110243.

[9] M Kumwenda, 'How Microgrids Can Facilit https://www.theigc.org/blog/how-microgrids-can-facilitate-energy-access-and-electrify-rural-africa accessed 6 March 2025.

[10] ibid.

[11] RES4Africa, 'Africa's Energy Future is Renewable' (15 June 2023) https://res4africa.org/wp-content/uploads/2023/06/Africas-Energy-Future-is-Renewables-Flagship2023.pdf accessed 4 September 2024.

[12] Mini-Grids Partnership (Sustainable Energy for All, Transforming Energy Access, and UK Aid), 'State of the Global Mini-Grids Market Report 2024' (Mini-Grids Partnership, August 2024) https://minigrids.org/global-market-report-2024/ accessed 28 June 2025.

[13] Africa Minigrid Developers Association and Odyssey Energy Solutions, 'Benchmarking Africa's Minigrids Report 2024' (3rd edn, AMDA 2024) https://www.africamda.org/wp-content/uploads/2025/04/Benchmarking-Africas-Minigrids-Report-2024-Online-version.pdf accessed 28 June 2025.

[14] ADB and IsDB Joint Report, 'Unlocking Islamic Climate Finance in Joint ADB and IsDB Countries' (November 2022) https://events.development.asia/materials/20221109/unlocking-islamic-climate-finance-joint-adb-and-isdb-countries accessed 31 January 2024.

[15] Global Ethical Finance Initiative, Greenpeace MENA, and Ummah for Earth, 'Islamic Finance & Renewable Energy' (Greenpeace MENA 2024) https://www.greenpeace.org/static/planet4-ummah-stateless/2024/11/d63785ad-iffe_report_en-.pdf accessed 6 January 2025.

remarkable growth, with green sukuk issuances amounting to USD 9.9 billion in the first half of 2024 alone, on par with the total for the entire year of 2023.[16] Strategically allocating 5 per cent of Islamic finance assets towards renewable energy initiatives by 2030 can generate an estimated additional $400 billion.[17]

This chapter examines the legal and regulatory reforms necessary to encourage and secure financing to facilitate DER integration. The chapter proceeds in six sections. After this introductory section, Section 2 discusses the role of DERs in sustainable development. Section 3 discusses the role of Islamic financing as a sustainable financing model. Section 4 analyses the regulatory and policy barriers to sustainable DER integration in Africa. Section 5 discusses the recommendations for addressing these barriers. Section 6 is the concluding section.

II. The Role of DERs in Sustainable Development

As discussed in Chapter 1 of this book, addressing energy poverty by diversifying the electricity mix is one of the key drivers of the rapid adoption of DER models across the world, especially in African countries which remains home to some of the world's energy poorest people. For example, currently, due to electricity outages, 40 per cent of all electricity consumed in Nigeria is produced from backup generation,[18] leading Nigerians to spend over USD 22 billion per year on fossil-fuelled generators, made worse by the removal of fuel subsidies in Nigeria in July 2023, which led to a 175 per cent spike in petrol prices.[19] Between 2021 and 2022, electricity costs for those dependent on diesel generators increased between 220 per cent and 260 per cent.[20] Utilities across the region, from Malawi to South Africa, have implemented indefinite load shedding. Distribution firms in Ghana and Kenya suggest rate increases of as much as 20 per cent to maintain funding for inconsistent energy service delivery.[21] Businesses experience efficiency loss, and additional costs for backup electricity could add up to USD 1.3 trillion by 2030.[22] Aside from the unstable cost, emissions from these generators also cause 33 million metric tons of CO_2 per year.[23]

Mini-grids can utilize diesel fuel or renewable energy sources (solar photovoltaic, hydro, wind, and biomass) or function as hybrids combining renewable energy with diesel.[24] Africa possesses 60 per cent of the world's best solar resources, yet has only 1 per cent of the total installed solar PV capacity.[25] Achieving universal access requires a considerable change in

[16] ibid.

[17] ibid.

[18] International Energy Agency, 'Unlocking Smart Grid Opportunities in Emerging Markets and Developing Economies' (IEA 2023) https://www.iea.org/reports/unlocking-smart-grid-opportunities-in-emerging-markets-and-developing-economies accessed 6 January 2025.

[19] C Dadzie and others, 'Scaling Utility-Enabled Distributed Energy Resources in Nigeria: A Roadmap to Boost Distribution Company Revenues and Improve Power Availability and Reliability' (RMI, June 2024) 17 https://rmi.org/insight/scaling-utility-enabled-distributed-energy-resources-in-nigeria/ accessed 23 December 2024.

[20] IEA (n 5).

[21] T Lee, K Jean-Pierre, and H Carr, 'Distributed Energy Resource Regulatory Building Blocks: A Framework for African Markets' (CrossBoundary Energy, March 2024) 2 https://www.crossboundary.com/publications/ accessed 23 December 2024.

[22] IEA (n 5).

[23] Dadzie (n 19) 17.

[24] African Development Bank Group, 'Green Mini-Grid Market Development Programme: GMG MDP Document Series No 1' (AfDB 2017) 3 https://greenminigrid.se4all-africa.org accessed 4 September 2024.

[25] IEA and AfDB (n 1) 26.

the types of projects that are financed. The IEA forecasted that by 2040, standalone systems and mini-grids would generate 70 per cent of the new rural electricity supply in Africa.[26] In Ghana, decentralized mini-grids have emerged as the most cost-effective option for providing electricity to remote villages.[27] There is a growing consensus within Nigeria's power sector that DERs offer a unique and scalable solution that can be tailored to meet specific customer needs.[28] The deployment of decentralized mini-grids gained momentum in Liberia following the adoption of the Renewable Energy Strategy and Master Plan.[29]

South Sudan presents a unique example of how DERs can prove beneficial in conflict areas. Despite abundant energy resources, substantial energy infrastructure is lacking due to historical and ongoing conflict, while the government struggles to control large swaths of its territory. Issues related to electricity supply are strongly linked to conflict as energy systems during and immediately after conflict face further complications due to the interconnected impacts of economic instability and the failure to invest in, construct, and repair energy infrastructure, which hinders affordable electricity access and economic development.[30]

Despite the targets set by the South Sudan Development Plan (SSDP), there has been a slight improvement in the energy sector since the nation gained independence in 2011. More than half of the population resides in these rural areas, where the cost of available electricity is prohibitively high, leading to a diminished standard of living and heightened social inequality.[31] Patankar and others outline a model energy system for South Sudan that considers the potential for the ongoing conflict, concluding that solar energy can hedge against such risks, diminishing the system's vulnerability to disruptions if a key generator or grid shuts down.[32] Their analysis of energy systems in fragile states shows that distributed generation is attractive, pointing to the added challenges of attracting and protecting investments in large-scale facilities, the capability to serve remote and conflict-affected areas, and the necessity to modify technology and financing.[33] Nhial Tiitmamer and others state that solar energy is the most promising and cost-effective route for South Sudan to achieve sustainable energy transition, given that extending the national grid to scattered rural communities would take decades, assuming all financial resources were available.[34]

Renewable energy technologies, especially solar power, provide electricity at a lower cost than grid-supplied electricity. DERs can offer greater reliability than the national grid when paired with storage solutions. The pace and extent of Africa's shift towards a sustainable economy will ultimately depend, however, on how policymakers adjust to accommodate DERs[35] and deal with regulatory and financing challenges.[36]

[26] AfDB (n 24) 3.
[27] Kumwenda (n 9).
[28] Dadzie (n 19) 9–10.
[29] Nyarko (n 7) 9.
[30] Thiak (n 4) 109.
[31] ibid.
[32] N Patankar and others, 'Building Conflict Uncertainty into Electricity Planning: A South Sudan Case Study' (2019) 49 Energy for Sustainable Development 53 https://doi.org/10.1016/j.esd.2019.01.003 accessed 5 March 2025.
[33] ibid.
[34] N Tiitmamer and JG Anai, 'Transitioning to Renewable Energy: An Analysis of Energy Situation in Juba, South Sudan' (2018) https://africaportal.org/publication/transitioning-renewable-energy-analysis-energy-situation-juba-south-sudan/ accessed 6 March 2025.
[35] Lee (n 21) 3.
[36] AfDB (n 24) 11.

III. Role of Islamic Financing in Sustainable DERs

In recent years, African nations have faced a series of interconnected challenges, including the COVID-19 pandemic, energy, and food crises resulting from Russia's invasion of Ukraine, and escalating climate threats. To meet the region's energy development and climate objectives, investments in energy must more than double from the current USD 90 billion by 2030, with nearly two-thirds of this expenditure allocated to clean energy.[37]

According to the World Bank, to provide electricity to 380 million people in Africa by 2030, over 160,000 mini-grids will be needed, requiring $91 billion.[38] However, concessional funding from major donors, including the World Bank and the African Development Bank, is around $9 billion.[39] Despite an increase in funding for the mini-grid sector over the last fifteen years, a notable financing gap remains.

While DERs provide a decentralized and resilient approach to addressing Africa's energy access issues, DER implementation still faces the same structural financing obstacles that challenge utility-scale clean energy projects. High initial costs, limited access to debt financing, political and currency uncertainties, and weak balance sheets of utilities all compromise the bankability of DERs.[40] To enable substantial private investment in the widespread deployment of DERs, it is increasingly important to utilize targeted concessional financing, mobilize local capital, and establish public green finance mechanisms.

Islamic finance presents a compelling alternative to traditional debt-focused financing approaches due to its focus on equity-oriented frameworks, shared risks, and ethical investment practices. Financial instruments like green sukuk, waqf-based infrastructure funds, and Islamic microfinance can bridge the capital shortages while being consistent with the principles of social equity and community empowerment that DER initiatives aim to promote.

A. Principles of Islamic Finance

In the global economic landscape, sustainable development is crucial in safeguarding environmental integrity and addressing social and economic impacts. The social dynamics within the Islamic community revolve around the definition of moral obligations and community involvement aimed at building a virtuous society.[41]

A founding principle of the Islamic worldview is its presentation of an interconnected and holistic perspective. Therefore, a modern interpretation of Maslaha (public interest) could lead to a broader theoretical understanding of morality. Islam embodies the natural state (Fitra), which signifies an inherent goodness. This natural state completely aligns with nature, humanity, and the constructed environment. It also denotes a full awareness of and acknowledgement of humanity's role as a caretaker and witness (Khalifa).[42] As vicegerents

[37] IEA and AfDB (n 1) 17.

[38] Africa Minigrid Developers Association (n 13).

[39] ibid.

[40] IEA and AfDB (n 1) 42.

[41] NM Abdul Aziz and F Azni Ahmad, 'The Islamic Moral Responsibility and Sustainable Social Engagement in Islamic Institutions' (2019) 2(2) International Journal of Islamic Economics and Finance Research 45.

[42] ibid 55.

on earth, humans have the responsibility (Amana) to inhabit and build a civilization for the good of all humanity and the environment.[43]

Islam places significant emphasis on two fundamental principles within the dynamic nature of social life: first, the efficient use of the resources bestowed by God to humanity and the surrounding environment, and second, the fair usage and distribution of these resources while fostering all human interactions based on the ideals of rights and justice.[44] Access to clean water and sanitation, affordable energy, infrastructure, and adequate housing are all included in these, and there is an opportunity within Islamic frameworks to make an impact through innovative financial solutions.[45]

Attention to the disadvantaged and marginalized through sharing resources, coupled with the financial contributions made through waqf (endowment) and zakat (charity), is a crucial concept in Islam that must be developed through institutional innovation (ijtihad) and governance reform. Within Islamic management entities, emphasis is placed on waqf and zakat institutions, which serve as intermediaries in collecting and distributing waqf and zakat funds. Zakat is considered one of the five fundamental pillars of Islamic teachings. It plays a vital role in enhancing the nation's social welfare, alleviating poverty, promoting equitable wealth distribution, and boosting the economic well-being of the Muslim community.[46]

Financial innovation can pave the way for sustainable and inclusive economic and social development. Accordingly, there is great potential to enhance the involvement of capital markets in developing inventive methods to fund projects focused on environmental, social, and sustainable development initiatives.[47]

B. Viability of Islamic Financing

The global Islamic finance market experienced an annual growth rate of 10 to 15 per cent from 2000 to 2012, appealing to Muslim and non-Muslim countries.[48] From 2009 to 2011, Shari'ah-compliant assets rose by over 160 per cent, and by 2016, the sector held approximately USD 1.6 trillion in banking assets.[49] By 2022, the global Islamic finance industry expanded by 11 per cent to USD 4.5 trillion.[50] The Islamic Finance Development Report 2022 reflected the global Islamic finance sector's resilience and growth, surpassing pre-COVID levels with a 17 per cent increase, totalling USD 4.0 trillion, thereby expanding its influence and size. The largest sector, Islamic banking, which represents 70 per cent of the Islamic

[43] S Sikandar Shah Haneef, 'Principles of Environmental Law in Islam' (2002) 17 Arab Law Quarterly 3, 241.
[44] Abdul Aziz (n 41) 53.
[45] S Jahan, A Muneeza, and SH Baharuddin, 'Islamic Capital Market for Social Development: Innovating Waqf Mobile Sukuk in Sub-Saharan Africa' (2021) 8(2) International Journal of Management and Applied Research 145 https://doi.org/10.18646/2056.82.21-009 accessed 4 January 2025.
[46] Abdul Aziz (n 41) 48.
[47] Jahan (n 45) 145.
[48] United Nations Environment Programme (UNEP), 'State of Financing for Nature: Report' (2021) 34 www.unep.org/resources/state-finance-nature-2021 accessed 6 February 2024.
[49] M Calderone and others, 'Investing in Financial Inclusion for Climate Resilience and Adaptation: The Role of Islamic Financial Services' Briefing Note for UN Climate Summit 2019 by ODI and Islamic Relief Worldwide (IRW) (May 2019) 4.
[50] Islamic Corporation for the Development of the Private Sector and London Stock Exchange Group, 'OIC Megatrends Report 2024' (2024) https://icd-ps.org/uploads/files/ICD_LSGE_Megatrends_Report_20241715586850_9187.pdf accessed 6 January 2025.

finance industry, grew by 17 per cent, reaching USD 2.8 trillion, partly driven by the sustained demand for Islamic banking, even from areas outside of traditional Islamic finance markets, reflecting the emergence of new markets as the sector stabilizes in more established areas.[51] The Islamic FinTech industry is estimated at USD 138 billion, constituting 1.2 per cent of the global FinTech market for 2022/23.[52]

A key factor in its rise and growth is its basis in equity-focused principles, which include sharing risk and profit and loss between the capital provider and the recipient, transforming fund suppliers into investors rather than debtors. Other factors include asset-backed financial transactions, directly connecting financing to genuine assets and real economic activities; a ban on interest and speculative actions, which prevents hoarding and eliminates excessive uncertainty, risk, and gambling; and the importance of maintaining contractual integrity and safeguarding property rights.[53]

Islamic finance promotes the financing of investment activities and projects in accordance with the tenets and procedures of Shari'a law.[54] Islamic financial instruments promote risk sharing and equitable profit-and-loss distribution through various financing methods, such as profit-and-loss sharing partnerships (Mudarabah), profit-and-loss sharing joint ventures (Musharakah), and Shari'ah-compliant lending instruments (sukuk).[55] This core concept of risk sharing is consistent with Islamic finance principles, emphasizing shared accountability and encouraging fair distribution of profits and losses, supporting ethical financial practices.

African countries have also implemented sustainability guidelines aligning Islamic finance with international sustainability initiatives. South Africa has achieved notable scores in CSR and ESG disclosures, particularly due to the transparency exhibited by organizations such as Sabana Industrial Real Estate and Al Baraka Bank South Africa.[56] As of 2021, the total Islamic Finance Assets in Sub-Saharan Africa exceeded USD 10 billion, including USD 3 billion in Islamic Banking Assets and USD 3 billion in Outstanding Sukuk Value.[57] In North Africa, the strong demand for Islamic financial products in Algeria prompted several conventional banks, such as Banque Nationale d'Algérie, Banque Extérieure d'Algérie, and Banque de Développement Local, to establish Islamic windows. The Islamic finance sector in Algeria is anticipated to expand with the forthcoming introduction of takaful products that will enhance Islamic banking offerings.[58]

1. Sukuk

The Islamic finance sector can contribute significantly to green economy goals. A clear indicator of the growing impact of the Islamic finance sector on the green economy is the increase in the issuance of green sukuk. Green sukuk is a category of securities categorized

[51] ICD and Refinitiv, 'ICD-Refinitiv Islamic Finance Development Report 2022' (2022) 6 https://icd-ps.org/uploads/files/ICD%20Refinitiv%20ifdi-report-20221669878247_1582.pdf accessed 6 January 2025.

[52] ICD and LSEG (n 50).

[53] M Iqbal and P Molyneux, *Thirty Years of Islamic Banking: History, Performance and Prospects* (Palgrave 2005).

[54] D Aassouli, 'Mobilizing and Leveraging Islamic Climate Finance in the MENA Region: The Potential Role of National Green Participative Banks' in D Olawuyi (ed), *Climate Change Law and Policy in the MENA Region* (Routledge 2022).

[55] N Saleh, 'Definition and Formation of Contract under Islamic and Arab Laws' (1990) 5 Arabic Law Quarterly 101–16.

[56] ICD and Refinitiv (n 51).

[57] ibid.

[58] ibid.

under ESG sukuk (which also includes social sukuk, aimed at projects with a positive social impact; sustainability sukuk, designed to fund both green and social initiatives; and transition sukuk, which targets the reduction of carbon emissions).[59] Sukuk is the second-largest asset sector in the global Islamic finance industry, growing by 14 per cent to USD 713 billion in sukuk outstanding.[60] Sukuk has become a key instrument governments and corporations use to address their financing needs and broaden their funding sources in support of economic and infrastructure development.[61]

Malaysia is the largest ESG sukuk market globally, representing USD 11.6 billion, or one-third of all ESG sukuk issued from 2017 to 2023. In 2014, Malaysia's Securities Commission introduced the Socially Responsible Sukuk (SRI) Guidelines to aid in the country's goal of becoming a centre for green and sustainable financing.[62] In 2017, Malaysia launched the world's first sustainable and responsible investment (SRI) initiative. Malaysia's Quantum Solar Park issued the largest solar project linked to green sukuk, which powered more than 90,000 homes, generated over 3,000 jobs, and cut emissions by 193,000 tons of carbon each year.[63] Sukuk has also been utilized to fund hydropower projects, including Telekosang Hydro One Sdn Bhd and RP Hydro (Kelantan) Sdn Bhd.[64]

Indonesia soon followed suit, issuing the first sovereign green sukuk, amounting to USD 1.25 billion and extending its reach beyond Shari'ah investors, with 29 per cent new green investors.[65] It secured USD 2 billion in private funding and supported twenty-three public green and sustainable initiatives.[66] Green sukuk has been promoted as one of the Islamic financial instruments capable of addressing sustainability deficits in achieving the SDGs, a financing tool for green projects in the wind energy sector in Italy,[67] for financing sustainable waste management initiatives in Indonesia,[68] as a tool to fill the financing gap for green projects to mitigate climate change and achieve the SDGs in sub-Saharan Africa,[69] and to be considered along with cash waqf to help finance afforestation programs by the Organization of Islamic Cooperation countries.[70]

In addition to these primary sukuk markets, Egypt and Nigeria issued sukuk in 2021 within Africa. In 2022, the Financial Regulatory Authority in Egypt introduced ten new

[59] ICD and LSEG (n 50).
[60] ICD and Refinitiv (n 51).
[61] Jahan (n 45) 145.
[62] ICD and LSEG (n 50).
[63] Securities Commission Malaysia and World Bank Group, 'Islamic Green Finance: Development, Ecosystem, and Prospects Report, based on the Conference on Harnessing Islamic Finance for a Green Future, Held from 14 to 15 May 2018, in Kuala Lumpur' (March 2019).
[64] ICD and LSEG (n 50).
[65] D Siswantoro and HV Surya, 'Indonesian Green Sukuk (Islamic Bond) of Climate Change: A Revisited Analysis' (2021) 716 IOP Conference Series: Earth and Environmental Science 1 https://iopscience.iop.org/article/10.1088/1755-1315/716/1/012044 accessed 6 February 2024.
[66] G Kala Buana and M Didi Hardiana, 'Islamic Finance's Answer to SDGs and Climate Change' (3 July 2019) www.undp.org/blog/islamic-finances-answer-SDGs-and-climate-change accessed 6 February 2024.
[67] D Campisi, S Gitto, and D Morea, 'Shari'ah-Compliant Finance: A Possible Novel Paradigm for Green Economy Investments in Italy' (2018) 10 Sustainability (Switzerland) 11, 1 https://doi.org/10.3390/su10113915 accessed 6 February 2024.
[68] HF Hariyani and H Kusuma, 'Green Sukuk-Based Project on Sustainable Waste Management in Indonesia' (2020) 12 Al-Iqtishad: Jurnal Ilmu Ekonomi Syariah 2, 165 https://doi.org/10.15408/aiq.v12i2.15129 accessed 6 February 2024.
[69] Aassouli (n 54).
[70] SI Abdullahi, 'Financing Afforestation in the Organization of Islamic Cooperation Countries: What Role for Islamic Economics and Finance?' (2020) 32 JKAU: Islamic Econ 2, 161 www.researchgate.net/publication/339791255_Financing_Afforestation_in_the_Organization_of_Islamic_Cooperation_Countries_What_Role_for_Islamic_Economics_and_Finance accessed 6 February 2024.

requirements for sukuk issuance.[71] Egypt has already set up a sukuk program worth USD 5 billion and launched its inaugural sukuk in 2023, amounting to USD 1.5 billion.[72] Green sukuk, which emphasizes asset-backed financing and risk-sharing, is appropriate for capital-intensive DER projects like mini-grids or solar installations in rural and underserved African regions.

2. Waqf and Zakat

Redistribution frameworks like zakat and waqf also serve as practical instruments for Islamic social financing for sustainable development. According to UNDP and the Islamic Development Bank (IsDB) estimates, zakat alone may generate between USD 200 billion and USD 1 trillion annually to support the SDGs.[73] The Organisation for Economic Cooperation and Development reports that the value of Awqāf assets ranges from USD 410 billion to USD 3 trillion, potentially yielding annual returns of approximately USD 150 billion.[74]

The link between waqf and zakat is essential for the redistribution of wealth in Islam. While zakat is a mandatory form of charity, waqf represents a voluntary contribution, and both are vital for enhancing social justice and mitigating inequality.[75] Waqf is often regarded as philanthropy or Sadaqah Jariyah (ongoing charity). However, the notion of waqf differs slightly from 'charity' in that the waqf property (asset) remains intact. At the same time, only its benefits are utilized per the donor's wishes for specific individuals or groups identified by the donor.[76] 'Waqf' refers to an asset designated so that its benefits are dedicated indefinitely to a charitable purpose. It literally means to hold or restrict the movement or transfer of a particular thing.[77]

Another prevalent misconception is that waqf exists solely for religious purposes, such as building mosques and Qur'an memorization schools. This is incorrect, as waqf has historically played a role in the provision of infrastructure, services, and social programs, encouraging entrepreneurship, job growth in communities,[78] cultural and religious preservation, and has supported marginalized groups such as women and people experiencing poverty.[79] It is a long-established religious institution aimed at enhancing social solidarity and empowering communities.[80] Waqf can advance long-term social, economic, and environmental concerns within communities. It has the potential to finance projects related to community and economic development, as well as infrastructure, while also supporting

[71] ICD and Refinitiv (n 51).

[72] ICD and LSEG (n 50).

[73] UNDP and Baznas, 'Unlocking the Potential of Zakat and Other Forms of Islamic Finance to Achieve the SDGs in Indonesia' (2018) www.undp.org/indonesia/publications/unlocking-potential-zakat-and-other-forms-Islamic-finance-achieve-SDGs-Indonesia accessed 6 February 2024.

[74] Organisation for Economic Cooperation and Development (OECD), 'How Islamic Finance Can Help Achieve the Sustainable Development Goals' (OECD 2020), OECD Development Cooperation Policy Papers, 30.

[75] M Hariri, 'Economic Potential of Waqf for Social Welfare: Patterns and Contrasts in MENAP and Countries in Sub-Saharan Africa' (2023) 4(1) Sukuk and Waqf Research 1 https://www.cribfb.com/journal/index.php/IJSWR/article/view/2344 accessed 4 February 2025.

[76] P Stibbard, D Russell, and B Bromley, 'Understanding the Waqf in the World of the Trust' (2012) 18(8) Trusts & Trustees 785.

[77] ibid.

[78] Hariri (n 75) 1.

[79] ibid.

[80] I Munshi and R Davids, 'Sustainable Giving through Waqf: A View from South Africa' (2018) 23(3) Alliance Magazine https://www.alliancemagazine.org/feature/sustainable-giving-through-waqf-a-view-from-south-africa/ accessed 4 February 2025.

initiatives focused on conservation and environmental protection.[81] Waqf serves as a versatile tool within Islamic finance and economics[82] and can be created for many initiatives; a notable example is the work of Finterra, a startup that combines blockchain technology with waqf, among other initiatives.[83]

Recently, there has been a global movement towards reviving and promoting the institution of waqf in modern societies. There is widespread recognition that this revival should foster a culture of giving as an integral part of sustainable community development. This initiative is gaining traction in South Africa, marked by a rise in new Awqāf. At the forefront of this movement is the National Awqāf Foundation of South Africa (Awqāf SA).[84] Many African nations have utilized the waqf system to fund social initiatives. For centuries, waqf institutions in Egypt have provided complementary social services. Egypt has successfully overseen waqf assets, creating revenue for social and religious initiatives and playing a crucial role in delivering public services. In particular, the waqf has historically supported Al Azhar University in Egypt. Sudan's waqf has played a role in financing education, agriculture, and shelters for the disadvantaged.[85]

In Nigeria, Land waqf is intended to deliver public benefits and humanitarian support. Waqf has played a vital role in facilitating broad religious activities by supporting the construction and upkeep of mosques.[86] Proposed waqf-based agricultural financing models in Nigeria illustrate how utilizing waqf land can help reduce poverty and assist farmers, stressing that establishing waqf land in Nigeria could generate economic opportunities and resources to improve living standards.[87] It has also been proposed that waqf resources be used to support initiatives that foster entrepreneurship through training and create jobs within communities.[88]

3. Islamic microfinancing and crowdfunding

Waqf designates land for activities that serve the public good, while waqf Microfinance is a perpetual social fund offering financial services to disadvantaged communities. This method is a sustainable, cost-effective, and easily accessible strategy with significant social and economic influence on local communities.[89] Strategies to enhance financial inclusion can empower low-income individuals, independent producers, women's co-operatives, and small to medium-sized enterprises (SMEs) to access and utilize formal financial services.[90]

These include Islamic microfinance alternatives such as Qard Hassan (interest-free loans), Ijara (leasing agreements), and Takaful (mutual guaranteed insurance), co-operative

[81] H Gabil and others, 'The Need for Shari'ah Compliant Awqāf Banks' (2020) 13(4) Journal of Financial and Risk Management 1 https://doi.org/10.3390/jrfm13040076 accessed 4 February 2025; Q Yulia, 'The Context of the Waqf in the Philosophy of Islamic Law Based on the Qur'an and Hadith' (2016) 1(1) Lambung Law Review Mangkurat 15 https://www.academia.edu/download/46977688/02.pdf accessed 4 February 2025.

[82] Hariri (n 75) 10.

[83] H Simisola Fa-Yusufa, S Abdullahi Busari, and B Lawal Shuaibu, 'Waqf Effectiveness in Nigeria: Problems and Solutions' (2021) 10(2) Journal of Islamic Finance 79–80.

[84] Munshi and Davids (n 80).

[85] Hariri (n 75) 8.

[86] Fa-Yusufa (n 83) 80–81.

[87] Hariri (n 75) 5.

[88] HT Haruna and AS Ibrahim, 'The Contribution of the Waqf Institution as a Financial Tool to Address Poverty Reduction: Evidence from the Literature' (2021) 2(2) AZKA International Journal of Zakat and Social Finance 151 https://doi.org/10.51377/azjaf.vol2no2.64 accessed 4 February 2025.

[89] ibid 1.

[90] Calderone (n 49) 62–63.

models such as mutual insurance based on a group of participants donating funds into a pool that members can then use in the event of specified unfavorable circumstances. The IsDB has acknowledged this and urged Islamic finance specialists to innovate new semi-formal Islamic microfinance offerings tailored for entrepreneurs and SMEs.[91]

Islamic finance is now becoming increasingly recognized as a distinct approach for both Muslims and non-Muslims globally, serving as a viable alternative to address the funding challenges faced by SMEs on a global scale.[92] More so in developing nations where traditional SMEs contribute significantly, accounting for approximately 45 per cent of overall job creation and around 33 per cent of gross domestic product (GDP), with a significant portion of these positions likely to be found in Asia and Sub-Saharan Africa.[93] Adams' cross-sectional study, which included 354 owners and managers of SMEs located in northern Nigeria, examined how Islamic finance affects the performance of SMEs in Nigeria, positing that Islamic financing positively influences the success of SMEs.[94]

Local businesses, particularly SMEs, are expected to be crucial in facilitating the implementation of decentralized solutions. The scale of these projects is considerably less than that of significant energy developments and many other end-use investments, while the risks tied to the end consumer can be significantly greater. Conventional methods of energy financing are not ideally suited to support these smaller, riskier projects or to provide funding for SMEs and local startups.[95] African countries can look to Islamic financing, structured around providing interest-free loans, sharing profits and losses, owning assets, actively engaging in trading, funding, and forming collective bargaining agreements between the investor and the entrepreneur[96] to be used for sustainable development and to uplift people from energy poverty.

Harnessing fintech can enhance the practice of online cash waqf, which is made possible when crowdfunding is considered.[97] With the development of the Integrated Islamic Crowdfunding Microfinance (IICM) model, it is evident that Islamic crowdfunding and microfinance can effectively complement one another. Microfinance reduces poverty and stimulates the economy, while Islamic crowdfunding can offer financial support for microfinance initiatives.[98]

There are examples of crowdfunding platforms being utilized for the management of waqf, zakat, and CSR initiatives. This process can be facilitated by collaboration with pertinent entities such as Islamic financial institutions, foundations, and religious organizations. For example, Global Sadaqah, an Islamic FinTech based in Malaysia, enables users to donate online to charity sadaqah and zakat. It supports payments via various channels,

[91] ibid.

[92] A Saidi Adedeji, 'Islamic Financing Effects on Small and Medium Enterprises Competitiveness' (2021) 1(1) Journal of Islamic Economic and Business Research 100–113 https://doi.org/10.18196/jiebr.v1i1.11753 accessed 4 January 2025.

[93] ibid.

[94] BS Adams, 'The Influence of Islamic Financing on SMEs Performance in Nigeria' (2020) 4(9) International Journal of Research and Innovation in Social Science 33.

[95] IEA and AfDB (n 1) 44–45.

[96] Adedeji (n 92) 101.

[97] Suhaili Alma'amun and others, *Waqf-Based Crowdfunding: A Case Study of Waqfworld.org* (2018) 1(1) International Journal of Islamic Economics and Finance Research 13–14.

[98] M Asmy bin Mohd Thas Thaker and others, 'A Proposed Integrated Islamic Crowdfunding–Microfinance (IICM) Model for Poverty Reduction: A Case Study of Indonesia' (2021) 4(1) International Journal of Islamic Economics and Finance Research 15–25.

including Bitcoin. FinTech companies can act as a financing solution for businesses. Through crowdfunding platforms, small enterprises can access affordable financing using innovative technologies, while lenders can earn appealing returns as these businesses grow. Islamic FinTech platforms, such as Saudi Arabia's Raqamyah, offer services in this domain. The platform merges technology and financial assessment to choose qualified SMEs for investment opportunities.[99]

Regarding Islamic microfinance, Tunisia saw a remarkable rise in its Other Islamic financial institutions (OIFIs) assets, which include financial technology companies, investment firms, financing companies, leasing, and microfinance firms, soaring from under US$1 million to US$60 million within a year. In Egypt, the Financial Regulatory Authority (FRA) issued the first license to a financial institution to provide Islamic microfinance services. Yonna Islamic microfinance launched operations as The Gambia's inaugural fully fledged Islamic microfinance institution.[100]

In Sudan, models of microfinance based on waqf have been introduced to offer interest-free loans to farmers and entrepreneurs. Sudanese microfinance, in which income generated from the Diwan Al Zakah funds interest-free loans for farmers affected by drought and a lack of fertilizers. The Muzara'ah initiative by the Sudan Islamic Bank has also proven effective for small-scale farmers in Sudan. However, there is still much room for improvement through the modernization of waqf management in Sudan.[101] Tanzania has also implemented the Muzara'ah financing framework, grounded in waqf and zakat, to assist impoverished farmers. In Zanzibar, a proposed model known as the Waqf-Muzara'ah-Supply Chain Model (WMSCM) suggests that a waqf institution can support agriculture-related inputs and logistics.[102]

Although there is growing potential for Islamic finance to enable investment in clean energy, challenges still hinder the potential of DERs in Africa. Legal and regulatory frameworks are necessary to ensure the successful deployment of DERs and attract investors. The following section discusses the current policy and regulatory challenges faced by integrating DERs in the African market.

IV. Institutional and Legal Barriers to Islamic Financing

In Africa, the cost of capital for clean energy initiatives is at least two to three times greater than in developed nations and China.[103] Consequently, this discourages private sector participation, making projects seemingly unaffordable or unviable for investors. This traps countries in a cycle of increased risks, elevated costs, energy shortages, and heightened dependence on fossil fuels, which usually require lower initial investments but expose consumers and governments to ongoing fluctuations in fuel prices and environmental hazards.[104]

[99] ICD and LSEG (n 50).
[100] ICD and Refinitiv (n 51).
[101] <IBTHariri (n 75) 6.
[102] ibid 7.
[103] IEA and AfDB (n 1) 28.
[104] ibid.

A. Regulatory Gaps in DER Financing

Mobilizing financing for DERs through Islamic financing will require addressing regulatory gaps that hinder third party financing for DER projects. Some African markets are hindered by regulatory approaches that prevent them from fully capitalizing on the advantages of DER technology and accessing financing. Even in cases where an independent energy regulator may exist, there is a high degree of utility influence in energy policymaking.[105] Regulations are seldom drafted and implemented based on thorough economic analysis or extensive market assessments, instead dealing with the various challenges facing centralized energy utilities.[106] This results in regulatory frameworks that discourage third-party and non-traditional financing methods, such as Islamic finance, and restrict the expansion of DERs.

Senegal is one such example, where the Ministry of Energy in Senegal is implementing a temporary halt on third-party financing for renewable DER projects, leaving customers no choice but to self-finance for on-site generation. A new set of decrees and Ministerial orders is anticipated in 2025 to clarify the terms under which independent power producers (IPPs) can provide renewable energy to private clients.[107]

The Mozambican government implemented significant reforms in the primary legislation. The 2022 Electricity Law aligns with the national Five-Year Plan (2020-2024), which proposed a strategy to enhance renewable production capacity by encouraging public and private investments in new generation infrastructure and power transmission.[108] However, since the Electricity Law primarily targets utility-scale projects, it imposes burdensome, costly, and restrictive concession conditions on renewable energy projects relying on third-party ownership. Thus, Mozambique's lack of a dedicated DER policy prevents third-party funding of distributed energy resources (DERs).[109]

Ghana's National Clean Energy Access Programme (NCEP) leverages results-based financing and carbon market tools (as outlined in Article 6 of the Paris Agreement) to promote solar photovoltaic systems.[110] The framework includes concessional loans and performance-based incentives but does not mention sukuk, waqf, zakat, or any other Islamic finance methods. In 2023, South Africa's energy sector regulator, NERSA, increased the licensing threshold for DERs due to a generation deficit and the ongoing necessity for nationwide 'load shedding'. In the amendments to Section 3 of Schedule 2 of the Electricity Regulation Act 4 of 2006, the requirement to secure a license was removed entirely.[111] However, the amendment does not include provisions or mechanisms for financing DERs.

The gaps in legal frameworks, or restrictions and limitations of third-party funding, hinder the viability of many Islamic Finance models. Islamic financial instruments like Ijara (leasing), Murabaha (cost-plus financing), and Musharaka or Mudaraba (equity partnership) depend on third-party ownership rights or collaborative investment arrangements.

[105] Lee (n 21) 17.
[106] United Nations Industrial Development Organization (UNIDO), 'Module 3: Introduction to Energy Regulation' (2009) https://www.unido.org/sites/default/files/2009-02/Module3_0.pdf accessed 6 January 2025.
[107] Lee (n 21) 18.
[108] ibid.
[109] ibid.
[110] IEA and AfDB (n 1) 122.
[111] Lee (n 21) 27.

Therefore, a waqf-based solar microgrid project intended to provide electricity to a school or village would face significant legal obstacles.

B. Lack of Standardized Islamic Finance Legislation and Policies

One of the significant obstacles to the widespread implementation of Islamic finance is the lack of clear and standardized legal frameworks and policies that address the unique nature and risks associated with the Islamic finance sector.[112] Islamic banks face unique risks that complicate their legal and regulatory structures.

The industry has two primary standard-setting bodies: the Accounting and Auditing Organization for Islamic Financial Institutions (AAOIFI), which focuses on Shari'ah accounting and auditing standards, and the Islamic Financial Services Board (IFSB), which concentrates on regulatory and supervisory standards. The implementation of these standards, however, varies significantly from one country to another. The inconsistent application of AAOIFI and IFSB standards, particularly in nations with active Islamic banking, threatens to diminish transparency and opens the door for 'regulatory arbitrage'. There is a need for greater regulatory clarity in banking laws and regulations, along with improved communication between Islamic standard-setting bodies and national regulators.[113]

There is also a lack of comprehensive and standardized Shariah frameworks to regulate Islamic finance.[114] Despite attempts to establish standards for Islamic finance, many nations regulate the industry using a framework intended for conventional finance, which does not account for the distinctive features of Islamic finance.[115] Additionally, the industry operates in an environment where legal and tax regulations, financial infrastructure, and access to financial safety nets are either lacking or, when present, do not sufficiently address the unique characteristics of Islamic finance.[116]

There is also a wide range of approaches regarding the regulatory framework for Islamic banking. A significant challenge in establishing and expanding Islamic banking services in Africa is the absence of a Shariah framework within the governance of many African nations. In Algeria, the lack of legislative laws regarding Islamic banking is one of the main obstacles to the establishment of an Islamic banking system in the country.[117] Kenya's Central Bank Act presents significant challenges to the development of Islamic banking, failing to define, acknowledge, or cover the existence of Islamic banks. Additional challenges include a lack of co-ordination between the Central Bank and Sharia Boards, the absence of a Sharia-Compliant Legal Framework, ineffective prudential regulation, and the reliance on conventional banking benchmarks. Despite the high demand for Islamic financial products in Kenya, there is an insufficient number of financial institutions providing

112 Aassouli (n 54).
113 ibid.
114 M Kabir Hassan, A Muneeza, and KNM Sonko, 'Introduction' in M Kabir Hassan, A Muneeza, and KNM Sonko (eds), *Islamic Finance in Africa: The Prospects for Sustainable Development* (Edward Elgar Publishing 2022) 8.
115 ibid 8.
116 ibid.
117 H Yaacob, R Mat Zin, and Q Ali, *Islamic Finance, Governance and Regulation: Global Perspectives* (Routledge 2025) 131.

Shariah-compliant services to meet customer needs. The primary reason is the inadequate support from the legal and regulatory framework that governs Kenya's banking sector.[118]

In Tanzania, the tax framework implemented by the Tanzanian Revenue Authority (TRA) does not create an equitable environment for Islamic banks, particularly concerning the potential double-tax implications of murabaha transactions. Tanzania also lacks dedicated regulations for takaful, meaning it falls under the regulations that govern conventional insurance.[119] Therefore, it is essential to examine the current legislative and regulatory landscape to develop a policy framework for Islamic banks.

Despite potential, the overall issuance of sovereign sukuk in Africa remains relatively low. In 2023, Egypt, Nigeria, and South Africa were the only three countries to issue public sukuk, which accounted for just 2 per cent of the global sukuk total. This was deemed due to the complexities and changing Islamic legal requirements associated with sukuk issuance, which hinder the effectiveness of the instrument.[120]

Nigeria is paving the way for emerging state-level regulators to build upon, refine, and enhance national regulations concerning DERs. The Constitutional Amendment of 2023 in Nigeria has opened the door for electricity planning and regulation at the state level. States such as Lagos and Oyo are leading the way in setting priorities and creating plans for improvements in utility-scale generation, transmission, distribution, and off-grid electrification.[121] Mini-grids provide electricity to approximately five million individuals in Nigeria due to a blend of electricity reforms, specialized government bodies like the Rural Electrification Agency promoting a robust involvement of development organizations and donors, and the capacity to secure funding through the Nigeria Electrification Project.[122] However, there is still no explicit inclusion of financing or Islamic financing instruments.

Lack of a strong legislative framework for waqf remains a challenge in Nigeria, as it is in many African countries.[123] In Nigeria, the governance of waqf is managed by trustees, unlike in countries like Malaysia, where it is governed by law. There is currently no Federal legislation specifically addressing the management and administration of waqf. The only mention of waqf in the Federal constitution is that all matters related to it should be governed by Shariah law. In the northern regions of Nigeria, the waqf institution has not fully realized its potential despite regulations governing its operations, and it continues to be overshadowed mainly by conventional practices.[124] The challenges facing the waqf institution in Nigeria include insufficient protection for donated land,[125] and the presence of unregistered waqf assets.[126]

Common challenges regarding waqf land ownership include insufficient clarity concerning land ownership rights, inadequate regulation and oversight, and a lack of transparency in managing waqf land. Furthermore, certain countries encounter corruption-related challenges and the misappropriation of income produced by waqf properties. It is crucial

[118] ibid 137.
[119] ibid 151–52
[120] M Parker, 'Slow but steady growth of sukuk in Africa' (African Business, 4 April 2024) https://www.african.business/2024/04/african-banker/slow-but-steady-growth-of-sukuk-in-africa accessed 30 June 2025.
[121] ibid 30–31.
[122] IEA and AfDB (n 1) 48.
[123] Hariri (n 75) 5.
[124] Fa-Yusufa (n 83) 80–81.
[125] Hariri (n 75) 5.
[126] Fa-Yusufa (n 83) 80–81.

to enhance transparency and accountability in the management of waqf land, enforce stringent regulations to safeguard property rights, and implement strict measures to combat corruption. Raising awareness and educating communities about the advantages of waqf can foster increased community involvement and support in preserving and developing waqf lands.[127]

For Islamic financing to thrive and be available to provide much-needed financing of DER infrastructure in Africa, there must be strong, standardized, and comprehensive laws and policies established. Governments need to create supportive legal and regulatory frameworks that will enhance the recognition, funding, and implementation of Islamic financing.

V. Legal and Regulatory Pathways for Sustainable DERs in Africa

Establishing a conducive policy and regulatory framework is crucial for encouraging private investment in infrastructure initiatives that may have low returns but significant social benefits. These policies must balance safeguarding and attracting investments while fulfilling the responsibility to advance social and economic growth among the impoverished rural population.[128]

Broader legal systems impact risk perceptions, and the lack of specific regulations for clean energy initiatives can deter investors in those markets.[129] Co-ordinated efforts are necessary to foster a conducive environment for DERs and non-traditional forms of funding, such as Islamic financing. The approach will differ by country but should encompass harmonization within government bodies to enact necessary reforms, given the unique characteristics of DERs and Islamic financing.

A. Strengthen Legal Frameworks for Islamic Financing of DERs

To promote DER infrastructure in Africa using Islamic financing, governments need to address the existing legal and regulatory gaps identified in this chapter by adopting a good governance approach to Islamic green financing. This approach involves the thorough integration of Islamic law principles into laws, policies, programs, budgets, and funding mechanisms.[130] While some countries have DER-specific regulations, some, like Mozambique, lack clear and specific DER regulations,[131] let alone ones that specify non-traditional forms of financing. DER-specific regulatory and policy frameworks that include financing models serve as a driving force in enhancing the development of off-grid power systems in emerging nations.

[127] Hariri (n 75) 9.

[128] NJ Williams and others, 'Enabling Private Sector Investment in Microgrid-Based Rural Electrification in Developing Countries: A Review' (2015) Renewable and Sustainable Energy Reviews http://dx.doi.org/10.1016/j.rser.2015.07.153 accessed 4 September 2024.

[129] IEA and AfDB (n 1) 36–37.

[130] Aassouli (n 54).

[131] Lee (n 21) 18.

Regulators should develop a comprehensive framework for renewable energy, which includes technical design specifications, tariff settings, established capacity goals for renewable energy, independent power producer (IPP) frameworks, along with well-defined power purchase agreements (PPAs), which are also measures that can be taken to mitigate risk, particularly regulatory risks.[132] These should also include non-traditional forms of financing, such as Islamic financing tools, including sukuk and waqf-based models.

To address these gaps, governments should revise energy and DER regulations to explicitly allow the use of Islamic financial instruments like green sukuk in DER initiatives. Standardized legal frameworks for Islamic public-private partnership contracts, particularly for microgrid projects and off-grid communities, should be developed, and Islamic microfinance institutions should be considered as eligible financiers in tenders for mini-grid developers.

B. Building Shariah-Compliant Financial Systems

For the advancement of the Islamic financial sector, several factors are essential, including dedicated regulations and policies, sound governance, effective supervision, well-trained professionals in Islamic banking, Shariah supervisors, and experts in Shariah audits.[133] Several African countries have made significant strides to develop their regulations; however, regional interaction and knowledge sharing between countries with experience and practice, such as Malaysia, can promote expertise on how Islamic green financing can best serve DER infrastructure.

Globally, Malaysia is recognized as having the best legal and regulatory framework for Islamic finance. There are three regulatory bodies overseeing financial operations in Malaysia, including the Central Bank of Malaysia, which regulates Takaful and Islamic banking, and the Securities Commission that oversees the Islamic Capital Market and other securities issuances. The Islamic Banking Act 1983 regulates Islamic banking[134] and a centralized Shariah Advisory Council (SAC) under the Central Bank acts as the final authority on Shariah issues within Islamic finance.[135] The Financial Services Act 2013 (FSA) and the Islamic Financial Services Act 2013 (IFSA) represent one of the most comprehensive legislative measures for Islamic finance in the world. It requires complete Shariah compliance for Islamic financial institutions, covering governance, operations, and product categories. The Act also introduced banking products such as mudharabah, musharakah, ijarah, etc.[136] The State Islamic Religious Council (SIRC) is responsible for managing waqf property, acting as the sole trustee of waqf assets.[137]

African countries should align their national frameworks with AAOIFI and IFSB standards[138] as well as UN SDGs and the Paris Agreement, for ESG-linked Islamic instruments.

[132] IEA and AfDB (n 1) 51.

[133] A Elzahi Saaid Ali, *Empowering the Poor through Financial and Social Inclusion in Africa: An Islamic Perspective* (Palgrave Macmillan 2022) 110 https://doi.org/10.1007/978-3-031-00925-9 accessed 30 June 2025.

[134] ibid 11–12.

[135] ibid 10.

[136] ibid 13–14.

[137] Alma'amun (n 97) 12.

[138] S Conteh and others, 'Regulatory Framework of Islamic Finance in Selected West African Countries' in Hassan, *Islamic Finance in Africa* (n 114) 63.

They should have comprehensive and dedicated Islamic Finance legislation[139] that recognizes Islamic banking, Takaful, sukuk, Islamic microfinance, etc, and defines key Shariah-compliant contracts (murabaha, mudarabah, ijara, musharakah, etc) and distinguishes Islamic financial institutions (IFIs) from conventional ones like Malaysia's Islamic Financial Services Act 2013 (IFSA).

An alternative to enacting comprehensive legislation like Malaysia's IFSA, which would require considerable time for review before approval, is for African nations to implement central bank guidelines that establish clear policies for regulating Islamic banking and finance practices. This has been Nigeria's approach and is currently being pursued by The Gambia. This approach is strategic given that Nigeria, Guinea, and The Gambia have delegated authority over their Islamic banking and finance matters to their respective central banks. Among these nations, Nigeria stands out as the only one that has put forth dedicated sets of guidelines to address various Islamic finance contracts and products. The challenge for Nigeria has been related to the enforcement of these guidelines rather than their establishment.[140]

Establishing a Centralized Shariah Governance Framework with a national Shariah Advisory Board[141] under the central bank or financial authority like Malaysia's Shariah Advisory Council could standardize and regulate financial products, guide Shariah boards, and require IFIs to perform internal Shariah audits and publish compliance reports. Establishing this board at the central bank level would create uniformity in Islamic finance practices nationwide, reducing conflicts that may arise during financial interactions if each bank's Shariah committee is given discretion to make its own Shariah rulings, thereby standardizing the Islamic finance sector, encompassing not only Islamic banking, capital markets, and takaful insurance but also extending to all finance-related areas, including Islamic social finance (like microfinance and waqf) and other Shariah-compliant ventures.[142] To support the Shariah governing bodies, especially those engaged in financial transactions, it is crucial to establish additional supportive organizations like academic institutions and research centres dedicated to Shariah laws.[143]

Islamic banks in Kenya have formed a Shariah Advisory Board to ensure that the products and services of Islamic banking adhere to Shariah principles and conduct audits to ensure that all Islamic financial institutions comply with Islamic standards.[144] The Central Bank of Nigeria Act 2007 was implemented on 25 May 2007, replacing the Central Bank of Nigeria Act 1991. Empowered by section 33(b), the Central Bank issued Guidelines on Shariah governance for non-interest financial institutions in Nigeria, establishing a clear mandate for creating a Shariah Advisory Council.[145]

Sudan is considered a pioneer in contemporary takaful practices. The Insurance Commission of Sudan regulates the insurance market under the governing law of the Insurance Act of 2003.[146] Sudan has also created a conducive environment for the establishment of Islamic financial institutions and a structured framework for microfinance.[147]

139 ibid 62.
140 ibid 63.
141 ibid.
142 ibid 63–64.
143 Conteh (n 138) 63–64.
144 Yaacob (n 117) 135–36.
145 ibid 140–41.
146 ibid 147.
147 ibid 7.

In 2007, the Central Bank of Sudan established the Islamic Microfinance Unit to promote microfinance, and in 2011, it issued a revised policy and regulations concerning Islamic microfinance.[148] It mandates that Islamic banks allocate up to 12 per cent of their reserves for microfinance purposes and authorized a specialized bank—the Sudanese Family Bank—to provide Shariah-compliant microfinance solutions.[149] This improved Islamic financial infrastructure has created a supportive environment for Islamic microfinance and the associated institutions, such as waqf and zakah, enabling them to fulfill their essential roles in assisting the less fortunate in Sudan.[150] Whilst the practical implementation has stagnated given the conflict, the legal framework is one that other African countries can adopt to enable Islamic financing of DER initiatives.

Harmonization with existing legal frameworks, for example, regarding land and ownership laws, is essential in Islamic Finance, as many transactions involve asset-based arrangements where contracts such as sales and leases are commonly employed.[151] Tax codes should be amended to ensure Islamic contracts are not penalized, such as the double tax on murabaha. South Africa has positioned itself as a potential centre for Islamic banking in the region. In 2010, the South African government revised its tax laws to attract foreign investment by offering Shariah-compliant options and incorporated legal definitions for musharaka and mudaraba in the draft Taxation Laws Amendment Bill.[152]

Regional collaboration is also beneficial. Due to the Ivory Coast's participation in regional organizations like the West African Economic and Monetary Union (WAEMU) and the Central Bank of West African States (BCEAO), the country has experienced significant improvements in its regulatory framework to support Islamic Finance. In its efforts to boost Islamic Finance in the region, the BCEAO established a collaboration agreement with the Islamic Development Bank (IDB), dedicated to making the necessary reforms better to address the specific characteristics of the Islamic financial sector.[153]

The Tanzanian Insurance Regulatory Authority is currently working on legislation to enable conventional insurance companies to create distinct structures for offering Takaful products.[154] The Central Bank of Tanzania is collaborating with the World Bank to create a supervisory, regulatory, and legal framework for Islamic banking. The World Bank's task team is actively engaging with local financial authorities to establish reporting requirements and to form an autonomous Shariah supervisory function. Current regulations indicate a need for a structured internal auditing system that includes an independent evaluation of control measures tailored to the unique nature of Islamic finance.[155]

The establishment of dedicated regulatory frameworks and governance systems that align with international prudential standards, emphasizing transparency and reducing Shariah-compliant risks, while also accommodating the unique institutional structures and market conditions of each country, would address the primary concerns of regulators, investors, and other stakeholders, ultimately fostering the growth of the Islamic Finance

148 ibid 80–81.
149 ibid 9.
150 ibid 76.
151 Conteh (n 138) 62.
152 Yaacob (n 117) 145–46
153 Conteh (n 138) 57.
154 ibid 151.
155 Yaacob (n 117) 151–52.

industry. This should be a collaborative effort, including active engagement from all stakeholders in the market, regulators, investors, and financial institutions.[156]

VI. Conclusion

Distributed Energy Resources (DERs) in Africa can significantly benefit various market actors. With a supportive regulatory framework and efficient processes for permitting and licensing, DERs can be deployed more rapidly than utility-scale projects, often at little or no financial burden to the government. Renewable DERs can contribute to achieving national renewable energy objectives, and as industries gain access to affordable, dependable, clean energy, they enhance manufacturing output, increase local job opportunities, and raise tax revenues.[157] Nevertheless, despite their promise, the growth of DERs in Africa encounters considerable financing hurdles, exacerbated by fragmented regulations, inadequate legal frameworks, and poor institutional co-ordination.

Africa, home to approximately 1.2 billion individuals, with nearly 40 per cent identifying as Muslim, has been referred to as a sleeping giant in the development of Islamic finance, which has awakened to contribute to the advancement of the industry actively.[158] Islamic finance represents a feasible and underutilized approach to closing Africa's funding gap for distributed energy. Based on principles of social equity, risk-sharing, and ethical investment, mechanisms like green sukuk, waqf-based infrastructure funds, and Islamic microfinance can provide long-term financing for DER projects. With significant expansion in Islamic finance worldwide and a growing interest in aligning it with sustainable development, considerable potential exists to mobilize funds for clean energy initiatives that support environmental and community upliftment goals.

African governments must establish a conducive environment through legislative frameworks and DER regulations that facilitate the implementation of DERs and include Islamic financing methods. The creation of comprehensive and dedicated Islamic Finance legislation, centralized Shariah frameworks, and governance structures that are in line with global standards, regulatory harmonization, and regional collaboration, which would tackle the main issues faced by regulators, investors, and other stakeholders, would thereby promote the development of the Islamic Finance sector and the implementation of more widespread DER infrastructure.

[156] Conteh (n 138) 50.
[157] Lee (n 21) 12.
[158] Hassan (n 114) 2.

13

Demand-Side Aggregation of Distributed Energy Resources

David Robinson

I. Introduction

This chapter explains the growing importance of demand-side aggregation in electricity systems that increasingly rely on Distributed Energy Resources (DERs) such as electric vehicles, heat pumps, and stationary batteries. It examines the challenges of developing demand-side aggregation within a decentralized, digitalized, and decarbonized power sector, focusing on its potential, the regulatory hurdles, and ways to overcome them.

Two fundamental shifts are reshaping the electricity sector: the rapid growth of intermittent renewables and the rise of behind-the-meter Distributed Energy Resources (DERs). Together, they alter the sector's technical and economic foundations. Flexibility, once provided mainly by dispatchable generation, is declining as renewables displace thermal and hydro plants. At the same time, controllable DERs now allow consumers to supply flexibility, making the demand side not only complementary but also a competitor to large-scale generation.

Unlocking this new role for demand requires efficient management of DERs. While individuals can react to tariff signals, competing directly with generation at wholesale level requires co-ordination—made possible through demand-side aggregation. Aggregators rely on consumer participation and supportive regulation to compete in electricity markets. The debate on their regulation began in the US and continues in the EU and Great Britain, with outcomes that will shape how consumers and DERs compete in future systems.

The chapter is structured in four parts beyond this introduction and the conclusions. Part II explains the technical and economic importance of decarbonized flexibility in systems with high renewable penetration. It highlights the potential of DERs—such as batteries, solar panels, EV chargers, and heat pumps—and introduces two of the potential business models:[1] 'implicit' flexibility from consumer price response, and 'explicit' flexibility sold in markets. It also explains why aggregation is essential to capture fully these benefits and why this chapter focuses on demand reduction competing with generation.

[1] A third potential model to exploit DSF is for a regulated utility or system operator to remotely control DERs, offering consumers financial compensation. Hybrids combining the models is also an option. For more on DSF models, see D Robinson and X Li, 'How Consumers can help to decarbonize the electricity system: the role of electricity demand-side flexibility (DSF) and policy implications for China' (2021) pp 7-8 and 23 <https://climatestrategies.org/wp-content/uploads/2021/10/DSF-China-Policy-Brief-ENG_v2.pdf> accessed 18 August 2025.

David Robinson, *Demand-Side Aggregation of Distributed Energy Resources* In: *Sustainable Distributed Energy Resources*.
Edited by: Íñigo del Guayo, Damilola S. Olawuyi San, Louis de Fontenelle, Milton Fernando Montoya, and José Juan González,
Oxford University Press. © International Bar Association 2026. DOI: 10.1093/9780198991946.003.0015

Part III compares US and EU regulation of aggregated demand response in wholesale markets. The US applies a clear 'net benefits' test to ensure demand reduction improves social welfare and pays it the same price as generation. The EU and GB lack such a formal test, creating uncertainty and inviting resistance from retail suppliers, who face costs when aggregators reduce their customers' demand. Resistance comes also from conventional generators that compete with DSF.

Part IV compares France and Great Britain (GB). GB regulation supports independent aggregators, while French regulation creates barriers. Each system has regulatory gaps.

Part V makes recommendations to address these gaps, promoting demand aggregation that improves consumer welfare and fairly compensates retail suppliers.

Finally, the chapter uses the terms demand-side flexibility (DSF), demand-side response (DSR), and demand response (DR) interchangeably, consistent with EU legislation: 'the change of electricity load by final customers from their normal or current consumption patterns in response to market signals . . . whether alone or through aggregation.' All three terms refer to increasing or reducing demand, though here DR often refers specifically to demand reduction.

II. The Flexibility Imperative and the Potential Role for DSF

Deep penetration of intermittent renewable electricity requires an increase in flexibility to ensure safe and efficient network operation. This has been referred to as the 'flexibility imperative'.[2] According to the European Commission's Joint Research Centre (JRC) report, the flexibility necessary to ensure the proper functioning of the system will grow from 11 per cent in 2021 to 24 per cent in 2030 and 30 per cent in 2050 of the projected and growing total electricity demand.[3]

There are three central questions that we face with respect to DSF in power systems. First, what role can DSF play in a system with deep penetration of intermittent renewable generation? Second, what is the value of DSF? Third, why are markets and intermediaries, especially aggregators, important for achieving the goals of system security and economic and environmental sustainability?

A. What Role Can DSF Play in the Future Decarbonized Electricity System?

Flexibility will increasingly come from large scale storage (eg batteries), networks (including interconnectors), and DSF. These resources will be able to replace fossil fuels to manage short term renewable intermittency, manage network congestion, and provide ancillary services, such as frequency control and voltage regulation. They are also able to provide resource adequacy for many hours. By replacing fossil fuels, these flexible resources

[2] ibid, pp 5–8.

[3] D Koolen, M De Felice, and S Busch, 'Flexibility requirements and the role of storage in future European power systems' (Publications Office of the European Union 2023) p 26 <eu-flexibility-requirements-and-the-role-of-storage.pdf> accessed 18 October 2025.

contribute to environmental sustainability and system stability.[4] However, for very long-duration challenges (spanning several weeks or seasons), addressing intermittency will necessitate long-duration storage and decarbonized generation solutions (eg pondage and pumped hydro) to supplement the important roles networks, storage, and DSF can play in this time dimension.

It is important to distinguish between electricity demand reduction and demand increase. While both contribute to system flexibility, DSF demand reduction delivers most cost savings by lowering the need for investments in generation, networks, and storage.[5] It also cuts emissions by consuming less, particularly during periods of peak demand or supply shortage.

Demand increase plays an increasingly important, but secondary, role in terms of costs savings, by absorbing renewable energy that would otherwise be curtailed and stabilizing prices when supply exceeds demand. It can help balance the system when reducing generation is not feasible.

Historically, DSF meant infrequent, often predictable, industrial demand cuts. Today, the need for frequent, less predictable real-time DSF—especially at the local level—has grown. Digitalization now enables DSF to respond dynamically to system needs. Realizing the full potential requires market structures and regulations that make DSF financially viable.

B. What is the Value of DSF?

The *potential economic value* of DSF is significant due especially to the growing penetration of digitalized DERs. A report by DNV and smarten estimated that by 2030, DSF could generate a direct benefit of €71 billion annually and an indirect benefit of over €300 billion per year in the EU.[6]

This flexibility will become increasingly important for *system security*.[7] First, the incidence of operations under conditions of reserve scarcity can be reduced by consumers willing to modify their demand at a price below the implied cost of reducing needed generation reserves to serve that demand. Second, system resilience improves with the growth of DERs because these local resources provide an alternative to system-wide security of supply. Third, local systems can aid in rebooting the central grid during extreme weather events. Fourth, blackouts or high prices can happen when the grid has insufficient capacity or access to flexible generation. Traditionally, the solution has been to invest in more infrastructure—like power lines and substations—to handle extreme weather and increased demand. But this approach is costly and raises network expenses. As a result, grid planning is shifting towards smarter, less infrastructure-heavy solutions. Instead of building more

[4] 'Power System Flexibility for the Energy Transition; Part 1: Overview for Policy Makers', IRENA, November 2018 < https://www.irena.org/-/media/Files/IRENA/Agency/%20%20%20%20Publication/2018/Nov/IRENA_Power_system_flexibility_1_2018.pdf> accessed 16 August 2025.

[5] HM Reeve and others, *Distribution System Operator with Transactive (DSO+T) Study Main Report: Volume 1* (Pacific Northwest National Laboratory 2022) < https://doi.org/10.2172/1842485> accessed 13 May 2025.

[6] *Demand-side flexibility: Quantification of benefits in the EU*, smartEn and DNV, 2022. <https://www.dnv.com/Publications/demand-side-flexibility-quantification-of-benefits-in-the-eu-232342/ > accessed 14 May 2025. See also Regulatory Assistance Project: Flex-ability for all, January 2024 for information on the potential for DSF for different technologies < https://www.raponline.org/knowledge-center/flex-ability-for-all-pursuing-socially-inclusive-demand-side-flexibility-europe/ > accessed 14 May 2025.

[7] *Overview of Flexibility Platforms*, 4E TCP, February 2025, at > accessed 16 August 2025.

physical assets, operators are looking for more flexible generation and increased demand flexibility in constrained load pockets.

DSF also contributes to *environmental sustainability*.[8] By increasing demand when renewables are operating, it avoids curtailment of those resources, and, more importantly, reduces the need for energy during scarcity periods. By reducing demand when renewables are not available, it avoids emissions from fossil fuels. Furthermore, DSF helps to lower investment needs for networks, thereby reducing costs and environmental impact.

C. What are the Roles of Prices, Markets, and Aggregators?

There are two ways to tap DSF benefits through markets and price signals: implicit and explicit flexibility. Both benefit the system and flexible consumers. 'Implicit' DSF is when consumers adjust behavior based on price signals like Time-of-Use (ToU) tariffs. These can provide useful signals but rarely reflect real-time market prices. 'Explicit' DSF involves consumers or their representatives selling flexibility in wholesale or balancing markets, with firm commitments to change demand accordingly.

Economists widely agree on the value of implicit DSF reflecting real-time pricing as a means of reducing system costs.[9] This form of DSF has been in use for decades, for instance in France, the UK, and many US states.

Explicit DSF differs in that it competes with generation on the supply side of markets, with its own offer price; so, the offer of 1 MWh demand reduction competes with the offer of 1 MWh increase in generation. When the offer price to reduce demand is below the offer price to increase generation, the market clearing price is lower and all consumers benefit provided some of the savings are passed on to them.

Yet, explicit DSF is difficult without demand aggregation—especially for residential consumers. This is for four reasons. First, while large users have sold their flexibility for years, small consumers face regulatory and transaction cost barriers that aggregators help to overcome, for instance by remotely managing resources and meeting capacity requirements to participate in markets. Second, from a system perspective, operators need certainty in delivery. Aggregators help meet reliability requirements, usually facing penalties for non-delivery. Third, residential DERs, like EVs or heat pumps, can potentially provide cheaper flexibility than large central assets built for the system (eg generators and storage) because consumers have already invested in them (and have low incremental costs to provide flexibility with those resources); but for these DERs to be competitive in markets requires aggregation. For instance, one study shows that Virtual Power Plants (VPPs), which aggregate DERs, can provide flexible resource adequacy at 40–60 per cent the cost of gas turbines or large-scale batteries, with additional societal benefits.[10] Finally, business models matter. Traditional retailers and distributors have little incentive to reduce demand, especially if

[8] Energy Transition Commission, 'Demand side flexibility—unleashing untapped potential for clean power' (February 2025) < https://www.energy-transitions.org/publications/demand-side-flexibility> accessed 16 August 2025.

[9] See S Borenstein. 'The Long-Run Efficiency of Real-Time Electricity Pricing' (2005) 26(3) The Energy Journal 93.

[10] R Hledik and K Peters, 'Virtual Power Plants (VPPs) Could Save US Utilities $15-$35 Billion in Capacity Investment Over 10 Years', Report prepared by The Brattle Group for Google, 2 May 2023 <https://www.brattle.com/insights-events/publications/real-reliability-the-value-of-virtual-power/> accessed 18 October 2025.

they are vertically integrated with generation. Demand-side aggregators, by contrast, are aligned with lower demand and sustainability goals.

III. Regulation of DSF and Demand-side Aggregators

This part summarizes and compares the legislative treatment of aggregated demand response in the US and in the EU.

A. US Federal Legislation for Demand Response

The US adopted policies to support explicit DSF in three key pieces of federal legislation.[11]

First, FERC Order 719 (2008) marked the first significant step towards integrating demand response (DR) into US wholesale electricity markets by requiring Regional Transmission Organizations (RTOs) and Independent System Operators (ISOs) to accept demand response bids when they could balance supply and demand cost-effectively.[12] However, the order left unresolved the crucial question of how such resources should be compensated, leading to inconsistent treatment across markets.

Second, FERC Order 745 (2011)[13] addressed this gap by establishing that demand response must be compensated at the full locational marginal price (LMP) whenever it can displace generation at equal efficiency and passes a net benefits test. In doing so, Order 745 put demand response on economic parity with generation, creating a stronger and more uniform incentive for its participation in wholesale markets. This decision recognizes that DR, which refers here to reducing demand, is equivalent to generation and competes with it. It enables monetary compensation for reducing electricity demand. This FERC Order is especially important for our analysis of European and British legislation and so the debate and decisions about compensation deserve further explanation.

Under FERC Order 745,[14] demand aggregators participate in the energy market by selling demand reduction. The FERC grappled with how to prevent demand response from being overcompensated relative to its value. Two main approaches were debated: (a) paying demand response the full locational marginal price (LMP[15]); or (b) compensating demand response at 'LMP minus g', where *g* represents the generation portion of the retail electricity price.

Supporters of option (a) argued that from the perspective of system balancing and economics, the reduction of demand was equivalent to an increase in generation and that both should be paid the same price, namely the LMP, which is the marginal cost of supplying, at least cost, the next increment of electric demand at a specific location on the transmission

[11] See Chapter 12 in this book by Lee Paddock, that examines the role of fiscal incentives such as tax reforms in promoting sustainable financing for DERs in the US.

[12] FERC Order No 719 (2008): Wholesale competition in regions with organized electric markets < https://www.ferc.gov/media/order-no-719> accessed 18 August 2025

[13] FERC Order No 745 (2011), Demand Response Compensation in Organized Wholesale Energy Markets, issued 15 March 2011 <https://www.ferc.gov/sites/default/files/2020-06/Order-745.pdf> accessed 13 May 2025.

[14] ibid.

[15] US Energy Information Administration. Glossary Electricity. https://www.eia.gov/tools/glossary/?id=electricity, accessed 7 December 2025.

system, taking into account generation marginal costs, the physical aspects of the transmission system, and system operating constraints. Supporters included aggregators and Kahn, who had been the Chairman of the New York Public Service Commission and a leading author on economic regulation. He argued that demand response is economically equivalent to supply response (generation) and that economic efficiency requires that it be regarded and rewarded equivalently to generation in competitive power market, namely at the market clearing price (which in FERC regulated wholesale markets is the Locational Marginal price, LMP).[16]

Supporters of option (b) included some incumbent utilities and Hogan, another leading expert on economic regulation. Hogan argued that paying LMP enables the consumer to avoid paying the energy component of the retail price (g) and that earning full LMP (which includes the cost of producing energy) would lead to overpayment and to inefficiently high levels of DR. His logic was that the flexible consumers were being rewarded twice for reducing demand: once by not paying for generation they were avoiding, and a second time for selling the reduction of demand.[17]

To respond to concerns about overcompensation, the FERC chose to adopt LMP as the correct compensation for demand response, subject to a 'net benefits' test. That test assesses whether the benefit to the system in terms of reduced prices for all consumers resulting from demand flexibility exceeds the cost of paying demand response resources at the LMP. The FERC 745 Ruling[18] argued that, when reductions in LMP from implementing demand response results in a reduction in the total amount consumers pay for resources that is greater than the money spent acquiring those demand response resources at LMP, such a payment is a cost-effective purchase from the customers' standpoint. They defined a net benefits test that would determine whether this cost effectiveness condition was met.[19]

Calculating net benefits in real time would require a hypothetical run of the day-ahead auction to determine the clearing price in the absence of any demand response bids. Instead, to implement the net benefits test, FERC directed RTOs and ISOs to determine a mechanism to approximate a price level above which DR would be cost-effective. Then, based on historical data as a starting point and updated for changes in relevant supply conditions such as changes in fuel prices and generator unit availability, they should determine a monthly threshold price corresponding to the point along the supply stack at which the overall benefit from the reduced LMP resulting from dispatching demand response resources exceeds the cost of dispatching those resources. They should accept DR bids and pay them LMP when market prices were equal to or above the threshold.[20]

[16] FERC Order No 745 (2011), Demand Response Compensation in Organized Wholesale Energy Markets, issued 15 March 2011 pp 17–18. https://www.ferc.gov/sites/default/files/2020-06/Order-745.pdf; accessed 7 December 2025.

[17] ibid, pp 21–22.

[18] Ibid. p 41.

[19] Ibid, p 41. FERC Order No 745 offered a worked example. Assume a market of 100 MW, with a current LMP of $50/MWh without demand response, and an LMP of $40/MWh if 5 MW of demand response were dispatched. Total payments to generators and load would be $4,000 with demand response compared to the previous $5,000. Even though the reduced LMP is now being paid by less load, only 95 MW compared to 100 MW, the price paid by each remaining customer would decrease from $50/MWh to $42.11/MWh ($4,000/95). Therefore, the payment of LMP to demand resources is cost-effective.

[20] FERC outlines the net benefits test and the threshold price level in FERC Order 745-A, 2011 (p 34): <https://www.ferc.gov/sites/default/files/2020-04/E-4_16.pdf> accessed 11 September 2025.

What explains the FERC's decision on Order No 745? The Commission—under Chairman Jon Wellinghoff, a strong advocate for demand response—recognized in its decision that the greater risk to consumers in paying LMP is not overpaying and over-stimulating demand response but under-paying and under-stimulating it, given the substantial entry barriers and market distortions (eg, LMP suppression due to out-of-market payments via capacity mechanisms in which only a small subset of economic DSF can qualify to participate) adversely affecting what DSF providers can realize in other markets relative to the system value they can provide. The net benefits approach is not perfect, but it strikes a useful balance between accessing as much DSF as possible while imposing some level of discipline on what is paid. That approach and the debate about what the aggregator should be paid are central to the debate in the EU and GB.

Third, in 2020, FERC Order 2222[21] extended the compensation rule of paying LMP for a wide range of distributed energy resources, including solar panels and batteries. However, in this case, there is no net benefits test because these distributed services act more like generators (or Virtual Power Plants, VPP), increasing or reducing production, and are assumed to compete with them in all markets. If VPP are chosen to be in the merit order, this is assumed to be beneficial for the system.

To summarize, the FERC Orders 745 and 2222 conclude that demand response and DER participation should be compensated at the market clearing price, LMP. For demand response (reduction), participation in the market (and subsequent payment of LMP) is subject to a net benefits test. For payment to a VPP, there is no need for a net benefits test because it should be treated in the same way as a generator.

B. Comparison of US and EU Legislation Related to Aggregated Demand Response

In 2019, the EU adopted legislation that requires Member States to allow and foster participation of the demand-side in all markets, in competition with generation. Before looking at the EU legislation in detail, it is worth providing some reasons why the EU approach differs from the US approach.

First, the two regions' approaches are shaped by different balancing regimes, which in turn reflect how their day-ahead (DA) and real time markets are structured. In the EU, the market operates under a self-dispatch model: the Transmission System Operator (TSO) only takes control at gate closure.[22] By contrast, the US follows a central dispatch model, where the ISO or RTO controls the market from the DA stage through to real-time operations.[23]

[21] FERC Order No 2222, Participation of Distributed Energy Resource Aggregations in Markets Operated by Regional Transmission Organizations and Independent System Operators, issued 17 September 2020 <https://www.ferc.gov/sites/default/files/2020-09/E-1_0.pdf> accessed 13 May 2025.

[22] See M Roggenkamp for a summary of network balancing and energy security under EU legislation, 'How Net Zero Goals Impact Electricity Systems and Supply Security: An Analysis of the EU Legal Framework', in *Net Zero and Natural Resources Law: Sovereignty, Security, and Solidarity in the Clean Energy Transition* (OUP 2024).

[23] Federal Energy Regulatory Commission (FERC), 'An Introductory Guide to Electricity Markets Regulated by the Federal Energy Regulatory Commission' <https://www.ferc.gov/introductory-guide-electricity-markets-regulated-federal-energy-regulatory-commission> accessed 16 August 2016.

Consequently, in the US, the ISO (or RTO) effectively acts as the only Balancing Responsible Party (BRP). Electricity injections and withdrawals are scheduled, and the ISO manages dispatch in real time to address any net deviations from the schedule. This leads to a final market-clearing settlement price, which determines what net buyers pay and what net sellers receive for any energy that was not scheduled in advance. The cost of these deviations from the schedule is shared across market participants in the ISO-managed markets; there is no need to compensate retail companies.

In the EU, by contrast, commitments are made in advance and the BRPs (and ultimately retail suppliers) may face a financial cost resulting from the aggregator's activation of demand response among the supplier's consumers, depending on how this activation is accounted for in the BRPs' positions—namely, with or without a 'perimeter correction', an option left up to Member States to adopt or not.[24] That explains why the EU legislation also allows Member States to grant compensation for retail companies or their BRPs that face financial costs due to the activation of demand response by independent aggregators. It also explains why the key debate related to aggregation in the EU is about who should pay for such compensation awarded to retail suppliers. Should the aggregators compensate the retailers, or should the costs be shared by all the retailers (hence ultimately their consumers) that benefit from lower wholesale market costs resulting from the aggregators' actions?

Second, the EU legislation is probably the result of a compromise that satisfied different parties, those that favoured demand-side aggregation and those that did not, with the opponents primarily the retail suppliers and generators whose current profits were threatened by aggregation. In the US, the FERC chose to pay the wholesale market price for DR (reduction) subject to the net benefits test, and to pay that price for VPP services without that test. The EU legislation recognizes the importance of net benefits but does not require a net benefits test, and the result is that the test is generally not applied. However, as explained in more detail below, the EU legislation[25] is clear about how such net benefits should be considered when applied.

Third, EU legislation[26] involves two options for system balancing related to DSF and aggregation. The first option, which is the pre-existing model, involves no correction of the electricity supplier's day-ahead commitment (ie the 'uncorrected model'). Under it, when consumption is reduced due to independent aggregation, the supplier for those participating consumers will be credited for a positive imbalance, which will be paid for as part of the imbalance financial settlement run by the system operator, at the balancing price (which may be higher or lower but is usually higher than the day ahead price). In other words, the suppliers receive financial compensation for the positive imbalances as part of the financial imbalance settlement.[27] However, the legislation allows Member States to adopt models where perimeter corrections are introduced. In the 'corrected' model, the

[24] The concept of a perimeter correction is explained below in this section.

[25] Article 17-4 of Directive (EU) 2019/944 of the European Parliament and of the Council of 5 June 2019 on common rules for the internal market for electricity and amending Directive 2012/27/EU (recast) (text with EEA relevance). https://eur-lex.europa.eu/legal-content/EN/TXT/?uri=CELEX:32019L0944

[26] Recital 39 of EU Directive 2019/944 allows the Member States to choose between a corrected perimeter model and an uncorrected perimeter model. 'Member States should be free to choose the model that fits best with their market organisation, for example a model where the supplier's perimeter is corrected or a model where it is not.' However, Recital 39 informs how Article 17 should be interpreted and implemented but does not create a binding obligation to establish compensation rights.

[27] If the balancing price is higher than the DA price, the supplier makes a profit; it is lower than the DA price, it faces a financial cost.

supplier/BRP will be deprived of financial compensation for the positive imbalance, since the imbalance disappears due to the correction. It is left with a cost related to the difference between the day ahead commitment and its final sales. The supplier may be entitled to financial compensation, but this is up to the Member State to decide. Both France and GB adopted the corrected model, and in each country, suppliers are entitled to financial compensation. However, as explained in Section IV, those countries differ as to who should bear the burden of the compensation granted to suppliers.

Fourth, the calculation of net benefits of demand response is different. In the US, this is calculated by reference to the reduction in prices for all consumers compared to the cost of paying for demand response; this requires hypothetical scenarios but is relatively easy to put into practice. In the EU, the net benefits test that makes economic sense is the benefits to all suppliers (and eventually to their consumers) due to lower wholesale prices, compared with the financial costs borne by those suppliers (and eventually their consumers) resulting from activation of demand response by aggregators. The net benefits of this calculation are likely to be positive, especially when market prices are high[28].

C. Independent Aggregators in EU Legislation

The relevant EU legislation for independent aggregators is part of the Clean Energy Package, published on 14 June 2019 in the Official Journal of the EU, includes the Directive on electricity market design, EU 2019/944[29] and the Regulation on electricity markets, EU 2019/943.[30]

The directive defines 'aggregation' as a function performed by a natural or legal person who combines multiple customer loads or generated electricity for sale, purchase, or auction in any electricity market. It defines an 'independent' aggregator as a market participant engaged in aggregation who is not affiliated to the customers' supplier. Retail supply companies and energy communities may also engage in aggregation and would be considered as independent aggregators when they operate demand response for consumers who are not their customers for the supply business.

Article 17 of the directive defines the regulation of demand response through aggregation. The text establishes that EU Member States must enable demand response participation through aggregation, ensuring final consumers can compete fairly with producers in all electricity markets. It also requires a regulatory framework that grants independent aggregators direct market access without needing consent from other participants, while obliging all market players to be financially responsible for imbalances they create in the system.

Article 17 also deals with the possibility of financial compensation. It permits Member States to require compensation from demand response participants (electricity

[28] Article 17-4 of Directive (EU) 2019/944 provides for such specification. A study prepared by Compass Lexecon (for DR4EU) on demand response in Europe estimated that the net benefits (savings to suppliers/consumers minus the costs to them) would be approximately €3.3 billion between July 2022 and June 2023 <https://dr4eu.org/wp-content/uploads/2022/12/CL-DR4EU-DR-benefits-2022_2023.pdf> accessed 13 May 2025.

[29] Directive (EU) 2019/944 (n 26).

[30] Regulation (EU) 2019/943 of the European Parliament and of the Council of 5 June 2019 on the internal market for electricity (recast) (Text with EEA relevance).

undertakings or participating final customers) if their actions directly affect other market players or their BRPs. However, this compensation must not hinder market entry or flexibility, and the calculation method should be strictly limited to covering actual costs incurred during demand response activation. Furthermore, Article 17-4 defines conditions under which demand response participants may not be required to pay compensation: 'The method for calculating compensation may take account of the benefits brought about by the independent aggregators to other market participants and, where it does so, the aggregators or participating customers may be required to contribute to compensation but only where and to the extent that the benefits to all suppliers, customers and their balance responsible parties do not exceed the cost incurred.'[31]

IV. Independent Aggregation in France and Great Britain (GB)

France and GB both adopted the 'corrected perimeter' model, which left suppliers with financial costs related to demand reduction activated by aggregators. These countries took different approaches to the issue of compensating the suppliers.

A. France

France was a pioneer in promoting DR in Europe,[32] using dynamic regulated tariffs since the 1960's. However, with liberalization, the end of some regulated tariffs led to a reduction in DR. Between 2003 and 2013, market mechanisms were opened to the participation of DR, with industrial consumers active participants in the balancing market (BM).

A new regulatory mechanism for DR came into force in January 2014. NEBEF (*Notification d'Échange de Blocs d'Éffacement*)[33] enabled third parties (including independent aggregators) to participate in the electricity markets, including the wholesale Day Ahead Market (DAM) and the Balancing Market (BM). DR was valued on the energy markets and treated like a supply resource, analogous to the treatment under FERC legislation in the US as summarized earlier.

Supplier compensation was included in the NEBEF rules for DR sold in the DAM. It required independent aggregators to compensate electricity suppliers for the volumes bought in the markets (whether volumes of electricity or demand response) but not paid for by their customers during aggregator activation of downward demand response. The compensation is calculated by reference to the applicable tariff for the relevant customer, with the aim of reimbursing the supplier for the energy-related revenue they would have received had there been no activation of downward demand response. Compensation was also introduced in the balancing market, where there had been no compensation until then.

[31] Article 17-4 of Directive (EU) 2019/944 (n 26).

[32] For details on the participation of DR in the French wholesale market, see Enefirst, 'Participation of Demand Response in French Wholesale Electricity Market' https://enefirst.eu/wp-content/uploads/4_PARTICIPATION-OF-DEMAND-RESPONSE-IN-FRENCH-WHOLESALE-ELECTRICITY-MARKET.pdf accessed 7 December 2025.

[33] RtE, 'Participate in the NEBCO mechanism' <https://www.services-rte.com/en/learn-more-about-our-services/participate-nebef-mechanism> accessed 25 July 2025.

The issue of compensation to suppliers in France has been the subject of a legal battle between the supplier EDF and Voltalis, an independent DR aggregator, over the impact of the approach and whether it acts as a barrier to aggregation. This battle may have influenced the drafting of the EU legislation summarized above, with the obligation not to create a barrier to DR, and with aggregators pressing for references to net benefits being introduced to strengthen the case for paying DR the market clearing price and the case for aggregators not compensating retail suppliers.

As noted earlier, under Directive 2019/944, Article 17, Member States are allowed to require compensation for suppliers when the country has adopted a 'corrected perimeter' model. However, the legislation says that such compensation must not create a barrier to market entry for aggregators or be a barrier to flexibility.

France's compensation model has discouraged DSF activation, and especially in periods when the energy component of tariffs is higher than wholesale spot prices. When wholesale electricity prices are very high compared to tariffs, aggregators earn enough in the market to cover compensation based on regulated tariffs, and DR activity increases. This was true in 2022 due to a combination of rising costs of gas and electricity related to the Russian invasion of Ukraine, and much lower retail tariffs. However, in subsequent years, the volumes of DR activation fell back significantly because wholesale prices fell, while tariffs rose.

In the context of the European Commission's review of France's implementation plan (pursuant to Article 20/5 of Regulation (EU) No 2019/943), the Commission commented on the introduction of a capacity mechanism, which raised questions concerning State Aids and barriers to activation of DSR.[34] The Commission urged France to review whether its market design, particularly balancing activation and remuneration, hinders demand response from developing under fair market conditions. It noted that demand response should primarily be driven by electricity prices without entry barriers. France is also asked to monitor energy activation volumes, as persistently low levels could indicate ongoing market obstacles.

That Commission document was written in 2021. In March 2024, experts from the French Agency for the Ecological Transition (Ademe) confirmed the concerns expressed by the European Commission, in Recommendation 17,[35] noting that the current supplier reimbursement price for undelivered energy was a barrier to demand response. Since April 2023, spot market prices had not exceeded this compensation level, making demand response unrecoverable. They stressed the need to revise both the calculation method and its evaluation frequency.

Overall, the conclusion is that under some market conditions, the French compensation mechanism can consume much or all the revenues available to aggregators, undermining the economics of explicit demand response and diminishing the associated potential benefits to customers. However, the analysis also suggests the need to define a market price threshold above which demand response generates net benefits for the system.

[34] Commission Opinion of 28 August 2021 pursuant to art 20(5) of Regulation (EU) No 2019/943 on the implementation plan of France, English version, p 12.

[35] Flexibilité due système électrique, Ademe, Mars 2024; <https://librairie.ademe.fr/index.php?controller=attachment&id_attachment=3151&preview> accessed 25 July 2025.

B. Great Britain

Until 2023, GB did not permit consumers and Virtual Lead Parties (VLP) to participate in the wholesale DAM. The category of VLP includes independent aggregators, which do not belong to retail suppliers, and the aggregation business belonging to the same group as a retail supplier, with unbundling requirements to ensure effective competition among VLP.[36]

In October 2023, in its P415 modification of the Balancing and Settlement Code (BSC) the energy regulatory, Ofgem, approved consumer participation via VLPs in the wholesale market, with effect from November 2024.[37] In its decision, Ofgem concluded that retail suppliers should be compensated for financial costs associated with activation of DSF and adopted the so-called 'mutualization' approach for this, as explained further below.

P415 was a proposed change to the Balancing and Settlement Code (BSC) intended to improve access to the wholesale electricity market for flexibility provided by VLPs. Under the system existing before the Ofgem decision, electricity consumers who could adjust their usage could not profit from that flexibility in the wholesale DAM unless they had a specific agreement with their supplier. VLPs at the time could support consumers in offering their flexibility within the Balancing Mechanism (BM), but they were not permitted to do so in the wholesale DAM. Allowing VLPs access to this market would remove that barrier and give consumers more direct control over how they offer and monetize their flexibility.

The modification proposed by P415 introduced a new concept known as 'deviation volumes', which would allow VLPs to trade consumer-driven changes in demand or generation in the wholesale electricity market. However, if a VLP reduced a consumer's demand, the supplier may be left with surplus volumes (ie a positive imbalance), having already purchased those volumes in the DAM (be they provided by electricity producers or demand aggregators). The volumes of DR bought in the market, as they do not come from electricity production but from demand curtailment, cannot be billed to consumers, and are not accounted for as a positive imbalance given the perimeter correction imposed on retailers, which could result in financial loss.

Two compensation models were put forward: (1) the Proposed Solution, under which suppliers who suffered financial losses due to VLP actions would be compensated, with the cost of compensation shared among all suppliers through a mutualization process; or (2) the Alternative Solution, under which VLPs would compensate affected suppliers, based on the deviation volumes they cause. This second model is like the model adopted in France.

A cost-benefit analysis found that P415 could deliver major benefits under both solutions. According to the analysis, the Proposed Solution (the mutualized approach, where costs are spread across all suppliers) lowered the variable costs for VLPs, making it more feasible for them to enter and participate actively in the market.[38] This greater participation

[36]

[37] Ofgem decision P415 'Facilitating Access to Wholesale Markets for Flexibility Dispatched by VLPs', 6 October 2023 <https://www.ofgem.gov.uk/decision/ofgem-decision-p415-facilitating-access-wholesale-markets-flexibility-dispatched-vlps> accessed 16 August 2025.

[38] Ofgem, Balancing and Settlement Code (BSC) P415: Facilitating access to wholesale markets for flexibility dispatched by Virtual Lead Parties (P415), 6 October 2023, < https://www.ofgem.gov.uk/sites/default/files/2023-10/Ofgem%20decision%20P415%20%27Facilitating%20Access%20to%20Wholesale%20Markets%20for%20Flexibility%20Dispatched%20by%20VLPs_0.pdf> accessed 16 August 2025.

of aggregators, Ofgem concluded, was likely to produce greater benefits for the system overall, notably with respect to flexibility, when compared to the Alternative Solution.

Ofgem recognized that there were concerns about the fairness of mutualization. Since the costs of VLP activity are distributed among all suppliers, mutualization created a potential imbalance where suppliers bear financial burdens that VLPs avoid. Despite this, Ofgem argued that the benefits from increased flexibility—such as lower wholesale electricity prices—were expected to improve market welfare overall and benefit all suppliers by reducing their sourcing costs. Ofgem committed to monitoring how these benefits and cost impacts unfolded for suppliers and whether they affected end users.

Ofgem's decision was made in October 2023 and took effect in November 2024.[39] It is too early to analyze evidence of VLP participation in the wholesale market. However, an indication of the effect of the decision is that Voltalis, a European leader in demand response services, has announced its intention to invest up to £1 billion to develop up to 5 GWp of demand response from residential and commercial buildings in the energy market by 2030.[40]

Another indication of the DSF potential in GB is the evidence of consumer participation in the Demand Flexibility Service (DFS),[41] managed directly by the National Electricity System Operator (NESO), which buys the flexibility from customers directly or possibly through aggregators and retail companies. The DFS is designed to help households and businesses participate in the electricity market by providing incentives for reducing or shifting demand. In 2022, NESO, operating then as the Electricity System Operator (ESO) launched the DFS for the winter period, rewarding homes and businesses for shifting their electricity use during DFS events. Over 1.6 million households and businesses participated in the service, helping to balance the national electricity network at peak times. DFS returned for winter 2023/24, and saw even more homes and businesses get involved, with 2.6 million participating.

However, Ofgem has more recently decided to introduce compensation to retailers in the BM[42] following a very different approach than it elected for the wholesale market, closer to the French model. This decision is not discussed here but the section on recommendations would, in principle, also apply to the BM.

V. Recommendations to Fill Regulatory Gaps

This chapter outlined several reasons why demand aggregation is central to the economic viability, environmental sustainability, and energy security of increasingly decentralized electricity systems reliant on intermittent renewables. Both GB and French legislation allow demand aggregators to participate in wholesale markets and compensate retail suppliers for the financial costs of activating demand-side flexibility (DSF). Some argue compensation

[39] ibid.

[40] A Rolt, 'Grid flexibility specialist Voltalis unveils BPS 1 bn UK Investment Plan', 29 January 2025. <https://www.businessgreen.com/news/4403400/grid-flexibility-specialist-voltalis-unveils-gbp1bn-uk-investment-plan> accessed 25 July 2025.

[41] NESO, 'Demand Flexibility Service Explained' <https://www.neso.energy/industry-information/balancing-services/demand-flexibility-service/demand-flexibility-service-explained> accessed 25 July 2025.

[42] Ofgem, 'Balancing and Settlement Code (BSC) Modification P444' https://www.ofgem.gov.uk/decision/balancing-and-settlement-code-bsc-modification-p444 accessed 25 July 2025.

is unnecessary since suppliers can also offer DSF services and benefit from lower prices through aggregation. Yet, in systems with corrected models, DSF activation typically imposes some financial cost when demand is reduced. While not obligatory, compensation appears both fair and helpful in facilitating the shift to aggregation. The challenge lies in designing compensation without creating barriers to entry.

The review of US, GB, EU, and French legislation highlights five key regulatory gaps. This section offers recommendations to address them: (A) adopt a net benefits test; (B) introduce mutualized financial compensation; (C) enable Virtual Lead Parties in the EU; (D) design capacity markets that do not discourage demand response activation; and (E) allow demand response to stack participation across multiple markets.

A. Net Benefits Test

In the US, FERC defined a clear net benefit test to assess whether demand response improves consumer welfare. Neither GB nor the EU have a clearly defined net benefit test. This raises doubts about whether, or more precisely when, aggregated DR is improving consumer welfare. The FERC test compared the total cost of compensating DR with the benefit resulting from the reduced system cost due to DR; where the net benefit was positive, there was an improvement in consumer welfare. It simplified this test by defining a market price threshold above which net benefits were positive.

In GB and the EU, the net benefits test must be different from that in the US because the balancing systems are different. In GB and the EU, under 'corrected' systems, balancing responsible parties (in particular the retail suppliers) bear the financial cost of deviations between their day ahead commitments to buy electricity and the final consumption of their customers. A net benefits test in these countries requires a comparison of the wholesale market cost reduction due to DR with the financial costs borne by the retailer suppliers whose customers reduced their demand due to aggregator activation of DR and the related perimeter 'correction'. Where the lower wholesale market costs due to aggregated DR are greater than the financial costs for affected suppliers, net benefits are positive. The test could be based on a monthly definition of a threshold DAM price, above which net benefits are positive, following the FERC guidelines, or on other agreed criteria.

The recommendation is that both GB and the EU define and implement net benefits tests and use these tests before determining who should pay compensation. In the EU, the key reform is to require a net benefits test rather than making it an option.

B. Compensation

GB has adopted mutualized compensation arrangements for suppliers affected by the activation of DR by Virtual Lead Parties (VLP, especially independent aggregators) for sales in the wholesale market. Under this arrangement, the retail supply companies that have been commercially impacted because of the VLP activation of DR, and subsequent perimeter correction, will be compensated for their financial costs, and the costs will be mutualized amongst all suppliers. This approach was adopted by Ofgem, following a cost-benefit analysis that concluded that the mutualization solution allowed for greater volumes of flexibility deployment due to the lower variable costs for deployment of flexibility by VLPs and therefore greater benefits for consumers compared to the alternative.

The EU legislation could be interpreted to favour the mutualization approach. Where net benefits are taken into account, Article 17-4 requires aggregators and their customers to contribute to compensation only when the activation of DR has not generated a net benefit. However, without a net benefits test, there is no way to determine whether the aggregation has indeed generated a net benefit. Furthermore, as illustrated in the case of France, requiring aggregators to compensate retail suppliers raises their costs and can act as a barrier to entry or at least to the activation of DR.

The recommendation is to amend Article 17 to ensure that mutualization in some form is adopted when net benefits are proven. Mutualization could follow the GB approach. Alternatives that reflect the contribution of demand response to meeting environmental and system security objectives and that involve socialization of the costs include introducing an electricity system levy, or cost recovery through the tax system.

C. Virtual Lead Parties

In the UK, the Virtual Lead Party (VLP) does not supply electricity but is able to participate in wholesale electricity markets by providing demand side response or other flexibility services to customers supplied by another company. Customers can stay with their retail supplier while contracting separately with a VLP for flexibility services. This separation of supply from flexibility services enables fair competition between independent aggregators and retail companies in the aggregation market, broadening participation, reducing costs, and strengthening system flexibility.

In the EU, while independent aggregators already can participate in markets without supplier consent, it leaves the revenue-sharing and operational details to Member States. There is no distinct, standardized market actor like the VLP. Introducing a VLP concept could create a clear, harmonized role for independent aggregators and facilitate retail suppliers to develop aggregation affiliates to target consumers outside their own supply portfolios. It would also support a level playing field in the market for aggregation services, encouraging wider deployment of DSF.

The recommendation is that the EU amend EU Directive 2019/944 to include the figure of Virtual Lead Party (VLP) and draw on GB legislation to define it.

D. Capacity Markets

In the US, GB, and the EU, demand response may participate in capacity markets, and make an important contribution to meeting system security of supply. However, the experience in France raises concerns about the value of paying a capacity payment and especially the legality of state aid. This explains the concerns expressed by the European Commission and the French Agency for the Ecological Transition (Ademe).

The proposal is not to eliminate capacity payments for DR in GB or the EU, but to ensure that where capacity payments apply, they do not act as a barrier to activation of demand response. This is best achieved by following the recommendations concerning the adoption of a net benefits test and mutualized compensation since this would eliminate the barrier to activating DR. However, there may be additional legislation required.

E. Multi-Market Access for Aggregators

One of the key conclusions of the Ofgem decisions on P415 was that, by reducing the costs faced by VLPs, mutualized compensation in the wholesale DAM would pave the way for greater participation of flexibility, increased load shifting opportunities, and increased welfare compared to the alternative.[43] These benefits correspond not only to the wholesale DAM market because investment in DR will enable that flexibility in balancing, capacity, and other markets.

Demand flexibility resources—such as batteries—often need to participate in multiple markets (eg wholesale, balancing, capacity, ancillary services) to earn enough revenue to make aggregation economically viable. However, regulatory frameworks sometimes create conflicting rules: meeting the conditions to provide flexibility in one market can restrict or exclude participation in another, reducing revenue opportunities and discouraging investment.

The solution and the proposal is to design coherent, market-wide rules that allow flexibility providers to stack revenues across markets; while ensuring they are not double-paid for the same service.[44] This requires harmonized participation criteria, clear definitions of responsibilities, and settlement systems that recognize contributions across different markets without creating barriers.

VI. Conclusions: Aggregation at the Heart of a New Sustainable, Flexible, and Decentralized Electricity System

The energy transition demands a radical shift in how electricity systems are managed, moving from centralized, supply-focused models to ones that actively integrate demand. As this chapter has shown, DSF—particularly to reduce demand through aggregation when prices are high—has the potential to provide cost-effective, low-carbon flexibility across multiple markets. By unlocking the flexibility in DERs, aggregation can reduce overall system costs and enhance reliability, while supporting the integration of renewables and environmental sustainability.

Legal and regulatory frameworks are crucial in determining whether this potential can be realized. This chapter has argued that GB's regulation of demand response in the wholesale DAM promotes the participation of independent aggregators (as VLP) and that the French model often discourages that participation. It identifies regulatory gaps in the GB and EU regulatory frameworks and makes recommendations to fill these gaps. The most

[43] Ofgem, 'Balancing and Settlement Code (BSC) Modification P444' <https://www.ofgem.gov.uk/decision/balancing-and-settlement-code-bsc-modification-p444> accessed 25 July 2025.

[44] A Danthine and A Zerain, 'What are the Economic Risks Associated With Investing in Energy Storage, and How Can They Be Mitigated?' in 'Powering the Future: Energy Storage in Tomorrow's Electricity Markets' (2024) 140 Oxford Energy Forum (April 2024) https://www.oxfordenergy.org/wpcms/wp-content/uploads/2024/04/OEF-140-Powering-the-Future.pdf accessed 29 October 2025.

important recommendations are the introduction of a clear net benefit test in GB and the EU to help assess whether DR is welfare improving; and the adoption by the EU of the GB model of mutualized funding (or socialization) of compensation to retail suppliers when the net benefits test has been passed.

This chapter has focused on electricity demand reduction, where it is usually clear that the effect is to reduce market prices and improve welfare, subject to the net benefits test and that some of the benefits are passed on to consumers through competitive pressure. Demonstrating improvement in social welfare resulting from a demand increase is more challenging if the effect is an increase in market prices. However, an increase in price may well still be a social welfare improvement if the alternative is even more expensive. The implication is that policy makers need to introduce net benefits tests for both demand increase and demand decrease to confirm that DSF yields an improvement in social welfare and that mutualization is the preferred compensation mechanism when one is needed.

The emerging policy consensus in both the EU and the US supports enabling demand response, directly and through intermediaries, to compete on an equal footing with generators in electricity markets. As electricity systems become more decentralized, policies that reward consumer participation and innovation—while maintaining system balance and fairness—will be central to the success of the energy transition.

14

Digitization of Networks and Data Protection

Lee Godden

I. Introduction

This chapter examines the data protection and consumer privacy challenges that arise with the digitization of energy systems under a Distributed Energy Resources (DER) model. In many jurisdictions globally, the progressive 'digitization' of energy systems over several decades has been driven by the technological reform of energy production and distribution that has both facilitated the adoption of DER and enhanced the digitalization of the energy sector. Digitization in energy systems is associated with decarbonization due to the shift to renewable energy generation, and with the decentralization of energy distribution. There is an accompanying rationale to digitalization in that such technological transformation among other benefits can promote sustainable energy systems and address climate change.[1] Alongside the greater penetration of digitization in energy systems, however, there has been a rise in public and regulatory concerns about privacy and protection of personal data.[2]

Accordingly, this chapter first examines the move to DER within nations (see chapter 1) and internationally (see chapter 2), in light of the accompanying challenges that arise with the digitization of energy networks. While outlining the technological changes brought by energy digitization (see also chapter 15), it focuses on efforts to ensure data protection and integrity in the context of energy data access and shared use alongside the sustainability benefits delivered by DER.

Attention to individual privacy and data protection has formed part of wider regulatory platforms designed around the need to secure personal data against its potential misuse by government and corporate organizations accessing personal data.[3] Energy utilities (whether publicly or privately owned) typically are included in regulatory efforts to ensure the security of the personal data of consumers.[4] More recently, the concern around energy digitization has turned to ensuring data sovereignty, given the recognition that there is a public or national interest in data security in the energy sector.[5] These technological changes to the foundations of energy production and distribution systems alongside the rise of 'prosumers'[6] are

[1] S Lavrijssen, B Espinosa Apráez, and T ten Caten, 'The Legal Complexities of Processing and Protecting Personal Data in the Electricity Sector' (2022) 15 Energies 1088, 1088.

[2] Lavrijssen (n 1) 1088.

[3] See P Balboni and K Elizabeth Francis, *Data Protection as a Corporate Social Responsibility* (Edward Elgar 2024).

[4] Many jurisdictions now require energy companies and energy regulators to have a data strategy.

[5] S Misra, K Barik, and P Kvalvik, 'Digital Sovereignty in the Era of Industry 5.0: Challenges and Opportunities' (2025) 254 Procedia Computer Science 108, 108.

[6] See Z Rehman and others, 'Prosumer Based Energy Management and Sharing in Smart Grid' (2018) 82 Renewable and Sustainable Energy Reviews 1675.

Lee Godden, *Digitization of Networks and Data Protection* In: *Sustainable Distributed Energy Resources*. Edited by: Íñigo del Guayo, Damilola S. Olawuyi San, Louis de Fontenelle, Milton Fernando Montoya, and José Juan González, Oxford University Press.
 DOI: 10.1093/9780198991946.003.0016

components of a wider transformation in industrial and economic sectors which increasingly depend on the technological utilization of data.[7]

Drawing lessons from the European Union and Australia,[8] the chapter in sections II, III, and IV outlines the tensions involved in facilitating shared data access and use, while enabling effective data protection and consumer agency as renewable energy is integrated into previously centralized energy models. Within a market-based regulatory environment such as Australia, policy and legal developments around data protection and digitization are predicated upon a dialogue of consumer/prosumer utility. These policy platforms obscure what is, in effect, a trend away from the first phase of small-scale DER. Small-scale solar and wind and now electric vehicles have been promoted directly by governments but are now being rapidly overtaken by large wind and solar projects where the opportunities for consumer participation are more constrained.

In the two final sections, by reference to Australian energy law and policy, the chapter considers how liberal, market-based energy systems are responding to digitization challenges, with particular emphasis on how data protection, and indeed data sovereignty, may be effectively regulated.[9] Having surveyed the challenges of ensuring energy data protection as DER models become integrated more fully into energy markets, the chapter concludes by exploring the feasibility of tighter regulation in respect of data protection and data sovereignty within the wider rubric of facilitating DER under energy legislation and relevant data protection regulatory schemes.

The chapter argues that while the DER policy and regulation are now reasonably effective in facilitating renewable energy integration as a component of decarbonization and energy transition the emphasis on data privacy protections, while important, needs to be extended to deal with the broader data security challenges of digitalization in energy markets that are masked by a promotion of consumer agency. Accordingly, the chapter provides some key points as the basis for ensuring the pivotal integrity of data as these transformations occur.

A. Australia as a Case Study

The chapter uses Australia as an exemplar to examine the implementation of DER, digitization, and data, given the particular complexity of policy, technology, and legal factors at play in the country. Australia has had a national energy security planning process since 2017 that is designed to enhance DER integration in energy markets. The first phase is to be complete by 2025.[10] Further, there are exceptionally high levels of small-scale rooftop

[7] Misra (n 5) 108.

[8] The European Union has been proactive in data protection. Regulation is comprised of the General Data Protection Regulation, the Law Enforcement Directive, and the Data Protection Regulation for EU institutions, bodies, offices and agencies: European Commission, 'Legal Framework of EU Data Protection' <https://commission.europa.eu/law/law-topic/data-protection/legal-framework-eu-data-protection> accessed 28 July 2025; for Australia, see Energy Security Board, 'Data Strategy Consultation Paper' (October 2020) <https://www.datocms-assets.com/32572/1657766904-esb-data-strategy-options-consultation-paper-october-2020.pdf> accessed 25 August 2025.

[9] F Dammann and G Glasze, 'Governing Digital Circulation: The Quest for Data Control and Sovereignty in Germany' (2023) 11 Territory, Politics, Governance 1100, 1102.

[10] Energy Security Board, 'Data Strategy: Final Recommendations' (July 2021) 3 <https://esb-post2025-market-design.aemc.gov.au/32572/1630275857-esb-data-strategy-july-2021.pdf> accessed 19 July 2025.

solar in domestic electricity provision in the country that enhance energy sustainability but pose issues for data and grid management.[11] The penetration of DER in energy markets was facilitated by the relatively early introduction of smart (i.e. digitally based) energy mechanisms, such as smart grids and smart meters by subnational governments.

Against the backdrop of these factors, the Australian Energy Security Board in 2021 noted that digitalization and changes in the energy market due to larger volumes of renewable energy generation are driving the need for more flexible and timely access to data. The Board indicated that '[g]reater variability, diversity and fragmentation in both supply and demand is increasing the challenge of coordination . . . Data volumes are growing exponentially, creating ongoing challenges for data management. Data privacy, protections and security concerns are also under pressure.'[12]

II. Digitization and Distributed Energy Resources

Clearly, DER models rely heavily on the integration of digital technologies into energy systems. Such resources, to adopt the definition of an International Energy Agency report, comprise 'small-scale, clean installations located behind the consumer meters, such as photovoltaic panels, energy storage and electric vehicles, are increasingly widespread and are already transforming our energy systems.'[13] As chapter 1 of this book notes, DER can comprise a wider group of energy system attributes, 'as it is not limited to generation but also includes other electricity resources that are also available to consumers (such as storage).'[14] The objective is that DER will allow for more informed demand response, as consumers use data technologies to make efficient energy consumption decisions and increase energy end user self-sufficiency.[15] Indeed, DER models such as small-scale renewable energy generation and distribution could not feasibly exist without the digital technologies that allow decentralized 'sources' of energy production such as solar photovoltaics at household level to be digitally managed remotely, and for remote digital systems to facilitate grid access and distribution for the energy that is produced.

In this manner, DER can assist energy democracy and diversification by advancing citizen-led energy solutions to renewable energy challenges such as electricity supply flexibility, and capacity demand management. Correspondingly, as the analyses in several other chapters in this book[16] contend, DER models offer opportunities to transition from centralized fossil-fuel based energy systems, but the intermittent nature of renewable energy and the diffuse scale of much renewable energy production (which can be highly disaggregated)

[11] Australian states have the highest levels of rooftop solar in the world. See A Akimov and others, 'Are Australian Regulatory and Policy Frameworks Ready for a Distributed-Energy-Resources-Led Energy Transition?' (2023) 82 Australian Journal of Public Administration 529, 545.

[12] Energy Security Board (n 10) 3.

[13] International Energy Agency, *Unlocking the Potential of Distributed Energy Resources: Power System Opportunities and Best Practices* (May 2022) 3 <https://iea.blob.core.windows.net/assets/3520710c-c828-4001-911c-ae78b645ce67/UnlockingthePotentialofDERs_Powersystemopportunitiesandbestpractices.pdf> accessed 25 August 2025.

[14] Íñigo del Guayo noted in chapter 1 of this book that DER refers to the small-scale generation, transmission, and storage of electricity through microgrids, backup batteries, fuel cells, and virtual power plants that are often at, or closer to, the point of consumption (for example, solar panels in the consumer's home).

[15] See chapter 2 of this book.

[16] See chapters 2, 3, 7, 9, 10, 14, 15, and 17 of this book.

poses challenges for effective co-ordination and regulation of these energy resources. Alongside these technical challenges is the need to ensure the integrity of data within these energy models while also realizing the sustainability potential of DER. That potential for DER, as the central thematic of this book explores, is predicated on principles denoting decarbonization, decentralization, digitalization, and democratization.

A. Digitization and DER in Energy Transformation

Digital technologies and manipulation of data have simultaneously transformed many processes associated with energy production and consumption in industry, households, and the public sector. The forerunner to the recent wave of technological changes was the introduction of smart grids, which are designed as a power system with integrated data/communication components to enable a two-way exchange of energy and information about energy use. In turn, the introduction of hybrid data 'tools' such as smart meters enabled remote monitoring and provision of consumer information about their energy use patterns to encourage better demand management. These two measures link to a global trend in the rapid expansion in 'Smart' digital technologies in energy systems.[17] As the US Department of Energy noted in 2017, '[p]rosumers are growing in the energy space as more Americans generate their own power from distributed energy resources . . . Gone are the days when electricity consumption was a one-way street. Today's electric grid is blurring the lines between power generation and consumption.'[18]

These technologies, in association with DER transition models, allow a more sustainable and efficient energy system that facilitates active consumer involvement.[19] Such objectives align with new generation data technologies that are shaping many dimensions of human-technology interactions in the 'phase 5.0 Industrial Age'. 'Industry 5.0 is characterized by human-centric, resilient, and sustainable systems, with digital sovereignty as a key factor in enabling secure and autonomous innovation.'[20] Digital sovereignty implies, 'the capacity of governments, organizations, and individuals to autonomously manage their digital resources, infrastructure, data, and technology, assuring alignment with local values, legislation, and priorities'.[21]

Notwithstanding the optimistic view of human-technology interactions and the vision of digital sovereignty as conferring the autonomy to manage digital resources consistently with local values, there is growing acknowledgement of the significant power and influence (and wealth) associated with controlling 'data' in an era characterized by a rapid rate of digital transformation. In turn, the capacity for third parties to control large amounts of data has precipitated concerns about personal privacy. Indeed, despite the large amounts of personal data that are being collected across all aspects of society, individuals may not own or be in a position to have effective control over the data that is collected about them.

[17] See F Wen and others (eds), *Proceedings of 2024 International Conference on Smart Electrical Grid and Renewable Energy (SEGRE 2024)* (Springer 2025) vol 2.

[18] Office of Energy Efficiency and Renewable Energy, 'Consumer vs Prosumer: What's the Difference?' (*US Department of Energy*, 11 May 2017) <https://www.energy.gov/eere/articles/consumer-vs-prosumer-whats-difference> accessed 19 July 2025.

[19] Rehman (n 6) 1676.

[20] Energy Security Board, *Data Strategy* (n 10) 109.

[21] ibid 112.

'Data governance, cybersecurity management, and privacy protection are essential for controlling who has access to how this data is used and for protecting one's digital assets and identity.'[22]

B. Privacy Law, Data Regulation, and Energy Law

What then is the interface between privacy law, data regulation and energy law where the latter promotes uptake of DER? Digitization processes that 'promote the energy transition that rely on the processing of consumer's data cannot be viewed in isolation from the protection of personal data.'[23] The protection of data was not a prominent part of the earlier phase of systemic energy law transformation. Commencing in the early 1990s, there was a significant movement driven by market economic agendas to 'liberalize' energy systems in market economies.[24] Existing state-based 'monopoly' systems were characterized as overcapacity and inefficient. The market-based reforms sought to disrupt centralized energy production and distribution by breaking up energy systems such as electricity supply into several component functional units, in order to introduce competition and efficiency objectives. These liberal, market-based energy reforms were largely still predicated upon negligible household electricity production. Now, the liberalized energy market system is itself undergoing radical change, and the polices that were embedded around earlier technologies are being scrutinized. In market-based energy systems, DER present opportunities but challenge the adequacy and direction of regulatory frameworks which should respond to such developments.[25] There is now an increasing level of policy and regulatory activity across the interface of digitization, data protection, and energy systems.

In many of the legal and regulatory contexts that operate at the interface of privacy, data protection, and energy transition, much of the language remains grounded in policy and governance objectives from an earlier era of energy transformation. In those policy settings, the discourse still tends to emphasize concepts such as sustainability, renewable energy transition, and decarbonization.[26] The potential gap between the expansion of digitization as a pervasive force in energy systems, and the gaps in data security suggest the need for more comprehensive legal and regulatory attention to data protection and data sovereignty in energy law and regulation. The corresponding rapid growth in DER suggests that data protection in energy markets needs to become more comprehensive, and to be expanded beyond individual 'tools' such as smart metres to embrace more systemic measures that can safeguard data protection and enhance digital sovereignty.

The European Union is the jurisdiction that has the most comprehensive law and regulation that sits at the interface of data protection and energy transition. Even so, there are some competing policy objectives in governing regulation. For example, there are energy-related measures that seek to expand the utilization of data to support demand driven sustainability.

[22] ibid 109.
[23] Lavrijssen (n 1) 1089.
[24] See generally, D Zillman and others, 'Introduction: How Technological and Legal Innovation are Transforming Energy Law' in D Zillman and others (eds), *Innovation in Energy Law and Technology: Dynamic Solutions for Energy Transitions* (OUP 2018).
[25] Akimov (n 11) 531.
[26] See, for example, the scope of responsibilities articulated for the Australian Renewable Energy Authority under the Australian Renewable Energy Agency Act 2012 (Cth).

An EU Directive (EU 2019/944) seeks to advance the use of smart metres to provide data for energy consumers to better enable households to track the energy consumption and production in order to incentivize decreased energy demand. Simultaneously, the data is identifiable as personal data and attracts protection under the EU General Data Protection Regulation. Protection of personal data has the status of a fundamental right as identified in Article 8 of the Charter of Fundamental Rights of the European Union and Article 16 of the Treaty of the Functioning of the European Union. As the right is embedded in the governing architecture of the European Union, it highlights the significance attached to data protection and its corresponding regulation. The right is given operational impetus by Regulation 2016/679 (GDPR) which protects the personal data of natural persons as a core objective. This protection is in some tension with another core objective of the GDPR—that of the free movement of personal data. The GDPR applies to the processing of personal data in all sectors and situations, with some relatively circumscribed exceptions.[27] Personal data comprises 'any information relating to an identified or identifiable natural person'.[28] Data attributed to companies such as energy utilities or providers thus would not fall within the ambit of the GDPR. Similarly, data pertaining to natural persons that is completely anonymized does not attract GDPR application. Even so, there is a vast amount of personal data that is likely to be associated with natural persons in energy systems such as the liberalized energy markets of the EU that rely heavily on digitization. This raises the question of how the two legal regimes, one prioritizing data protection and privacy and the other that is encouraging consumer initiatives such as 'prosumers' coexist.[29] Much of this incentivization is to be found in the EU Recast Electricity Directive.

In a comprehensive analysis of the interface between the data protection rules of the GDPR and the rights concerning 'Consumer Data' in the EU Recast Electricity Directive, Saskia Lavrijssen and others note that the Recast Electricity Directive, 'does not elaborate on the exact relationship between the GDPR and the processing of (including access to) consumer data in the electricity sector'.[30] They highlight how the Recast Directive places emphasis on facilitating access to data by consumers themselves—which might be explained by reference to the fact that this measure is related to consumers obtaining information from smart meter systems.[31] Further, the Recast Directive promotes consumer data access to allow energy consumers to participate in energy services beyond the traditional supply of electricity such as in aggregation contracts.[32] These emergent services are supported by a right to access consumer energy data at no cost. On the other hand, the Recast Directive does reflect the data processing principles of the GDPR which constrain requests for personal data, for example, in energy switching contexts and there are obligations of transparency about the collection and processing of personal data.

In summary, the EU regimes of data protection clearly impose specific obligations of data protection in energy settings, but such obligations exist within a more dynamic position that supports data movement and access that underpins the emergent technologies of DER

[27] Lavrijssen (n 1) 1094.
[28] Regulation (EU) 2016/679 of the European Parliament and of the Council of 27 April 2016 on the protection of natural persons with regard to the processing of personal data and on the free movement of such data, and repealing Directive 95/46/EC (General Data Protection Regulation) [2016] OJ L119/1, art 4(1).
[29] Lavrijssen (n 1) 1089.
[30] ibid 1102.
[31] ibid.
[32] ibid.

and the promotion of the individual consumer as an active agent in the management of data and energy consumption. Issues of digital sovereignty and the public interest are mediated through the privatized lens of liberal energy markets and to be displaced by a discourse of economic agency. A similar dynamic tension is to be found in the Australian case study context, although that jurisdiction does not have the benefit of an overarching right to data protection and the substantial regulatory infrastructure associated with privacy and data protection in the EU. While there is regulatory attention to data it too emphasizes data management to support new technologies, forms of production, and services.

III. Australia as a Digitization and DER Case Study

A. Australia's Decarbonization Transformation

Australia, as a market-based developed economy and democratic polity, has moved to adopt DER at multi-scales, across its national and subnational jurisdictions to fulfill significant decarbonization functions in its energy transition. Over almost two decades, there has been strong promotion of renewable energy at the provincial and local government levels. In recent years in Australia's national political and legal landscape, the climate change debate also has shifted markedly to include a clear decarbonization agenda based around greater renewable energy generation in the national electricity grid.[33] The Western Australian energy market, once a climate change outlier is now embedding DER initiatives. Households and businesses are shaping 'their own energy futures by investing in solar, batteries and electric vehicles'.[34] Effectively, they have the potential to help lower costs for all consumers by offsetting the need for an additional $4.1 billion of grid-scale investment.[35] Australia has been an early adopter of the smart technologies such as smart metres in the renewable energy space.[36]

The Australian domestic decarbonization agenda is strongly aligned with the trend to digitization noted above. Decarbonization has a local democracy orientation too, in the growth of a new 'Teal' democracy movement that advocates for stronger measures to address climate change.[37] The Teal democracy movement has given rise to a group of women who are independent political candidates elected to Parliament in the last two national elections. The Teal movement initially captured the sense of civic protest around Australia's previously weak Nationally Determined Contribution emission reduction targets under the Paris Agreement. The Teal Independent politicians have since forged a wider energy

[33] See generally, Australian Energy Market Operator, *2024 Integrated System Plan for the National Electricity Market: A Roadmap for the Energy Transition* (26 June 2024) <https://www.aemo.com.au/-/media/files/major-publications/isp/2024/2024-integrated-system-plan-isp.pdf> accessed 25 August 2025.

[34] Australian Energy Market Operator, *AEMO's 2024 Integrated System Plan* (26 June 2024) <https://www.aemo.com.au/-/media/files/major-publications/isp/2024/2024-integrated-system-plan-overview> accessed 20 July 2025.

[35] ibid.

[36] L Godden and A Kallies, 'Smart Infrastructure: Innovative Energy Technology, Climate Mitigation, and Consumer Protection in Australia and Germany' in D Zillman and others (eds), *Innovation in Energy Law and Technology: Dynamic Solutions for Energy Transitions* (OUP 2018).

[37] R Millar, 'Inside the Teal Wave: How the Independent Revolution Happened' *The Sydney Morning Herald* (Sydney, 22 May 2022) <https://www.smh.com.au/politics/federal/inside-the-teal-wave-how-the-independent-revolution-happened-20220522-p5ani0.html> accessed 19 July 2025.

transition policy platform,[38] which includes DER elements around energy transition. Nonetheless, the overall political position on climate change measures and small-scale renewable energy infrastructure in Australia remains politically volatile.[39]

While adverse political rhetoric may suggest otherwise, the changes in Australia's energy mix have been profound. The nation is reasonably well advanced along the energy transition pathway in respect of renewable power generation and the integration of distributed renewable energy into electricity grid systems. Digitalization and data management are critical foundations to that transition.[40] The national energy markets have experienced steady growth in participation by wind, solar, and batteries, which have increased renewable market share and reduced greenhouse gas emissions.[41] There are predictions that there will be a renewable resource dominated system, firmed by battery storage with a small amount of gas to meet peak electricity demand within fourteen years.[42] Similar to international trends though, the phenomenon of the increasing contribution of distributed renewable energy is happening alongside increased electricity demand in specific commercial and social sectors.[43] Moreover, it is unlikely that there will be a substantial decrease in Australia's major fossil fuel exports given current policy settings.

In the Australian DER setting with respect to the four vectors of decarbonization, decentralization, digitalization, and democratization, there is increasing convergence around digitization, data, and democratization. Such interactions are reshaping the understanding of 'sustainability' in the governance of energy in the nation. Historically, the intersections between data protection and sustainable energy management have not been a strong focus in the Australian policy and regulatory settings around DER. That has been changing rapidly since the Finkel Review in 2017.[44] What is apparent though is that the synergy of decarbonization, decentralization, digitization, and democracy as it evolves in Australia through the expansion of DER presents regulatory problems beyond operational adjustment. More than a sustainability 'tweak' is required to address the 'wide range of environmental, social and economic challenges ... which if not holistically addressed could limit the full potential of DERs to advance a secure and sustainable energy future'.[45]

Moreover, while the DER pathway is firmly established in Australia, the sustainability agenda is being co-opted in various ways to advance a 'Digitization future'. It is acknowledged in Australian energy policy that innovation and transformation of consumer energy

[38] C Long, 'The Independents in the 2025 Election, Their Electorate, and What They Stand for' *ABC News* (4 April 2025) <https://www.abc.net.au/news/2025-04-04/independent-mp-parliament-politics-teal-federal-election-2025/104990946> accessed 19 July 2025.

[39] In May 2025, the political opposition party endorsed a plan to promote nuclear energy as an alternative pathway to meet Australia's Net Zero obligations.

[40] Energy Security Board (n 10) 5.

[41] Australian Energy Market Operator, 'Quarterly Energy Dynamics Q1 2025' (7 May 2025) 4 <https://www.aemo.com.au/-/media/files/major-publications/qed/2025/qed-q1-2025.pdf> accessed 25 August 2025 (*QED Q1 2025*).

[42] D McConnell, 'Coal-Free in 14 Years as Renewables Rush in: New Blueprint Shows How to Green the Grid—Without Nuclear' *The Conversation* (26 June 2024) <https://theconversation.com/coal-free-in-14-years-as-renewables-rush-in-new-blueprint-shows-how-to-green-the-grid-without-nuclear-232985> accessed 18 July 2025.

[43] B Kizhakedath, 'Renewable Energy Contribution to Electricity Generation in 2024: IEA' *Greentech Lead* (24 March 2025) <https://greentechlead.com/renewable-energy/renewable-energy-contribution-to-electricity-generation-in-2024-iea-49256> accessed 25 August 2025.

[44] A Finkel and others, *Independent Review into the Future Security of the National Electricity Market: Blueprint for the Future* (Commonwealth of Australia, June 2017) <https://www.dcceew.gov.au/sites/default/files/documents/independent-review-future-nem-blueprint-for-the-future-2017.pdf> accessed 25 August 2025.

[45] Akimov (n 11) 534.

services and production is central to the data-led energy transition.[46] The Australian Energy Security Board's 2021 Data Strategy was designed to provide direction for managing changing data needs in energy transition and to optimize the long-term interests of energy consumers in a digitalized economy.[47] As the Board recognizes, consumers can now access specialized energy products, and, if regulatory reform ensues, may obtain energy from multiple service providers—all dependent on emergent data technologies.[48] Such providers will 'benefit by improved access to customer's data'.[49] While there is concurrent acknowledgement of the need for data protection, with a consumer data right in energy regulation,[50] the existing data protections are regarded as barriers to facilitating access in a rapid technological transformation.[51] Like the EU, there are inherent tensions around the rights to data protection in the energy sector. Moreover, given the scale and scope of the changes that reach across customers, service providers, network operators, and market entities (regulation, development, planning, and operations) what is occurring is a paradigm shift in Australian energy systems and law.

The messaging around digitization and data, however, while nominally consumer oriented around DER, also provides a general mandate for augmenting data capability and data services across energy systems.[52] In turn, these future systemic attributes, that the Data Strategy terms 'Needs Tomorrow' act to enhance the acceptability of the increased penetration of digitization into the Australian energy market. These measures not only support future consumer energy outcomes but advantage multinational capital and the large-scale renewable energy projects that are well positioned to leverage 'new data technologies'. By contrast, consumer-led localized prosumer and aggregation models for energy supply are not strongly in evidence yet, despite the preparatory policy and legal settings being progressively developed.[53] Australian authorities continue to enthusiastically embrace the prosumer model—predicting strong growth in the extent to which battery storage will heighten the capacity for households installing solar panels to be both electricity consumers and prosumers.

The Australian energy markets have witnessed large investments in 'digitization' across the generation, grid distribution, and retail electricity sectors[54]—partly by regulatory authorities and partly by a range of multinational companies expanding their operations in the Australian energy sector. Thus digitization, while furthering decarbonization and decentralization, offers less support for democratic participation and consumer agency at the level of the individual user/producer, despite the advent of the political 'Teal' style political representation. Older models of data protection no longer are held to be appropriate. These tensions are part of competing visions for future directions for distributed energy in Australia. There remains a disconnect between the scenario of a prosumer and community-led upsurge in digitized DER with the financial structure of the market-based

[46] Energy Security Board, *Data Strategy* (n 10) 17.
[47] ibid 5.
[48] ibid.
[49] ibid.
[50] ibid 19.
[51] ibid 17, stating that 'requirements to protect data can be the largest barriers'.
[52] ibid.
[53] Akimov and others (n 11) 550.
[54] Y Strengers and others, 'Digital Energy Futures: Future Home Life' (Monash University, July 2021) <https://www.monash.edu/__data/assets/pdf_file/0011/2617157/DEF-Future-Home-Life-Full-Report.pdf> accessed 20 May 2025.

energy framework that is characterized by the large-scale investment that now predominates in the roll out of renewable energy resources over much of the nation. The large-scale renewable energy projects and the public infrastructure that enables them also have necessarily invested heavily in digitization.

In Australia, the exponential increase in renewable resources such as small-scale photovoltaics, wind, and battery storage has been promoted and subsidized by subnational governments over the last decade or so. Yet it may not translate into greater local sustainability of energy resources. For example, increasing temperatures in southern Australia linked to climate change have lifted peak energy demand. These localized increases in household energy consumption are not fully offset by the penetration of DER in the electricity system,[55] particularly given the relatively limited adoption of 'prosumer' household energy production beyond battery storage.[56]

Further, while DER now has a firm policy direction in Australia, the intensity and scale of decentralized renewable energy roll out remains variable across the electricity supply system, and consequently, in terms of the needs for data management. Further, there are specific concentrations in the spatial location of DER, with for example, high levels of penetration of small-scale residential solar energy systems in Australia's metropolitan areas.[57] More centralized energy generation and distribution is still prevalent outside urban metropolitan areas. The skewed distribution of DER introduces equity anomalies not only in consumer/prosumer access, but it has ramifications for consumer knowledge and information about data protection in energy markets, particularly in regard to data sharing and consumer data rights.

B. Consumer Data Protection and DER Future Directions

The Australian Energy Security Board over several years developed a policy and regulatory platform for energy data.[58] The focus for the data strategy was to review the current regulatory energy and privacy legal frameworks to identify barriers that limited effective data management, and which it was argued limited consumer's long-term interests. Key restrictions of the existing framework were identified ascomplexity, an unworkable public interest test (notably though data sovereignty was not invoked), and constraints in the management of privacy and commercial sensitivities.[59] While not providing direct evidence, the data strategy indicated that such difficulties were resulting in planning inefficiencies and duplication of costs. The data strategy did note that these barriers were adopted for sensible reasons at the time, due to a conservative approach to consumer (data) protection or to limit regulatory burden.[60] The strategy noted that as data sets increase in scale and value, the current regulatory model increased costs to consumers. Even so, what is not investigated is why it should be consumers that bear those costs rather than the energy sector entities best able to take financial advantage of the enhanced value brought about by growth in data sets.

[55] Australian Energy Market Operator (n 41) 10.
[56] Akimov (n 11) 547–49.
[57] ibid 535.
[58] Energy Security Board, *Data Strategy* (n 10) 18.
[59] ibid.
[60] ibid.

Instead, the strategy advocates the reduction of risks aligned to 'limited data visibility misaligned standards or protection settings'.[61] In short, the paradigm that is advocated is one of further advancing the national liberalized energy market via well-established competition and consumer style measures supported by data regulation reforms.[62] The final strategy is rationalized as keeping the system secure given the intermittent nature of small-scale renewable generation and use, while 'making sure customers can get the best value from reduced energy bills, more innovative products, and simple easy to access choices'.[63]

The specific regulatory reforms advocated for in the strategy are nominally aligned to changing consumer expectations (no evidence is provided as to their derivation). It is hard though not to see the classic neoliberal open markets and consumer preference model being reinvented in the energy data sphere in the strategy analysis. The other vector of classic neoliberal economic models—that of the purported transparency of government and business—similarly is invoked in the strategy. This package of economic aspirations pertaining to future energy data are also substantially aligned in the policy underpinning the wider national structural reforms around data access and management. Those structural reforms in Australian national data management have three core tenets: open data, consumer data right, and data availability and transparency legislation.[64]

C. Consumer Data Right and Data Availability and Transparency Act

A national consumer data right was introduced in Australia in 2020. It articulates with the Data Availability and Transparency Act 2022 as a component of a broader data sharing and data protection regulatory framework. The Act is not energy system specific, but a large number of energy organizations and companies participate in the scheme of data sharing, collection, and use established under the legislation.[65] The data sharing scheme established under the Act authorizes 'Commonwealth bodies' (ie national government departments and agencies, including energy regulatory authorities) to share the public sector data that they hold with accredited users.[66] Data may be shared directly with an accredited user, or through an intermediary (accredited data service provider). Users such as energy retailers which are accredited through the consumer data right process are authorized to collect and use the data, albeit in a controlled context prescribed by the statute.

There are a number of safeguards in that the sharing, collection, and use of data must be part of a project for one or more of the defined data sharing purposes.[67] Sharing and the collection and use of data must be done consistently with the statutory data sharing principles and under a registered data sharing agreement.[68] Privacy protections apply to the sharing of personal information,[69] that articulate with the requirements under the Privacy Act

[61] ibid.
[62] ibid 19.
[63] ibid 8.
[64] ibid. The Data Availability and Transparency Act 2022 (Cth) subsequently came into effect.
[65] Data Availability and Transparency Act 2022 (Cth) s 15. Data sharing purposes comprise delivery of government services; informing government policy and programs; and research and development.
[66] ibid ss 13–13A.
[67] ibid s 11.
[68] ibid s 11A.
[69] ibid s 16A.

1988 (Cth). The Act establishes the office of a National Data Commissioner and a National Advisory Board.[70]

The ostensible orientation of the legislation is primarily to govern public sector data sharing for research and policy development. Nonetheless, the measures for data sharing and the collection and use of data by accredited users such as corporations in the energy context make the system integral to the further digitization of energy systems in Australia. Accordingly, given the nexus between digitization and DER, these models will also underpin future planning and evolution of the Australian energy system in respect of DER.

The embedding of the specific modality of data access, sharing, collection, and use as developed in the data availability legislation is in line with the economy-wide digitization advocated for in the Energy Security Board data strategy. Each system is characterized by a similar set of data principles.[71] Those principles, according to the Energy Data Strategy, should align with energy market objectives, ensure appropriate privacy and security safeguards are maintained, and be in line with wider national reforms on data.[72]

Significant in the list of principles is the trend to align energy regulation with wider structural reforms in the national finance and energy sectors that are built around digitization technologies and data sharing platforms. Such models are exemplified by roll out of the consumer data right in the energy sector.[73] In effect, the consumer data right is an authorized, automated data sharing platform between consumers and providers of services that is administered by an Australian government agency.[74] To date, the process is only available within the banking and finance and energy sectors in Australia, but the plan is to extend it across the economy.

The consumer data right is advertised as an 'opt in service', designed to give consumers 'greater choice and control through the convenience of a simple, easy-to-use process'.[75] The safeguard for personal data is provided in that the data transfer is completed between providers, but 'the Australian Government has designed and oversees the system to ensure it is safe and secure for consumers'.[76] The providers must undergo an accreditation system. There is a consent regime and privacy safeguards.[77] The Australian Competition and Consumer Commission along with the Information Commissioner have responsibility for managing the consumer data right process. The Competition Commission has wider responsibilities for general competition and market regulation in Australia, although there are energy specific national agencies, such as the Australian Energy Regulator. Interestingly, even the purpose of the regulator now is designated as, 'to make energy consumers better off, now and in the future'.[78] While this agency is an independent regulator, the refrain of

[70] ibid ch 4.
[71] Energy Security Board (n 10) 20.
[72] ibid.
[73] Australian Government, 'Your Rights' (*Consumer Data Right*) <https://www.cdr.gov.au/your-rights> accessed 19 July 2025.
[74] Australian Government, 'What is CDR?' (*Consumer Data Right*) <https://www.cdr.gov.au/what-is-cdr#goto-introduced-in-banking-and-energy> accessed 19 July 2025.
[75] ibid.
[76] ibid.
[77] The privacy safeguards are found in the Competition and Consumer Act 2010 (Cth) which sets out thirteen legally binding obligations on businesses collecting and handling the data. Australian Government, 'What is CDR?' (n 74).
[78] Australian Energy Regulator, 'About Us' <https://www.aer.gov.au/about> accessed 19 July 2025.

concern for consumer interests may reflect the current political sensitivities of high energy prices. It also reflects a similar objective to that of the Energy Security Board's strategy.

In summary, there have been far-reaching reforms around data management in the energy sector in Australia based on enhanced digitization and the alignment of energy consumer interests with data management. Many reforms have dual objectives in terms of simultaneously facilitating data sharing and use with protection of personal data — all in the context of extending the neoliberal fabric of Australian energy markets. The suite of regulatory changes is relatively recent, but the compliance around the consumer data right and energy companies has been studied.[79] The Office of the Australian Information Commissioner in 2023 found that the consumer data right policies of three large energy retailers demonstrated what was termed strong compliance, meeting an average of 82 per cent of the minimum requirements under Privacy Safeguards and the Consumer Data Rules 2020.[80] Whether the dual objectives of extending data protection while facilitating more expansive 'sharing' can continue to be synchronized over a longer time frame in the expansion of DER in the energy system is a matter for future consideration.

To return to an earlier theme though, to what extent are the reforms around digitization and DER aligning to realistic consumer protection and prosumer outcomes when the policy vectors of decarbonization and decentralization are now pulling in different directions to the diffuse, small-scale energy settings around DER? Which raises a consequent query about what democracy and sustainability mean in these evolving energy settings? These questions are explored in the next section.

IV. Renewable Energy Planning and DER

To provide some tentative conclusions to the foregoing questions, this final section considers the renewable energy planning underway in Australian energy markets that is at odds with the promotion of consumer/prosumer agency. In 2024, the Australian Energy Market Operator's Integrated System Plan identified renewable energy distribution zones as a central mechanism to support the twin vectors of decarbonization and decentralization. The Plan was part of a renewable energy transition roadmap for the National Electricity Market. The plan identifies the essential infrastructure required to meet Australia's future energy needs, by detailing the generation, storage, and network investments to transition to net zero by 2050, and by reference to policy settings designed to deliver 'significant net market benefits for consumers'.[81] Yet these initiatives include planning for grid and transmission facilities that tend to support large-scale wind and solar farm project development more so than local scale community or individual household energy production and individual consumer grid access. Access to the grid for consumers (prosumers) is still mediated principally through energy retailers rather than direct consumer access to the electricity grid. Thus, much of the new capacity in solar and wind and batteries is coming from very large-scale renewable projects, which are at 'grid-scale'.[82] Typically, these are not local community

[79] Office of the Australian Information Commissioner, 'Openness and Transparency: Energy Retailers' Management of Consumer Data Right Data' (6 September 2024) <https://www.oaic.gov.au/consumer-data-right/consumer-data-right-assessments/> accessed 19 July 2025.

[80] ibid.

[81] Australian Energy Market Operator (n 41) 10.

[82] ibid 40.

'bottom up' prosumer projects, but projects requiring substantial capital investment, which is often drawn from international equity sources.[83] These projects are being strongly supported by subnational governments in Australia—the very governments who supported the first waves of small-scale DER. Thus, there is a gap between the rhetoric of DER and democracy advanced by national energy agencies, and what is being supported through renewable energy infrastructure development at a subnational level.

A. Subnational Drivers of DER and Grid-Scale Renewable Energy

Australia is a federal nation, so sub-national jurisdictions play an important role in developing law and policy for DER. State governments, such as Victoria, have sovereign powers over energy resources, but also legislate across critical industry sectors such as energy, transport, agriculture, and commercial organizations as well as holding approval powers for energy projects. In Victoria, the Climate Change Act 2017 (Vic) is the primary mechanism to deliver sustainability outcomes under a 'sector-based pledging system'.[84] This system spreads responsibility for climate change actions across spheres of government, and it incorporates climate change into decision-making across different policy areas.[85]

This package of legislation also sits alongside the omnibus of the regulatory framework of the national energy market. The legislation contains 'bottom up' consultative measures such as the 'principle of community engagement'.[86] The legislation does not comprise a specific system for DER, nor does it require statutory agencies to directly regulate for emissions reductions. Instead, the Act leaves much of the planning for and implementation of DER to executive policy or 'default' regimes such as planning and development laws. While there has been significant decarbonization through retirement of Victorian coal-fired power stations, the pattern of government engagement around decentralization of renewable energy and 'direct democracy' of consumer agency has been variable.

There is a long history of community-based challenges to wind farm and solar farm projects in Victoria. Those challenges have related to a wide range of project impacts on communities which include the sheer scale of many projects over vast spaces, the location of projects on sensitive coastal regions, noise impacts on nearby residential communities, and biodiversity losses.[87] Victoria is developing offshore wind facilities at the moment, with a significant degree of community concern about the impacts of extremely large wind turbines in sensitive marine areas that overlap with cultural heritage sites.

The local community impact of projects is the 'physical' reality of what is happening in much of Victoria in terms of renewable energy. These large projects[88] differ from urban 'small-scale' prosumer models of DER that supposedly are the objective for the digitalization

[83] ibid.

[84] Climate Change Act 2017 (Vic) pt 5 div 3.

[85] A Calabro, S Niall, and A Skarbek, 'The Victorian Climate Change Act: A Model' (2018) 92 Australian Law Journal 814, 816–17.

[86] Climate Change Act 2017 (Vic) s 27.

[87] L Godden and J Peel, 'The Environment Protection and Biodiversity Conservation Act 1999 (Cth): Dark Sides of Virtue' (2007) 31 Melbourne University Law Review 106.

[88] For example, the Golden Plains projects comprise up to 228 turbines across 16,739 hectares: 'About the Project' (*Golden Plains Wind Farm*) <https://goldenplainswindfarm.com.au/about-golden-plains-wind-farm/> accessed 18 July 2025.

reforms. In turn, for several years there has been a scaling back of measures supporting small-scale residential solar and reductions in the tariffs paid for consumers 'feeding back' into the grid. Instead, there is a focus on electrification of household domestic appliances.[89] Battery storage at residential scale still attracts some subsidization. In summary, investment in 'grid scale' energy resources is rapidly eclipsing the contributions of individual energy consumers to rapid energy transition in Victoria while energy market operators still need to deal with the fluctuations in supply due to myriad small-scale DER.

B. DER, Digitization, and Large Energy Projects in Australian Energy Markets

Given the scale of many Victorian wind and solar projects, there has been an inflow of off-shore investment into the projects, with multinational companies developing large projects. Such companies are required to meet community consultation requirements.[90] Even so, these trends underscore that energy systems tie domestic and regional energy systems, including DER, to international trade and finance modalities. These companies will utilize much energy data going forward but there are limited opportunities for community-based (democratic) input into the data governance of such projects. This situation can be contrasted with the small-scale community wind farms that first were developed in Victoria twenty years ago. In this sense, democracy and data management in the decarbonization and decentralization of energy systems has been diminished by stronger regulatory engagement with digitization, markets, and alignment with the national financial sector that is overtaking DER, but which is shielded by the consumer-oriented rhetoric of digital-enabled consumer agency.[91]

Thus, while relevant legal and political frameworks allocate roles and responsibilities between the State and the markets in the Australian electricity sector, the balance seems to be tilting towards markets and the financial sector. Clearly, following the Paris Agreement in 2015 there is a stronger presence of the commercial and business sector in strategies to address climate change,[92] and to advance energy transition.[93] Neoliberal economic theory has been significant in advancing these strategies.

[89] Department of Energy, Environment and Climate Action Victoria, 'Cheaper, Cleaner, Renewable: Our Plan for Victoria's Electricity Future' (The State of Victoria Department of Energy, Environment and Climate Action 2024).

[90] Environmental Effects Act 1978 (Vic) and Environmental Protection and Biodiversity Conservation Act 1999 (C'th).

[91] The company developing Victoria's largest wind farm notes, '[a]s the world grapples with the twin challenges of ageing infrastructure and climate change, companies like ACCIONA, that deliver sustainable solutions at scale will be increasingly in demand': M Law, 'ACCIONA: A Sustainable Future Through Digital Transformation' *Technology Magazine* (Norwich, 6 September 2024) <https://technologymagazine.com/company-reports/acciona-a-sustainable-future-through-digital-transformation> accessed 10 May 2025.

[92] ML Banda, 'The Bottom-Up Alternative: The Mitigation Potential of Private Climate Governance after the Paris Agreement' (2018) 42 Harvard Environmental Law Review 325.

[93] J Peel and others, 'Governing the Energy Transition: The Role of Corporate Law Tools' (2019) 36 Environmental and Planning Law Journal 459.

V. Recommendations

Pursuant to neoliberal theory, as market actors, consumers do not effectively incorporate externalities such as energy demand reduction into their decisions because they have imperfect information. As a market failure, the remedy must be market-based too, namely providing consumers with more accurate information to perfect markets, and to make energy data more 'accessible' and 'shared' to promote economic efficiency. The belief is that, if consumers receive accurate energy data, it will build DER models such as aggregation into overall market preferences.[94] This model reinforces the primacy of data and digitization to guide consumer behaviour under a DER model. It has driven many elements of the current energy data access and shared use regulatory initiatives in Australia.

The consumer/prosumer agency and user choice paradigms in energy policy that seemingly characterize DER are now strongly aligned to market-based measures and financial transactional security via regulatory initiatives such as the consumer data right. These policy platforms are largely framed within the structural preference for economic utility measures over social inclusion (participatory democracy) which informs much of the Australian energy transition law and policy. There is a need to re-balance the data access and transactional emphasis in current energy data strategies and associated regulation to strengthen the protection and value basis of data in the legislative frameworks associated with energy data management.

The focus for current protections is around safe-guarding individual privacy through prohibitions in the Privacy Act 1988 (C'th) on disclosure of personal information by government agencies and energy companies. While important, this focus does not fully address the risks associated with energy data access and sharing at the consumer/prosumer level as envisaged under the Energy Data Strategy as central to a digitalization future. Moreover, the Privacy Act only has application to organizations with a turnover of more than $3 million per annum—small-scale DER aggregation and prosumer activity is unlikely to be of this level. Prosumer governance models typically adopt co-produced modes of energy policy making and rule setting characterized by normative models that utilize dialogue and deliberation and devolved decision making. In such information-based regulation, the public (and the national interest in data sovereignty) are typically only able to participate as 'consumers' of energy sector services.

Further, critiques of information-based measures point to the narrowness of the focus on markets, given embedded assumptions around consumer self-interest. Markets are not particularly open to participants on a public interest basis—other than as consumers requiring certain protections. There is a lack of attention in the current policy and regulatory framings to potential data access and sharing by community-based collectives which would transcend the current consumer data right model. Nonetheless, the current model of data-based consumer/prosumer models assumes the self-containment of markets.[95] These measures embed the information asymmetries between the energy corporate sector and the individual consumers with consequent power disparities between the two groups.[96] These

[94] B Neville and J Peel, 'Transforming the Systemic Role of Corporate Law for the Climate Emergency' (37th EGOS Colloquium, Amsterdam, SWG 15, 9 July 2021).

[95] JE Cohen, *Between Truth and Power: The Legal Constructions of Informational Capitalism* (OUP 2019) 3.

[96] B Christophers, 'Climate Change and Financial Instability: Risk Disclosure and the Problematics of Neoliberal Governance' (2017) 107 Annals of the American Association of Geographers 1108.

issues of market power in data access and sharing arrangements also require further regulatory attention.

Finally, the neoliberal market orientation of current energy regulation and policy has constrained attention to energy data sovereignty predicated on a public/national interest model. Energy security, however, has been identified as a matter of regulatory attention in market design—that is, a concern primarily with managing vagaries of supply and demand in energy markets rather than concerns about potential breaches of the underpinning data systems that operate the market. As global geopolitical conflicts intensify, these aspects may become more pressing.

VI. Conclusion: A Digital Energy Future

The Australian Energy Security Board's data strategy has been a prominent platform for the development of future energy markets. As the strategy notes, '[g]rowing data capabilities and digital tools will be vital to managing future energy markets with greater variability diversity and fragmentation'.[97] These techno-economic paradigms echo the predominant models for Australian economic development and policy with a similar orientation in the national Climate Change Act 2022 (Cth). That orientation is mediated in relatively modest ways by data protections, privacy laws, sustainability, and Teal-style democracy (as aligned to climate change outcomes rather than data protections). Again, these patterns have parallels in market-based energy technology and energy security measures to be found in transnational settings, such as the European Union.[98]

Policy platforms, legislation, instruments, and codes for climate action may reference sustainability objectives, but the legal modality of decarbonization in many market-based energy settings is distributed through the practices of self-regulation, markets, data production, and 'consumer agency'. These approaches are normative as well as regulatory. The clean energy transition globally has brought a plethora of new laws and policies pertaining to data management as well as market and technology shifts for the energy and finance sectors as major funders of energy projects.[99] These trends influence the effective policy response to renewable energy and decentralized modes of electricity production and distribution.[100]

Most national governments when responding to the imperative of energy transition initially relied upon public law. By contrast, in recent energy regulation, the DER vision is that of an energy system that synchronizes data technologies and renewable energy to deliver digitized energy choice to consumers, and which facilitates prosumers as independent energy managers.[101] 'With the proliferation of DERs . . . , market constructs are evolving

[97] Energy Security Board (n 10) 8.

[98] K Huhta, 'Trust in the Invisible Hand? The Roles of the State and the Markets in EU Energy Law' (2020) 13 The Journal of World Energy Law and Business 1, 2.

[99] Taskforce on Climate-Related Financial Disclosures (TCFD), *Final Report: Recommendations of the Task Force on Climate-Related Financial Disclosures* (June 2017) 7–12 <https://assets.bbhub.io/company/sites/60/2021/10/FINAL-2017-TCFD-Report.pdf> accessed 25 August 2025.

[100] A Foerster, K Sheehan, and D Parris, 'Investing for a Safe Climate?' (2021) 44 UNSW Law Journal 1409.

[101] D Prasad Mishra and others, 'Smart Grid and Energy Management Systems: A Global Perspective' in S Reddy Salkuti (ed), *Energy and Environmental Aspects of Emerging Technologies for Smart Grid* (Green Energy and Technology, Springer 2024) 629.

to allow DERs to participate in multiple possible markets, at different levels of grid hierarchy.'[102] This signals a move to govern energy futures by more direct reference to contractual and private law paradigms. As the International Energy Agency notes, '[n]ew, diverse technologies are helping consumers be more proactive and are prompting new players to enter power markets, such as aggregators who pool together small-scale resources and act on their owners' behalf'.[103] Such endorsements resonate with neoliberal economic assumptions around deepening the market and enhanced opportunities for energy entrepreneurship.[104] These features are clearly evident in the Australian case study.

The centring of the individual consumer/actor as purportedly synchronizing with a property-democracy based model is apparent in that '[t]he primary beneficiaries of DERs are the consumers who own them'.[105] DER are advanced as synonymous with civic agency, whereby '[c]onsumers are increasingly able to take control of their own energy demand through a complex web of interactive smart energy devices'.[106] Democracy is being reframed as consumer energy agency.

Yet, there is an emerging disjunct between this form of DER and the roll out of renewable energy in market-based jurisdictions, as exemplified by the Australian energy system. The policies and regulatory model for energy transition increasingly prioritize large-scale projects with direct access to the national electricity grid.[107] Currently, there is some realignment underway to provide a stronger regulatory basis to DER energy consumer models that is predicated now on financial and data management structures such as the data availability laws and a consumer data right. Further, the language of DER energy transition is now also enmeshed in neoclassical economic assumptions about individual agency and preferences and organizational data management behaviour, that are to be channelled via digitization to drive effective economic, consumer, and regulatory outcomes. Tensions still manifest in the legislative and regulatory spheres between the imperatives to facilitate greater 'controlled' access to energy data, and the regimes of privacy and data protection. These tensions are likely to continue to exist as the digitization of energy systems becomes even more pervasive.

Finally, DER modes have moved from being a 'disruptor' to the previously highly centralized energy systems, to now being positioned as one of the main platforms of an emerging generation of energy infrastructure and data technologies that can both address climate change and sustainability imperatives while enabling energy markets to diversify across demand and supply structures.

[102] International Energy Agency (n 13) 7.

[103] ibid.

[104] See B Holligan, 'Property and Environmental Markets' in N Graham, M Davies, and L Godden (eds), *The Routledge Handbook of Property Law and Society* (Routledge 2022) 264, 265.

[105] International Energy Agency (n 13) 7.

[106] S Skillings and P Clutton-Brock, 'Digital Innovation for a Net Zero Electricity Grid' (May 2022) E3G Briefing Paper <http://www.jstor.org/stable/resrep41536> accessed 18 July 2025.

[107] The Government has committed $23 billion to modernize Australia's electricity grid: Australian Government, 'Powering Australia' (Department of Climate Change, Energy, the Environment and Water, 24 May 2024) <https://www.dcceew.gov.au/energy/strategies-and-frameworks/powering-australia> accessed 25 August 2025.

PART III
SUSTAINABLE DEMOCRATIZATION OF ENERGY SYSTEMS

15

The Role of Active Consumers in Decentralized Energy Systems

Kaisa Huhta and Emma Merikanto*

I. Introduction

The low-carbon energy transition requires that the means of producing and consuming energy change fundamentally. This process of transition is well-known to take place in different ways, one of which is the change of emphasis from a centralized energy system to a decentralized, distributed one where energy is not only produced but also transmitted and stored in smaller-scale.[1] The rise of distributed energy resources (DERs) has been recognized as an important development in the decarbonization of energy systems worldwide, but also because of its potential to improve access to energy and to engage individual consumers to become more active in the energy system.[2] However, DERs also face a different set of social, environmental, economic, and technical challenges compared to a centralized energy system, as demonstrated throughout the chapters of this book. This chapter focuses on the social and economic risks that energy consumers face when they become more active in the energy markets.

The increase of DERs changes the role of the consumer from passive recipients of energy to active components of the energy system, challenging established legal and institutional frameworks that have been constructed with the centralized model in mind. While activating energy consumers and engaging the demand-side more generally has long been identified as a necessity in the low-carbon energy transition,[3] new risks and challenges of such consumer activation have also been identified, as increasing consumers' opportunities and choices in the energy markets also inevitably allocates more responsibility and more risk on consumers.[4] To begin with, the financial opportunities for adopting DERs

* Funded by the European Research Council (INTEL 101160605). Views and opinions expressed are however those of the author(s) only and do not necessarily reflect those of the European Union or ERCEA. Neither the European Union nor the ERCEA can be held responsible for them. This work was also supported by the Research Council of Finland (361603).

[1] A Savaresi, 'The Rise of Community Energy from Grassroots to Mainstream: The Role of Law and Policy' (2019) 31 Journal of Environmental Law 487.

[2] See, eg M Funsho Akorede, H Hizam, and E Pouresmaeil, 'Distributed Energy Resources and Benefits to the Environment' (2010) 14 Renewable and Sustainable Energy Reviews 724; B Zakeri and others, 'Centralized vs Distributed Energy Storage—Benefits for Residential Users' (2021) 236 Energy 121443.

[3] S Lavrijssen, 'Power to the Energy Consumers' (2017) 26 European Energy and Environmental Law Review 172; I J Perez-Arriaga and C Batlle, 'Impacts of Intermittent Renewables on Electricity Generation System Operation' (2012) 1 Economics of Energy & Environmental Policy 3; H Christian Gils, 'Assessment of the Theoretical Demand Response Potential in Europe' (2014) 67 Energy 1.

[4] K Huhta, 'Unleashing Consumer Potential in the Energy Transition: A Reflection of the Transforming Role of the EU Consumer' (2019) 17 OGEL Energy Law Journal <https://www.ogel.org/article.asp?key=3832> accessed 23 March 2025.

Kaisa Huhta and Emma Merikanto, *The Role of Active Consumers in Decentralized Energy Systems* In: *Sustainable Distributed Energy Resources*. Edited by: Íñigo del Guayo, Damilola S. Olawuyi San, Louis de Fontenelle, Milton Fernando Montoya, and José Juan González, Oxford University Press. © International Bar Association 2026. DOI: 10.1093/9780198991946.003.0018

are disproportionately distributed between consumers that have the opportunity and the financial capacity to invest in energy self-production and those that do not. In fact, it has been widely documented that there are social disparities in the adoption of DERs.[5] Those consumers that are able to participate in the financing and operation of DERs face risks and responsibilities related to the safe and financially rational operation, maintenance, and governance of DERs.[6] In particular, they are subject to new kinds of technical, financial, and legal obligations and responsibilities that have been previously handled—in centralized energy systems—by professional energy actors specialized in compliance with the requirements of the energy sector. Furthermore, those consumers that are disconnected from the grid and entirely dependent on decentralized self-production of energy in the form of DERs, are also more vulnerable to the adverse effects of supply disruptions. It is in this context of the new type of environment for the energy consumer that this chapter seeks to explore how legal frameworks can address the challenges related to active energy consumers in a decentralized energy system by using the European Union (EU) law as a case study.

Governing these new kinds of challenges and potential solutions to these challenges through law has been discussed in literature from various points of view.[7] Existing literature has identified general risks associated with increased consumer participation in distributed energy systems and identified mechanisms to address those risks.[8] It has also explored some of the legal challenges related to active consumers in the energy sector in general and in EU law particular.[9] However, a combined understanding of the new risks and responsibilities faced by active energy consumers in decentralized energy systems on the one hand and the legal solutions that can be used to address those risks, is missing. Furthermore, with the rise in DERs and the most recent legislative developments in EU law adopted to respond to that rise, no up-to-date analysis on how the EU legal system addresses the new challenges of increased consumer participation in decentralized energy systems is available. To address these gaps in existing scholarship, this chapter contributes to the growing body of legal literature by exploring the role of active consumers in decentralized energy systems from a legal point of view and in the context of EU energy law to showcase concrete legal approaches to transform the role of the consumer in the energy transition. The chapter argues that while the legal framework in the EU gives consumers new opportunities to engage in activities in the energy sector, it also places more responsibilities and risks on active energy consumers that require corresponding legal mechanisms to protect those consumers. An analysis of EU law as a regional legislative structure, with its progressive and globally unique energy governance framework, provides an excellent legislative example

[5] AM Brockway, J Conde, and D Callaway, 'Inequitable Access to Distributed Energy Resources Due to Grid Infrastructure Limits in California' (2021) 6 Nature Energy 892.

[6] C Decker, 'Protecting Consumers in Digitized and Multi-Source Energy Systems' (2021) 16(11–12) Energy Sources, Part B: Economics, Planning, and Policy 1127.

[7] R Monaco and others, 'Digitalization of Power Distribution Grids: Barrier Analysis, Ranking and Policy Recommendations' (2024) 188 Energy Policy 114083; S Lavrijssen, B Espinosa Apráez, and T ten Caten, 'The Legal Complexities of Processing and Protecting Personal Data in the Electricity Sector' (2022) 15 Energies 1088; J Jaria-Manzano, 'Energy Transition and Democratic Challenges: The Contribution of EU Law' (2024) 33 European Energy and Environmental Law Review 254; RJ Heffron and L de Fontenelle (eds), *The Power of Energy Justice & the Social Contract* (Springer Nature 2024) <https://directory.doabooks.org/handle/20.500.12854/132000> accessed 19 March 2025.

[8] C Decker, 'Protecting Consumers in Digitized and Multi-Source Energy Systems' (2021) 16(11–12) Energy Sources, Part B: Economics, Planning, and Policy 1127.

[9] K Cseres, 'The Active Energy Consumer in EU Law' (2018) 9(2) European Journal of Risk Regulation 227–44.

with valuable insights for not just European energy law professionals but also global energy law scholarship. While the chapter acknowledges, especially given the fast-paced change in the development of DERs, that the example provided by the EU is not by far a perfect one, it does provide a globally unique way of governing the role of the consumer in increasingly decentralized energy systems.

The objective of this analysis is to demonstrate how the trade-offs between encouraging the consumer to participate in the energy sector on the one hand and protecting the consumer from the risks associated with such participation on the other can be balanced in a legal framework. With this objective in mind, the analysis in this chapter focuses on the legal governance of the social and economic risks that energy consumers face when they become more active in the energy markets (rather than the environmental questions, for example). In line with the chapter's objective, traditional consumer rights provisions, such as the right to billing information, also fall outside the scope of this chapter.[10]

This chapter is structured in the following way. It first explores existing conceptualizations of energy citizens and active energy consumers by reviewing the relevant literature, focusing on legal scholarship and the ways in which energy consumers have been conceptualized in the energy transition (Section II). It then analyses the key legal approaches to activate energy consumers in EU energy law (Section III) and explores the risks of increased consumer engagement before presenting legal solutions to manage these challenges in the EU legal system (Section IV). In line with the chapter's objective and the literature gap it aims to address, Section IV will first identify, by using the relevant literature, the social and economic risks faced by active energy consumers, and then analyse, by using the EU as a case study, how law allocates this risk to consumers—and even consolidates it—and how the law also responds to and protects the consumer from these risks. Finally, conclusions are drawn with insights for jurisdictions beyond the EU (Section V).

II. Conceptualizations of Active Energy Consumers in the Energy Transition

Traditionally, energy markets have been dominated by a few vertically integrated energy companies who have had a monopoly over the whole value chain: energy production, transportation, and supply.[11] In this type of centralized energy market structure, the consumer played a passive role unable to choose between different energy suppliers. Due to the liberalization and deregulation of energy markets globally, consumers have become more able to choose between different suppliers, accelerating competition and contributing to more efficient markets.[12] In addition to this earlier development, DERs and the growing role of energy efficiency in the energy policy have created new opportunities for the consumer to operate in the energy markets. Legislative and policy frameworks have transformed the consumer's role from a passive recipient of energy to an active market participant, allowing the consumer to produce, store, share, and sell energy. The diversified

[10] Electricity Directive, art 18.

[11] S Milčiuvienė and others, 'The Role of Renewable Energy Prosumers in Implementing Energy Justice Theory' (2019) 11 Sustainability 5286.

[12] Cseres (n 9) 231.

roles of the consumers have also changed the perception of energy from a mere commercial commodity to a necessity or right, which the consumer should have.[13]

Active energy consumers have been given many definitions in literature and in legal and policy discourses.[14] The concept is also sometimes used interchangeably with energy citizens, active customers, prosumers, and self-consumers and often goes hand-in hand with conceptualizations of energy communities and consumer engagement.[15] Typically, all these forms of active consumers are discussed under the umbrella of decentralization or distributed energy, which—in this book—are collectively discussed under the concept of DERs.[16] Despite this inclusive terminology, this chapter also acknowledges that different forms of active consumer engagement (eg individual, collective, or commercial) often imply different DER configurations—and sometimes conflicting governance models, which is well demonstrated throughout the chapters of this book.

In EU energy law alone, a diversity of concepts and their definitions can be found. In the Electricity Directive, an 'active customer' is defined as 'a final customer, or a group of jointly acting final customers, who consumes or stores electricity generated within its premises located within confined boundaries or self-generated or shared electricity within other premises, or who sells self-generated not constitute its primary commercial or professional activity'.[17] In the Renewable Energy Directive, a 'renewables self-consumer' is a final customer operating within its premises located within confined boundaries or, where permitted by a Member State, within other premises, who generates renewable electricity for its own consumption, and who may store or sell self-generated renewable electricity, provided that, for a non-household renewables self-consumer, those activities do not constitute its primary commercial or professional activity.[18] A 'renewable energy community' is defined as a legal entity composed of natural persons, small and medium-sized enterprises, or local authorities that own and develop renewable energy projects in their vicinity with the primary purpose of providing environmental, economic, or social community benefits for its shareholders or members or for the local areas where it operates, rather than financial profits.[19] These definitions partially overlap and demonstrate the diversity of approaches to active energy consumers.[20]

[13] G Jiglau and others, 'Energy and the Social Contract: From "Energy Consumers" to "People with a Right to Energy"' (2024) 32 Sustainable Development 1321, 1322.

[14] T Van Der Schoor and B Scholtens, 'Power to the People: Local Community Initiatives and the Transition to Sustainable Energy' (2015) 43 Renewable and Sustainable Energy Reviews 666; C Inês and others, 'Regulatory Challenges and Opportunities for Collective Renewable Energy Prosumers in the EU' (2020) 138 Energy Policy 111212.

[15] L Diestelmeier, 'Regulating Residential Prosumers' in MM Roggenkamp, KJ De Graaf, and RC Fleming (eds), *Energy Law, Climate Change and the Environment* (Edward Elgar Publishing 2021) 729–40; Van Der Schoor (n 14); Inês (n 14); MA Heldeweg and S Saintier, 'Renewable Energy Communities as "Socio-Legal Institutions": A Normative Frame for Energy Decentralization?' (2020) 119 Renewable and Sustainable Energy Reviews 109518. Several of these concepts are also discussed elsewhere in this book. For energy citizens, for instance, see Chapter 2.

[16] AJ Wieczorek and others, 'Citizen-Led Decentralised Energy Futures: Emerging Rationales of Energy System Organisation' (2024) 113 Energy Research & Social Science 103557; Savaresi (n 1); R Leal-Arcas, N Akondo, and J Alemany Rios, 'Energy Decentralization in the European Union' (2019) Queen Mary School of Law Legal Studies Research Paper No 307/2019, Available at SSRN: https://ssrn.com/abstract=3333694.

[17] Directive (EU) 2019/944 of the European Parliament and of the Council of 5 June 2019 on common rules for the internal market for electricity and amending Directive 2012/27/EU (recast) [2019] OJ L158/125 (Electricity Directive), art 2.

[18] Directive (EU) 2018/2001 of the European Parliament and of the Council of 11 December 2018 on the promotion of the use of energy from renewable sources (recast) [2018] OJ L328/82 (Renewable Energy Directive), art 2.

[19] Renewable Energy Directive, art 2.

[20] Inês (n 14).

This chapter adopts a generalist definition of active energy consumers. It refers to them as individual or organized groups of non-professional energy consumers that participate in energy activities beyond merely passively receiving energy from a centralized energy production and transmission system. This definition encapsulates all the above legal definitions but also extends to individual consumers who, for example, voluntarily enter into dynamic price contracts with their supplier and thereby offer demand side flexibility to the energy system.

An active energy consumer defined in this way can also operate within a centralized energy system. In fact, it is not possible to clearly distinguish when an active consumer participates in a centralized system as opposed to a decentralized one. An active consumer can be connected to and supplied by a highly centralized energy system, while also actively participating in a decentralized community energy initiative. The distinction between two systems is therefore not clear cut. Instead, it is more like a continuum where at the opposite ends of the spectrum there is a centralized system with customers that passively receive energy and a decentralized system where all energy is produced, supplied, and consumed in a distributed way. In this way, all increase in the activity levels of individual energy consumers or energy communities contributes to the decentralization rather than centralization of an energy system.

In addition to the definition of active energy consumers alone, the different roles of the active energy consumer have been discussed in the legal scholarship.[21] In this body of literature, decentralized energy systems are recognized as systems where consumers manage their own energy consumption while being able to (co-) produce energy.[22] Consumers are empowered to own, run, and develop energy solutions, technologies, projects, and facilities.[23] However, it is widely acknowledged that consumers cannot compete on an equal footing with big market players requiring empowerment and special protection from the legislators.[24] Ideally, legislative and policy frameworks should be supportive enough for a consumer to participate in the energy markets independently.[25] It has also been recognized that, for the consumers to actively participate in the energy markets, their role in relation to other market players should be clearly regulated, such as the co-operation with distribution system operators to enable energy sharing.[26] The concepts and regulatory issues related to the production, storing, sharing, and selling of energy by the consumer have also been discussed extensively.[27] For instance,

[21] See eg Milčiuvienė (n 11); J Grabiel Luis Cordova, K Byttebier, and Y Delgado Triana, 'Decentralized Energy Generation for Sustainable Energy Development in EU' (2023) 32 European Energy and Environmental Law Review 164.

[22] Cseres (n 9) 228.

[23] D Olawuyi, 'From Energy Consumers to Energy Citizens: Legal Dimensions of Energy Citizenship' in K Huhta and others (eds), *Sustainable Energy Democracy and the Law* (Brill 2021) 101–23.

[24] L Diestelmeier, 'Changing Power: Shifting the Role of Electricity Consumers with Blockchain Technology—Policy Implications for EU Electricity Law' (2019) 128 Energy Policy 189, 196.

[25] ibid 195.

[26] M Krug and others, 'Implementing European Union Provisions and Enabling Frameworks for Renewable Energy Communities in Nine Countries: Progress, Delays, and Gaps' (2023) 15 Sustainability 8861, 8880.

[27] L Diestelmeier and V Cappelli, 'Conceptualizing "Energy Sharing" as an Activity of "Energy Communities" under EU Law: Towards Social Benefits for Consumers?' (2023) 12(1) Journal of European Consumer and Market Law 15–24; J Lowitzsch, 'Investing in a Renewable Future—Renewable Energy Communities, Consumer (Co-)Ownership and Energy Sharing in the Clean Energy Package' (2018) 9 Renewable Energy Law and Policy Review 14.

the regulatory framework should incentivize consumers to store, share, and sell surplus energy.[28]

The relevant legal literature acknowledges that becoming an active energy consumer through community energy projects requires risk-taking, financial resources, skills, and knowledge.[29] Therefore, an active consumer must be able to navigate complex information about energy markets, funding, and technical, legal, and policy issues.[30] Currently, it seems that the co-production of renewable energy easily falls under the control of for-profit and commercial entities that have resources.[31] This sort of development of decentralization of energy markets falls short in activating consumers to increase citizen-led projects.

The different vulnerabilities and barriers affecting the consumer's active participation in energy prosumership have also been discussed in the literature.[32] Such vulnerabilities can relate to individual characteristics, discriminating structures, and policymaking.[33] For example, there is still a major gender disparity among the participants in renewable energy projects—men constituting the majority of participants.[34] To make energy transition more inclusive, the applicable literature has recognized that the increase in DERs should be closely linked to national energy poverty policies.[35] The Member States should ensure that there are suitable funding mechanisms available for citizen-led energy projects and establish accessible information dissemination policies to activate energy consumers.[36] Strategic investor support and allowing community energy projects to enter into partnerships with municipalities can help citizens to establish joint energy projects.[37] Furthermore, existing research suggests that the increasing use of auctions instead of feed-in tariffs to finance renewable energy projects favors larger projects that have more resources.[38] This leads often to the exclusion of community-based participants.[39] If Member States do not establish supportive structures to activate consumers from different socio-economic backgrounds, there is a risk that active energy consumers remain a homogeneous group of consumers.

[28] B Bonfert, '"We like Sharing Energy but Currently There's No Advantage": Transformative Opportunities and Challenges of Local Energy Communities in Europe' (2024) 107 Energy Research & Social Science 103351, 103354.

[29] CE Hoicka and others, 'Implementing a Just Renewable Energy Transition: Policy Advice for Transposing the New European Rules for Renewable Energy Communities' (2021) 156 Energy Policy 112435, 112440.

[30] V Brummer, 'Community Energy—Benefits and Barriers: A Comparative Literature Review of Community Energy in the UK, Germany and the USA, the Benefits It Provides for Society and the Barriers It Faces' (2018) 94 Renewable and Sustainable Energy Reviews 187, 193.

[31] L Horstink and others, 'Collective Renewable Energy Prosumers and the Promises of the Energy Union: Taking Stock' (2020) 13 Energies 421, 429.

[32] See eg J Lowitzsch and F Hanke, 'Consumer (Co-)ownership in Renewables, Energy Efficiency and the Fight Against Energy Poverty—A Dilemma of Energy Transitions' (2019) 9(3) Renewable Energy Law and Policy Review 5–21; F Hanke, R Guyet, and M Feenstra, 'Do Renewable Energy Communities Deliver Energy Justice? Exploring Insights from 71 European Cases' (2021) 80 Energy Research & Social Science 102244.

[33] See eg F Hanke and J Lowitzsch, 'Empowering Vulnerable Consumers to Join Renewable Energy Communities—Towards an Inclusive Design of the Clean Energy Package' (2020) 13 Energies 1615.

[34] L Horstink and others, 'Collective Renewable Energy Prosumers and the Promises of the Energy Union: Taking Stock' (2020) 13 Energies 421, 4.

[35] M Feenstra and others, 'Humanising the Energy Transition: Towards a National Policy on Energy Poverty in the Netherlands' (2021) 3 Frontiers in Sustainable Cities 1.

[36] Hanke (n 33) 1615.

[37] Lowitzsch (n 27) 16.

[38] M Segura, M Clayton, and L Manuel, 'State Aid Rules for Environmental Purposes: An Effective Instrument for Implementing EU Policy Priorities?' (2021) 21 ERA Forum 667, 677.

[39] OW Fitch-Roy, D Benson, and B Woodman, 'Policy Instrument Supply and Demand: How the Renewable Electricity Auction Took over the World' (2019) 7 Politics and Governance 81, 86.

From the perspective of just energy transition and leaving no one behind, the different capabilities of individuals to become active consumers in the energy market must be taken into consideration by legislators and policymakers. Ownership in energy production can support the consumer to consume less energy, implement additional energy efficiency measures, reduce the energy bills, and provide a second source of income when the excess energy is sold back to the grid.[40] These benefits of DERs should be distributed equally in the society, and legal frameworks can play an important role in achieving this goal. The EU example is next studied to demonstrate how this plays out.

III. Key Legal Approaches to the Active Energy Consumer in EU Energy Law

A. The Evolution of the EU's Legal Framework for Active Energy Consumers

Consumer protection has been included in EU energy law at least since the first energy package in the mid-1990s.[41] However, placing the consumer in the centre stage of the energy transition and facilitating active consumer engagement in the energy market has only gradually gained larger foothold first in EU energy policy and then in EU energy law.

The Clean Energy for All Europeans package, which entered into force in 2018–2019 clearly put the consumer in the centre of the energy transition and aimed to 'provide a fair deal for consumers'.[42] Similarly to the past three energy packages from 1996, 2003, and 2009, the Clean Energy for All Europeans package was founded on the idea that a well-functioning, competitive, and integrated energy market is the best instrument with which to pursue the objectives of EU energy policy, including consumer activation.[43] One of its key functions was to put the consumer in the centre of the energy transition, allowing consumers to take ownership of it.[44]

Since the Clean Energy for All Europeans package, the EU's legislative framework for energy has evolved rapidly. The Fit for 55 package proposed in 2021,[45] the EU hydrogen and gas decarbonization package proposed in 2021,[46] and the RePowerEU initiative adopted in response to the Russian invasion of Ukraine in 2022[47] include several legal instruments that

[40] Lowitzsch (n 32).

[41] Directive 96/92/EC of the European Parliament and of the Council of 19 December 1996 concerning common rules for the internal market in electricity [1997] OJ L27/20.

[42] Commission, 'Communication from the Commission to the European Parliament, the Council, the European Economic and Social Committee, the Committee of the Regions and the European Investment Bank, Clean Energy for All Europeans' COM (2016) 860 final, 3.

[43] On the functioning of the market, see K Huhta, 'Too Important to Be Entrusted to Neighbours? The Dynamics of Security of Electricity Supply and Mutual Trust in EU law' (2018) 43 European Law Review 920.

[44] COM (2016) 860 final. In literature, L Diestelmeier, 'Citizen Energy Communities as a Vehicle for a Just Energy Transition in the EU—Challenges for the Transposition' (2021) 1 OGEL Energy Law Journal.

[45] K Huhta, 'The "Fit for 55" Package in the Context of EU Energy Law and Policy' (2022) OGEL Energy Law Journal 1.

[46] For a summary, see European Commission, 'Hydrogen and decarbonised gas market' <https://energy.ec.europa.eu/topics/markets-and-consumers/hydrogen-and-decarbonised-gas-market_en#related-links> accessed 16 January 2025.

[47] Commission, 'Communication from the Commission to the European Parliament, the European Council, The European Economic and Social Committee and the Committee of Regions, RePowerEU' COM (2022) 230 final.

alter the legislative setting for energy consumers in the EU energy transition and have all gradually entered into force.

EU energy law now includes several legal instruments with direct relevance with regard to the focus of this chapter. While many key energy law instruments include consumer protection provisions that lay out consumer rights, the majority of the provisions that specifically seek to activate and empower consumers provisions are included in the Electricity Directive and Renewable Energy Directive, which has also been updated to include provisions with relevance to consumers' market participation, particularly from the point of perspective of distributed renewable generation. Therefore, they are the main focus of the following analysis. Furthermore, the Electricity Regulation, last revised in 2024, includes provisions on price formation and market-based operation of the energy sector in a way that are relevant to identifying and addressing the risks associated with activating consumers in the energy markets.[48] The following sections focus on the existing legislative framework in force and identify two key trends in particular that are enforced by the existing rules: active consumers offering demand-side flexibility to the electricity markets and active consumers collectively producing energy.

B. Active Consumers Offering Demand-side Flexibility to the Electricity Markets

The EU legislative framework applicable to energy consumers aims to transform consumers from a passive receiver of energy to an active resource that offers demand-side flexibility and is able to balance some of the intermittency that necessarily increases as the share of wind and solar renewable generation increases in the energy mix.[49] To this end, all final customers should be entitled to not just consume electricity but also generate, store, and sell self-generated electricity in all organized markets either individually or through aggregators.[50] They should not be subjected to disproportionate or discriminatory technical requirements, administrative requirements, procedures, and charges, and to network charges that are not cost-reflective.[51]

Member States are obliged to allow final customers, including those offering demand response through aggregation, to participate alongside producers in a non-discriminatory manner in all electricity markets.[52] In this context, the requirement for non-discrimination should be understood as requiring supply and demand side resources to compete on an equal basis. Furthermore, Member States are to ensure that active customers that own energy storage have the right to a grid connection, are not subject to any double charges or disproportionate licensing requirements, and are allowed to provide several services simultaneously, if technically feasible.[53] Overall, Member States are obliged to ensure that all

[48] Regulation (EU) 2019/943 of the European Parliament and of the Council of 5 June 2019 on the internal market for electricity (recast) [2019] OJ L158/54 (Electricity Regulation).

[49] N Soininen and K Huhta, 'Law for Decentralized Electricity Systems—Lessons from Complexity Theory' in R Fleming, K Huhta, and L Reins (eds), *Sustainable Energy Democracy and the Law* (Brill 2021) 78.

[50] Electricity Directive, art 15.

[51] ibid.

[52] ibid art 17.

[53] ibid art 15.

customers are free to purchase and sell electricity services independently from their electricity supply contract and from an electricity undertaking of their choice.[54]

Empowering consumers in the energy sector is first and foremost based on the removal of obstacles that hinder consumer activation. Many of the provisions are worded as obligations to Member States to ensure and guarantee that nothing prevents active consumers from partaking in energy sector activities, giving consumers access to more room for manoeuvre in the energy markets. Some of the provisions that aim to remove these obstacles have already been in place in the previous legislative frameworks, such as ensuring that customers are free to purchase electricity from the supplier of their choice and easily switch their suppliers.[55] The number of these types of provisions has increased, providing consumers with diverse opportunities to profit on their activities and hence incentivizing consumer activities.

In addition to removing obstacles to consumer participation, there are a number of provisions that seek to add further incentives to ensure that consumers use the new opportunities allowed for them within the legal framework. For instance, the Renewable Energy Directive obliges Member States to ensure that self-consumers of renewable energy are entitled to receive remuneration for the self-generated renewable electricity they feed into the grid, while maintaining their rights and obligations as final consumers.[56] This remuneration should reflect the market value of electricity.[57] Energy customers have also been guaranteed the right to energy sharing.[58] Member States must ensure that all households and small enterprises and medium-sized enterprises have the right to participate in energy sharing as active customers in a non-discriminatory manner. They must also ensure that active customers are entitled to share renewable energy between themselves based on private agreements or through a legal entity.[59]

Finally, there are clear efforts to incentivize consumer activation through increasing the predictability and clarity of the legal framework. Member States are obligated to adopt and maintain legal frameworks, which contain transparent rules clearly assigning roles and responsibilities to all market participants.[60] While increasing predictability and clarity in the legal framework alone might not seem like a concrete measure to empower consumers, it has been identified as an important element of improving the legitimacy and efficacy of law.[61]

C. Active Consumers Collectively Producing Energy

In addition to activating individual consumers in the energy sector, the EU's legislative framework includes provisions to engage groups of active consumers to collectively take

[54] ibid art 13.
[55] ibid art 2.
[56] Renewable Energy Directive, art 21.
[57] ibid.
[58] Diestelmeier (n 27).
[59] Electricity Directive, art 15a; L Diestelmeier and D Kuiken, 'Is Sharing Caring? "Energy Sharing" within Energy Communities under EU Law' in R Fleming and others (eds), *A Force of Energy: Essays in Energy Law in Honour of Professor Martha Roggenkamp* (University of Groningen Press 2022) 275–82.
[60] Electricity Directive, art 17.
[61] On the criteria of legitimacy of law, see LL Fuller, *Morality of Law* (rev edn, Yale UP 1969) 33–41 and its application in EU law; N Soininen and others, 'The Impact-based Regulatory Strategy in Environmental Law: Hallmark of Effectiveness or Pitfall for Legitimacy?' (2023) 35 JEL 185.

part in energy activities either as citizen energy communities or renewable energy communities.[62] Both the Electricity Directive and the Renewable Energy Directive contain provisions to this end.

Article 16 of the Electricity Directive contains rules relevant to citizen energy communities. It obliges Member States to provide an enabling regulatory framework for citizen energy communities, ensuring that participation in a citizen energy community is open and voluntary, participants of a citizen energy community do not lose their rights and obligations as household customers or active customers, and that these communities are given fair compensation. Furthermore, citizen energy communities should be subject to non-discriminatory, fair, proportionate, and transparent procedures and charges, including with respect to registration and licensing, and to transparent, non-discriminatory, and cost-reflective network charges.[63] Member States are to ensure that citizen energy communities are able to access all electricity markets in a non-discriminatory manner and are treated in a non-discriminatory and proportionate manner.[64]

The Renewable Energy Directive includes provisions with similar aims. Article 21 of the Directive obliges Member States to ensure that consumers are entitled to become renewables self-consumers and that they are entitled to generate renewable energy, including for their own consumption, store and sell their excess production of renewable electricity. With regard to renewable energy communities, Member States are obligated to ensure that final customers, in particular household customers, are entitled to participate in a renewable energy community while maintaining their rights or obligations as final customers.[65] Renewable energy communities must be given the rights to produce, consume, store, and sell renewable energy as well as share renewable energy within the community and access all suitable energy markets both directly or through aggregation in a non-discriminatory manner.[66] Member States are additionally obligated to assess existing barriers and potential of development of renewable energy communities in their territories and provide an enabling framework to promote and facilitate the development of renewable energy communities.[67] In other words, EU law explicitly highlights the need to support the participation of vulnerable and low-income households in the co-production of renewable energy.[68]

IV. Risks and Risk Management of Increased Consumer Engagement in EU Energy Law

A. Exposure to Market Prices, Investment Risks, and Producer Responsibility

The EU's legal framework on active energy consumers gives consumers new opportunities to benefit from the market developments and from the low-carbon energy transition. However, the increase in these opportunities also brings with it new responsibilities to

[62] Inês (n 14).
[63] Electricity Directive, art 17.
[64] ibid art 16.
[65] Renewable Energy Directive, art 22.
[66] ibid.
[67] ibid.
[68] ibid art 22(4)(f).

which previously passive consumers have not been exposed. This section identifies three categories of new risks that face active energy consumers: the exposure to market prices; the investment risks of becoming self-producers of energy; and the responsibilities that come with becoming an energy producer. In fact, the analysis of the legislative framework shows that EU law allocates these responsibilities to active energy consumers and, in doing so, consolidates the existence of these new risks.

First, the fact that active energy consumers are entitled to a dynamic price contract and to offer their demand side flexibility through aggregation means that they become interactive players in the electricity system. This also means that active consumers are exposed to market prices, which may become more volatile as the share of intermittent energy sources increases in the energy mix.[69] Overall, the EU legal framework for electricity is built on the premise that market prices should be formed on the basis of supply and demand, providing the most economically efficient balance.[70] Furthermore, any actions that prevent this price formation or constitute a disincentive to the development of more flexible generation or demand or low carbon generation should be avoided.[71]

The rationale for establishing further provisions to facilitate market-based price formation lies in the existing market conditions. In this regulatory setting, high prices are needed to signal scarcity. While this is necessary to incentivize new investment, individual consumers' exposure to these scarcity prices presents an economic risk that electricity consumers are not traditionally used to. Price caps or State interventions may prevent prices from signalling scarcity, which, in turn, hinders market-based investment decisions and can lead to resource adequacy concerns.[72] Market-based price signals should also provide necessary incentives on flexible resources, such as demand side response by consumers.[73] The increased emphasis on consumers and the resulting new consumer opportunities are a step towards the right direction in creating a more flexible demand side in the electricity market.

The emphasis on incentivizing the consumers to become active market participants is necessary to increase the flexibility of the electricity market, particularly in the context of the energy transition and to balance the increasing variable generation in the overall energy mix. However, increasing market opportunities and choices also inevitably allocate more responsibility and more risk on consumers. The element of risk is particularly visible in the context of exposing consumers to dynamic pricing.[74] Linking wholesale prices to retail prices has the potential to deliver consumers with significant savings. However, it also exposes retail customers to the price spikes of the wholesale markets. This is particularly harmful to those groups of consumers who cannot shift their demand to cheaper hours. The most vulnerable groups of energy consumers are the ones already struggling to

[69] European University Institute, 'Recent Energy Price Dynamics and Market Enhancements for the Future Energy Transition' (Publications Office 2022) <https://data.europa.eu/doi/10.2870/641134> accessed 23 March 2025; P Pereira Da Silva and P Horta, 'The Effect of Variable Renewable Energy Sources on Electricity Price Volatility: The Case of the Iberian Market' (2019) 38 International Journal of Sustainable Energy 794.

[70] Electricity Regulation, art 3; in literature, see K Huhta, 'Trust in the Invisible Hand? The Roles of the State and the Markets in EU Energy Law' (2020) 13(1) Journal of World Energy Law & Business 1.

[71] Electricity Regulation, art 3.

[72] Recast Electricity Regulation, 3; for an understanding of how price formation and price caps operate, see Leigh Hancher, A de Hauteclocque, K Huhta, and M Sadowska (eds), *Capacity Mechanisms in the EU Energy Markets: Law, Policy, and Economics* (2nd edn, OUP 2022).

[73] Electricity Regulation, recital 22.

[74] S Numminen, S Ruggiero, and M Jalas, 'Locked in Flat Tariffs? An Analysis of Electricity Retailers' Dynamic Price Offerings and Attitudes to Consumer Engagement in Demand Response' (2022) 326 Applied Energy 120002.

cover their monthly costs.[75] For these groups of consumers, mistakenly increasing electric heating during a price spike hour may already be economically devastating.[76]

Second, becoming an energy self-producer or a participant of an energy community also means that consumers have to invest in the energy production technology they wish to use. This means financial contribution to acquire the solar panels, for example and already rules out those groups of consumers who are not in the socio-economic position to make such an investment. Investment recovery, ie when and if the production of energy from these panels becomes profitable, depends on market price developments, which the investor cannot predict with full certainty.[77] For a large energy utility, the uncertainty of recovery and the broader investment risk is a known part of the business, but for an individual consumer, the awareness level of this kind of risk varies depending on qualities such as educational background and previous familiarity with existing rules and market.

Third and finally, the existing legislative framework includes new responsibilities for active energy consumers. For example, Article 16 of the Electricity Directive requires that citizen energy communities are financially responsible for the imbalances they cause in the electricity system and that they must bear the balancing responsibility. Furthermore, Member States must ensure that renewables self-consumers contribute in an adequate and balanced way to the overall cost sharing of the system when electricity is fed into the grid.[78] These are important elements, as decentralization necessarily increases the number of players in the energy sector and therefore contributes to increasing complexity of the energy system.[79] However, it also means that consumers engaging in these activities have more responsibility than what they did as mere passive recipients of centrally produced energy.

The identified three categories of risk that energy consumers face when becoming more active in the energy sector are, to a large extent, inherent when consumers are given more opportunities and cannot be entirely removed. However, their adverse effects can be mitigated and managed through appropriate governance structures that offer protection and safeguards for active energy consumers. These kinds of provisions can be identified throughout the EU legislative framework, which are identified and analysed in the next sub-section.

B. Protecting and Safeguarding Active Energy Consumers

1. Clear identification of affected consumer groups

Managing the risks resulting from exposing consumers to market prices and other risks associated with consumers engaging in energy activities begins with identifying the relevant risk groups: not all consumers or groups of consumers necessarily need the same level of protection and the same safeguards. To identify and delineate these groups, Member States

[75] S Borenstein, 'Effective and Equitable Adoption of Opt-In Residential Dynamic Electricity Pricing' (2013) 42 Review of Industrial Organisation 127, 131; more recently G Makridou, K Matsumoto, and M Doumpos, 'Evaluating the Energy Poverty in the EU Countries' (2024) 140 Energy Economics 108020.

[76] A Faruqui, 'The Ethics of Dynamic Pricing' in FP Sioshansi (ed), *Smart Grid: Integrating Renewable, Distributed and Efficient Energy* (Academic Press 2011) 61–82, 75–76.

[77] V Casalicchio and others, 'From Investment Optimization to Fair Benefit Distribution in Renewable Energy Community Modelling' (2022) 310 Applied Energy 118447.

[78] Renewable Energy Directive, art 21.

[79] Soininen (n 49) 77.

are obligated to define the concept of vulnerable customers which may refer to energy poverty and, inter alia, to the prohibition of disconnection of electricity to such customers in critical times.[80] The definition may be based on elements such as income levels, the share of energy expenditure of disposable income, the energy efficiency of homes, or critical dependence on electrical equipment for health reasons among other elements.[81] In addition to defining the concept, Member States must take all appropriate measures to protect these identified groups of vulnerable consumers.[82] Member States must also ensure that vulnerable customers and customers affected by energy poverty are fully protected from electricity disconnections.[83]

As part of their integrated national energy and climate plans, Member States have to assess the number of households in energy poverty taking into account the necessary domestic energy services needed to guarantee basic standards of living in the relevant national context, existing social policy, and other relevant policies.[84] When assessing the number of households in energy poverty, Member States are also obligated to establish and publish a set of criteria, which may include low income, high expenditure of disposable income on energy, and poor energy efficiency.[85] However, the identification of vulnerable consumers does not mean that these groups could not be offered dynamic price contracts. Instead, it means that additional safeguards to protect these groups may be adopted within the EU legal framework.[86]

2. Financial incentives

In addition to identifying the most vulnerable consumer groups, the legislative framework includes measures to manage the identified risks. Importantly, the Electricity Directive requires Member States to guarantee universal service and a supplier of last resort.[87] Universal service refers to the right to be supplied with electricity of a specified quality within their territory at competitive, easily and clearly comparable, transparent, and non-discriminatory prices, whereas a supplier of last resort is a designated supplier that must offer consumers electricity contracts to ensure continuity of supply for household customers.

Under the existing electricity market rules, the protection of vulnerable consumers has also been possible through retail price regulation, although the legislative framework's general approach to price regulation is rejective. Nevertheless, retail price regulation has often been justified on the basis of Article 9 of the Electricity Directive, which governs public

80 Electricity Directive, art 28.
81 ibid art 28; in literature, see Diestelmeier (n 44).
82 Electricity Directive, art 28.
83 ibid art 28a.
84 Regulation (EU) 2018/1999 of the European Parliament and of the Council of 11 December 2018 on the Governance of the Energy Union and Climate Action, amending Regulations (EC) No 663/2009 and (EC) No 715/2009 of the European Parliament and of the Council, Directives 94/22/EC, 98/70/EC, 2009/31/EC, 2009/73/EC, 2010/31/EU, 2012/27/EU, and 2013/30/EU of the European Parliament and of the Council, Council Directives 2009/119/EC and (EU) 2015/652 and repealing Regulation (EU) No 525/2013 of the European Parliament and of the Council [2018] J L328/1 (Governance Regulation), art 3. This has been done by Member States in their integrated national energy and climate plans. For example, see Germany: Final updated NECP 2021-2030, available at <https://commission.europa.eu/publications/germany-final-updated-necp-2021-2030-submitted-2024_en> accessed 4 August 2025 and Spain: Final updated NECP 2021-2030, available at https://commission.europa.eu/publications/spain-final-updated-necp-2021-2030-submitted-2024_en.
85 Electricity Directive, art 29.
86 Huhta (n 4).
87 Electricity Directive, art 2, art 27a.

service obligations.[88] According to the provision, Member States are allowed to impose public service obligations on undertakings which, among other things, may relate to quality and price of supplies. However, this type of price regulation must comply with the rules on market-based supply prices, which require that suppliers are free to determine the price at which they supply electricity to customers.[89] The Directive also explicitly underlines that energy poor or vulnerable customers must be protected by means other than the State-driven price-setting for the supply of electricity, such as instruments of social policy or the social security system.[90]

Some exceptions to the prohibition of State-driven price-setting are specifically allowed. As an exception to the above rule, Member States can apply public interventions in the price setting for the supply of electricity to energy poor or vulnerable household customers if they fulfil certain criteria.[91] Among other criteria, the interventions must pursue a general economic interest. They must be proportional, be clearly defined, transparent, non-discriminatory, and verifiable and guarantee equal access for Union electricity undertakings to customers.[92] These criteria are nearly a mirror image of the conditions established by the Court of Justice of the EU in 2010 in the case of *Federutility*.[93]

3. The extension of protective provisions from consumers to energy communities

The protection of vulnerable or poor energy consumers is clearly meant to apply to individuals. However, other mitigation measures that are meant to protect active consumers engaging in energy communities can also be found. First and foremost, the Renewable Energy Directive states that participants in renewable energy communities maintain their rights (as well as obligations) as final customers and, therefore, benefit from the protective provisions identified above.[94] Furthermore, articles 21 and 22 of the Renewable Energy Directive include provisions that aim to ensure non-discriminatory opportunities for renewable energy communities and self-consumers, and to clarify their right to remuneration and support schemes.[95] The legal framework also acknowledges the new supplier risks of increased community energy activities. Article 18a of the Electricity Directive posits that Member States should endeavour to ensure the accessibility of hedging products for citizen energy communities and renewable energy communities to manage supplier risks.[96]

[88] L Rimšaitė, 'Public Services in the Modern Economy: The Evolution and Future' in *The Crossroads of Competition Law and Energy Regulation* (Springer 2024); H Schweitzer, 'Services of General Economic Interest: European Law's Impact on the Role of Markets and of Member States' in M Cremona (ed), *Market Integration and Public Services in the European Union* (OUP 2011) 11-62; N Fiedziuk, 'Services of General Economic Interest and the Treaty of Lisbon: Opening Doors to a Whole New Approach or Maintaining the "Status Quo"' (2011) 36 European Law Review 226.

[89] Electricity Directive, art 9, art 5.

[90] ibid art 5.

[91] ibid.

[92] ibid art 5.

[93] Case C-265/08 *Federutility and Others* [2010] ECR I-3377. Also see Case C-121/15 *ANODE* [2016] Judgment of the Court (Fifth Chamber) of 7 September 2016, published in the electronic Reports of Cases.

[94] Renewable Energy Directive, art 22.

[95] For more on support schemes, see MM Sokołowski, 'Renewable and Citizen Energy Communities in the European Union: How (not) to Regulate Community Energy in National Laws and Policies' (2020) 38 Journal of Energy & Natural Resources Law 289.

[96] Electricity Directive, art 18a.

V. Conclusions: More Opportunity, More Responsibility

This chapter has explored the EU's legal framework as an illustrative and globally unique model on a governance structure in order to provide a legal account of active energy consumers. In particular, the analysis has addressed the trade-offs between empowering the consumer to participate in the energy markets on the one hand and protecting the consumer from the risks associated with such participation on the other. In doing so, it has contributed to the theme of the book by analysing the legal mechanisms needed to ensure socially and economically sustainable consumer engagement in decentralized energy systems.

Two key components can be identified in the EU's legal framework for active energy consumers. The first component is composed of provisions that seek to engage consumers to offer demand-side flexibility to the electricity markets. The second component focuses on collective action in the form of renewable energy communities or citizen energy communities. Both forms of consumer empowerment carry new kinds of risks. This analysis identified three key categories of these new risks that the EU legal framework acknowledges and consolidates: the exposure to electricity market prices; the financial risk associated with investing in self-generation, such as solar panels; and the increased responsibilities that arise from consumers becoming producers of electricity.

Based on this analysis, the chapter also identified three legal approaches within the EU legal system that could provide useful benchmarks for other jurisdictions seeking to develop legislative solutions to govern the intersection between DERs and the protection of active energy consumers. The first group of provisions focuses on the clear and appropriate identification and acknowledgement of the affected consumer groups. Not all active energy consumers require the same level of protection, while others are clearly in a more vulnerable position. The second group relates to the mitigation of the financial risks associated with increased activities for consumers and can be addressed through different kinds of financial incentives spanning from providing initial investment support to hedging price variations. Third and finally, the provisions extend the protection of active energy consumers to energy communities, which ensures that collective operation of DERs is not left outside the scope of the relevant provisions.

Given the fast-paced development of the energy transition, the applicable EU legislative frameworks have been adapted and revised several times in the most recent decade or so. Research is needed to understand and monitor whether, and to what extent, these legislative frameworks are de facto able to address the increased risks and responsibilities of increased consumer engagement. The rapid development of DERs and the diversifying role of active consumers also means that legislative frameworks need to be adapted to respond to future needs.

The chapter has demonstrated how the legal framework provides consumers with more opportunities but also places more responsibility onto consumers. While these responsibilities and risks are inherent when opportunities diversify, legal frameworks play an increasingly important role in managing and mitigating the increased risks of active consumer engagement.

16

Sustainable Energy Democratization Through Community-Centric Models

Nadia B Ahmad, Kierra Peoples, Evangeline Fletcher, and Brianna Pruitt

I. Introduction: Community-Based Energy Systems and Sustainable Energy Democratization

The global transition towards distributed energy resources (DERs) has revealed a fundamental tension between technological advancement and social equity, as community-centric energy models struggle with entrenched regulatory frameworks designed to preserve centralized utility control. Across jurisdictions worldwide, communities seeking to develop microgrids, community solar gardens, and other participatory energy initiatives face persistent barriers, including restrictive ownership regulations, inadequate financing mechanisms, complex permitting processes, and utility resistance to distributed generation that threatens traditional business models.[1] These challenges are particularly acute in the context of sustainable DERs, which are decentralized energy systems that integrate the social (people), environmental (planet), and economic (profit) pillars of sustainable development while advancing energy citizenship through meaningful community participation in energy governance and ownership.[2]

This chapter focuses on the legal and regulatory landscape surrounding two promising community-centric sustainable DER initiatives in Florida, namely microgrids and community solar gardens. Through an analysis of Florida's regulatory barriers and a comparative examination of supportive frameworks in other jurisdictions, this chapter offers comprehensive legal and policy recommendations aimed at creating more supportive environments for community-centric sustainable DERs that foster energy democratization while advancing environmental justice and resilience. The concept of energy democratization, as understood through the lens of sustainable DERs, extends far beyond physical infrastructure changes to encompass a fundamental reconfiguration of power relationships within energy systems that have historically prioritized corporate interests over community needs.[3] Energy citizenship, which refers to the active engagement of citizens in energy production, distribution, and governance, represents a cornerstone of this transformation,

[1] E Obuseh and others, 'A Systematic Review of Barriers to Renewable Energy Integration and Adoption' (2025) 9 Journal of Asian Energy Studies 26.

[2] Drawing from concepts outlined in DS Olawuyi, 'International Law on Distributed Energy Resources' in Chapter 2 of this book.

[3] S Axon and J Morrissey, 'Just Energy Transitions? Social Inequities, Vulnerabilities and Unintended Consequences' (2020) 1 Buildings and Cities 393.

Nadia B Ahmad, Kierra Peoples, Evangeline Fletcher, and Brianna Pruitt, *Sustainable Energy Democratization Through Community-Centric Models* In: *Sustainable Distributed Energy Resources*. Edited by: Íñigo del Guayo, Damilola S. Olawuyi San, Louis de Fontenelle, Milton Fernando Montoya, and José Juan González, Oxford University Press.
DOI: 10.1093/9780198991946.003.0019

enabling communities to participate meaningfully in energy decisions that directly affect their environmental conditions, economic opportunities, and social well-being.[4] As Araújo notes, energy transitions involve 'complex socio-technical changes that co-evolve and include changes in user practices and institutional (eg, regulatory) structures as well as technological dimensions', highlighting the multifaceted legal and policy transformations required to create more democratic and sustainable energy systems.[5]

Florida presents a particularly compelling and urgent case study for examining these dynamics, given the convergence of abundant renewable resources, acute climate vulnerability, and deeply entrenched regulatory barriers to community energy development.[6] Despite ranking third nationally for solar energy potential and experiencing an increasing frequency of extreme weather events that highlight the need for resilient distributed energy systems, Florida lags significantly behind comparable states in community energy deployment.[7] Recent policy developments have further hindered Florida's renewable energy progress. In 2023, Governor DeSantis vetoed $346 million in federal energy efficiency programs that would have provided rebates for heat pumps, electric vehicle chargers, and energy-efficient appliances for Florida residents.[8] This decision particularly impacts low-income communities and communities of colour, who face the highest energy burdens and would benefit most from efficiency programs.[9] The state's pattern of rejecting federal clean energy funding demonstrates how political resistance to renewable energy development actively undermines community-centric sustainable DER initiatives.[10] The state's energy democracy challenges are compounded by persistent energy burden inequities affecting low-income communities and communities of colour, regulatory frameworks that systematically favour investor-owned utilities over community initiatives, and the limited legal recognition of co-operative and community ownership models for energy infrastructure.

While existing scholarship has extensively examined the technical aspects of DER integration and analysed individual state policy approaches to community energy, significant gaps remain in understanding how legal frameworks can systematically advance sustainable DERs that integrate environmental, social, and economic objectives while promoting energy citizenship and democratic participation. Most legal analyses focus narrowly on technical interconnection standards or economic compensation mechanisms, failing to address the broader governance transformations necessary for genuine energy democratization.[11]

[4] Olawuyi (n 2).

[5] K Araújo, 'The Emerging Field of Energy Transitions: Progress, Challenges, and Opportunities' (2014) 1 Energy Research & Social Science 112.

[6] J Lopez, 'The Climate Is Changing and so Must We: The Need to Prioritize at-Risk Communities and Ecosystems' (2024) 74 Washington University Journal of Law and Policy 44.

[7] OO Babayomi, DA Dahoro, and Z Zhang, 'Affordable Clean Energy Transition in Developing Countries: Pathways and Technologies' (2022) 25 iScience 104178.

[8] 'DeSantis Veto Cost $346 Million in Energy-Saving Programs for Florida' Tampa Bay Times (17 August 2023) <https://www.tampabay.com/news/florida-politics/2023/08/17/desantis-veto-energy-savings-program-air-conditioner/> accessed 5 September 2023.

[9] M Perry, 'Kathy Castor Still Upset over DeSantis Veto That Cost FL $350 Million in Energy Funds' Florida Phoenix (4 August 2023) <https://floridaphoenix.com/2023/08/04/kathy-castor-still-upset-over-desantis-veto-that-cost-fl-350-million-in-energy-funds/> accessed 4 August 2023.

[10] 'Florida Misses out on Millions of Dollars in Federal Energy Rebate Programs' ABC Action News (17 August 2023) <https://www.abcactionnews.com/news/state/florida-misses-out-on-millions-of-dollars-in-federal-energy-rebate-programs> accessed 9 October 2023.

[11] J Radtke, 'Understanding the Complexity of Governing Energy Transitions: Introducing an Integrated Approach of Policy and Transition Perspectives' (2025) 35 Environmental Policy and Governance 595 https://doi.org/10.1002/eet.2158 accessed 25 August 2025.

Furthermore, limited research has examined how regulatory frameworks can proactively address energy justice concerns by ensuring equitable access to community energy benefits, particularly for historically marginalized communities that face the greatest energy burdens but encounter the most barriers to participation.[12]

This chapter addresses these gaps by arguing that the successful implementation of community-centric sustainable DERs requires comprehensive legal frameworks that simultaneously: (1) establish clear ownership and governance structures enabling genuine community control over energy assets; (2) create supportive regulatory environments that value the full range of community energy contributions, including resilience, environmental benefits, and social equity; (3) develop innovative financing mechanisms and participatory processes that ensure meaningful access for diverse communities; and (4) embed energy justice principles throughout regulatory design to address historical inequities while building more inclusive energy systems. Through Florida's experience, the chapter demonstrates that piecemeal regulatory adjustments are insufficient; transformative legal frameworks must address the interconnected technical, economic, and social dimensions of sustainable energy democratization.

This chapter is organized into six main sections that systematically examine the legal and policy pathways for advancing community-centric sustainable DERs. Following this introduction, Section II develops a conceptual framework explaining energy democratization and community ownership within Florida's regulatory context, establishing connections between sustainable DER principles and energy citizenship concepts. Section III provides a comprehensive analysis of Florida's current legal and regulatory barriers to community energy development, organized thematically to examine ownership restrictions, interconnection challenges, financing limitations, and equity gaps. Section IV examines transformative legal models from other US states and international jurisdictions, identifying innovative approaches that successfully integrate sustainable DER principles while promoting energy citizenship and democratic participation. Section V synthesizes these comparative insights to present comprehensive legal framework recommendations for advancing community-centric sustainable DERs, structured as lessons learned that can inform policy development in other jurisdictions seeking to democratize their energy systems. Section VI provides a conclusion to assess the impacts of improved legal frameworks for clean energy policy.

II. Conceptual Framework: Community-Centric Sustainable DERs and Energy Democratization

A. Defining Community-Centric Sustainable DERs

Community-centric distributed energy resources represent a distinctive subset of sustainable DERs characterized by collective ownership structures, shared governance mechanisms, and localized benefit distribution patterns that integrate the three pillars of sustainable development while advancing energy citizenship.[13] These systems operate at an intermediate scale, larger than individual households but significantly smaller than

[12] ibid.
[13] Olawuyi (n 2).

utility-scale installations, making them particularly suitable for neighbourhood, municipal, or institutional applications that can democratize energy ownership and decision-making.[14] Sustainable DERs, as conceptualized within international frameworks, balance social (people), environmental (planet), and economic (profit) objectives through decentralized energy systems that promote community participation while addressing environmental and economic sustainability issues.[15] The social pillar emphasizes addressing social exclusion and human rights considerations that may result from energy system design, ensuring meaningful participation by historically marginalized communities.[16] The environmental pillar focuses on reducing greenhouse gas emissions and environmental pollution while minimizing adverse effects on ecosystems. The economic pillar emphasizes diversification opportunities and cost-effective energy solutions that provide economic benefits to the participating communities.[17] Community-centric models advance these sustainable development objectives by embedding energy citizenship and the active engagement of citizens in energy production, distribution, and governance throughout their design and implementation.[18] This approach transforms passive energy consumers into active participants with genuine agency in energy decision-making processes, creating pathways for broader participation and benefit-sharing that can democratize energy systems according to local needs, values, and priorities, rather than profit-driven imperatives alone.

B. Energy Democratization Through Community Ownership Models

Energy democratization represents a fundamental paradigm shift in how energy systems are governed and who benefits from them, a shift of particular relevance to Florida's hierarchical utility structure, dominated by investor-owned utilities.[19] This multidimensional concept encompasses several key interrelated elements: distributed ownership of energy infrastructure enabling community wealth building, participatory governance mechanisms that provide a genuine voice to historically excluded stakeholders, equitable distribution of both economic and environmental benefits, and accessible decision-making processes that prioritize transparency and community input over corporate interests.[20] At its core, energy democratization seeks to transfer power from centralized institutions to diverse communities, enabling them to shape energy systems according to local needs, values, and priorities.

The historical development of Florida's energy landscape reflects broader patterns observed throughout the United States, where conventional energy systems are characterized by centralized production, top-down decision-making processes that exclude community input, and ownership structures concentrated among powerful corporate entities with substantial political influence.[21] This entrenched model has systematically excluded

[14] J Mattes, A Huber, and J Koehrsen, 'Energy Transitions in Small-Scale Regions—What We Can Learn from a Regional Innovation Systems Perspective' (2015) 78 Energy Policy 255.
[15] Olawuyi (n 2).
[16] Radtke (n 11).
[17] Olawuyi (n 2).
[18] ibid.
[19] Anthony Giancatarino and Donna House, *Energy Democracy: Reparation, Regeneration, Reinvestment* (Emerald Cities Collaborative 2021), <https://emeraldcities.org/wp-content/uploads/2021/04/Energy-Democracy-FINAL_Flipbook.pdf>.
[20] Axon and Morrissey (n 3).
[21] Florida Administrative Code r 25-6.0423; Florida Statutes § 704.07 (Solar Easements).

communities, particularly economically disadvantaged and minority populations, from meaningful participation in energy governance decisions that directly affect their well-being, environmental conditions, and economic opportunities. As Axon and Morrissey observe, 'Energy justice focuses on the fair distribution of both environmental benefits and burdens associated with energy production and consumption', a consideration particularly relevant in Florida, where climate vulnerability and energy burden disproportionately affect low-income communities and communities of colour.[22]

C. Two Models of Community-Centric Sustainable DERs

This chapter examines two prominent models of community-centric sustainable DERs with relevance to Florida's unique geographic, demographic, and climate conditions: community microgrids and community solar gardens. These models exemplify how sustainable DER principles can be operationalized through community ownership and governance structures that advance energy citizenship while addressing local resilience and equity requirements.

1. Community Microgrids

Community microgrids represent sophisticated localized energy systems that can operate either connected to the main grid or independently in isolation (island mode), providing enhanced resilience during increasingly frequent and severe weather events that plague Florida, including hurricanes, tropical storms, and flooding.[23] These systems typically integrate multiple generation sources (particularly solar in Florida's sun-rich environment), advanced storage capabilities essential for maintaining power during outages, and intelligent load management systems, making them exceptionally valuable for enhancing local resilience in Florida, a state particularly vulnerable to climate change impacts and associated power disruption.[24]

Community microgrids advance sustainable DER objectives by simultaneously addressing all three pillars. The environmental pillar is served through the integration of renewable energy sources and the reduction of greenhouse gas emissions. The economic pillar is advanced through community ownership models that enable local wealth building and reduce energy costs. The social pillar is addressed through enhanced energy security and resilience for vulnerable communities, particularly during climate emergencies when centralized infrastructure is compromised.[25] The state's experiences with devastating hurricanes that caused widespread power outages lasting weeks in some areas highlight the critical need for resilient energy systems that can operate independently when the centralized infrastructure fails.[26]

[22] ibid 400.

[23] Babayomi (n 7).

[24] SC Bronin, 'Curbing Energy Sprawl with Microgrids' (2010) 43 Connecticut Law Review 547; 'Microgrids Can Provide Resiliency During Extreme Weather Events' (*Power*, 1 February 2023) <https://www.powermag.com/microgrids-can-provide-resiliency-during-extreme-weather-events/> accessed 20 August 2025.

[25] Olawuyi (n 2).

[26] 'Attaining Energy Resilience during Hurricanes: PowerSecure' (*PowerSecure*, 26 November 2024) <https://powersecure.com/blog/microgrids-the-backbone-of-resilient-energy-infrastructure-during-hurricanes> accessed 20 August 2025.

2. Community Solar Gardens
Community solar gardens represent innovative shared solar facilities that allow multiple community members to benefit collectively from a single solar installation, a model with tremendous potential in Florida, where abundant sunshine provides ideal conditions for solar generation throughout the year. Participants may either own portions of the system directly or subscribe to receive benefits from the energy produced, creating pathways to solar access for Floridians who cannot install rooftop systems because of financial constraints, unsuitable roof conditions, rental status, or homeowners' association restrictions.[27] This model particularly advances the social and economic pillars of sustainable DERs by addressing energy equity concerns in Florida's diverse communities, from densely populated urban centres such as Miami and Orlando to retirement communities and rural areas, where residents may lack the financial resources or suitable property for individual solar installations. Community solar gardens enable energy citizenship by providing accessible pathways for community participation in renewable energy ownership and governance, regardless of individual property characteristics or the financial capacity of the participants.[28]

D. Integration with Sustainable Development and Energy Citizenship

Both community microgrid and community solar garden models offer promising pathways to democratize Florida's energy systems by broadening ownership beyond powerful utility companies, enhancing meaningful community participation in energy decisions that directly affect local residents, and creating a more equitable distribution of benefits across diverse socioeconomic groups. These models operationalize energy citizenship by creating institutional structures that enable communities to actively participate in energy system governance while building local capacity for energy self-determination.[29] However, their successful implementation depends heavily on transforming Florida's legal and regulatory frameworks, which currently prioritize utility interests over community energy development and democratic participation. The following section examines the specific legal and regulatory barriers that impede the development of community-centric sustainable DER models in Florida's current institutional environment.

III. Florida's Current Legal and Regulatory Landscape: Barriers to Sustainable Energy Democracy

Florida's energy governance structure creates a complex and often inhospitable environment for community-centric sustainable DER initiatives, largely reflecting the substantial political influence of the state's investor-owned utilities (IOUs). The Florida Public Service Commission (FPSC) serves as the primary regulatory body overseeing these powerful IOUs (Florida Power & Light, Duke Energy Florida, Tampa Electric Company, and Gulf

[27] RL Ottinger and J Bowie, 'Innovative Financing for Renewable Energy' (2015) 32 Pace Environmental Law Review 701.
[28] Mattes (n 14).
[29] Axon and Morrissey (n 3) 405.

Power), while the state's broader energy policy is shaped by multiple entities, including the Department of Agriculture and Consumer Services' Office of Energy, the Governor's office with its shifting political priorities, and the state legislature, where utility interests maintain a significant lobbying presence.[30] This fragmented governance structure, characterized by overlapping jurisdictions and occasionally conflicting mandates, has resulted in policy inconsistencies and regulatory gaps that complicate the development and implementation of community energy projects in the state. With the right political will, energy policy education, and prioritization of sustainable development, these barriers can be overcome.

A. Lack of Comprehensive Renewable Energy Mandate

Unlike twenty-nine other states with progressive renewable portfolio standards (RPS) or clean energy standards, Florida lacks a comprehensive renewable energy mandate requiring utilities to procure specific percentages of electricity from renewable sources.[31] This policy gap signals the state's limited commitment to transitioning away from fossil fuel dependence, despite its extraordinary renewable energy potential, ranking third nationally for solar energy potential with an average of 237 days of sunshine annually.[32] The absence of mandatory renewable procurement requirements is compounded by recent policy setbacks that actively impede clean energy development. In 2023, Governor DeSantis vetoed $346 million in federal energy efficiency programs, costing Florida residents significant savings on energy bills and undermining efforts to reduce energy burdens in vulnerable communities.[33] Instead, Florida relies primarily on voluntary utility initiatives, often designed to maintain utility control over renewable development, limited and frequently changing incentive programs without long-term stability, and piecemeal regulatory adjustments to drive renewable energy adoption. The absence of mandatory renewable procurement requirements under state law means that community-centric sustainable DER projects cannot rely on guaranteed market demand or stable regulatory support for long-term planning and financing of such projects. This inconsistent and inadequate policy environment has profound implications for community-centric energy projects, which systematically struggle to secure the necessary regulatory support, interconnection arrangements, financing mechanisms, and fair compensation structures required for financial viability and long-term sustainability.[34] This pattern of rejecting federal clean energy funding demonstrates how political resistance translates into concrete barriers for community energy initiatives.[35]

[30] Araújo (n 5).
[31] C Christ, 'Florida's Energy Policy: A Model for the Nation' (2009) 39 Environmental Law Reporter 10061.
[32] Florida Solar Energy Center, 'Solar Energy in Florida' (University of Central Florida 1989).
[33] 'DeSantis Veto Cost $346 Million' (n 8).
[34] F Bradley-Wright, 'Outdated Energy Efficiency Policies Still Holding Florida Back' (Clean Energy for America, 5 April 2023) <https://www.cleanenergy.org/blog/outdated-energy-efficiency-policies-still-holding-florida-back/> accessed 1 August 2025.
[35] Perry (n 9).

B. Legal Uncertainty Regarding Community Ownership Structures

Florida law provides limited clarity regarding legal ownership structures for community microgrids and solar gardens, creating substantial uncertainty and risk for potential developers, community organizations, and investors. Unlike states such as New York or California, which have established explicit frameworks authorizing and defining community energy ownership rights and responsibilities, Florida's regulations leave it fundamentally ambiguous whether non-utility entities can legally own and operate energy systems that serve multiple customers across property boundaries without being classified as regulated utilities.[36]

This legal ambiguity stems from Florida's broad statutory definition of 'public utility', which potentially encompasses any entity providing electricity service to multiple customers for compensation. Classifying it as a regulated utility would impose prohibitive regulatory burdens on community-scale projects, including rate regulation, service territory restrictions, and comprehensive PSC oversight designed for large-scale utility operations rather than community-owned sustainable DER initiatives. This regulatory framework was designed to protect utility monopoly interests rather than facilitate innovative community energy solutions that are responsive to local needs and resilience concerns.[37]

C. Inadequate Interconnection Standards for Community-Scale Projects

Although Florida has established basic interconnection standards for individual customer-owned generation systems, these regulations were never designed with community-scale microgrids in mind and fail to address their unique operational characteristics. The existing regulatory framework lacks clear, standardized provisions addressing the distinctive technical and operational requirements of microgrids that serve multiple customers across multiple properties and potentially operate in island mode during emergencies, capabilities that are particularly important in hurricane-prone Florida.[38] Instead, communities seeking to develop microgrids face a complex, time-consuming, and often prohibitively expensive process of negotiating custom interconnection agreements with reluctant utilities that frequently impose excessive technical requirements, demand costly system upgrades, or impose operational limitations that undermine the microgrid functionality. This bureaucratic process, governed primarily by utility-specific tariffs rather than standardized state regulations, discourages all but the most well-resourced and determined communities from pursuing microgrid development, despite the clear resilience benefits these systems could provide in Florida's vulnerable coastal communities.

[36] Axon and Morrissey (n 3).
[37] ibid.
[38] Mattes (n 14).

D. Restrictive Third-Party Sales Prohibitions

Florida's regulatory interpretation, reinforced by judicial decisions favourable to utility interests, effectively prohibits third-party power purchase agreements (PPAs), a critical financing mechanism that has enabled community solar development in numerous other states.[39] This prohibition stems from an expansive interpretation of 'public utility' that classifies any entity selling electricity to multiple customers as a regulated utility subject to comprehensive PSC jurisdiction, a designation that renders independent community solar projects legally and financially infeasible. The restrictive interpretation of third-party sales rules effectively prevents non-utility entities from selling power directly to multiple customers, which is a critical component of many community microgrid and community solar business models. This restriction severely limits the development of diverse community solar models that have flourished elsewhere, particularly those that rely on third-party ownership and operation while allowing community governance and benefit distribution.

E. Limited Virtual Net Metering Authorization

While Florida has established basic net metering regulations allowing individual customers with renewable energy systems to receive credit for excess generation, these provisions remain fundamentally inadequate for community solar applications, where generation occurs at a shared facility and benefits must be distributed among multiple participants at different locations. The current regulatory framework conspicuously fails to authorize virtual net metering, a critical policy tool implemented in numerous other states that enables the allocation of credits from a shared solar facility to multiple subscribers' bills based on their ownership share or subscription level.[40] Without this essential mechanism, community solar projects in Florida face significant administrative barriers and economic disadvantages compared to individually owned systems, effectively limiting participation to those with suitable properties for on-site installation and sufficient financial resources for ownership. Community-centric sustainable DERs are designed to overcome these barriers to advance energy citizenship and democratic participation.[41]

F. Insufficient Financing Mechanisms and Accessibility Barriers

The development of community-centric energy projects in Florida is further complicated by substantial financing obstacles and accessibility limitations that disproportionately affect communities with limited resources. Unlike progressive states such as Connecticut, New York, and Massachusetts, which have established dedicated green banks or specialized clean energy funds with explicit mandates to support community-based energy projects through innovative financing tools, Florida conspicuously lacks comprehensive public

[39] Ottinger and Bowie (n 27).
[40] ibid.
[41] Olawuyi (n 2).

financing mechanisms specifically designed to address the unique challenges faced by community-centric renewable energy initiatives.[42]

This institutional gap leaves community projects dependent on conventional financing sources that frequently fail to understand or appropriately value their distinctive characteristics, including longer development timelines, complex ownership structures, and multiple benefit streams that extend beyond simple financial returns. Existing financing models and program designs for renewable energy in Florida systematically exclude low-income communities, communities of colour, and historically marginalized groups through multiple reinforcing mechanisms: stringent credit requirements, substantial upfront cost obligations, technical complexity requiring specialized expertise, and program designs that fundamentally fail to address the specific needs of diverse populations.[43]

G. Inadequate Compensation for Grid Services and Resilience Benefits

The current utility rate structures and compensation mechanisms in Florida systematically undervalue or entirely ignore the substantial resilience benefits and grid services provided by community-centric, sustainable distributed energy resources (DERs), creating fundamental economic obstacles to project development. Unlike states that have implemented value-stacking mechanisms to compensate for distributed energy resources for multiple benefits, including capacity value, ancillary services, resilience enhancement, and environmental attributes, Florida's regulatory framework narrowly focuses on basic energy compensation while failing to monetize the full range of values that sustainable DERs provide to both participants and the broader electric system.[44]

This economic disincentive significantly undermines the financial viability of community energy projects, particularly those serving vulnerable or low-income communities, where the energy burden is already disproportionately high, but where resilience benefits would be especially valuable. The absence of fair compensation mechanisms effectively transfers value from communities to utilities without appropriate remuneration, reinforcing existing power imbalances in the energy system and perpetuating the inequitable distribution of both costs and benefits.[45]

The regulatory ambiguities, policy inconsistencies, and legal uncertainties described above collectively create a challenging investment environment characterized by significant uncertainty for potential investors in Florida's community energy projects. This uncertainty translates directly into higher capital costs, risk premiums demanded by investors, reluctance from traditional lenders, and reduced availability of financing for community-based projects compared to conventional energy investments or utility-controlled developments.[46]

[42] Ottinger and Bowie (n 27).
[43] Axon and Morrissey (n 3).
[44] Mattes (n 14).
[45] Axon and Morrissey (n 3).
[46] Ottinger and Bowie (n 27).

IV. Transformative Legal Models for Community-Centric DERs: Lessons from Other Jurisdictions

The development of supportive legal frameworks for community-centric sustainable DERs requires comprehensive approaches that address the interconnected technical, economic, and social dimensions of energy democratization. Several progressive jurisdictions have developed innovative legal frameworks that successfully integrate sustainable DER principles while advancing energy citizenship, offering valuable lessons for policymakers seeking to create more supportive environments for community energy development in Australia.[47]

A. Comprehensive Legal Recognition of Community Energy Models

New York has established a nationally recognized legal framework for community energy that demonstrates how comprehensive legislation can simultaneously advance all three pillars of sustainable distributed DERs while promoting energy citizenship. The state's community solar program, established through the Clean Energy Standard and implemented via detailed regulations, provides clear legal recognition of diverse project ownership structures, including community-based organizations, co-operatives, and municipal entities.[48] This legal clarity directly addresses one of the primary barriers to community energy development: regulatory uncertainty regarding ownership structures and operational authority. New York's framework advances the social pillar of sustainable DERs through meaningful low-income participation provisions that enhance energy equity, including dedicated funding mechanisms, enhanced incentives for projects serving disadvantaged communities, and simplified participation processes designed to address the historical barriers to clean energy access.[49] The environmental pillar is served through requirements for renewable energy generation and environmental benefit verification, while the economic pillar is advanced through subscriber protection provisions and transparent benefit-sharing arrangements that ensure the fair distribution of project value among participants.

Successful community energy legislation must provide explicit legal recognition of diverse ownership models while establishing clear operational guidelines that protect both community interests and system reliability. This step requires moving beyond technical interconnection standards to address governance structures, subscriber management, and benefit distribution mechanisms.

B. Value-Based Compensation Mechanisms

Minnesota's approach to community solar compensation offers a sophisticated legal precedent for recognizing the full value that community-centric sustainable DERs provide to both participants and the broader energy system. The state's value-of-solar methodology, established through comprehensive legislation and detailed regulatory proceedings,

[47] Olawuyi (n 2).
[48] NY Public Service Commission, 'Order Adopting a Clean Energy Standard' Case 15-E-0302 (1 August 2016).
[49] ibid.

calculates compensation based on the full range of values that distributed solar energy provides, including energy value, generation capacity value, transmission and distribution capacity value, environmental benefits, and avoided fuel price risk.[50] This comprehensive valuation methodology advances sustainable DER objectives by properly monetizing environmental benefits (environmental pillar), ensuring fair economic returns to community participants (economic pillar), and creating incentives for distributed generation that enhance system resilience and equity (social pillar). The legal framework establishes clear methodologies for calculating these values, while providing transparency and predictability for project developers and community participants.[51]

Effective legal frameworks must move beyond basic net metering to establish sophisticated compensation mechanisms that recognize the multiple values of community energy projects. This requires detailed regulatory proceedings to establish valuation methodologies and ongoing mechanisms for updating these values as technology and grid conditions evolve further.

C. Equity-Centred Policy Design

Colorado's Community Solar Gardens Act demonstrates how legal frameworks can embed energy justice principles throughout program design while advancing sustainable DER objectives. The Act includes innovative legal requirements that community solar projects reserve minimum percentages of capacity specifically for low-income subscribers, creating enforceable obligations rather than voluntary commitments to do so.[52] This legal approach directly addresses the social pillar of sustainable DERs by ensuring that community energy projects contribute to energy justice rather than exacerbating existing inequities. Colorado's framework also establishes clear legal protections for community energy participants, including subscriber rights, transfer provisions, and dispute resolution mechanisms that build trust in community energy models while protecting vulnerable populations from predatory practices.[53] These provisions demonstrate how legal frameworks can balance innovation and consumer protection while advancing energy citizenship through meaningful community participation.

Legal frameworks for community-centric sustainable DERs must proactively address equity concerns through enforceable requirements rather than voluntary guidelines. This includes establishing minimum participation thresholds for disadvantaged communities, consumer protection provisions, and accessible dispute-resolution mechanisms.

D. Institutional Innovation for Energy Citizenship

Germany's legal framework for energy co-operatives provides a valuable precedent for institutionalizing energy citizenship through the explicit legal recognition of

[50] Minnesota Statutes § 216B.164 subd 10 (2013).

[51] Minnesota Public Utilities Commission, 'In the Matter of the Application of Xcel Energy for Approval of a Solar Garden Program' Docket No E002/M-13-867 (2014).

[52] Colorado Revised Statutes § 40-2-127 (2010).

[53] ibid.

community energy organizations. The German Renewable Energy Sources Act (EEG) and Cooperative Act provide clear legal status for energy co-operatives, establishing their rights, responsibilities, and protections within the energy system while creating supportive regulatory frameworks for their operation and growth.[54] This legal approach has enabled remarkably widespread community participation in Germany's renewable energy transition, with energy co-operatives operating throughout the country, collectively managing significant renewable capacity, and involving hundreds of thousands of citizen members in democratic energy governance.[55] The legal framework successfully integrates all three pillars of sustainable DERs: environmental objectives through renewable energy deployment, economic benefits through community ownership and local value creation, and social objectives through democratic participation and community control.

Institutionalizing energy citizenship requires the explicit legal recognition of community energy organizations as legitimate actors within the energy system. This includes establishing clear legal frameworks for co-operative formation, operation, and governance while providing supportive regulatory environments that enable community energy organizations to participate effectively in energy markets.

E. Integrated Support Systems

Scotland's Community and Renewable Energy Scheme (CARES) demonstrates how comprehensive legal frameworks can integrate multiple policy tools to create supportive ecosystems for community energy development. The program combines legal authorities for community energy development, dedicated financing mechanisms, technical assistance programs, and capacity-building initiatives within a single integrated framework.[56] This approach recognizes that successful community energy development requires addressing multiple barriers simultaneously rather than pursuing piecemeal solutions. Scotland's framework advances sustainable DER objectives through integrated approaches that address the technical, financial, and social barriers to community energy development. The legal framework establishes clear pathways for community energy project development while providing supporting institutions and financing mechanisms that enable communities to participate effectively in the transformation of energy systems.[57]

Effective legal frameworks for community-centric sustainable DERs require integrated approaches that simultaneously address multiple barriers. This includes establishing not only basic legal authorities but also supporting institutions, financing mechanisms, and capacity-building programs that enable communities to participate effectively in energy system transformation.

[54] Erneuerbare-Energien-Gesetz [Renewable Energy Sources Act] 2017 (Germany).
[55] Axon and Morrissey (n 3).
[56] Scottish Government, 'Community and Renewable Energy Scheme (CARES)' (2024).
[57] Mattes (n 14).

F. Key Elements of Supportive Legal Frameworks

A comparative analysis of progressive jurisdictions reveals several key elements that must be integrated into comprehensive legal frameworks for community-centric sustainable DERs. First, legal frameworks must explicitly recognize diverse community energy ownership models while establishing clear operational guidelines that protect community interests and ensure system reliability. Second, compensation mechanisms must recognize the full value that community energy projects provide to participants and the broader energy system, moving beyond basic energy compensation to value-stacking approaches. Third, legal frameworks must proactively address equity concerns through enforceable requirements that ensure meaningful participation by disadvantaged communities are ensured. Fourth, successful frameworks institutionalize energy citizenship through the explicit legal recognition of community energy organizations and their rights within the energy system.[58] These elements must be integrated within comprehensive legal frameworks that address the interconnected technical, economic, and social dimensions of sustainable energy democratization while providing the legal certainty and institutional support necessary for effective community energy development.

V. Legal Framework Lessons for Community-Centric Sustainable DERs

Florida's experience with community-centric energy development reveals fundamental insights into the legal and regulatory reforms necessary to advance sustainable DERs while promoting energy citizenship and democratic participation. The barriers identified in Florida's current framework and the transformative models observed in other jurisdictions offer valuable lessons for policymakers seeking to develop supportive legal environments for community energy development that integrate environmental, social, and economic objectives.[59]

A. Comprehensive Legal Recognition of Community Energy Models

Successful implementation of community-centric sustainable DERs requires comprehensive legislation that explicitly recognizes diverse community ownership models while establishing clear operational guidelines that protect both community interests and system reliability.

Florida's experience demonstrates that regulatory ambiguity regarding community ownership structures creates insurmountable barriers to developing projects. Legal frameworks must address the ownership structure uncertainties identified in Florida by providing explicit statutory recognition of community microgrids and community solar gardens as distinct categories of energy infrastructure with clear provisions regarding ownership structures, operational requirements, and the right to provide services to multiple customers without automatic classification as regulated utilities.[60] Drawing from New York's

[58] Olawuyi (n 2).
[59] ibid.
[60] Axon and Morrissey (n 3).

comprehensive community solar framework, effective legislation should establish clear legal pathways for diverse ownership models, including community-based organizations, co-operatives, and municipal entities, and provide operational guidelines that ensure system reliability and participant protection.[61] This legal clarity directly addresses the social pillar of sustainable DERs by enabling meaningful community participation, while supporting the economic pillar through clear investment frameworks and the environmental pillar through the deployment of renewable energy.

Effective implementation requires statutory definitions that distinguish community microgrids and community solar gardens from regulated utilities, providing clear legal pathways for development while avoiding burdensome utility regulations. Legal frameworks must establish explicit authorization for diverse ownership structures, including co-operatives, municipal entities, and community-based organizations, each with appropriate governance mechanisms that ensure democratic participation and transparent decision-making. Operational guidelines should balance community autonomy with system reliability requirements, establishing technical standards that protect grid stability while enabling community control over energy assets. Additionally, comprehensive participant protection provisions must ensure transparent governance structures and fair benefit distribution mechanisms that prevent exploitation of community members while promoting genuine energy citizenship.[62]

B. Standardized Interconnection Procedures for Community-Scale Projects

Community-centric sustainable DERs require specialized interconnection standards that address their unique technical characteristics while streamlining approval processes.

Florida's inadequate interconnection standards for community-scale projects create substantial technical and economic barriers that discourage community energy development in the state. Legal frameworks must address these challenges by establishing standardized interconnection procedures specifically designed for community microgrids and community solar gardens that address their distinctive characteristics, including islanding capabilities, multiple generation sources, and potential grid services.[63] Effective frameworks should establish clear technical standards while streamlining approval processes to reduce soft costs and development timelines, thereby making projects more viable for community groups with limited resources. This addresses the economic pillar of sustainable DERs by reducing development costs and supporting the social pillar through enhanced accessibility for diverse communities.

Effective legal frameworks must establish standardized technical requirements for community-scale interconnection that address unique operational characteristics while incorporating streamlined approval processes with clear timelines and transparent cost structures. Regulatory standards should address technical requirements for islanding capabilities and grid services that enable community systems to operate independently during

[61] NY Public Service Commission Order (n 48).
[62] KEH Jenkins and others, 'Towards Impactful Energy Justice Research: Transforming the Power of Academic Engagement' (2020) 67 Energy Research & Social Science 101510.
[63] Mattes (n 14).

emergencies while providing grid support functions. Jurisdictions should establish fast-track procedures for standardized community energy configurations, allowing projects that meet predetermined specifications to proceed through expedited approval processes without extensive custom reviews or utility-specific negotiations that create barriers for community organizations.

C. Value-Based Compensation Mechanisms

Sustainable economic viability for community-centric DERs requires compensation mechanisms that recognize the full range of values these systems provide to participants and the broader energy system.

Florida's narrow focus on basic energy compensation while failing to monetize resilience benefits, grid services, and environmental attributes creates fundamental economic obstacles to community energy development. Drawing from Minnesota's value-of-solar methodology, legal frameworks should establish mechanisms to fairly compensate community energy projects for the full range of services they provide, including enhanced resilience, grid support functions, and environmental benefits.[64] This comprehensive valuation approach advances all three pillars of sustainable DERs: properly monetizing environmental benefits (environmental pillar), ensuring fair economic returns (economic pillar), and creating incentives for enhanced resilience in vulnerable communities (social pillar).

Establishing value-stacking methodologies that recognize multiple benefit streams from community energy projects, including energy generation, capacity value, grid services, environmental benefits, and resilience enhancement, is more beneficial than limiting compensation to basic energy production alone. These frameworks should incorporate regular valuation updates that reflect changing grid conditions, evolving technology costs, and shifting market dynamics to ensure fair compensation over project lifespans. Enhanced compensation mechanisms should specifically address resilience benefits in climate-vulnerable areas like Florida, where community energy systems provide critical backup power during hurricanes and extreme weather events that increasingly threaten centralized infrastructure. Additionally, jurisdictions must establish transparent calculation methodologies with meaningful stakeholder input processes that enable community representatives to participate in valuation proceedings and ensure that compensation frameworks accurately reflect the full range of values that community energy projects provide to both participants and the broader energy system.

D. Authorization of Virtual Net Metering and Third-Party Ownership

Community solar development requires explicit legal authorization of virtual net metering and clarification that third-party-owned community projects do not constitute regulated utilities.

[64] Minn Stat § 216B.164 (n 50).

Florida's prohibition of virtual net metering and restrictive interpretation of third-party sales create insurmountable barriers to diverse community solar models in the state's energy landscape. Legal frameworks must explicitly authorize virtual net metering to enable credit allocation among multiple subscribers, while clarifying that third-party-owned community solar projects serving multiple subscribers do not constitute regulated utilities.[65] This legal clarity enables the innovative financing models necessary for broad community participation while ensuring that community solar gardens can serve their intended function of expanding access to renewable energy benefits regardless of individual property characteristics or financial capacity.

Successful community solar development requires explicit statutory authorization of virtual net metering that enables credit allocation among multiple subscribers at different locations, removing the geographic constraints that limit renewable energy access to property owners with suitable rooftops. Legal frameworks must provide clear exemptions for community energy projects from public utility regulation, establishing that third-party-owned community solar gardens serving multiple subscribers do not constitute regulated utilities subject to comprehensive oversight designed for large-scale utility operations. Comprehensive subscriber management requirements should protect participant interests through transparent governance structures, fair benefit distribution mechanisms, and accessible dispute resolution processes that prevent predatory practices while building trust in community energy models. Additionally, regulatory frameworks should accommodate flexible ownership models that address diverse community needs, including co-operative structures for grassroots organizations, municipal ownership for local governments, and hybrid models that combine community governance with professional management to ensure both democratic participation and operational effectiveness.

E. Specialized Financing Mechanisms and Institutional Support

Community-centric sustainable DERs require specialized financing institutions and mechanisms designed to address their unique characteristics and development challenges.

Florida's lack of specialized financing mechanisms for community energy projects creates substantial accessibility barriers, particularly for underserved communities in Florida. Drawing from successful green bank models in Connecticut and New York, effective frameworks should establish dedicated financial institutions focused on clean energy deployment that can provide specialized financing products, including pre-development loans, credit enhancements, and construction financing.[66] Additional mechanisms should include the expansion of property-assessed clean energy (PACE) financing to explicitly include community-owned energy projects, enabling long-term financing secured by property assessments that can make projects more financially viable for diverse communities.

Addressing financing barriers requires establishing dedicated clean energy financing institutions that provide specialized financing products for community energy development, including pre-development loans, credit enhancements, and construction financing that accommodates complex ownership structures. Property-assessed clean energy (PACE)

[65] Ottinger and Bowie (n 27).
[66] ibid.

program expansion should explicitly include community-owned projects, enabling long-term financing secured by property assessments that make projects more financially viable for diverse communities. Additionally, comprehensive technical assistance programs must provide project development expertise, regulatory navigation support, and capacity building services that enable community organizations to participate effectively in energy system transformation despite limited technical and financial resources.

F. Mandatory Equity and Inclusion Provisions

Community energy programs must include enforceable equity provisions to ensure benefits reach vulnerable and historically marginalized communities.

Florida's systematic exclusion of low-income communities and communities of colour from renewable energy benefits demonstrates the necessity of proactive equity. Drawing from Colorado's innovative approach, legal frameworks should include minimum participation requirements for low-income households, rental communities, and environmental justice areas, with enhanced incentives to ensure that these projects are financially viable despite potentially higher development costs.[67] Program designs should incorporate specific targets for reducing the energy burden in economically vulnerable communities while prioritizing community energy development in areas with historical environmental burdens or particular vulnerability to climate impacts.

Community energy programs must include minimum participation thresholds for disadvantaged communities to ensure that clean energy benefits reach historically marginalized populations rather than exacerbating existing inequities. Enhanced incentives should specifically target projects serving environmental justice areas, providing additional funding, streamlined permitting, or preferential treatment that makes community energy development financially viable in communities that face the greatest energy burdens but encounter the most barriers to participation. Energy burden reduction targets with robust monitoring and verification systems should establish measurable goals for decreasing energy costs in vulnerable communities while tracking progress through transparent reporting mechanisms. Additionally, accessible participation processes must be designed for diverse communities, incorporating multilingual outreach, flexible meeting schedules, culturally appropriate engagement strategies, and simplified administrative procedures that enable meaningful participation regardless of educational background, immigration status, or previous experience with energy system planning.

G. Integrated Stakeholder Engagement and Capacity Building

Successful energy democratization requires systematic capacity building and meaningful stakeholder engagement processes that enable communities to participate effectively in energy system governance.

[67] Colo Rev Stat § 40-2-127 (n 39).

Florida's utility-dominated regulatory processes demonstrate the need for innovative approaches to public participation that ensure diverse voices influence energy decisions. Legal frameworks should incorporate intervenor funding, community representation requirements, and accessible hearing formats, while providing dedicated technical assistance to community organizations seeking to develop energy projects.[68] Meaningful stakeholder engagement requires intervenor funding and community representation requirements in regulatory proceedings to ensure that community voices can effectively participate in complex administrative processes despite limited financial resources and technical expertise. Dedicated technical assistance programs for community energy development should provide project development support, regulatory navigation guidance, and feasibility assessment services that enable community organizations to advance energy initiatives without relying solely on expensive private consultants. Knowledge sharing networks and peer learning platforms must connect successful community energy projects with emerging initiatives, facilitating the transfer of lessons learned, best practices, and practical implementation strategies across diverse geographic and socioeconomic contexts. Leadership development programs focusing on energy literacy and project development skills should build local capacity for energy citizenship by training community members in energy system fundamentals, project finance, regulatory advocacy, and collaborative governance models that enable sustained community participation in energy system transformation.

VI. Conclusion

The legal framework lessons derived from Florida's experience demonstrate that the successful advancement of community-centric sustainable DERs requires integrated approaches that address technical, economic, and social barriers simultaneously while embedding energy citizenship principles throughout the regulatory design. Effective frameworks should also establish formal networks for knowledge sharing among community energy projects to accelerate learning and build collective expertise in the community energy sector. However, the transformation of energy systems towards greater democracy and sustainability cannot rely solely on regulatory reform; it demands active engagement from communities, businesses, and advocacy organizations working collaboratively to build political power for systemic change.

Community members and project planners bear crucial responsibilities for advancing energy democracy through grassroots organizing, knowledge sharing, and innovative financing mechanisms that complement regulatory advocacy efforts. Successful community energy development requires communities to build organizational capacity for energy literacy, project development, and political advocacy, moving beyond passive energy consumption to active participation in energy system governance.[69] This transformation demands sustained community organizing that connects energy issues to broader concerns

[68] Axon and Morrissey (n 3).

[69] L Ross and M Day, 'Community Energy Planning: Best Practices and Lessons Learned in NREL's Work with Communities' (National Renewable Energy Laboratory 2022) NREL/TP-6A50-82937.

about economic justice, environmental health, and climate resilience, creating powerful coalitions capable of challenging entrenched utility interests.[70]

Knowledge-sharing networks among community energy projects serve as essential infrastructure for scaling successful models across diverse geographic and socioeconomic contexts. Communities that have successfully navigated regulatory barriers and developed thriving energy projects can provide valuable mentorship and technical assistance to emerging initiatives, creating peer learning platforms that strengthen the broader energy democracy movement.[71] These networks enable communities to share lessons learned about effective governance structures, financing strategies, stakeholder engagement processes, and regulatory advocacy tactics that can be adapted to local conditions and needs.

Crowdfunding and community investment co-operatives represent emerging models that can democratize access to clean energy capital while building community wealth and local ownership. Community solar gardens, energy efficiency co-operatives, and share ownership models demonstrate the viability of democratic energy alternatives while creating political constituencies for broader policy change.[72] These demonstration projects provide concrete examples of energy democracy in action, making abstract policy concepts tangible for community members and policymakers while generating economic benefits that remain within local communities.

The business community also bears responsibility for supporting just energy transitions through innovative financing mechanisms, technical assistance partnerships, and advocacy for supportive policy frameworks. Rather than viewing community energy development as competition, businesses can partner with communities to develop shared ownership models, provide technical expertise, and advocate for regulatory reforms that benefit both community interests and sustainable business development. Social impact investing, patient capital providers, and mission-driven financial institutions can play crucial roles in addressing the financing gaps that currently limit community energy development, particularly in historically marginalized communities.

Significant research gaps remain in understanding how regulatory frameworks most effectively advance energy justice goals while promoting community ownership and democratic participation. Comparative studies of different state approaches to community energy regulation could provide valuable insights for jurisdictions seeking to develop supportive frameworks, particularly regarding the relationship between regulatory design and equity outcomes.[73] Future research should examine the long-term impacts of community energy projects on local economic development, social cohesion, and environmental justice outcomes, assessing whether community ownership models deliver on their promises of democratic participation and equitable benefit distribution.[74]

Additional research is needed on the intersection of climate vulnerability and community energy resilience, particularly in coastal states like Florida, where sea-level rise and extreme weather events threaten centralized energy infrastructure. Understanding how

[70] J Mijin Cha and M Pastor, 'Just Transition: Framing, Organising, and Power-Building for Decarbonization' (2022) 90 Energy Research & Social Science 102588.

[71] RE Shelton and H Eakin, 'Who's Fighting for Justice?: Advocacy in Energy Justice and Just Transition Scholarship' (2022) 17 Environmental Research Letters 063006.

[72] Ottinger and Bowie (n 27).

[73] S Carley and DM Konisky, 'The Justice and Equity Implications of the Clean Energy Transition' (2020) 5 Nature Energy 569.

[74] Jenkins (n 62).

community-scale energy systems can enhance climate adaptation while advancing social equity could inform both energy policy and climate resilience planning.[75] Longitudinal studies should track how community energy projects evolve over time, identifying design features that promote sustained community engagement and prevent elite capture of community energy benefits. The role of emerging technologies, including energy storage, smart grid systems, and distributed ledger technologies, in enabling community energy governance deserves further investigation. Research should examine how these technologies can enhance democratic participation in energy system management while ensuring that technological complexity does not create new barriers to community engagement or exacerbate digital divides that exclude vulnerable populations from energy decision-making processes.[76]

By implementing comprehensive legal framework reforms while building community capacity for advocacy and project development, jurisdictions can create supportive environments for community energy development that advance environmental sustainability, economic equity, and democratic participation in energy system governance. The transformation towards energy democracy requires both regulatory innovation and grassroots organizing, combining policy expertise with community power to challenge entrenched interests and build more just and sustainable energy systems for all.[77]

[75] X Wang and K Lo, 'Just Transition: A Conceptual Review' (2021) 82 Energy Research & Social Science 102291.

[76] BK Sovacool and others, 'Decarbonization and Its Discontents: A Critical Energy Justice Perspective on Four Low-Carbon Transitions' (2019) 155 Climatic Change 581.

[77] Olawuyi (n 2).

17

The Role of Energy Communities in a New Decentralized Energy Model

Legal Progress and Challenges in Colombia

*Milton Fernando Montoya and José Vicente Zapata**

I. Introduction: The Need for a Decentralized Solution in the Context of the Energy Transition

This chapter addresses the emergence of energy communities in Colombia as a transformative element in the shift towards a decentralized energy model. It evaluates the prospects and legal challenges that energy communities face, and two potential pathways for addressing the existing challenges: aligning incentives and the legal framework with the private sector and restoring investor's confidence.

One of Colombia President Gustavo Petro's main goals ever since his presidential campaign has been to radically change the energy matrix towards cleaner sources, but this has proven difficult as Colombia's economy heavily depends on the extraction of natural resources, with activities such as the exploitation of coal and minerals.[1] During his presidential campaign, Mr Petro vowed to end oil exploration and fracking projects,[2] but once in office, his approach became more moderate[3] as the transition had to be gradual to maintain economic stability, energy self-sufficiency, and avoid international investment disputes.[4] This was especially important given Colombia's post-pandemic context: rising inflation, high unemployment, and significant debt despite tax reforms.[5] Achieving Mr Petro's

* Authors appreciate the research support from Maria Paulina Rendón and Ernesto Guzmán.

[1] Presidency of the Republic of Colombia, 'President Gustavo Petro's Speech at the Economic Reactivation Forum: Confidence for Growth' (9 August 2024) https://www.presidencia.gov.co/prensa/Paginas/Intervencion-del-presidente-Gustavo-Petro-durante-el-Foro-de-Reactivacion-Econ-mica-Confianza-para-crecer-240809.aspx accessed 25 July 2025.

[2] Inter-American Dialogue, 'Expecting Major Changes: Petro and Colombia's Energy Policy' (The Dialogue, 7 July 2022) https://thedialogue.org/analysis/a-la-expectativa-de-grandes-cambios-petro-y-la-politica-energetica-de-colombia?lang=es accessed 28 July 2025.

[3] L Abu Shihab Vergara, 'Gustavo Petro's record on energy and environment analyzed' (Dialogue Earth, 18 July 2023) https://dialogue.earth/en/climate/390308-gustavo-petros-record-on-energy-and-environment-analysed/ accessed 28 July 2025.

[4] T Alcazar and L Victoria, 'Colombia on the Brink of Oil Decline Due to Petro's Policies' (Global Affairs, University of Navarra, 19 July 2023) https://en.unav.edu/web/global-affairs/colombia-al-borde-del-declive-petrolero-por-las-politicas-de-petro accessed 25 July 2025.

[5] J Bonet-Morón and others, 'Impacto económico regional del COVID-19 en Colombia: un análisis insumo-producto' (Banco de la República 2021) https://www.banrep.gov.co/es/impacto-economico-regional-del-covid-19-colombia-analisis-insumo-producto accessed 25 July 2025.

Milton Fernando Montoya and José Vicente Zapata, *The Role of Energy Communities in a New Decentralized Energy Model*
In: *Sustainable Distributed Energy Resources*. Edited by: Íñigo del Guayo, Damilola S. Olawuyi San, Louis de Fontenelle, Milton Fernando Montoya, and José Juan González, Oxford University Press. © International Bar Association 2026. DOI: 10.1093/9780198991946.003.0020

energy and climate goals is largely dependent on obtaining sufficient funding and securing investor's trust, which has been a challenge for the President. This has created a dilemma for the government, who has decided to launch programs with the goal of meeting supply targets in remote areas and empower communities to use natural resources before other structural issues are resolved.

To address these challenges and move towards sustainable energy, Colombia launched its Energy Plan 2050 in 2016, aiming to diversify energy resources and ensure reliable supply by incorporating wind, solar PV, and geothermal technology into the national mix.[6] In this context, the National Development Plan 2022–2026 (NDP), drawing on the Energy Plan, created the concept of energy communities (ECs).[7] They are defined in Article 235 of the NDP as 'organizational and/or associative forms that collectively seek the generation, commercialization, and/or efficient use of energy through the use of non-conventional renewable energy sources and distributed energy resources'.[8]

ECs function in organized structures where there are defined responsibilities and collective ownership of energy assets for the decentralized use of energy resources. A group of citizens—a community of neighbors, for example—organize and benefit from the shared management of the assets rather than relying solely on traditional utility companies, while promoting energy democracy through collective decision making.[9] In that way, energy communities have potential for socio-environmental transformation in remote places of the country, where there are structural problems that have hindered a reliable supply.

ECs empower local users to generate and consume renewable energy: for example, pilot projects in La Guajira and Medellín, have shown these initiatives can lower energy costs amidst the rise of prices of traditional energy. However, scaling up is challenging due to limited financing, technical complexity, and rising demand that current projects cannot fully meet.[10] For example, a recently documented experience in Palmor, a town in the Sierra Nevada de Santa Marta in Colombia, shows that 'Initial estimates for the plant's capacity were based on a census that underestimated the demand for electricity—when asked, Palmoreños expressed lower energy needs based on their previous energy realities, not their future consumptions.'[11]

While international experience shows these models can improve resilience and reduce emissions, developing countries still face significant regulatory, administrative, and infrastructure challenges, as we will explain later in this chapter. In Colombia,

[6] Mining and Energy Planning Unit, 'National Energy Plan 2020–2050: Consultation Document' (UPME 2016) https://www1.upme.gov.co/DemandaEnergetica/PEN_documento_para_consulta.pdf accessed 24 July 2025.

[7] National Planning Department, 'National Development Plan 2022–2026: Colombia, World Power of Life' (2023) p 176 https://colaboracion.dnp.gov.co/CDT/Prensa/Publicaciones/plan-nacional-de-desarrollo-2022-2026-colombia-potencia-mundial-de-la-vida.pdf [1] accessed 29 July 2025.

[8] Ministry of Mines and Energy, 'ABC of Energy Communities 2024' (2024) https://www.minenergia.gov.co/documents/11739/ABC-ComunidadesEnergeticas-2024.pdf accessed 24 July 2025.

[9] EIA (2023), 'Rethinking energy communities' EIA Working Paper https://www.eia.edu.co/wp-content/uploads/2023/10/1.EIA-WorkingPaper-Rethinking-Energy-Communities-2023.pdf accessed 24 July 2025.

[10] REScoop.eu and Ecopower, 'Financing Guide for Energy Communities: SCCALE 203050' (2023) https://managenergy.ec.europa.eu/document/download/04a6a3d1-5289-40f9 9ef4a8ad87823416_en?filename=SCCALE203050_financingguide_energycommunities.pdf accessed 29 July 2025.

[11] SV Zapata Alfonso and others, 'Associative Energy: Palmore's Community-Led Hydroelectric Success' in S Biswas and others (eds), *Let Communities Lead* (2024) www2.energia.coop/brasil/wp-content/uploads/2025/06/letcommunitieslead_2024.pdf accessed 19 October 2025.

energy communities are seen as a way to address energy poverty and unreliable supply, supported by recent regulations that promote collective self-generation and renewables. However, further adaptation and research are needed to ensure these models are viable and effective in the local context.[12] Particular business and technical skills are still necessary for the effective implementation of such projects in Colombia, as evidenced by documented experiences. The communities' capacity to overcome these challenges is rooted in the interplay between newly acquired competencies and the broader knowledge base of the community, which ranges from organizational processes traditionally related to coffee production and everyday collective practices. Fundamentally, documented experiences show that successful energy communities require the active participation of a community, not only in correctly assessing how much energy is needed but maintaining and managing the assets over long periods of time, which can be complex.[13]

From a practical point of view, EC projects face three main challenges that we would like to highlight: the first one has to do with decision making, as assessing accurately and planning accordingly for future electricity demand is difficult. Initial estimates can often underestimate growth, leading to insufficient capacity, blackouts, and the need for rationing; second, ensuring regular payment for electricity remains problematic, since some community members may resist paying for what they perceive as a public gratuitous service, resulting in financial strain and even conflict over enforcement. Financial strain can also come from difficulties in receiving subsidies: some EC that qualify for subsidies may not actually obtain this economic aid due to government inefficiencies: 'Electropalmor does not receive any government subsidies (despite having the legal right to receive them) and relies on bill payments for its day-to-day operations.'[14] Third, building effective relationships with external stakeholders, such as public institutions, is often hampered by bureaucratic delays and legal uncertainty, limiting access to financial support. In recent years, Colombia has progressed in implementing energy communities that are more cost-effective and reliable, reducing dependence on a carbon-based economy.

In this chapter, we will critically analyze the emergence of energy communities in Colombia given the context and legal challenges: while energy communities can offer significant potential for decarbonizing the economy and decreasing reliance on traditional utility companies, it is clear there are many obstacles before they can be become a widespread and investable model. Some of these obstacles are regulatory, administrative, financial, and technical barriers. We argue that the legal framework still needs tweaks in order to enable EC to unlock their full potential, and that targeted fiscal incentives and simpler administrative procedures are still necessary to ensure scalability and financial viability. This chapter looks at legal reforms, policy instruments and documented experiences and makes recommendations for overcoming legal and financial barriers.

[12] MF Medina-Reyes, JG Fajardo-Cuadro, and JC Martinez-Santos, 'Driving the development of energy communities in Colombia: challenges and opportunities for a decentralized energy transition' (2025) Habitat Substantiable V.15, N.1 15(1) Universidad Tecnológica del Bolivar https://doi.org/10.22320/07190700.2025.15.01.01 accessed 23 July 2025.

[13] Zapata Alfonso (n 11).

[14] ibid.

II. Energy Communities in Colombia and the National Development Plan

Energy communities can be defined as a legal entity composed of a group of people that generate and consume their own energy. They can be powerful tools in contributing to local development and solidarity between members of a community, while producing clean and sustainable energy in remote areas where there are often difficulties in having stable energy provided to them. ECs are sustainable energy participation from the community and through a system that enables decision making from the people that benefit from the energy being produced.[15] Energy communities usually take the form of nonprofit organizations, co-operatives, associations, or partnerships, allowing the members to make democratic decisions when operating energy assets.[16]

The main activities of ECs include generating renewable energy, providing energy efficiency services, producing, supplying, consuming, storing, and distributing clean energy, as well as offering electric mobility. Economic benefits are allocated to members' savings and the social development of the community. All members—citizens, companies, and local authorities—should actively participate and exercise control over the assets for proper control.

As for the assets, solar and wind energy are common options for residential self-consumption; small scale wind farms are also some of the most affordable options. Rural communities can replace traditional boilers with biomass boilers, using agricultural and forest waste while promoting forest cleaning, fire prevention, and rural development.

The Colombian Ministry of Mines and Energy (MME) has announced that it received over 18,000 energy communities (ECs) applications from all regions of Colombia.[17] After completing the technical process, the communities that meet the criteria established in MME Resolution 40137 of 2024, are in theory accepted for receiving subsidies. However, as we mentioned before there is a problem with governmental efficiency that can hinder projects. The criteria considered include energy poverty of the ECs region, multidimensional poverty, gender equality conditions, membership in PDET[18] or ZOMAC[19] territories. The fact that there are so many applications demonstrates there is a lot of interest in these associative structures as well as significant need and pressing demand for energy reliability across Colombia.

[15] D Olawuyi, 'From Energy Consumers to Energy Citizens: Legal Dimensions of Energy Citizenship' in K Huhta and others (eds), *Sustainable Energy Democracy and the Law* (Brill 2021) 101–23.

[16] European Commission, 'Energy Communities' https://energy.ec.europa.eu/topics/markets-and-consumers/energy-consumers-and-prosumers/energy-communities. accessed 9 August 2025.

[17] 'This program is part of the national strategy for Energy Communities. As of today, there are over 18,000 Energy Communities submitted, and more than 400 are already in the implementation phase.' Ministry of Mines and Energy, 'Government of Change Begins Delivery of Educational Energy Communities' (2024) https://www.minenergia.gov.co/es/sala-de-prensa/noticias-index/gobierno-del-cambio-inicia-con-la-entrega-de-comunidades-energ%C3%A9ticas-educativas/ accessed 29 July 2025.

[18] Programa de Desarrollo con Enfoque Territorial (Development Program with a Territorial Focus).

[19] Zonas Más Afectadas por el Conflicto Armado de Colombia (municipalities considered as the most affected by the armed conflict).

A. Legal and Regulatory Developments

1. Background

One of the defining principles of the Colombian constitution is being a 'social rule of law'. The Colombian Constitution provides in article 1 that one of the essential purposes of the government is serving the community, promoting general prosperity, and ensuring participation in decisions affecting citizens' lives, meaning democratic participation is a cornerstone the government should constantly strive for. 'In the legal system, evidence of this principle of Social Rule of Law as a duty to guarantee some minimum means to allow people exercise their rights can be found in the Constitution itself, which has an entire chapter devoted to social, economic, and cultural rights, and an article on the State intervention in the economy in order to guarantee the fair distribution of opportunities.'[20] Moreover, article 365 states that public services are governed by law and may be provided by the State, organized communities, or individuals. Both these dispositions serve as the constitutional basis for precisely organized communities such as energy communities.

On the other hand, Colombia also has a robust legal framework for a state obligation to advance the energy transition and decarbonize the economy. The country has entered into many international commitments that require progress towards these goals: in 2015, when Colombia joined the Paris Agreement, the government committed to reducing greenhouse gas (GHG) emissions and pursue the Sustainable Development Goals (SDGs), which include universal energy access (Goal 7) and climate action (Goal 13). In consequence, we updated in 2020 our Nationally Determined Contribution (NDC), pledging a 51 per cent reduction in GHG emissions by 2030 and launched numerous programs towards those goals.

Since then, technological advances have transformed the energy sector, and Colombia has been on a path towards sustainable sources such as solar, wind, hydrogen, biomass, and geothermal energy.

Law 2294 of 2023, as part of the National Development Plan 2022–2026, formally introduced the concept of Energy Communities in Article 235. When structuring the NDP, ECs were conceived as part of the energy transition: they were expected to provide efficient energy services while expanding renewables at a cost-effective level, and ECs were also envisioned with a focus on areas with high poverty, conflict, and weak institutions, hence why this law explicitly requires that the design and development of ECs consider the autonomous initiatives of small farmers and ethnic groups, tailored to each territory's needs.

Decree 2236 of 2023 further regulates Article 235, allowing individuals and legal entities to register as an EC and participate in the electricity value chain by using non-conventional renewables with the goal of 'generating, commercializing and using energy efficiently' (Article 5). Energy communities are framed as social initiatives that can be urban or rural where organizations of people manage energy assets, legally constituted as either natural or legal persons created by a private contract between the members of the community. This already poses the need for legal assistance in the redaction of the contract.

[20] CA Pérez-Garzón, 'Unveiling the Meaning of Social Justice in Colombia' (2018) 10(2) Mexican Law Review 27 https://www.scielo.org.mx/scielo.php?script=sci_arttext&pid=S1870-05782018000100027 accessed 10 August 2025.

This model should theoretically enable users to become producers, consumers, and even sellers of surplus energy, moving beyond passive consumption and contributing to solutions for sustainability.

2. Normative and regulatory developments since the NDP

To further enhance the role of Energy Communities, the National Development Plan outlines specific objectives and benefits. These communities are established as one of the new legal instruments suitable for harnessing local energy resources, allowing demand to play a more active role in the energy service supply chain, and providing access to electricity services for economically vulnerable populations. By empowering these populations through collective association schemes, Energy Communities become key legal instruments to ensure equitable access to electricity services and promote the efficient use of energy.

In this regard, and considering that with the adoption of Law 1715 of 2014, the integration of renewable energies into the electricity sector is declared of national interest, the role of self-generation is redefined from being exclusively for the consumption of a user to primarily allowing the possibility of delivering surplus energy to the system. The law also establishes small-scale self-generation, large-scale self-generation, and distributed generation as new activities in the service chain, with specific regulatory conditions to facilitate their integration into the system and provide new instruments that enable decision-making by the demand side. This means a more active demand in the development of an efficient and competitive market.

In the development of this mandate, the Ministry of Mines and Energy (MME) issued several decrees establishing: (i) the symmetry policy for Large-Scale Self-Generation (AGGE) and (ii) a differential treatment for Small-Scale Self-Generation (AGPE) and Distributed Generation (GD). This was interpreted and developed in the current regulatory framework issued by The Energy and Gas Regulation Commission (CREG), through Resolution 24 of 2015 and Resolution 174 of 2021. Additionally, the Mining and Energy Planning Unit (UPME) issued a resolution determining the capacity limit for Small-Scale Self-Generation (AGPE) through Resolution 281 of 2015.

On another hand, Resolution CREG 101 072, from early 2025 'By which the regulation for the integration of energy communities into the National Energy System is harmonized, and other provisions are enacted', supposedly marks a regulatory breakthrough for energy communities. However, CREG made the choice to change the pricing mechanism for the production of energy for energy communities, carrying substantial economic implications:[21] communities will receive payments based on system operational costs, not market valuations—a distinction that has historically resulted in lower compensation levels, compromising the viability of projects yet again.

[21] Energy and Gas Regulation Commission (CREG), 'Resolution 101 072 of 2025'. https://gestornormativo.creg.gov.co/gestor/entorno/docs/originales/Resoluci%C3%B3n_CREG_101_072_2025/Resoluci%C3%B3n_CREG_101_072_2025.pdf accessed 29 July 2025.

III. The Integration of Sustainability in Existing Policies in Colombia: An Evaluation of the Practical Advancement of Distributed Energy Resources Programs

The integration of sustainability into Colombia's energy policies has become a central objective in the country's transition towards a decentralized and low-carbon energy model. This chapter examines how distributed energy resources (DER)—including distributed generation, energy storage, and demand response—are being incorporated into the national regulatory and policy framework. It evaluates the extent to which existing policies and recent legal developments have facilitated or hindered the practical advancement of DER programs, with a particular focus on the technical, legal, and financial challenges that persist.

This section considers the effectiveness of current incentives, the adequacy of legalization procedures for energy communities, and the targeting criteria used to direct public resources. By assessing these dimensions, we provide a critical overview of Colombia's progress in embedding sustainability within its energy sector and to identify the main barriers and opportunities for the further development of distributed energy resources and energy communities in the country.

A. Technical Limitations of DER

On this point, we should note that the technical limitations of Distributed Generation (DG) and distributed energy resources (DER) are one of the major challenges for their implementation. In this regard, the main challenges include: (i) high initial costs, (ii) overloading of electrical grids, and (iii) lack of adequate infrastructure for integration into the energy system.

Given that the initial cost of installing the energy grid can be thousands of dollars, although they can generate savings on energy bills in the long term, the purchase and installation of solar panels and other assets, is undoubtedly unaffordable for many communities. Under this premise, government incentives, such as tax credits and subsidies, play an important role, as they are essential to facilitate the adoption of these types of technologies and thus make them more affordable.

Now, regarding the other challenge of electrical grid overload, it is more evident since the distribution infrastructure was built in the twentieth century and was not designed at the time of construction for a bidirectional flow of energy. Consequently, today in Colombia, conventional grids are mainly designed to transport electricity from large power plants to consumers, but not in the opposite direction, ie, to receive energy generated by distributed systems. This problem can undoubtedly cause grid congestion and increase the risk of blackouts in certain areas of the country. To mitigate this, it is necessary to improve coordination between regulators, grid operators, and consumers, as well as to implement smart grid technologies that support this bidirectional flow of energy.

In addition to the above, another obvious technical limitations of DERs is the lack of a centralized information system, which in itself makes it difficult to plan and monitor the impact of these new resources on the grid. Furthermore, it should be noted that the communication and supply infrastructure is insufficient, so what is undoubtedly needed is greater investment in advanced metering infrastructure (AMI) and also consideration of communication systems that allow for more efficient control of distributed energy generation.

We cannot ignore another significant barrier that is the low integration of energy storage systems, an aspect that has not been thoroughly regulated in Colombia, which of course prevents their use as backup in critical situations and limits the optimization of energy consumption. This is compounded by difficulties in Distributed Generation (DG), mainly due to a lack of clarity in grid connection processes, the failure to consolidate a standardized format, the high costs of backup contracts, and the lack of adequate remuneration mechanisms for distributed generators.

Regarding Demand Response (DR), the smart metering infrastructure (AMI) has not yet been widely adopted, which prevents efficient measurement and control of variability in energy demand. Additionally, the absence of clear technical procedures represents an obstacle, as there are no standardized protocols for the installation, operation, and maintenance of distributed generation and energy storage systems.

With all this in mind, it is clear that on one hand the lack of economic and financial incentives discourages the expansion of DERs in Colombia. On the other hand, the absence of clear remuneration schemes means that investment in these technologies is not entirely attractive, which limits their growth and adoption in the national electricity market.

B. Legalization Procedures

In this regard, we find that Colombia has made progress in formalizing and legalizing energy communities, which is undoubtedly an essential process for guaranteeing and implementing their operation and access to the various legal and regulatory benefits currently in force. This process is very simple, as it is carried out through the energy communities Register (RCE), a tool managed and created by the Ministry of Mines and Energy (MME) through its Electric Power Directorate. The regulatory framework and the steps necessary for the registration of these energy communities are described below.

In line with the above, the RCE was created by Decree 2236 of 2023. This resolution regulates and establishes the conditions under which this registry must operate, thus ensuring that energy communities comply with the necessary requirements for theïr legalization and formalization at a legal and technical level.

At this point, it should be clarified that the resolution applies to all energy communities established in accordance with Decree No 2236 of 2023. It also applies or extends to National Interconnected System (SIN) Network Operators and Distributors in Non-Interconnected Areas (ZNI). All this is so that Collective Self-Generators (AC) or Collective Distributed Generators (GDC) can request connection to the grid through the Network Operator, in such a way that it is essential that the corresponding energy community is previously registered as an initial energy community.[22]

[22] It is important to clarify that, while the terms Collective Self-Consumption (Autoconsumo colectivo) and Energy Communities (CE) are closely related in the Colombian context, the legal and organizational figure of the Energy Community, as established in the National Development Plan and related regulations, serves as the broader conceptual and operational framework. CEs are associative entities that can undertake various activities within the energy value chain, including generation, commercialization, and efficient energy use. Within this framework, Collective Self-Consumption represents one of the key modes of participation and operation,

In cases where projects are not connected to the grid, supporting documents and information must be provided to evidence the functioning of the energy projects, through a commissioning act completed by the person responsible for administering, operating, and maintaining the community's technological solution.

On the other hand, a classification of energy communities was introduced in the regulations, as these can be classified as initial energy communities or operational energy communities.[23] As for initial energy communities, they have a period of twelve months, extendable for an additional six months, to complete the information necessary for their start-up or operation. If they do not submit all the required documents within this period, the application for registration will be considered withdrawn or abandoned, which would render it unfeasible until such registration is obtained.

The RCE allows for the modification of registered information in case of technical, social, legal, or environmental changes affecting the nature or purpose of the energy community. It is the responsibility of the community to keep the information in the register updated.

It should be noted that registration in the RCE registry does not constitute automatic approval of the energy projects submitted, but it is an essential regulatory requirement for legal operation, which makes it a legal prerogative and therefore an essential aspect to be considered by developers of this type of project. Furthermore, because the MME assigns a Unique National Recognition and Identification Number (NURIN) to registered communities during the registration process, duly accredited communities can subsequently access legal and regulatory benefits.

The above seeks to make it clear that the process of registering and legalizing energy communities in Colombia is essential to guarantee the transparency, efficiency, and sustainability of all energy projects in the country. Therefore, simplifying and streamlining these procedures, without neglecting control, could encourage greater participation and development of sustainable energy projects in the country, which is the goal sought by the country with the energy transition and the fulfilment of decarbonization targets.

From the above, the regulation details the 'what' and 'how' of the standard legalization procedure, but fails to explicitly address contingencies or alternatives for communities that cannot complete these procedures online. It is important that future support mechanisms, alternative processes (face-to-face or assisted) are contemplated to solve this barrier, as it is not yet clearly resolved within the framework for Energy Communities in Colombia.

allowing groups of users to collectively generate and consume their own energy. Therefore, Collective Self-Consumption is a fundamental activity that can be developed and managed by an Energy Community, but the CE figure encompasses a wider spectrum of potential functions and local energy services.

[23] According to Resolution 40136 of 2024 of the MME, Initial Energy Communities are those that are in a preliminary phase. They have expressed their intention to be formed and are in the process of gathering and submitting all the necessary documentation and information to formalize their registration and start operations and have twelve months to provide the information of the start-up, which may be extended for up to half of the initial time only once. Once the term of reference has expired and if the required documents are not submitted in their entirety, the application for registration will automatically be considered abandoned. Energy Communities in Operation are those that have already complied with all the registration requirements and are actively generating and/or managing energy.

C. Lack of Tax Incentives

When it comes to the lack of tax incentives, we should point out that the financial structure that permeates Colombian energy communities highlights a fundamental misconception about how these initiatives work. Although many cite Law 1715 or Law 2099 as proof of support for tax incentives, the truth is that these regulations envisage a completely different actor, as they were designed at the time for the typical large-scale renewable energy developer. So, for example, in the case of rural communities operating on a solidarity economy, accelerated depreciation as a tax incentive means nothing when there is hardly any taxable income, which is why it is said that these regulations are not the basis for tax incentives for energy communities. Added to this is the fact that the non-reimbursable resources of the Non-Conventional Energy and Efficient Energy Management Fund (FENOGE) are not promising, with requirements that seem designed to exclude rather than facilitate their adoption.

Digitalization discourse adds another layer of difficulty. While smart meters and monitoring systems could theoretically optimize community energy systems, we must assume that communities from remote parts of the country have yet to overcome the more basic challenge of securing grid connection. In regions where communities still struggle with intermittent electricity, discussions of real-time demand response and advanced aggregation platforms do not resonate with the members of the community. There is a clear disconnect between promoting high-tech solutions when there is a general lack of basic resources in a community, costs are high and people lack certainty that the project will meet their demands, especially when the government have failed to deliver subsidies and skill development resources for the community.

This technological optimism is constrained by the reality of financial exclusion. Sadly, when banks evaluate credit proposals from energy communities, they see informal associations with no credit history, operating in high-risk regions that propose business models that do not fit standard parameters. Traditional lenders expect individual property rights as collateral, predictable cash flows and corporate governance structures, precisely what community initiatives cannot and should not provide. Thus, the absence of adequate financial instruments or facilitators for this type is urgent.

The situation is even more concerning when the National Development Plan (Law 2294 of 2023) left energy communities without concrete implementation alternatives. On the one hand, one might think that energy communities have been used as symbolic gestures of inclusivity, rather than being a serious and practical solution for many communities. Meanwhile, the outlook on this issue is that until financial institutions can recognize that energy communities are profitable, the electricity sector will likely continue to struggle between ambitious climate goals and exclusive financial realities.

In summary, to improve the financial viability of DG and EC projects in Colombia, the government will need to develop and expand fiscal incentives, financing mechanisms, and skill local and regional CEs development programs. The digitalization of the energy system will also play a key role in the optimization and sustainability of these projects.

All of the above must go hand in hand with an analysis of incentives for users in the ZNI. This is mentioned because there is a lack of awareness that these populations are often located in rural areas with difficult access. Consequently, the lack of access to the electricity grid implies a dependence on other types of energy sources that tend to be expensive, inefficient, and polluting, such as kerosene or small diesel generators. In this sense, it is correct to say that the difficult access to bank deposits in many of these areas makes the implementation of distributed renewable energy systems, such as individual solar photovoltaic systems or mini-grids, financially unattainable or impossible, almost a utopia, without significant and real external support to address the needs described.

The socioeconomic context of off-grid users is also a factor justifying the need for differentiated and specific government financial support. It is not merely about incentivizing the adoption of clean energy, but about ensuring equitable access to an essential basic service. Programs of direct subsidies, financing schemes with highly favorable terms (low or zero interest rates, extended periods), adapted to the payment capacity of these communities, or non-reimbursable seed funding, are crucial mechanisms for overcoming the initial economic barriers preventing these populations from benefiting from distributed generation technologies.

This focused financial support not only alleviates an immediate economic burden for households (by reducing expenditure on fossil fuels) but also generates a broader positive socioeconomic impact. The availability of reliable energy in these communities improves living conditions, facilitates access to education (lighting for studying), boosts local productive activities (small businesses, refrigeration), improves health (reduction of indoor smoke), and contributes to closing gaps in territorial inequality. Therefore, government financial support for off-grid users must be understood as a strategic investment in inclusive social and economic development, going beyond the general incentives for renewable energy investment.

Despite progress in promoting energy communities, significant obstacles or hurdles remain to ensure adequate financing for these initiatives, primarily due to the fact that existing credit mechanisms are not designed to address the characteristics of distributed generation. As demonstrated by the 'Financing Guide for Energy Communities', which provides examples of how energy community projects have been financed, it is clear that for Latin American projects to be successful, regulators must define appropriate financing schemes to ensure that citizens can afford and benefit from such initiatives without placing undue burdens on public funds, as we see in Colombia. While government-led procurement schemes can facilitate more affordable prices for system components and enable economies of scale, it is key to identify suitable compensation mechanisms for on-grid solutions—such as net-metering, feed-in tariffs, or other remuneration methods—and establishing appropriate subsidies for electricity and equipment can help ensure that projects benefit the intended groups. Furthermore, mechanisms to finance capital expenditures (CAPEX) should ideally involve a mix of public and private funding to safeguard the long-term sustainability and scalability of energy communities. Public expenditure on these programs should be assessed within the specific regional context, balancing the need for economic efficiency with social equity.[24]

According to the International Energy Agency, the operational guidelines should be clear and straight-forward. While many communities are eager to take on these initiatives (hence the 18,000 applications), there is still a need to address practical challenges such as lack of technical

[24] International Energy Agency, 'Bottom-up energy transitions: Managing the rise of energy communities in Latin America' (IEA, 25 September 2024) https://www.iea.org/commentaries/bottom-up-energy-transitions-managing-the-rise-of-energy-communities-in-latin-america accessed 29 July 2025.

expertise, risks of financial mismanagement, and the logistics of ongoing maintenance. Clearly defining roles and establishing a solid financial plan are key components of the long-term success of energy communities. E-tools and digitalization play a vital role in enhancing the efficiency and traceability of these projects. By leveraging digital technologies, communities can improve the monitoring and management of their energy systems, ensuring reliable operation and facilitating transparent reporting and accountability.[25] In light of the above, it should be noted that Colombia currently has a significant delay in the adoption and integration of these technological tools, which of course remains an area of improvement for the development of energy communities and in order to fully exploit their potential, as previously mentioned.

Finally, as previously mentioned, the perception of risk in relation to social and regulatory factors should not be overlooked, as these aspects deter potential investors. Addressing these barriers is essential to harnessing the full potential of energy communities in Colombia and, in doing so, ensuring their contribution to the energy transition.

D. Targeting Criteria

Regarding the targeting criteria, it should be noted that they are a public policy tool created to effectively allocate state resources to energy infrastructure projects, specifically in Colombia's energy communities. In other words, according to Resolution 40509 of 2024 of the Ministry of Energy and Mines (MME), these criteria seek to ensure that public resources are appropriately prioritized for energy communities with certain characteristics. These characteristics correspond to energy communities in situations of vulnerability, energy poverty, or located in territories with special needs. According to Article 11 of the aforementioned Resolution, these benefits will only be allocated to energy communities that apply through an open call from the Ministry of Energy and Mines (MME). All of this is viewed favorably since government-owned energy communities and those arising from government commitments are excluded. Similarly, the targeting criteria are considered in a segmented manner, since they are divided into two categories: (i) regional targeting criteria, and (ii) population targeting criteria.

1. Regional targeting criteria

As already mentioned, one of the targeting criteria for Energy Communities in Colombia is the regional category. This category is primarily aimed at communities located in territories with special needs. The Multidimensional Poverty Indicator (MPI) is used to determine these needs and prioritize the allocation of state resources.

Thus, communities in municipalities where the index exceeds 50 per cent will be prioritized. To do so, the most recent and updated records from DANE (National Administrative Department of Statistics) must be used to identify these territories or municipalities. Subsidiarily and complementarily, Multidimensional Energy Poverty will also be considered as a criterion, prioritizing communities with energy deficits or lack of services. This will be done through the Multidimensional Energy Poverty Indicator (MPI), created by the Ministry of Energy and Mines (MEM).

[25] ibid.

Prioritization will also extend to Peace Territories, which encompass communities located in the Zones Most Affected by the Armed Conflict (ZOMAC) or within the Territorial Development Programs (PDET), in accordance with Law 2272 of 2022. Similarly, consideration will be given to Territories of the National Agrarian Reform and Rural Development System; that is, those communities within territories recognized by Law 160 of 1994 or related rural development regulations. Along these lines, Agro-food Campesino Territories (TECAM), defined as those traditionally inhabited by campesino families and organizations and delimited according to Decree 780 of 2024, will also be a focus of this targeting.

Special attention will call Reincorporation Territories, which include communities in the Former Territorial Training and Reincorporation Spaces (AETCR) and the Special Areas of Collective Reincorporation (AERC). Likewise, legally recognized Collective Territories of Indigenous or Afro-descendant Peoples form part of this prioritization strategy.

Finally, the category of Special Areas covers communities located in areas of lower rural development, subnormal neighborhoods (informal settlements), and areas of difficult coordination and oversight, as established in Decree 1073 of 2015. These regional criteria seek to ensure that support reaches the communities facing the greatest challenges and vulnerabilities within the national territory.

IV. Legal Approaches to Address the Challenges Regarding Energy Communities in Colombia: Lessons to Consider

To overcome the challenges in the implementation of energy communities in Colombia, it is essential that legislation explicitly incorporates them within the existing regulatory framework. Currently, the lack of clear and specific regulation hinders their development. Although energy communities have been mentioned in Law 2294 of 2023 and Decree 2236 of 2023, the regulations remain nascent and lack clear operational definitions. Furthermore, a specific legal entity that delineates their organization and functioning has not been established, creating uncertainty in their application. There is also a lack of harmonization between existing norms on distributed generation, self-generation, and energy commercialization, which complicates their effective implementation.

The regulatory treatment of energy communities in Colombia reveals an ecosystem of disadvantage. One example of this is the sale of surplus energy: it is clear that while individual prosumers can sell excess energy generation, energy communities face barriers that make collective self-consumption economically unviable or inappropriate. This reflects an institutional inability, from a regulatory perspective, to conceptualize energy systems beyond the duality between large generators and individual consumers. Another example is the absence of differentiated tariffs, given that communities often provide services in areas where traditional utility companies have failed to provide reliable or efficient service.

On the other hand, in practice, communities have experienced circumstances in which they spend eighteen to twenty-four months processing procedures that cost more than the initial investment in infrastructure, an adverse outcome for initiatives aimed at expanding access to electricity. There are even close cases where a proponent submits the same documentation to three different agencies, each claiming to be unaware of the others' requirements. This, aside from being inefficient and generating setbacks, demonstrates that this is a system designed to prevent existing utility companies from easily integrating into community energy initiatives.

While Law 1715 and Law 2099 created tax benefits, the truth is that many banks sometimes require guarantees that energy communities cannot offer or guarantee. When looking at development funds such as FENOGE, it is evident that criteria are applied to favor conventional projects. Even when energy communities are able to demonstrate their technical viability and social impact.

For its part, the 2022-2026 National Development Plan gives energy communities a special mention; it does not focus on establishing paths for actual implementation. Given this, it should be considered that the Ministry of Energy, the UPME, and the CREG are responsible for promoting the development of energy communities within their functions and responsibilities.

Given this scenario, there are several challenges and aspects to address. Regulation must be geared towards taking into account where and in what environment energy communities operate. In this sense, not only more realistic regulations are required, but also structural frameworks. Otherwise, these initiatives will remain trapped between the transformative potential and the systemic limitations described.

Considering the above, first, it is crucial to include affected neighbors in the financial participation and benefits of energy projects. This not only fosters social acceptance but also ensures that the benefits of the projects are equitably distributed among local communities. Energy communities can be beneficiaries of public resources for the financing of investment, operation, and maintenance of infrastructure, implying active community participation in the benefits.

Second is the need to address environmental impacts. To mitigate environmental impacts, better operational practices must be promoted. This includes implementing more efficient technologies and projects to reduce greenhouse gases. Additionally, it is essential to train communities in responsible use of hazardous materials and redesign operations to comply with environmental regulations. Energy efficiency and the use of non-conventional renewable energy sources (FNCER) are fundamental to reducing greenhouse gas emissions. The financing and sustainability of energy communities in Colombia depend on the existence of clear legal mechanisms that enable access to public and private resources, as well as specific tax incentives for renewable energy projects.

Regarding communication strategy, from a legal perspective, transparency and community participation are principles recognized in the Constitution and sectoral regulation (Articles 2 and 365 of the Constitution; Law 2294 of 2023). The regulations require energy communities to be established as legal entities, which ensures democratic participation and accountability. Regulation must guarantee that information and prior consultation processes are effective, especially in territories with ethnic communities or groups under special constitutional protection, in accordance with standards for prior consultation and citizen participation. With respect to performance indicators, regulation should establish clear criteria for measuring the efficiency, service quality, and environmental impact of energy communities. Resolution CREG 101 072 of 2025 introduces technical and commercial rules for integrating these communities into the electricity system, but further harmonization is needed to incorporate indicators of sustainability, economic efficiency, and emissions reduction, aligned with Colombia's international commitments (Paris Agreement, NDC).

Within the existing risk landscape for energy communities in Colombia, there is a lack of alignment between regulation and what is practical or truly necessary. Although Decree 2236 of 2023 established the Energy Communities Registry (RCE), this registry does not fully fulfill its purpose, as the promise of 'simplified procedures' remains unclear, as communities are faced with the reality of spending months on grid connection approvals, while companies with greater generation capacity take less time and have greater priority. Thus, it is evident that rather than truly mitigating risks, regulations expand obligations, forcing communities to operate under new parameters that are often unnecessary. The RCE, which is conceived as a solution tool, turns out to operate more as an administrative obstacle in an already complex scenario.

In terms of supply-demand balance and GHG mitigation, regulation should promote the adoption of efficient technologies and compliance with environmental standards. Law 1715 of 2014 and CREG regulations establish obligations for integrating non-conventional sources and reducing emissions, but it is necessary to strengthen monitoring and reporting mechanisms, as well as regulatory incentives for demand management and energy efficiency. GHG monitoring in energy communities should be governed by recognized international and national protocols, such as the GHG Protocol and ISO 14064, but Colombian regulation must advance in requiring periodic and verifiable reports, integrating these requirements into the registration and operation procedures for energy communities.

The results of the response to demand in Colombia reveal a regulatory gap. While it is true that the CREG has made progress in regulating standards for self-generation and distributed generation, it has not fully advanced specific mechanisms for the participation of energy communities. When reference is made to specific participation mechanisms, this means that there is a lack of clear criteria from participation to remuneration, which leaves energy communities in regulatory limbo. It should not be overlooked that the integration of storage is vital to demand response.

In this type of situation, Colombia's regulatory approach to energy communities lacks specificity in practice. The regulations list what should be defined, such as settlement methodologies, facility sharing protocols, or quality criteria, but they do not actually define them, and when they do, they often do so in a way that does not correspond to reality or the current situation.

To overcome them, it is necessary to establish a clearer and more flexible regulatory framework, with simplified administrative processes, financial incentives, and fair commercialization mechanisms. Additionally, greater integration with the energy transition strategy is essential so that these initiatives can contribute to the decentralization of the energy system and equitable access to energy in the country.

In summary, the development of energy communities in Colombia still requires a more coherent legal framework that addresses the reality we have explained. For example, it should be a priority to explicitly recognize energy communities as actors in the provision of public utility services, in accordance with the expected amendment to Law 142 so that they can access many more legal and economic benefits, and to strengthen contractual instruments and incentives for their viability.

V. Conclusions: Energy Communities to Achieve a Sustainable Decentralization

The development of energy communities in Colombia represents a promising yet challenging pathway towards achieving energy self-sufficiency and advancing the country's energy transition in a decentralized technical manner. Colombia's energy matrix remains heavily reliant on hydro (65 per cent), coal and natural gas (30 per cent), solar (3.7 per cent), and other minor sources (wind and biomass, 1 per cent).[26] Despite the country's significant capacity for renewable energy generation, including hydropower, wind, and solar, much of this potential remains underexploited. Political complexities and the need to maintain international investment further hinder efforts to address energy challenges, underscoring the importance of long-term, sustainable solutions.

Colombia has currently set itself some ambitious climate targets. It should be remembered that for some years now, the state has been aiming to reduce greenhouse gas emissions by 51 per cent by 2030 and achieve complete carbon neutrality by 2050, but the path to achieving these targets remains quite uncertain, and it is likely that they will not be met.[27] However, it is worth noting that energy communities have emerged as a key and fundamental part of the goals that have been set, although their role goes beyond simply reducing emissions if they are to contribute to the composition of Colombia's electricity generation matrix. Recent government measures, in particular the 2050 Energy Plan and Decree 2236 of 2023, suggest that their potential in this regard is recognized, but political statements and implementation show a different reality.

Taking the above into account, it can be said that the reality is that Colombia's energy community initiatives face a paradox: while policymakers focus on their transformative potential, actual implementation faces predictable challenges that have yet to be fully resolved. Additionally, it should be noted that there are few successful models to serve as a guide for new projects, and the regulatory framework remains incomplete despite efforts to update regulations. Above all, the technical challenge of grid integration persists. Perhaps the most important factor will always be co-operation and co-ordination between the government, private investors, and communities.

Looking forward, energy communities offer Colombia a compelling vision: locally controlled energy systems that could enhance both resilience and equity. But we should resist the temptation to view them as a panacea. Their success hinges not on grand pronouncements about Colombia becoming a 'leader in the global transition', but on mundane yet crucial factors—efficient delivery of subsidies, availability of financing mechanisms, and genuine community capacity building. The question is not whether energy communities can transform Colombia's energy landscape, but whether the country's institutions are prepared to transform themselves to make such change possible.

[26] Asociación Colombiana de Generadores de Energía Eléctrica ('Acolgen'), 'Installed Capacity in Colombia' https://acolgen.org.co/ accessed 2 August 2025.

[27] 'International relations will be oriented towards clean productive transformation, embracing the goals of the Paris Agreement. Under this accord, the country committed to reducing greenhouse gas (GHG) emissions by 51% by 2030, and to achieving carbon neutrality by 2050. These targets are linked to Colombia's Long-Term Climate Strategy E2050 and the Climate Action Law 2169 of 2021.' National Planning Department, National Development Plan 2022–2026: Colombia, World Power of Life (2023) p 176 https://colaboracion.dnp.gov.co/CDT/Prensa/Publicaciones/plan-nacional-de-desarrollo-2022-2026-colombia-potencia-mundial-de-la-vida.pdf accessed 29 July 2025.

18

Gender Justice and Decentralized Energy Resources

Perspectives from the United Kingdom and Nigeria

Pedi Obani and Adenike Akinsemolu

I. Introduction

Energy generation and supply are moving away from the traditional centralized models to more localized production and distribution systems. As discussed in Chapter 1 of this book, this paradigm shift provides significant benefits, including the increased adoption of renewable sources of energy such as solar, hydro, biomass, and wind energy, the promotion of suitable and appropriate energy storage technologies that support energy supply in peak demand and in the absence of other power sources, an increased energy supply, and reduced losses during energy transmission. However, energy decentralization presents some legal issues, particularly in addressing the inequities in the production and consumption of energy under the new energy systems.[1] Using two case studies, Nigeria and the United Kingdom, this chapter analyses the challenges and prospects of decentralized energy generation through a gender equality lens. Despite their divergent socioeconomic contexts, there are shared challenges in both nations' energy transition strategies, including the uneven distribution of resources, deteriorating infrastructure, and an inertia in energy systems that could be attributed to resource dependency and historical trajectories.

Overall, the crucial importance of gender justice in relation to decentralized energy resources is increasingly recognized more in Nigerian legal and policy frameworks and to a lesser but growing extent, the United Kingdom. However, significant gaps remain in practice, especially in Nigeria, where the issue is more urgent due to energy poverty and gender disparities in access, participation, and policymaking. Even in the United Kingdom where households typically enjoy better levels of energy access, women continue to be underrepresented in STEM disciplines and leadership positions that influence the green transition, and this can be traced back to the underrepresentation of girls in STEM education particularly from diverse communities.[2]

[1] N Soininen and K Huhta, 'Law for Decentralized Electricity Systems: Lessons from Complexity Theory' in R Fleming, K Huhta, and L Reins (eds), *Sustainable Energy Democracy and the Law* (Brill 2021) https://doi.org/10.1163/9789004465442_005; MA Heldeweg and S Saintier, 'Renewable Energy Communities as "Socio-legal Institutions": A Normative Frame for Energy Decentralization?' (2020) 119 Renewable and Sustainable Energy Reviews 109518 https://doi.org/10.1016/j.rser.2019.109518 accessed 11 September 2025.

[2] G La Malfa and M Jorgensen, 'Women in STEM: Bridging the Gender Gap' (Women in STEMM Season, King's College London, 2 July 2024) https://www.kcl.ac.uk/women-in-stem-bridging-the-gender-gap accessed 6 August 2025; L Hodge and others, 'Progression at Age 16 of Young People from Underrepresented Backgrounds

Pedi Obani and Adenike Akinsemolu, *Gender Justice and Decentralized Energy Resources:* In: *Sustainable Distributed Energy Resources*. Edited by: Íñigo del Guayo, Damilola S. Olawuyi San, Louis de Fontenelle, Milton Fernando Montoya, and José Juan González, Oxford University Press. © International Bar Association 2026. DOI: 10.1093/9780198991946.003.0021

This chapter examines the effectiveness of the energy laws and policies for promoting decentralized energy in Nigeria and the United Kingdom through a feminist perspective. For this purpose, the intersections of energy justice and feminist political ecology provide the conceptual foundations for this analysis. On the one hand, energy justice, as defined by Sovacool and Dworkin (2015) entails procedural justice (equitable participation in decision-making), distributive justice (fair allocation of energy benefits and burdens), and recognition justice.[3] On the other hand, feminist political ecology theory admonishes that the prevailing patriarchal societal structures influence resource access and environmental governance, disproportionately marginalizing women.[4] In Nigeria, for instance, land ownership, which is a quintessential prerequisite to the implementation of DER projects, as well as access to political power and influence, is skewed in favor of men, with women owning only 8 per cent of the crucial resource across 120 communities in the country.[5] This exclusion of women from owning or accessing resources leads to their systemic exclusion from participating in decisions related to environmental management.[6] These decisions include domestic energy use decisions and participation in DER projects in the community. Further, cultural factors limit women's access to education.[7] This creates a ripple effect that limits their ability to participate in decision-making with regard to DERs at the household, community, and governance or policy formulation levels by preventing them from taking up STEM courses, working in the energy sector, keeping up with trends in DER and related technologies, contributing to policies related to DER, establishing business in the energy sector, and accessing DER financing. This is evident in the limited participation of women in the energy sector, which has predominantly employed men, with emerging data indicating low representation of women in the renewable energy sector, too. In 2023, the International Renewable Energy Agency (IRENA) and the International Labour Organization (ILO) reported that the sector employed 13.7 million people worldwide with only 32 per cent being women.[8] In the United Kingdom, women represented 21 per cent of the offshore wind workforce and overall remain underrepresented in relevant training routes.[9] Within Nigeria and the global south context generally, women are also disproportionately impacted by energy insecurity though there is a dearth of data in most

Towards Careers in STEM' (EPI, CfEY and Mission 44 2024) https://epi.org.uk/wp-content/uploads/2024/11/EPI-CfEY_STEM-underrepresented-backgrounds_publication-version.pdf accessed 6 August 2025.

[3] BK Sovacool and MH Dworkin, 'Energy Justice: Conceptual Insights and Practical Applications' (2015) 142 Applied Energy 435 https://doi.org/10.1016/j.apenergy.2015.01.002 accessed 11 September 2025.

[4] S Vigil, 'Towards a Feminist Political Ecology of Migration in a Changing Climate' (2024) 155 Geoforum 104076 https://doi.org/10.1016/j.geoforum.2024.104076 accessed 11 September 2025.

[5] R Nyirongo, 'African Women Bear the Brunt of Climate Change Tracking the Impact of the Crisis in Nigeria and Mozambique' (2023) 91 New Agenda: South African Journal of Social and Economic Policy 58.

[6] M Raimi and others, 'Women Role in Environmental Conservation and Development in Nigeria' (2019) 1(2) Ecology & Conservation Science 0042. DOI: 10.19080/ECOA.2019.01.555558.

[7] R Onoshakpor and C Onoshakpor, 'Gender Inequality in the Nigerian Power Sector: The Case of Nigerian Electricity Companies' (2023) 12 American Journal of Engineering Research 21.

[8] International Renewable Energy Agency (IRENA) and International Labour Organization (ILO), 'Renewable Energy and Jobs: Annual Review 2023' (IRENA and ILO 2023) <https://www.irena.org/-/media/Files/IRENA/Agency/Publication/2023/Sep/IRENA_Renewable_energy_and_jobs_2023.pdf> accessed 29 March 2025.

[9] Department for Education, 'Skills England: Sector Skills Needs Assessments Clean Energy Industries' (DfE 2025) https://assets.publishing.service.gov.uk/media/683d6938f17469e343ebb98e/Sector_skills_needs_assessments_Clean_Energy_Industries.pdf accessed 6 August 2025

cases.[10] The foregoing underscores the imperative of gender justice to strengthen the social equity dimension of decentralized energy resources.

This chapter is divided into five sections. The introductory section establishes an overall link between energy decentralization and gender justice, laying the foundation for the exploration of distributed energy resources through a feminist lens. Following this introduction, the second section offers insights into the context and rationale for distributed energy resources. The third section explores the current state of energy justice in Nigeria and the United Kingdom to identify best practices and gaps in efforts to maintain efficiency in energy production and consumption while upholding gender justice. The section features an analysis of the existing laws/policies in the case study countries, through a feminist lens. The fourth section explores the gender dimensions and gaps, identifying barriers in the legal framework for DER legal and policy frameworks in both countries that impacts gender justice. The fifth section proffers recommendations for legal reforms to address the critical challenges of gender equity in the deployment of sustainable DERs in Nigeria and the United Kingdom. It is followed by a concluding section that features key highlights from the chapter and a call for action for both Nigeria and the United Kingdom to pursue further reforms to mainstream gender justice in their DER laws and policies.

II. The Justice Dimension of Distributed Energy Resources

Since the industrial revolution, natural resources extraction and the widespread use of fossil fuels have induced the current climate crisis.[11] The patterns of adverse climate change impacts also show disproportionate impacts on the poor, marginalized, and vulnerable populations. Hence, the current energy transition is informed by the need to move from carbon-intensive energy systems to low-carbon energy systems to reduce greenhouse gas emissions and address climate change and the related injustices. Rooted in the principles of sustainability and energy justice, the global shift towards energy decentralization, including the relevant legal frameworks, are underpinned by the urgent need to mitigate climate change, enhance energy security, and ensure equitable access to energy resources.[12] Decentralized renewable energy systems play an increasingly prominent role in the transition with prospects for diversifying countries' energy mix and promoting equitable energy security. To start with, decentralization has implications for the ownership structures of energy systems, establishing new hierarchies in operation, ownership, and distribution of energy infrastructure. From a feminist perspective, the establishment of these hierarchies necessitates the development of a fair, effective, and functional decentralized governance system, presenting the legal challenge of establishing a governance system that

[10] P Obani and A Akinsemolu, 'Editorial Comments: Special Issue on Gender Justice and Energy Transition in the Global South' (2025) 16(2) Journal of Sustainable Development Law and Policy xi.

[11] F Ahmed and others, 'The Environmental Impact of Industrialization and Foreign Direct Investment: Empirical Evidence from Asia-Pacific Region' (2022) 29(20) Environ Sci Pollut Res Int 29778. DOI: 10.1007/s11356-021-17560-w; F Hodu and others, 'Natural Resource Use, Industrialization and Climate Change in Africa: Blueprints for Sustainable Regional Development' (2024) 9 Research in Globalization 100245 https://doi.org/10.1016/j.resglo.2024.100245 accessed 11 September 2025; E Evans Osei Opoku and M Kofi Boachie, 'The Environmental Impact of Industrialization and Foreign Direct Investment' (2020) 137 Energy Policy 111178.

[12] Chapter 2 of this book.

upholds gender and energy justice while maintaining efficiency in energy production and consumption.[13]

For both Nigeria and the United Kingdom, there are critical legal issues requiring a direct focus in energy laws and governance systems, to ensure that women and other marginalized groups can access decentralized energy resources to address energy poverty and other environmental and equity concerns that are exacerbated by the prevalent centralized energy systems. Nigeria is one of the most vulnerable countries in terms of energy security. Historically, the country has an established centralized energy system, based on fossil fuels, which has also exacerbated gender inequality, environmental degradation, and other significant development challenges.[14] There have been several legislative reforms that are aimed at addressing the failings of the energy sector through decentralization and renewable energy. Similarly, despite its high ranking on the energy security index, the United Kingdom faces the challenge of effective deployment of decentralized energy.[15] Hence there is a need for critical analysis of the legal and policy frameworks for energy decentralization, particularly through a feminist lens.

III. Legislative and Policy Frameworks for Distributed Energy Resources

Decentralization could transfer authority from centralized utilities to local communities.[16] Burke and Stephens characterize this concept as 'energy democracy', wherein women and other marginalized groups possess greater influence over the use and benefits derived from energy infrastructures.[17] Legislative systems, inadvertently, often reflect power asymmetries and the legacy of centralized energy systems and disproportionate gender impacts continue to plague the energy transition movement. Against this backdrop, this section of the paper interrogates the gender justice implications of the legislative and policy frameworks for distributed energy resources in Nigeria and the United Kingdom.

A. United Kingdom Context

The legal and policy framework for decentralized energy resources in the United Kingdom is fragmented and includes a complex suite of national reforms and strategic initiatives, including financial incentives, rather than a single law. Against the backdrop of domestic

[13] P Obani, *Gender and Africa's Low-carbon Transition* (Green Finance Platform 10 June 2020) https://www.greenfinanceplatform.org/blog/gender-and-africa%E2%80%99s-low-carbon-transition accessed 12 June 2025.

[14] E Ekhator and P Obani, 'Women and Environmental Justice Issues in Nigeria: An Evaluation' in JJ Dawuni (ed), *Intersectionality and Women's Access to Justice in Africa* (Rowman and Littlefield 2024).

[15] B Woodman and P Baker, 'Regulatory Frameworks for Decentralised Energy' (2020) 36(12) Energy Policy 4527 https://doi.org/10.1016/j.enpol.2008.09.017 accessed 2 November 2025; MC Brisbois, 'Decentralised Energy, Decentralised Accountability? Lessons on How to Govern Decentralised Electricity Transitions from Multi-level Natural Resource Governance' (2020) 2 Global Transitions 16 https://doi.org/10.1016/j.glt.2020.01.001 accessed 11 September 2025.

[16] K Kaze, N Balta-Ozkan, and E Shrimpton, 'Connecting Power to People: Integrating Community Renewable Energy and Multi-level Governance Towards Low-carbon Energy Transition in Nigeria' (2025) 121 Energy Research & Social Science 103938 https://doi.org/10.1016/j.erss.2025.103938 accessed 11 September 2025.

[17] MJ Burke and JC Stephens, 'Political Power and Renewable Energy Futures: A Critical Review' (2018) 35 Energy Research & Social Science 78 https://doi.org/10.1016/j.erss.2017.10.018 accessed 11 September 2025.

and global energy security concerns vis-à-vis the Sustainable Development Goal target for universal access to clean and affordable energy by 2030 and the UK government's legally binding 2050 Net Zero target, the overriding focus of domestic energy policy has been on ensuring affordable access while meeting emission reduction targets, though there has been no direct focus on the gender dimensions. The Climate Change Act 2008 aims to reduce greenhouse gas emissions by 100 per cent by 2050 and mandates carbon budgets, which set a cap on the total amount of emissions for a five-year period. The Climate Change Act has been instrumental in driving policy and regulatory action to integrate decentralized renewables towards achieving net zero by 2050. However, despite its ambitious climate and emission aims, the Climate Change Act is gender blind. Similarly, the Energy Acts of 2013 and 2023 while gender blind promote energy production and security and the regulation of the energy market, with provisions such as Contracts for Difference (CfDs), Capacity Market, and energy market oversight, facilitating investment in renewables and smaller distributed generation.

Decentralized energy systems offer opportunities for accelerating emissions reductions and decarbonization and enhancing energy security. Beyond the environmental value, decentralized energy systems are seen to offer significant economic savings.[18] While decentralized energy systems could reduce the demand on the central energy generation and transmission system, current United Kingdom energy policy does not envisage decentralized systems replacing new large-scale centralized electricity generating facilities with the latter enabling advantages from the pooling of energy generation and demand.[19] The Electricity Act 1989 enables electricity licensing, generation, supply, and distribution, including frameworks for connecting decentralized resources to the grid and requirements for Ofgem regulation. It is projected that by 2030 decentralized generation would account for up to 30 per cent of total electricity generation capacity in the UK.[20] However, progress has been impeded by policy reversals and incoherent communications.[21] In the case of *R (Friends of the Earth) v Secretary of State for BEIS* (2022), the High Court deemed the UK's net-zero strategy unlawfully vague, emphasizing the need for inclusive policy design.[22]

With decentralization comes the risk of effective co-ordination of energy systems to achieve national objectives across all levels of governance, from the national to the regional and local levels. Nationally, as part of the Review of Electricity Market Arrangements Programme (REMA), the policy proposal includes market arrangements under Reformed National Pricing to incentivize the efficient siting of new energy infrastructure, which would strengthen decentralized generation systems across the country.[23] Also, from 2026

[18] M Jeremiah Barasa Kabeyi and O Akanni Olanrewaju, 'Economic Evaluation of Decentralised Energy Sources for Power Generation' (2025) 60 Energy Strategy Reviews 101774 https://doi.org/10.1016/j.esr.2025.101774 accessed 11 September 2025.

[19] Department for Energy Security & Net Zero (DENZ), 'Closed Consultation Draft: Overarching National Policy Statement for Energy (EN-1) Accessible Webpage' updated 19 May 2025 https://www.gov.uk/government/consultations/planning-for-new-energy-infrastructure-2025-revisions-to-national-policy-statements/draft-overarching-national-policy-statement-for-energy-en-1-accessible-webpage accessed 27 June 2025.

[20] National Energy System Operator, 'Future Energy Scenarios 2025: Pathways to Net Zero' (Vol 3, July 2025) <https://www.neso.energy/document/364541/download> accessed 22 July 2025.

[21] Climate Change Committee, '2024 Progress Report to Parliament' (CCC 2024) < https://www.theccc.org.uk/publication/progress-in-reducing-emissions-2024-report-to-parliament/> accessed 10 August 2025.

[22] [2022] EWHC 1841 (Admin).

[23] Department for Energy Security and Net Zero (DENZ), 'Policy paper—Review of electricity market arrangements (REMA): Summer update 2025' (10 July 2025) https://www.gov.uk/government/publications/review-of-electricity-market-arrangements-rema-summer-update-2025/review-of-electricity-market-arrangements-rema-summer-update-2025-accessible-webpage accessed 7 August 2025.

new and refurbished energy plants would need to be 'decarbonisation ready' which would support modular, decentralized energy options.[24]

Regionally, the recently introduced Regional Energy Strategic Plan (RESP) mandates strategic regional planning for energy distribution, co-ordinated by the independent National Energy System Operator (NESO), ensuring that regional stakeholders collectively set out how the energy system supports both local priorities and national decarbonization goals.[25] For this purpose, RESP mandates NESO to establish three pillars of governance, including: (a) structured and accessible engagement processes; (b) regional governance through Strategic Boards; and (c) national governance through a National Steering Committee.

Locally, Local Area Energy Planning (LAEP) is a key process for effective local action to contribute to the national 2050 net zero greenhouse gas emissions target. Despite the lack of formalization of the role of local authorities and the multi-level governance structure for net zero plans, LAEP is increasingly being adopted across the country.[26] Furthermore, feed-in tariffs and community energy projects have been fundamental in promoting local (and households) investments in decentralized energy resources in the United Kingdom.

1. United Kingdom feed-in tariffs scheme and small export guarantee

In 2010, the United Kingdom government introduced a Feed-in Tariffs (FIT) scheme to promote electricity generation from renewable and low-carbon sources. Under the FIT scheme, the owners of accredited installations generating electricity from solar photovoltaic (solar PV), wind, micro combined heat and power (Micro CHP), hydro, or anaerobic digestion (AD) would receive payments for the electricity generated from the accredited installations and exported. The payments were administered by a FIT licensee, a licensed electricity supplier, while Ofgem provided regulation. The FIT scheme was popular, though there were several legal challenges particularly regarding the government's decision to reduce the incentives for solar panels and the closure of the scheme to new applications.[27] FIT generators receive support for between ten and twenty-five years depending on technology type, capacity, when their installation was commissioned, and whether it was previously accredited under the Renewables Obligation scheme. The costs of the FIT scheme are distributed across all licensed electricity suppliers in Great Britain. The process, known as 'levelisation', is based on the market share figures for each electricity supplier.

Small Export Guarantee (SEG) for renewable generators began on 1 January 2020, providing guaranteed payments and market routes for small-scale and community generators exporting excess electricity to the grid. SEG which replaced FIT, is available for low-carbon technologies up to a capacity of 5 MW.[28] It mandates electricity suppliers with more than 150,000 domestic customers to offer tariffs for the surplus power that renewable energy

[24] DENZ, Closed Consultation Draft (n 19).

[25] Ofgem, 'Decision on the Regional Energy Strategic Plan Policy Framework' (April 2025) https://www.ofgem.gov.uk/sites/default/files/2025-04/RESP-policy-framework-decision.pdf accessed 17 June 2025.

[26] A Collins and A Walker, *Local Area Energy Planning: Achieving Net Zero Locally—Research Briefing* (UK Parliament POST 25 July 2023).

[27] P Bennett, 'Breaking News: High Court Rules that Government FiT Cuts are "Legally Flawed"' (Solar Power Portal, 21 December 2011) <https://www.solarpowerportal.co.uk/breaking_news_high_court_rules_that_government_fit_cuts_are_legally_flawed/> accessed 4 March 2025.

[28] S Hinson, 'Briefing Paper Number 8624: Support for Small Scale Renewables' (House of Commons Library 27 January 2020) https://researchbriefings.files.parliament.uk/documents/CBP-8624/CBP-8624.pdf accessed 2 February 2025.

generators export to the grid. Unlike FIT, that allowed estimates for payment, SEG requires exported power to be metered.

2. Community energy

There is no consensus on the meaning of community energy or community renewable energy. However, the common elements are community and renewable energy projects. Within this context, the community could include groups of people within a location or people with shared interests or objectives. In UK policy, community energy 'refers to energy projects that are wholly or partially owned and controlled by local communities. The projects can span generation—usually renewable or low carbon—or demand reduction such as energy efficiency.'[29] Community energy could also include collective action for energy purchase or infrastructure management. In addition to the ownership and control criteria, another important element of the definition of community energy is the benefit criterion. Community energy projects are designed to produce shared benefits and positive social outcomes, 'rather than only financial benefit for shareholders'.[30] There are diverse community energy ownership models, ranging from formal structures such as co-operatives, social enterprises, community charities, development trusts, and community interest companies, to informal structures for executing the projects. A major distinguishing characteristic of community energy projects is that they are driven by the local community and respond to the peculiar challenges regarding energy in the local area. Communities could also be involved in the conceptualization of the project or its management.[31]

Historically, government support for renewable energy in the United Kingdom has been weak at best with successive governments appearing to be heavily invested in domestic energy security albeit through the exploitation of nuclear, North Sea oil and gas, and other non-renewable energy sources.[32] Moreso, policy fluctuations, including the removal of the Feed-In-Tariffs for renewables have adversely affected the overall growth of the sector.[33] In 2022, in England, the Rural Community Energy Fund, the key source of development funding for urban projects in England, closed to applications.[34] This compounds the financial challenges of setting up a community energy project for communities without alternative funding for capital intensive local energy projects. Nonetheless, scaling up municipal and community energy is a priority for the current Labour government's local power

[29] B Mawhood and A Adcock, 'Empowering Community Energy Schemes' (House of Commons Library 2021) 2 https://researchbriefings.files.parliament.uk/documents/CDP-2021-0200/CDP-2021-0200.pdf accessed 8 February 2025.

[30] Department of Energy & Climate Change, 'Community Energy Strategy: Full Report' (Department of Energy & Climate Change 2014) 20.

[31] M Oteman, M Wiering, and JK Helderman, 'The Institutional Space of Community Initiatives for Renewable Energy: A Comparative Case Study of the Netherlands, Germany and Denmark' (2014) 4(1) Energy Sustainability and Society 1 https://doi.org/ 10.1186/2192-0567-4-11; FD Musall and O Kuik, 'Local Acceptance of Renewable Energy—A Case Study from Southeast Germany' (2011) 39(6) Energy Policy 3252 https://doi.org/ 10.1016/j.enpol.2011.03.017.

[32] HM Government, 'The UK Renewable Energy Strategy' (2019) https://assets.publishing.service.gov.uk/media/5a7c601ce5274a7ee501a8b6/7686.pdf accessed 7 August 2025; E Wifa, D Shapovalova, and P Obani, 'Balancing Environmental Protection with Offshore Wind and Petroleum Development Two Peas in a Pod?' in I Herrera Anchustegui and others, *Offshore Energy Law Challenges and Opportunities for Renewables and Hydrocarbons* (Bloomsbury 2025).

[33] Energy Saving Trust, 'Our Response to Barriers to Community Energy Projects Call for Evidence' (Energy Saving Trust 2025) https://energysavingtrust.org.uk/report/barriers-to-community-energy-projects-response/ accessed 7 August 2025.

[34] ibid.

plan.[35] Beyond the funding challenge, community energy projects also face significant barriers such as regulatory barriers to selling their energy within their communities; and difficulties connecting to the distribution network;[36] these need to be addressed for long-term sustainability of their operations.

B. Nigeria Context

Although similar to the United Kingdom, to the extent that the legal and policy framework for decentralized energy resources in Nigeria is also fragmented, some of the provisions in the Nigerian context directly address gender equality implications of energy governance unlike the United Kingdom (Section A). Decarbonization and energy efficiency are key government policy priorities, but the targets are off track and there is a general lack of implementation of relevant legal and policy provisions which should otherwise promote distributed energy. For instance, as part of the 2060 net-zero target, the Federal Government of Nigeria announced at the 26th Conference of Parties (COP26) that 7 GW of renewable energy would be added annually to the country's energy mix and 5 million households would benefit from decentralized solar solutions by 2030 and the renewable energy target of 23 per cent by 2025 was outlined in the Revised Renewable Energy Master Plan but only 4 per cent was achieved by 2023.[37] The NERC Mini-grid Regulations 2016 aims to increase electricity access for unserved populations, and encourage the use of renewables on a small scale.[38] In line with the objectives for promoting mini-grids, the Regulations provide that the Commission shall issue a permit to an applicant within thirty days from the date of receipt of complete documentation and for applications regarding a mini-grid of up to 100kW of distributed power, a developer can commence operations as a Registered Mini-Grid Operator Applications pending the grant of the permit. In reality, some applications are pending for months and there is no evidence of accommodation for mini-grids over 100 kilowatts.[39]

The Climate Change Act 2021, which sets the legal framework for the realization of low greenhouse gas emission, indicates a target of year 2050–2070 for achieving net zero greenhouse emission nationally.[40] Furthermore, the Nigeria Energy Transition Plan (NETP) launched in 2022, indicates 2060 as the target year for achieving the country's net zero target.[41] The NETP acknowledges that the initial 2050 net-zero target was unattainable and that the 2060 target offered a more feasible pathway to decarbonization. The key sectors for the emissions reduction are power, cooking, oil and gas, transport, and industry

[35] Labour, 'Make Britain a Clean Energy Superpower to Cut Bills, Create Jobs and Deliver Security with Cheaper, Zero-Carbon Electricity by 2030, Accelerating to Net Zero' (Labour 2024) https://labour.org.uk/wp-content/uploads/2024/03/Make-Britain-a-Clean-Energy-Superpower.pdf accessed 7 August 2025.

[36] C Kumar, 'Community Energy 2.0: The Future Role of Local Energy Ownership in the UK' (Green Alliance 2019).

[37] Kaze (n 16).

[38] Nigerian Electricity Regulatory Commission Regulation for Mini-grids 2016—Regulation No: NER/-R-110/17.

[39] Sustainable Energy for All, 'Achieving Economies of Scale in the Nigerian Solar Value Chain: Opportunities and Benefits of Upstream Localization' (2021) https://www.seforall.org/system/files/2021-02/Solar-Value-Chain-Project-SEforALL.pdf accessed 16 February 2025.

[40] Section 1(f).

[41] Nigeria's Energy Transition Plan https://www.energytransition.gov.ng/ accessed 7 August 2025.

which account for around 65 per cent of Nigeria's emissions.[42] Remarkably, the NETP priority sectors do not coincide with the mitigation sectors outlined in Nigeria's Nationally Determined Contributions—energy; agriculture, forestry, and land use; waste; industrial processes and product use; and short-lived climate pollutants.[43] The Climate Change Act 2021 does not indicate any specific sectors for emissions reduction. This undermines a clear policy direction for decentralized energy investments.

There is also fragmentation in the institutional and governance framework and apparent lack of compliance, which creates significant uncertainty for investors and other stakeholders. To start with, the Federal Ministries responsible for Environment and National Planning, respectively, are mandated to set the carbon budget and budgetary period and to periodically revise the budget in line with the country's NDCs and international climate change obligations.[44] Furthermore, the Secretariat of the National Council on Climate Change (established under the Climate Change Act 2021) has a duty to produce an Action Plan, in consultation with the federal ministries responsible for the environment, budget, and national planning.[45] The Action Plan should include an articulated carbon budget for the five-year cycle that is consistent with the carbon budget and an articulated annual carbon budget for each of the years that make up the five-year cycle among others, and the pilot Action Plan should have been produced not later than twelve months from the commencement of the law.[46] However, there is no indication that the carbon budget nor pilot Action Plan have been produced, as of July 2025, over three years since the commencement of the Act. Moreso, the failure to produce a carbon budget within the time outlined in the Climate Change Act 2021 has not been challenged through the courts despite this being an abrogation of a statutory duty.[47] Against this backdrop, the section explores the relevance of feed-in tariffs, fiscal incentives, and community energy projects for promoting decentralized energy in Nigeria.

1. Nigeria feed-in-tariffs

Feed-in-tariffs are an important element of the policy incentives for advancing Nigeria's net-zero emission development plans. This and other incentives for the uptake of renewables are particularly important to counter the misgivings about solar energy resulting from unsuccessful public solar street lights projects.[48] The Nigerian Electricity Regulatory Commission Regulations on Feed-In-Tariff for Renewable Energy Sourced Electricity in Nigeria which came into effect in 2015 provides the regulatory framework for the structure and obligations of feed-in-tariff systems for renewable electricity generation in Nigeria. By these Regulations, the off taker who purchases electricity for the purpose of selling the same to customers connected to the national grid or off-grid systems is obligated to 'connect to accredited plants generating electricity from eligible renewable energy sources, subject to

[42] ibid.

[43] Federal Ministry of Environment (FME), 'Nigeria's Nationally Determined Contributions—First Update' (Abuja 2021).

[44] Section 19(1)(b).

[45] Section 20.

[46] Climate Change Act (Nigeria), s 20.

[47] P Obani, 'Climate Change Litigation: Trends and Future Directions' (Natural Justice Workshop on Climate Change Litigation Prepping Lawyers for Climate Action in Nigeria for Female Lawyers, Lagos, June 2025).

[48] JD Nixon and others, 'Analysis of Standalone Solar Streetlights for Improved Energy Access in Displaced Settlements' (2021) 17 Renewable Energy 895 https://doi.org/10.1016/j.renene.2021.05.105 accessed 11 September 2025.

the costs being met by the developer'.[49] Furthermore, the off-taker is required to 'treat as "must buy" all electricity supplied by such renewable energy projects within the capacity limits defined in this regulation'.[50]

The Nigeria Bulk Electricity Trading Company (NBET) is obligated to purchase 50 per cent of the renewable energy electricity capacity limit while distribution licensees are required to take up the remainder.[51] Nigeria's Long-Term Low-Emission Development Strategy—2060 outlines incentives for competitive pricing and adoption of low emission and renewable energy technologies, such as solar and wind power, including tax credits, subsidies, or feed-in tariffs that ensure a guaranteed price for renewable energy generation.[52] Regarding the building and residential sector, Nigeria's Long-Term Low-Emission Development Strategy—2060 refers to 'green tariff systems' as part of the economic or regulatory policies and measures that would contribute to the funding of the implementation of low-emission development strategy, subsidized renewable energy materials and equipment and a pricing structure and feed-in-tariff (FiT) for household renewable energy equipment among measures for mitigation.[53] Despite the foregoing, there is no clear information about the operationalization of a pricing structure and FiT for households' renewable energy equipment. At the macro level, the operationalization of FiT and the overall development of solar projects have also been limited[54] due to a range of challenges, including economic monopoly, changes in the financial regulations and tariff structures, stakeholder distrust, and historical fossil-fuel subsidy regime. [55]

2. Fiscal incentives for renewable energy importers

The significant capital costs of renewable energy technologies is a barrier for adoption, particularly among women and other demographics with limited access to funding. In addition to the cost of development and operation, buyers in non-producing countries would also bear the cost of importation. It is estimated that a 200kW solar-diesel hybrid mini-grid could cost over $2 million to build and over $100,000 for the annual operational costs.[56] Despite the reduction in the cost of solar photovoltaic (PV) modules, the costs of solar systems are still high in Nigeria partly due to the dependence on foreign suppliers and the devaluation of the Naira.[57] Beyond the cost, the quality of imported renewable technologies is also a concern in Nigeria, due to lapses in the regulatory system for the technical standards of products.[58] The heavy dependence on imported technologies presents critical supply chain risks.

[49] Regulation 5(a).
[50] Regulation 5(b).
[51] Regulation 5(f).
[52] National Climate Change Council, 'Nigeria's Long-Term Low-Carbon Development Strategy' (NCCC 2023).
[53] ibid 138.
[54] F Adeniyi and A Isah, 'Unlocking Renewables Amid Rentierism: Market Constraints to Nigeria's Energy Transition' (2023) 104 Energy Research & Social Science 103248 https://doi.org/10.1016/j.erss.2023.103248 accessed 11 September 2025.
[55] ibid 9.
[56] Nigerian Customs Declare Solar Panels as Duty-Free (PVKnowHow.Com 11 September 2024) <https://www.pvknowhow.com/news/nigerian-customs-declare-solar-panels-duty-free/> accessed 19 February 2025.
[57] Sustainable Energy for All (n 39); S Stout and others, 'Landscape of Climate Finance in Nigeria 2024' (CPI 2024) https://www.climatepolicyinitiative.org/wp-content/uploads/2024/10/Landscape-of-Climate-Finance-in-Nigeria-2024.pdf accessed 7 August 2025.
[58] Sustainable Energy for All (n 39).

The Nigerian government has provided fiscal incentives for renewable energy technology, including import duty waivers or reduction from around 5 per cent to 20 per cent for solar panels and inverters, for example.[59] The amount of waiver is defined based on the category of the imported technology. For instance, there is 0 per cent import duty on solar panels under HS Code 8541, 5 per cent import duty on solar that are imported with other systems under HS Code 8502, while import duty of up to 20 per cent is imposed on all other solar components, including bulbs, fans, and television sets.[60] There is also an exemption of renewables from the payment of Value Added Tax, which further reduces the costs.[61] Furthermore, based on the accelerated depreciation allowance under the Companies Income Tax Act (CITA), companies who invest in renewable energy projects can depreciate their capital investments at an accelerated rate.[62] Accordingly, companies thereby accrue lower taxable income. There are no additional fiscal incentives targeted at female-owned businesses or investors in the renewable energy sector.

One of the main limitations of the current fiscal incentives for renewable energy in Nigeria is that most of the focus is on solar energy. There is limited emphasis on other sources of renewable energy, which could further diversify the country's energy mix. Moreso, the regulatory environment is complex, with most importers being charged higher import duties due to the classification of their goods under the wrong code.[63] There are also concerns over the risk of stifling local manufacturing through the import duty waivers and other fiscal incentives for the importation of renewable energy technologies.[64] To mitigate this, the local industry can be incentivized through introducing fiscal incentives for local manufacturers, setting a time limit on the fiscal incentives for imported technologies, and incorporating local content policies that require importers benefiting from the fiscal measures to invest in strengthening of the domestic sector through capacity building and technology transfer and gender transformative approaches.[65]

3. Community energy in Nigeria

Given the depth of Nigeria's energy crisis, and the limited funding and capacity of the centralized infrastructure for the delivery of energy services, particularly electricity, there is an urgent need for decentralized energy systems to be deployed. Nigeria's Rural Electrification Agency's Off Grid Electrification Strategy specifically focuses on the provision of clean and sustainable electricity using decentralized energy solutions for the benefit of households,

[59] GOLGA, Nigeria Country Brief (2024) https://gogla.org/wp-content/uploads/2024/11/Nigeria-Country-Brief.pdf accessed 5 March 2025.

[60] ibid.

[61] PWC, 'VAT Modification Order 2024—Incentivising Energy Transition in Nigeria' https://www.pwc.com/ng/en/assets/pdf/vat-odification-order-2024-incentivising-energy-transition-in-nigeria.pdf accessed 5 March 2025.

[62] EY, 'Nigeria Tax Act, 2025 has been Signed—Highlights' (EY Global 2025) < https://www.ey.com/en_gl/technical/tax-alerts/nigeria-tax-act-2025-has-been-signed-highlights> accessed 5 March 2025.

[63] Nigeria Customs Service, 'T&T/2024/Circular No.18: Clarification on the Classification of Solar Panels' https://www.energy-box.com/post/nigeria-customs-confirms-solar-panels-remain-duty-free accessed 5 March 2025.

[64] Renewable Energy Association of Nigeria (REAN) and ALL ON, 'Advocacy and Engagement on Fiscal Incentives & Importation Procedures for Renewable Energy Technologies in Nigeria' (REAN and ALL ON 2024).

[65] P Obani, 'Gender Equality in Local Content Policies' in EG Pereira, A Ingelson, and D Olawuyi (eds), *Local Content in Oil and Gas Law: A Practical Guide from Design to Implementation* (Routledge Abingdon 2025); DS Olawuyi and AJ Olaniyan, 'Local Content and the Sustainable Development of Oil and Gas Resources in Nigeria' in D Olawuyi (ed), *Local Content and Sustainable Development in Global Energy Markets* (CUP 2021).

communities, and businesses. Among the key objectives are by 2023 to develop 10,000 mini grids to provide power to 14 per cent of the population and to deploy 5 million solar standalone systems for residential and SMEs.[66] For this purpose, the energy database lists states and local government areas in relation to potential off grids.[67] However, there is no clear indication of the potential role of communities as part of the governance structure for achieving the objectives of the off grid electrification strategy.

There is limited coverage of community energy in the Sub-Saharan African region. Scholars argue that community renewable energy offers the region an opportunity to address climate change impacts, tackle poverty, and promote economic opportunities.[68] However, existing literature on community renewable energy that are predominantly based on European case studies fails to address the unique context of non-Western countries and the related factors that influence energy development.[69] Hence, the prospective environmental, social, and economic value of community renewable energy in the SSA region and the underpinning policy and legal framework requires analysis.

In Nigeria, the Electricity Act 2023 provides a framework for subnational governments to advance electricity generation, transmission, distribution, supply, and retail, but excludes roles for local government authorities, citizens, and communities.[70] Furthermore, the Act empowers states within Nigeria to regulate intrastate grids and offers incentives for renewable energy adoption. The decentralization of electricity provision offers a good basis for advancing community energy projects, with the support of states and local governments, though this is yet to be actualized and the Nigerian Electricity Act focuses predominantly on improving the traditional energy infrastructure, not prioritizing citizen engagement.[71]

The community energy model presents potential opportunities for advancing climate change targets and mitigating the lack of investments and infrastructure deficit in relation to electricity services.[72] There are however complex issues regarding land ownership and the socio-political environment which would need to be addressed for a successful community energy project, particularly in relation to energy generation in urban settings where most of the energy consumption occurs. The cultural norms and land tenure systems are such that while it is possible to acquire land for private and commercial uses, minority groups such as non-locals and women in the communities might experience additional restrictions to accessing land for energy projects.

IV. Gender Dimensions and Gaps

Women face unique barriers leading to their exclusion or underrepresentation in the energy sector. The following key barriers require direct interventions to challenge the stereotypes

[66] Rural Electrification Agency, 'Off Grid Electrification Strategy' https://rea.gov.ng/theagency/off-grid-electrification-strategy/# accessed 20 January 2025.
[67] Rural Electrification Agency, 'Energy Database' < https://rea.gov.ng/energydatabase/ > accessed 21 January 2025.
[68] Kaze (n 16) 5.
[69] ibid.
[70] Electricity Act 2023, s 2.
[71] Kaze (n 16).
[72] ibid.

and provide an enabling environment for women's full and effective participation through decentralized energy systems arrangements.

A. Lack of Dedicated Gender-focused DER Laws

With the exception of the Draft Energy Strategy and Just Transition Plan for Scotland which integrates capacity building and training for women to engage in UNFCCC discussions, in both United Kingdom and Nigeria, explicit legal frameworks to advance gender equality in the decentralized energy sector are lacking.[73] Rather than acknowledging their role and potential in leading and enabling energy sector initiatives, women's inclusion in the energy sector policies is characteristically associated with vulnerability, poverty, and rural areas.[74] Beyond highlighting the relevance of energy access for poverty reduction and the need for women's participation in the energy sector leadership for gender equality, the energy law and policy framework and decentralization reforms have not directly addressed the gender dimensions of energy justice.

B. Limited Access to Land

Globally, women account for less than 15 per cent of agricultural landholders.[75] Also, in Nigeria, only 8.2 per cent of women have sole ownership of land, compared to 34.2 per cent of men,[76] with fragile land tenure often dependent on marriage, inheritance, and other patriarchal factors. Particularly in Indigenous communities, women experience multiple forms of marginalization including forced displacement and eviction, adverse climate change impacts, poverty, low literacy levels, historical discrimination, and the arbitrary provisions of the Land Use Act which impact on tenure security.[77] Similarly, women in the United Kingdom also face barriers to acquiring land, particularly in the rural areas: 18 per cent of women are far more likely to be farm owners or part of a family partnership/company compared to 38 per cent of men and 19 per cent more likely to enter into agriculture through marriage compared to 1 per cent of men.[78] This hinders women's ability to obtain financing or equity investment opportunities needed for DER initiatives and reinforces gender disparities in energy entrepreneurship.

[73] REAN, 'Women in Renewable Energy Baseline Report' (REAN 2021) https://rean.org.ng/wp-content/uploads/media/REAN_Women_in_Renewable_Energy_Baseline_Report_2.pdf accessed 7 August 2025; A Soler, J Jæger, and D Lecoque, 'Women Entrepreneurs as Key Drivers in the Decentralised Renewable Energy Sector: Best Practices and Innovative Business Models' (Alliance for Rural Electrification 2020) https://www.ruralelec.org/wp-content/uploads/2023/11/Gender-Energy-Publication.pdf accessed 7 August 2025.

[74] FME (n 43) 28.

[75] FAO, 'FAO Policy on Gender Equality 2020–2030' (FAO 2020 Rome).

[76] I Gaddis, R Lahoti, and W Li, 'Gender Gaps in Property Ownership in Sub-Saharan Africa' Policy Research Working Paper 8573 (World Bank 2018).

[77] D Olawuyi, 'Gender, Indigeneity and the Search for Environmental Justice in Post-Colonial Africa', in C Gonzalez, S Seck, and S Attapatu (eds), *Cambridge Handbook of Environment Justice and Sustainable Development* (CUP 2021).

[78] A Kay, 'Exclusive Survey: Has Working in AG Become Easier for Women Since 2014?' (Farmers Weekly, 2024) https://www.fwi.co.uk/news/has-working-in-ag-become-easier-for-women-since-2014 accessed 12 August 2025.

C. Digital Divide and Limited Access to Finance

In low- and middle-income countries, and despite the widespread of digital technology women are 10 per cent less likely to own a mobile phone and 23 per cent less likely to access mobile internet, compared to men.[79] The resulting lack of access to information and networks undermines women's effective participation and influence in energy governance. Women also receive a limited amount of equity financing for complex reasons, including limited participation in the sector overall and the size of women-owned enterprises compared to men's.[80] In addition, other financial incentives including local content and procurement laws also do not prioritize gender transformation in most cases.[81]

D. Underrepresentation in STEM and Technical Roles, and Gender Pay Gap

The underrepresentation of women in STEM affects their capacity for technical roles in DER programmes.[82] Moreso, the existing data shows that women experience a gender pay gap. In the United Kingdom, the Equality Act 2010 prohibits discrimination on grounds of gender and employers with a minimum of 250 employees are obligated to report on their gender pay gap annually, yet the gender pay gap in the energy sector is as high as a 21.3 per cent mean gap and a 26.3 per cent median gap in the case of EDF Energy.[83] The Nigerian Constitution also protects against discrimination on the grounds of sex but in the absence of gender disaggregated data on the pay gap, for example, it is difficult to enforce this protection for women's rights in the energy sector.[84]

V. Addressing the Challenges: Lessons Learned

Although decentralized energy systems offer prospects for improving access to cleaner forms of energy, there are also concerns about the potential adverse impacts of renewable energy generation and off-grid energy infrastructure systems on women and marginalized populations. Furthermore, off-grid systems could increase the pressure on limited land and environmental resources for women while also increasing their workload if they are assigned responsibilities for the operation or management of the systems. It is therefore important to mainstream gender equality and women's rights at every stage of the lifecycle of energy decentralization projects. This chapter recommends four key legal/policy steps for mainstreaming gender equality in DER programmes pertaining to the prioritization of women's participation in governance, clear gender equality obligations for government, wholistic consideration of gender equality throughout the lifecycle of DER projects, and alternative inclusive financing structures for DERs.

79 FAO (n 75).
80 REAN (n 73).
81 Obani (n 65).
82 La Malfa (n 2).
83 EDF, 'Gender Pay Gap Report 2024' https://www.edfenergy.com/sites/default/files/2025-02/Gender%20Pay%20Gap%202024.pdf accessed 8 August 2025.
84 Constitution of the Federal Republic of Nigeria 1999, s 42.

First, given the peculiar constraints which women face, including economic, social, and political which constrain their opportunities to engage in energy governance, states and energy companies should prioritize women's participation, non-discrimination, and investment in capacity building to enhance the position of women as clean energy consumers, producers, and leaders in the sector. Drawing from the example set by the United Kingdom through Scotland's Gender Equality in Energy Transition Roadmap (2022), countries such as Nigeria can integrate gender quotas and mentorship programs into decentralized energy projects. For instance, community energy projects that prioritize women and other marginalized groups can be established to offer prospects for social cohesion, equitable distribution of costs and benefits, and legitimacy of the energy governance process. Also, for Nigeria, at the subregional level, the Economic Community of West African States (ECOWAS) Policy for Gender Mainstreaming in Energy Access (2017) promotes women's involvement in renewable energy initiatives and the ECOWAS Directive on Gender Assessment in Energy Projects which requires the development of a Gender Assessment, Gender Management Plan, and Gender Performance Monitoring for energy projects that meet the specified criteria and thresholds in terms of expected social impacts and benefits.[85] These offer good guidance for strengthening gender equality in the domestic energy sector.

Second, there should be a clear gender equality obligation in government's net zero laws, to ensure synergies between the energy decentralization and climate goals and gender equality outcomes. It is also important to ensure that fiscal and other incentives for decentralized energy are clearly integrated and effectively co-ordinated across all levels of governance, from the national to the local, and understood by all the regulators, enforcers and beneficiaries. The incentives should target a wide range of decentralized energy systems that are suited to the local context, while also promoting domestic production with the equal participation and fair compensation of female and male workers. In addition, there should be targeted education and capacity building opportunities to promote women and girls' participation on STEM and leadership roles and to build a wider supportive environment for women to thrive through mentoring and allyship. Furthermore, regulation and reporting requirements should be mainstreamed to incentivize action from governments and employers to address the existing gender pay and other gaps through targeted interventions in DER and the energy sector generally.

Third, there should be clear objectives for gender equality across all stages of energy policy, programme design, and implementation as well as in the governance of digitalization, energy financing, and land resources. This needs to go beyond recognizing the links between energy poverty and gender norms, the adverse impact of energy production activities on women's rights, or the general need to empower women for leadership in the energy sector. Decentralized energy projects should be underpinned by a holistic assessment of the intersections between women's rights and wellbeing and access to livelihood resources across all the stages of the project's lifecycle, from conceptualization to commercialization for domestic and industrial uses. The assessment of gender impacts and related gender targets for the energy sector should be clearly outlined in legal frameworks with mechanisms for enforcement and the necessary legal aid support provided for women who may be unable to afford the cost of legal advisory services and representation. Also, there should be

[85] M Maduekwe and A Gabriela Factor, 'Gender Assessment in Energy Projects: Perceptions, Practices and the Role of a Regional Directive in ECOWAS' (2021) 39(3) Impact Assessment and Project Appraisal 251.

targeted legislative reforms to address patriarchal traditions, and restrictive land use legislations and customs which hinder women's equal access to land.

Fourth, the financing structure for DERs needs to be flexible and to accommodate non-financial approaches. For community energy projects, for example, recognizing the non-financial contributions which women make to the design, development, and operation and maintenance of DERs could provide pathways for poor women and those with less disposable income to own DER and engage in energy entrepreneurship. Moreso, local content laws and fiscal incentives for DERs should be redesigned to advance gender transformative outcomes.

VI. Conclusion

The chapter highlighted the potential for energy decentralization to advance low-carbon transitions while also democratizing and enhancing energy access. This is in line with Net Zero obligations and states' commitment to energy security. At the same time, decentralized energy systems can perpetuate or worsen existing gender inequalities, depending on their design and the underpinning legal and policy frameworks. Through a comparative exploration of the United Kingdom and Nigeria, this paper shows that despite the close intersections between gender inequality and energy, there is minimal if any integration of gender justice as a guiding principle for energy decentralization. In both countries, decentralized energy laws and policies are advanced with significant progress in implementation. However, the legal frameworks fall short of addressing the structural barriers women face and systematically prioritizing women's effective participation and leadership, access to equitable benefit sharing, and protection from the adverse impacts of energy projects. The example of the regulatory framework for community energy projects in the United Kingdom potentially offers a template for cohesive local action for energy security and gender justice if underpinned by the relevant rights-based, community-centred, and gender-sensitive legal reforms. The recognition of the role of the decentralized energy for alleviating women's poverty within Nigerian energy sector laws is also remarkable. The mainstreaming of gender equality also needs to permeate fiscal incentives and other areas of energy governance to advance justice for all.

19

Eco-Entrepreneurship Law and Distributed Energy Resources

Kehinde Folake Olaoye and Josephine Ilesanmi

I. Introduction

Despite the energy democratization potential of distributed energy resources (DERs), several legal and regulatory barriers, including business facilitation, registration, and formalization requirements, continue to hinder the abilities of energy citizens and prosumers to commercialize DER innovation. Drawing lessons from Nigeria, this chapter examines the political, environmental, social, technological, and legal (PESTEL) gaps and challenges which act as barriers to distributed energy entrepreneurship and explores legal and policy initiatives for addressing such gaps.

An estimated 80 million of Nigeria's 201 million population do not have access to affordable and reliable electricity supply.[1] An indication of this perennial problem is the public outcry amidst grid collapses in April 2025, to a report that under the 2025 budget, the Federal Government of Nigeria had approved a ₦10 billion budget for a solar mini-grid to power the Aso Rock Presidential Villa—the official residence of the Nigerian President.[2] Over the last decades, there have been several challenges to achieving affordable and renewable energy in Nigeria. These include the recent removal of electricity subsidies, inflation, poor governance, and other policy changes which have resulted in litigation. In Nigeria, there has been heavy reliance on petrol and diesel-run generators which are unsustainable sources of energy.

Demand for alternative and cheaper sources of energy has created opportunities for eco-entrepreneurs who play an important role in Nigeria's energy supply chain. For developing countries like Nigeria, DERs like solar photovoltaics (PV), energy storage, electric vehicles (EVs), and smart buildings have the potential to solve a perennial energy crisis while contributing to the attainment of the sustainable development goals.[3] Citizen engagement in renewable DERs manifest as energy citizens, ecopreneurs, and energy communities which engage in eco-entrepreneurship and establish SMEEs.[4] Nigeria is a signatory to the 2025

[1] World Bank, 'NESI: Progress Report and Outlook' (2022) 4 <https://documents1.worldbank.org/curated/en/099062123033010348/pdf/P1747600ac17a20120af2f0d81585897cd4.pdf> accessed 28 April 2025.

[2] 'Aso Rock Opting for N10bn Solar Questions Presidency Belief in National Grid' (International Centre for Investigative Reporting, 22 April 2025) <https://www.icirnigeria.org/aso-rock-opting-for-n10bn-solar-questions-presidency-belief-in-national-grid/> accessed 23 April 2025.

[3] See Chapter 2 of this book.

[4] J Ilesanmi, 'Advancing Eco-Entrepreneurship in Nigeria through Transformative Implementation of Sustainable Business Model Innovations' (2025) 15(2) Journal of Sustainable Development Law and Policy 298–327.

Kehinde Folake Olaoye and Josephine Ilesanmi, *Eco-Entrepreneurship Law and Distributed Energy Resources* In: *Sustainable Distributed Energy Resources*. Edited by: Íñigo del Guayo, Damilola S. Olawuyi San, Louis de Fontenelle, Milton Fernando Montoya, and José Juan González, Oxford University Press. © International Bar Association 2026. DOI: 10.1093/9780198991946.003.0022

Dar es Salaam Energy Declaration under which it declares commitment to embracing distributed renewable energy and creating a supportive environment for private sector participation.

SMEEs generate eco-ventures, services, products, and green technologies promoting a decentralized electricity system powered by renewable energy resources.[5] As small and medium enterprises (SMEs) account globally for 50 per cent of employment and represent 90 per cent of businesses, they have a higher environmental footprint than large enterprises in some sectors and consume more resources. SMEs and energy communities which engage in eco-entrepreneurial activities can contribute significantly to economic development while promoting energy decarbonization and the energy transition.[6]

The small-scale characteristic of DERs means that small and medium enterprises inherently play a key function in implementation of DERs.[7] Although there is no accepted definition of SMEs, based on the Central Bank of Nigeria Guidelines for Small and Medium Enterprises Equity Investment Scheme, SMEs can be defined as registered companies with a maximum asset base of 500 million Naira excluding land and working capital.[8] Eco-entrepreneurs or green startups can be defined as newly established businesses that manufacture products and provide services with environmental and renewable energy benefits.

In Nigeria there has been an increase in off-grid renewable electricity generation through decentralized solutions. Nigeria is one of the world's largest hubs for renewable startups.[9] However, large corporations and businesses remain key players. For example, in 2024, all the nine licenses issued and renewed by the Nigeria Electricity Commission for off-grid solar generation were granted to one company—Daybreak Power Solutions.[10] In December 2022, Shell Overseas International, a wholly owned subsidiary of Shell plc, acquired Daystar Power Group. Daybreak Power Solutions Limited is a 100 per cent owned Nigerian subsidiary of Daystar Power Group which provides solar PV, hybrid energy, and storage solutions in Nigeria and Ghana with subsidiaries in Kenya and Zambia.[11] In 2024, Ignite Power acquired Oolu and its operations which extend to Nigeria, Senegal, Burkina Faso, and Cameroon. The merger has been investigated by the ECOWAS Regional Competition Authority.[12] While large DER enterprises like Daybreak Power Solutions Limited can

[5] D Mihailova and others, 'Exploring Modes of Sustainable Value Co-creation in Renewable Energy Communities' (2022) 330 Journal of Cleaner Production 129917.

[6] I Costa Melo and others, 'Sustainable Digital Transformation in Small and Medium Enterprises (SMEs): A Review on Performance' (2023) 9(3) Heliyon e13908; UN, '2025 Technology and Innovation Report Inclusive Artificial Intelligence for Development' (2025) 26.

[7] See 'European Covenant of Companies for Climate and Energy: Best Practices for SMEs in the Energy Sector' (2024) <https://www.smeunited.eu/admin/storage/smeunited/ccce-best-practices-for-smes-in-the-energy-transition-final.pdf> accessed 23 April 2025; World Bank, 'From Sun to Roof to Grid: Distributed PV in Energy Sector Strategies' (2021) 5 <https://documents1.worldbank.org/curated/en/532941634543900788/pdf/From-Sun-to-Roof-to-Grid-Distributed-PV-in-Energy-Sector-Strategies.pdf> accessed 23 April 2025.

[8] Central Bank of Nigeria, 'Guidelines for the Small and Medium Enterprises Equity Investment Scheme' (2005) para 4.0 <https://www.cbn.gov.ng/out/publications/guidelines/dfd/2005/smeeis%20guideline.pdf> accessed 15 April 2025.

[9] S Fankhauser and others, 'The Renewable Energy Entrepreneurs of the Global South' (2004) https://www.smithschool.ox.ac.uk/sites/default/files/2024-11/The-renewable-energy-entrepeneurs-of-the-Global-South.pdf accessed 15 April 2025.

[10] Nigerian Electricity Regulatory Commission (NERC), '2024 Annual Report and Accounts' (2025) 179 https://nerc.gov.ng/wp-content/uploads/2025/07/2024-Annual-Report-1.pdf accessed 15 June 2025.

[11] IFC Project Disclosures https://disclosures.ifc.org/project-detail/AS-ESRS/43623/daystar-power accessed 15 April 2025.

[12] K Koff, 'Filling of Merger to the ECOWAS Regional Competition Authority: Ingite Power Ltd and Engie Afrique SAS' (14 February 2025) <https://erca-arcc.org/filling-of-merger-notification-to-the-ecowas-regional-competition-authority-ingite-power-ltd/> accessed 15 April 2025.

navigate the complex process of doing business in Nigeria and obtain critical finance from international financial institutions,[13] smaller businesses continue to face hurdles in Nigeria. These include high import duties on essential raw materials, policy inconsistencies, stringent customs processes, and limited access to affordable financing.[14]

Despite growing awareness of the importance of eco-entrepreneurship in distributed energy resources, few existing academic studies examine the supportive policy, legal, and regulatory framework necessary for distributed energy business models in developing countries. This chapter aims to fill this research gap by identifying these policies. Using Nigeria as a case study, this chapter examines the PESTEL gaps and challenges which act as barriers for DER entrepreneurs. PESTEL is an analytical tool which evolved from the scholarly research developed by Francis Joseph Aguilar, an American scholar of strategic planning and general management in 1967.[15] This chapter contribution provides a contemporary account of the unique challenges of DERs in Nigeria combining entrepreneurial law studies and management studies. As a country with one of the world's largest energy problems, Nigeria is an important case study which provides comparative lessons for policy design and implementation in other developing countries.

This chapter is divided into five sections. The first section examines the unique energy challenges faced in Nigeria showing that eco-entrepreneurship and distributed energy resources eco-products and services are crucial for sustainable energy. The second section clarifies key concepts and examines existing scholarship on the connections between entrepreneurship law and distributed energy resources. The next section examines gaps which exist in Nigeria's energy and entrepreneurship laws making an argument for policy coherency. Section IV identifies legal approaches for addressing the gaps identified and provides recommendations for supporting SMEE generated distributed energy business models. The final section provides a summary and conclusion on the importance of eco-entrepreneurship and SMEEs as incubators for achieving the 4Ds of DERs and fulfilling the energy transition plan.

II. Eco-entrepreneurship and Energy Sustainability

It is universally recognized that entrepreneurship can address environmental challenges and promote energy sustainability through the introduction of new climate change mitigation and adaptation technologies.[16] It can deliver energy-efficient products and provide renewable energy equipment thereby promoting environmentally sustainable consumption and production patterns.[17] This key relationship between entrepreneurship and sustainable development is recognized in important documents like the 2022 United Nations General Assembly Resolution 77/160 on Entrepreneurship for Sustainable Development and the

[13] See T Elias, 'Financing Distributed Energy Infrastructure' in Chapter 12 of this book.

[14] W Alli, 'How Removal of Import Duties on Solar Raw Materials Can Boost Nigeria's Industrial Future' (*Business Day*, 15 April 2025) https://businessday.ng/companies/article/how-removal-of-import-duties-on-solar-raw-materials-can-boost-nigerias-industrial-future/ accessed 15 June 2025.

[15] FJ Aguilar, *Scanning the Business Environment* (Macmillan 1967).

[16] UNDP, 'From Pitch to Impact: Deep Dive into Climate Technology Startups' (2025) https://www.undp.org/sites/g/files/zskgke326/files/2025-03/climatech_report_update.pdf accessed 15 June 2025.

[17] UN General Assembly Resolution 77/160 on Entrepreneurship for Sustainable Development (14 December 2022); Report of the Secretary-General, 'Entrepreneurship for Sustainable Development' (22 July 2024).

European Union SME Strategy for a Sustainable and Digital Europe. Eco-entrepreneurship is also closely related to the four main vectors of transformation in DERs—decarbonization, decentralization, digitalization, and democratization.[18] This section of the chapter examines the relationship between eco-entrepreneurship and DERs.

A. Entrepreneurship Law and Start-Up Law

Although there is no consensus on the existence of entrepreneurship as a distinct field of law,[19] entrepreneurship law can be described as laws and regulations which promote and protect entrepreneurial activity. Different fields of law including company law, environmental law, employment law, intellectual property law, contract law, and energy law provide an enabling environment for entrepreneurial activity.[20] Entrepreneurial law is thus embedded in business, economics, and management studies.[21] Eco-entrepreneurial activity in the electricity sector can potentially transform the energy landscape of a jurisdiction, providing environmental, social, and economic benefits.[22] Environmental benefits include emission reduction and public health benefits. Social benefits include the alleviation of energy poverty, the promotion of self-generation, and energy autonomy.[23] Renewable distributed energy enables energy communities, especially low-income, vulnerable populations, and rural areas, to access electricity.[24] Also, eco-entrepreneurship implemented by energy communities and energy citizens provide economic empowerment and a source of livelihood especially for low-income or the bottom of the pyramid (BOP) communities.[25] Eco-entrepreneurship can advance innovation and differentiation in the electricity landscape through sustainable business model innovations that advance climate action while accelerating energy security.[26]

Despite growing awareness of the importance of eco-entrepreneurship in DERs, few existing academic studies critically examine the supportive policy, legal, and regulatory

[18] I del Guayo Castiella, 'Distributed Energy Resources as the New Electricity Paradigm: Legal Context and Principles' (in this volume).

[19] JM Lipshaw, 'Why the Law of Entrepreneurship Barely Matters' (2009) 31 Western New England Law Review 701.

[20] SF Reed, *Entrepreneurship Law: Cases and Materials* (Wolters Kluwer 2013) 1. See, AR Lucas and AE Ingelson, 'Entrepreneurial Legal Education and Energy Democratization: an Indigenous Renewal Distributed Energy Clinic' (in this volume).

[21] SC Parker, 'Law and the Economics of Entrepreneurship' (2006) 28 Comparative Labor Law & Policy Journal 695; A Travassos Rosário and J Figueiredo, 'Sustainable Entrepreneurship and Corporate Social Responsibility: Analysing the State of Research' (2024) 10 Sustainable Environment 1.

[22] JK Binder and F-M Belz, 'Sustainable Entrepreneurship: What It Is' in P Kyrö (ed), *Handbook of Entrepreneurship and Sustainable Development Research* (Edward Elgar 2015) 30.

[23] F Gianaroli and others, 'Development of Dynamic Sharing Keys: Algorithms Supporting Management of Renewable Energy Community and Collective Self Consumption' (2024) 311 Energy and Buildings 1.

[24] S Algarni and others, 'Contribution of Renewable Energy Sources to The Environmental Impacts and Economic Benefits for Sustainable Development' (2023) 56 Sustainable Energy Technologies and Assessments 103098.

[25] T Chikweche and others, 'Innovative New Product Adoption of Renewable Energy Mobility Products at the Bottom of the Pyramid: The Case of Hamba, An integrated Community-driven Solution' (2023) 35 Journal of International Consumer Marketing 194.

[26] J Ilesanmi, 'Advancing Eco-Entrepreneurship in Nigeria through Transformative Implementation of Sustainable Business Model Innovations' (2025) 16 The Journal of Sustainable Development Law and Policy 256.

framework necessary for DER entrepreneurs in developing countries. Existing legal studies on Nigeria have focussed generally on Nigeria's framework for off-grid renewable electricity.[27] Eco-entrepreneurship is an important tool for climate change action and low carbon transition in developing countries like Nigeria.[28] It provides a platform for citizen participation in electricity decentralization initiatives.[29] A related field of law which combines entrepreneurial innovation with sustainable development goals is the law of social enterprises which focuses on businesses which aim to solve societal problems like protection of the environment.[30]

In the last few years, some countries, a majority of which are developing countries, have enacted startup laws. Unlike older entrepreneurship laws, startup laws aim to take advantage of the fourth revolution, promote technological innovation, attract business in specific sectors, and facilitate the scaling up of businesses. Some countries like Germany and Seychelles have developed national startup strategies.[31] Jurisdictions which have recently introduced startup laws include Côte d'Ivoire, Tunisia, Brazil, and Nigeria.[32] While some of these laws recognize the contributions of entrepreneurial innovation to sustainable development, the direct link between energy transition, DERs, and entrepreneurship is missing from most laws and policies.[33] An exception is the Statement of the Objects and Reasons for Kenya's Startup Bill which references M-Kopa's 'pay-as-you-go' energy services for off-grid customers.[34] M-Kopa which was established in 2010 is a good example of a green startup which has scaled up and extended its operations to other African countries including Nigeria.

[27] N Chinwa Ole, 'The Nigerian Electricity Regulatory Framework: Hotspots and Challenges for Off-grid Renewable Electricity Development' (2020) 38 Journal of Energy & Natural Resources Law 367; N Chinwa Ole and others (eds), *Regulatory Support for Off-Grid Renewable Electricity* (Routledge 2023).

[28] J Ilesanmi, 'The Importance of Eco-Entrepreneurship as a Tool for Climate Action and Low Carbon Transition in Developing Countries: A Case Study of Nigeria' (2024) 15 Journal of Sustainable Development Law and Policy 298.

[29] DS Olawuyi, 'From Energy Consumers to Energy Citizens: Legal Dimensions of Energy Citizenship' in R Fleming, K Huhta, and L Reins (eds), *Sustainable Energy Democracy and the Law* (Brill 2021) 101.

[30] H Fleischer and M Pendl, 'The Law of Social Enterprises: Surveying a New Field of Research' (2024) 25 European Business Organization Law Review 269.

[31] Federal Ministry for Economic Affairs and Climate Action, 'Start Up Strategy of the Federal Government' (2022) <https://www.bmwk.de/Redaktion/EN/Publikationen/Wirtschaft/2022-startup-strategy-of-the-federal-government.pdf?__blob=publicationFile&v=5> accessed 15 April 2025; Republic of Serbia, 'Startup Ecosystem Strategy of the Republic of Serbia (2021–2025)' <https://nitra.gov.rs/images/ministarstvo/dokumenta/startup%20ecosystem%20development%20strategy.pdf> accessed 15 April 2025; Ministry of Entrepreneur Development and Cooperatives, 'Malaysia National Entrepreneurship Policy: An Outstanding Entrepreneurial Nation 2030' <https://www.kuskop.gov.my/admin/files/med/image/portal/NEP2030-Final.pdf> accessed 15 April 2025.

[32] Belarus, Law of the Republic of Belarus 22 April 2024 No 365-Z on Amending Laws on Entrepreneurial Activities (adopted by the House of Representatives on 21 March 2024, approved by the Council of the Republic on 10 April 2024); Brazil, Complementary Law No 182, of 1 June 2021 Establishing the Legal Framework for Startups and Innovative Entrepreneurship (signed on Complementary Law in 1 June 2021); Democratic Republic of Congo, Ordinance-Law No 22/030 of 8 September 2022 on the Promotion of Entrepreneurship and Start-ups (signed into law on 8 September 2022); Law of Georgia on Entrepreneurs (No 875-VRS-XMP of 2021) (adopted on 2 August 2021, effective 1 January 2022).

[33] Brazil Complementary Law no 182, of 1 June 2021 art 3(1); see Republic of Serbia, The Law on Social Entrepreneurship (passed on 4 February 2022) art 4(3).

[34] Republic of Kenya, The Startup Bill 2022 <http://www.parliament.go.ke/sites/default/files/2023-02/The%20Startup%20Bill%2C%202022.pdf> accessed 14 April 2025.

B. The PESTEL Analytical Framework and DER Entrepreneurship in Nigeria

PESTEL Analysis is important for examining and addressing the challenges of the law and policy of sustainable DER law and policy in Nigeria. Different theories and analytical frameworks have been used in scholarship to examine the relationship between eco entrepreneurship and sustainable development.[35] These include complex network theory,[36] theory of market failure,[37] institutional theory,[38] and regulatory focus theory.[39] Many existing studies use a case study analysis[40] to conduct management empirical studies.[41] While these theoretical frameworks are all important, PESTEL remains one of the most developed strategic management theories for a combined law and policy analysis. Although there are policy reports and documents on DERs, there are very few academic legal studies on DERs and entrepreneurship and no legal studies on entrepreneurship and DERs in Nigeria.[42]

The PESTEL analysis, encompassing political, economic, social, technological, environmental, and legal factors enables a comprehensive assessment of key indicators like the external macroeconomic environment of growth, interest rates, currency movements, input prices, regulations, and role of corporations in society.[43]

Variations of PESTEL acronym include PESTLE and PEST STEEPLE which includes corporate social responsibility and ethical considerations. The STEEPLED acronym incorporates an analysis of demographic factors. An increasing focus on the environmental and sustainability governance of businesses has also lead to the creation of the STEER analytical tool, which incorporates regulatory factors.[44] Although in practice the variations in acronyms are negligible in comparison to the equally common Strengths, Weaknesses, Opportunities and Threats (SWOT) analysis, PESTEL is an important analytical tool for entrepreneurial studies. It provides a holistic framework for analyzing DERs from

[35] P Muñoz and B Cohen, 'Sustainable Entrepreneurship Research: Taking Stock and Looking Ahead' (2018) 27 Business Strategy and the Environment 300; CG Crawford and others, 'Power Law Distributions in Entrepreneurship: Implications for Theory and Research' (2015) 30 Journal of Business Venturing 696; A Silvestri and S Veltri, 'Exploring the Relationships Between Corporate Social Responsibility, Leadership, and Sustainable Entrepreneurship Theories: A Conceptual Framework' (2020) 27 Corporate Social Responsibility and Environmental Management 585.

[36] D Dwivedi, P Kumar Yemula, and M Pal, 'Evaluating the Planning and Operational Resilience of Electrical Distribution Systems with Distributed Energy Resources Using Complex Network Theory' (2023) 46 Renewable Energy Focus 156.

[37] TJ Dean and JS McMullen, 'Toward a Theory of Sustainable Entrepreneurship: Reducing Environmental Degradation Through Entrepreneurial Action' (2007) 22 Journal of Business Venturing 50.

[38] M Wolsink, 'Distributed Energy Systems as Common Goods: Socio-political Acceptance of Renewables in Intelligent Microgrids' (2020) 127 Renewable and Sustainable Energy Reviews 1.

[39] D Fischer, R Mauer, and M Brettel, 'Regulatory Focus Theory and Sustainable Entrepreneurship' (2018) 24 International Journal of Entrepreneurial Behavior & Research 408.

[40] M Hellström and others, 'Collaboration Mechanisms for Business Models in Distributed Energy Ecosystems' (2015) 102 Journal of Cleaner Production 226; FM Belz and JK Binder, 'Sustainable Entrepreneurship: A Convergent Process Model' (2017) 26 Business Strategy and the Environment 1.

[41] M Asad and others, 'Green Entrepreneurial Leadership and Performance of Entrepreneurial Firms: Does Green Product Innovation Mediates?' (2024) 11 Cogent Business & Management 1; J Marzouk and R El Ebrashi, 'The Interplay Among Green Absorptive Capacity, Green Entrepreneurial, and Learning Orientations and Their Effect on Triple Bottom Line Performance' (2024) 33 Business Strategy and the Environment 1962.

[42] J Behrendt, 'Uncertainty-Related Transaction Costs in EU Law for Entrepreneurs Developing and Operating Microgrids' in K Baris Atici and others (eds), *Energy Entrepreneurship, Sustainability, Innovation and Financing: Practical Applications and Future* Directions (2025 Springer) 221.

[43] RS Kaplan and DP Norton, 'Mastering the Management System' (2008) 86 Harvard Business Review 62.

[44] S-B Tanya and D Galea, 'PEST Analysis' in *Wiley Encyclopedia of Management* (John Wiley & Sons 2014).

a multidisciplinary and comprehensive perspective without neglecting sustainability concerns.

Although PESTEL is widely accepted for its utility in anticipating risk and uncertainty, it is not a bullet-proof method as PESTEL factors can change very quickly due to the macroeconomics of doing business in developing countries.[45] In addition, interpretation of PESTEL analysis must take into account the distinct differences between developing and developed economies.

The political factor of PESTEL considers government policies and political institutions including predictability of business, type of government, taxation policy, and government regulation. In Nigeria, key political factors which affect entrepreneurs investing in DERs include change of government and change in the ruling party. For example, in 2025, the new Nigerian government introduced the Renewed Hope Solarization Project which aims to deliver 2 MW of clean, renewable energy to federal universities to ensure sustainable and reliable power. Also, reforms of Nigeria's tax law which will come into force in 2026 are part of the new government's economic policy on revenue.

The economic component of PESTEL examines how macroeconomics affects businesses. Some of the challenges faced by green entrepreneurs include high bank interest rates, hyperinflation, unemployment, gross domestic product, subsidies and global trade wars. Societal factors in PESTEL are demographic factors like change in income, change in population, and cultural attitudes towards DERs. For example, in Nigeria consumers believe that DERs are too expensive to obtain and maintain.

Technological factors create both opportunities and challenges for eco-entrepreneurs. New and emerging technologies like quantum computing and artificial intelligence are disrupting all sectors of the economy including renewable energy. Environmental factors in PESTEL are key considerations for attaining energy transition. This factor encompasses local and global climate change initiatives which could lead to changes in consumer behavior. Legal factors are the legislative and regulatory framework for doing business. For entrepreneurs these laws include environmental laws, competition law, labour regulations, health laws, technology laws, and entrepreneurship laws.[46]

As this chapter shows, there are several legal incentives and policies for DER entrepreneurs in Nigeria. However, the main challenge is poor implementation of laws and policy uncertainty. There is also limited availability of information and policy incoherence. Although the Nigerian government provides incentives for renewable energy, more can be done to provide accessible and effective incentives.[47] A market study conducted by the International Finance Corporation in 2021 showed that importers had been unable to secure 0 per cent duty on PV panels due to solar panels being imported as a combined unit.[48] In 2024, the Nigeria Customs Services clarified that in accordance with the ECOWAS Common External Tariff (CET) 2022-2027, solar panels assembled in modules or made up into panels are classified under HS Code 8541.42.00.00 and 8541.43.00.00 at 0 per cent duty and VAT-free.

[45] AE Henry, *Understanding Strategic Management* (5th edn, OUP 2025) 48.

[46] ibid 44.

[47] See L Paddock, 'Behind-the-Meter Distributed Generation: Lessons from the United States', in Chapter 11 of this book.

[48] IFC, 'Rooftop Solar Solutions Nigeria Market Assessment' (2021) <https://www.ifc.org/content/dam/ifc/doc/mgrt/nigeria-distrib-solar-market-assessment-fig-05042021.pdf> accessed 15 June 2025, 63.

Overall, a PESTEL analysis of the DER ecosystem in Nigeria shows that while several initiatives exist for DERs, in practice, they are not as effective. The next subsection makes proposals for supporting DER entrepreneurs in Nigeria. As the following sections show, it is important for developing countries like Nigeria to establish clear policy guidelines for DERs and startups.

III. Legal and Policy Gaps in Nigeria's Eco-entrepreneurial DER Framework

Nigeria's eco-entrepreneurial ecosystem for DERs is composed of different laws, policies, and regulations. Like most developing countries, Nigeria does not have a dedicated DER law or policy. Instead DERs are regulated by mini-grid regulations, policy documents, and contractual arrangements with international development partners.[49] This section of the chapter examines the main gaps in Nigeria's regulatory framework. These gaps are the absence of a dedicated DER policy and limited incentives for DER entrepreneurs.

A. Lack of Dedicated DER Law or Policy

The main regulatory gap in Nigeria is the absence of a dedicated DER law or policy document. On 9 June 2023, Nigeria's Electricity Act 2023 was enacted to replace the decades old 2005 Electricity and Power Sector Reform Act. Although the Electricity Act does not expressly refer to DERs, it introduces a regulatory framework firmly based on DERs by empowering state governments, companies, and individuals to generate, transmit, and distribute electricity from renewable energy sources including solar energy, wind, small hydro, and biomass.[50] The Act defines a mini grid renewable power system as 'a network of electricity supply from renewable energy technologies which is not connected to the grid' and an off grid renewable power system as 'electricity supply from renewable energy technologies which is not connected to the grid'.[51] This definition encapsulates Nigeria's policy shift towards decentralized renewable energy.

The 2023 Mini-Grid Regulations were established by the Nigerian Electricity Regulatory Commission in line with the revised electricity act.[52] The Mini-Grid Regulations define distributed generation as 'electricity generation from various small-scale, decentralized sources connected to the mini-grid or distribution system or directly to the point of use'. It recognizes diverse business models which are organized as sole proprietorships, partnerships, public limited liability companies, private limited liability companies, co-operative societies, and incorporated trustees. Although by recognizing diverse business models including sole proprietorships, the regulations indirectly support the development of DER

[49] See Nigeria Electricity Commission, 'Summit on Accelerated Scale-up of Renewable and Distributed Energy Resources in Nigeria, 22–23 July 2024 Agenda' <https://nerc.gov.ng/wp-content/uploads/2024/07/nerc-renewable-energy-summit-agenda.pdf> accessed 25 June 2025.

[50] Nigeria Electricity Act 2023 (signed 8 June 2023), art 80(1) and (2).

[51] ibid art 232.

[52] Nigerian Electricity Regulatory Commission, 'Mini-Grid Regulations 2023', Regulation no NERC R-116-2023 <https://nerc.gov.ng/wp-content/uploads/2024/01/MINIGRIDREGULATIONS.pdf> accessed 14 April 2025.

entrepreneurs, this does not correlate with the focus of the Nigerian Start-Up Act on only registered limited liability companies.

The Mini-Grid Regulations include detailed provisions on protection of the environment and sustainability. They provide that a breach of environmental laws could result in termination of or suspension of a permit. They also provide that where possible, a mini-grid permit developer shall remove, recycle, or dispose of equipment in an environmentally friendly manner. Schedule Six of the Mini-grid regulations provides for health and safety guidelines which include recommendations on environmental protection. Although the Mini-Grid Regulations don't refer to entrepreneurs, they provide a strong framework for licensing of small and medium DER businesses and encourage sustainable practices.

A second limitation of Nigeria's regulatory framework is the existence of multiple government departments and agencies in the energy sector. These agencies include the Federal Ministry of Power, Nigeria Electricity Regulatory Commission, Energy Commission of Africa, the Bank of Industry, and the Rural Electrification Agency (REA). The REA is an Agency under the Federal Ministry of Power established to provide electrification for Nigeria's unserved and underserved communities to catalyze economic growth and improve quality of life for Nigerians. REA implements several DER related projects including the Rural Electrification Fund (REF), Nigeria Electrification Project (NEP), Solar Power Naija (SPN), Energizing Education Programme (EEP), Energizing Economies Initiative (EEI), Energizing Agriculture Programme (EAP), and Energy for All—Mass Rural Electrification and Research and Innovation Hub.

The Standalone Solar Home Systems for Households and Micros Small Medium Enterprise aims to provide access to stand-alone systems provided by private sector companies. Nigerian states like Lagos State have also developed DER-related plans.[53] The REA is also a key stakeholder in DER projects like the Africa Mini-Grids Programme, Derisking Sustainable Off-Grid Lighting Solutions, and the Korean Energy Project which all target SMEs. The United Nations Development Program (UNDP) funded Derisking Sustainable Off-grid Lighting Solutions in Nigeria aims to develop a private sector-led technology value for making off-grid renewable energy technologies like solar lanterns. The Small and Medium Enterprises directorate of the Nigerian Bank of Industry provides finance for financing undeserved and budding sectors including solar. Under the National Renewable Energy and Energy Efficiency Policy, plans have been made to establish micro-credit facilities for entrepreneurs, especially women groups, for the establishment and operation of commercial solar energy facilities in remote and off-grid areas.[54]

The Solar Power Naija Programme was established in 2021 to champion and accelerate the development of public and private markets for off-grid solar power in Nigeria. The commercial workstream which is being implemented with technical support from the International Finance Corporation was established to develop Nigeria's market for commercial and industrial solar power. This component emphasizes captive, off-grid power and removing barriers to the flow of commercial finance through the solar value chain, by providing grants and support to solar companies. The consumer market component of

[53] Lagos State Ministry of Energy and Mineral Resources, 'The Lagos Off-grid Electrification Strategy and Action Plan' (2023) <https://lagosmemr.com/wp-content/uploads/2023/05/New-update-Double-Page-View-the-lagos-state-off-grid-electification-strategy-and-action-plan.pdf> accessed 14 April 2025.

[54] National Renewable Energy and Energy Efficiency Policy (2015) 19.

the programme awarded grants to businesses developing the markets for household solar systems and devices in Nigeria. These grants aim to help to mitigate capital constraints of SMEs.[55]

In 2017, after the approval of the Power Sector Reform Program (PSRP), the Nigerian government established the Distributed Access Renewable Energy Scale-up (DARES) which promotes DER objectives.[56] The Nigeria DARES project is the first joint World Bank attempt to create a private sector-led electrification. It is funded by the IFC (debt facility) and the World Bank (IDA financing) to create a platform for scaling up private sector-led electricity access solutions. Component 1 will focus on mini-grid expansion while Component 2 focuses on stand alone solar systems for electrification.

In 2016, the Nigerian government approved the Rural Electrification Strategy and Implementation Plan (RESIP) to facilitate private investment in expanding off-grid access to electricity rapidly in a cost-effective manner. The Nigeria Electrification Project was launched in 2018 as a private sector-led initiative. The Project is financed by US$550 million of public financing comprising US$350 million in credit from the World Bank and a US$200 million loan from the African Development Bank. It also involves private sector financing with the aim to connect more than 3.5 million people, 90,000 MSMEs, and public institutions.

Under the Nigerian Energy Transition Plan approved in 2022, the Nigerian government aims to achieve universal electricity access in the near term and a carbon neutral energy system by 2060. The Plan which was developed in 2021 with support from the COP26 Energy Transition Council (ETC) has the following DER related objectives: local manufacturing/assembly of key technologies such as solar panels, inverters, solar standalone systems, electric vehicles by 2025; technical assistance for skill development and knowledge transfer for the deployment of electric vehicles, establishment of a carbon market, and development of a just transition pathway beyond oil and gas; promotion of a just, inclusive, and equitable energy transition in Africa; and providing a conducive business and investment environment for the energy transition for deep decarbonization by 2060.[57] Some of the projects that have been developed include the 5 million solar projects which supported private distributors in deploying solar home systems and mini grids spread across geopolitical zones to electrify 5 million homes by 2023. The expected impact of this project is a 25,000 metric tons reduction of carbon emissions and empowerment of 200,000 SMEs.[58] Other initiatives are pilot rollout of electric vehicle supporting infrastructure and increased adoption of electric vehicles across commercial and residential segments.[59]

In Nigeria, a strong focus on solar technologies has limited the growth of other DERs.[60] It is clear that Nigeria has developed several DER-related strategies and initiatives which

[55] UK Foreign, Commonwealth Development Office (FCDO), 'Solar Nigeria Programme' <https://iati.fcdo.gov.uk/iati_documents/90000553.odt> accessed 15 April 2025.

[56] Rural Electrification Agency, 'Environmental and Social Framework for the Distributed Access Through Renewable Energy Scale Up Project' <https://ead.gov.ng/wp-content/uploads/2023/11/DARES-ESMF_Final-.docx9272023.pdf> accessed 14 April 2025.

[57] Nigerian Government, 'Investing in Nigeria's Energy Transition Opportunity' (March 2022) <https://www.energytransition.gov.ng//wp-content/uploads/2022/05/Investing-in-Nigeria-Energy-Transition.pdf> accessed 15 June 2025.

[58] ibid 4

[59] ibid 7.

[60] Renewable Energy Association of Nigeria, 'Advocacy & Engagement on Fiscal Incentives & Importation Procedures for Renewable Energy Technologies in Nigeria' (2024) <https://rean.org.ng/wp-content/uploads/2024/12/REAN-Report-Second-Review-with-updated-Table-of-Content.pdf> accessed 15 June 2025.

are beneficial to entrepreneurs. However, in the absence of a dedicated agency and database for entrepreneurs, it is difficult for small businesses to keep track of these initiatives. These challenges are closely linked to the political and legal components of PESTEL analysis.

B. Lack of National Strategy on Eco-entrepreneurship

Dedicated entrepreneurship law and distributed energy resources laws are a recent legislative development.[61] Traditionally, DERs have been regulated through contracts and private law.[62] In some US states draft DER legislation have been considered.[63] With the global rise in solar photovoltaic disputes, regulation is extremely important.[64] For example, in 2024, the Republic of Ireland's Supreme Court dismissed an appeal by a group of residents who sought to challenge the validity of a decision by Ireland's national independent planning body to grant planning permission for a photovoltaic solar farm.[65] While references to sustainability can be found in DER laws[66] and startup laws, express references to distributed energy resources in startup laws are rare. A good example is Korea's 2023 Special Act on Activation of Distributed Energy which came into force in 2024. It defines a distributed energy business as a small and medium-sized nuclear power generation business and small-scale electricity brokerage business.[67]

The main law which regulates entrepreneurship in Nigeria is the Nigeria Start Up Act 2022. The Act is applicable to companies incorporated under the Nigerian Companies and Allied Matters Act which have been granted the startup label and organizations whose activities affect the creation, support, and incubation of labelled startups in Nigeria. Under the Act, the startup label is granted to companies which have been in existence for less than ten years and which have the objects of innovation, development, production, improvement, and commercialization of a digital technology innovative product or process.[68] When compared with some laws in other jurisdictions which grant startup label for up to five years after incorporation, the ten-year period is realistic as it covers several DER entrepreneurs in

[61] For example, Republic of Korea Dispersed Energy Promotion Special Act 2023 (enacted 13 June 2023, entered into force 14 June 2024); Government of Western Australia, Electricity Industry Amendment (Distributed Energy Resources) Act 2024 (passed 7 March 2024); Republic of Seychelles Electricity Act 13 of 2023) Electricity (Distributed Generation System) Regulations, 2024 (signed 21 October 2024); Costa Rica Executive Decree No 43879-MINAE Regulations to the Law for the Promotion and Regulation of Distributed Energy Resources from Renewable Sources, No 10086 of 7 January 2022 (entered into force 1 February 2023, and repealed Executive Decree No 39220).

[62] Philippines, An Act Granting Solar Para Sa Bayan Corporation a Francise to Construct, Install, Establish, Operate and Maintain Distributed Energy Resources and Microgrids in the Remote and Unviable, or Unserved or Undeserved Areas in Selected Provinces of the of the Philippine to Improve Sustainable Energy 2019 (signed 23 July 2018).

[63] Eg New Hampshire Senate Bill 230 of 2025 Relative to Electric Utility Restructuring and Investment in Distributed Energy Resources; California Public Utilities Commission Decision Adopting Implementation Rules for Multi-Property Microgrid Tariffs and Other Matters (14 November 2024).

[64] Notable disputes include *Solar Century Holdings Ltd & Ors v Secretary of State for Energy and Climate Change* [2016] EWCA Civ 117.

[65] *Concerned Residents of Treascon and Clondoolusk v An Bord Pleanála, Ireland and the Attorney General and Elgin Energy Services Limited* (24 July 2024).

[66] Article 3(1)(c) of the Western Australian Electricity Industry Amendment (Distributed Energy Resources) Act 2024 provides that the state objective is to promote electricity efficiency in relation to the environment, including reducing greenhouse gas emissions.

[67] Act No 19437, 13 June 2023, art 2.

[68] Nigeria Start Up Act (signed 19 October 2022), art 15.

Nigeria and takes into consideration the average period of seven years during which startups can scale up and raise crucial funding.

As the long title of the Nigeria Start Up Act indicates, the law aims to create an enabling environment for technology-enabled startups and to position Nigeria as Africa's leading digital technology jurisdiction. The law does not contain any references to sustainability or renewable energy. However, as the law focusses on new technologies, distributed energy technologies which meet the requirements for labelling as startups can also benefit from the incentives provided under the startup law. This is reinforced by Policy Statement 4 of Nigeria's National Digital Innovation and Entrepreneurship Policy (2021–2023) which provides that the government shall stimulate more private sector contribution to power/energy and infrastructure.[69]

Under the Start Up Act, a startup label is a certificate issued to a limited liability company registered under the Companies and Allied Matters Act, which has been in existence for not more than ten years. Considering that the startup label is valid for ten years from the date of issuance, a DER start up could potentially take advantage of the incentives provided for no more than twenty years from the date of its incorporation.[70] To be eligible, at least one-third shareholding must be held by one or more Nigerians as founder or co-founder of the startup.[71] This criteria is very important for building Nigeria's capacity in the production of DERs as the biggest players in Nigeria are subsidiaries of foreign companies. A sole proprietorship or partnership can be granted a pre-label status for a period of six months to enable the sole proprietorship or partnership comply with the requirements for corporations.[72] By enacting a start-up law, Nigeria has shown a clear commitment towards supporting innovative businesses. To promote policy coherency and business facilitation, Nigeria should develop a comprehensive policy on DERs and entrepreneurship which consolidates all the policy developments occurring simultaneously in the different government agencies.

C. Limited Incentives and Subsidies for DER Entrepreneurs

DER startups are covered by several incentives in Nigeria under different policies and initiatives including the Start Up Act. However, there is no central policy document on existing incentives for DERs. As a result, DER entrepreneurs may find it difficult to access these initiatives. Under the Nigerian Electricity Act (2023), the REA's mandate includes advocating for tax incentives, investment capital allowance, and low interest loans to promote local production of renewable energy products.[73] In 2021, the REA unveiled the Research and Innovation Hub which includes a Start-Up Innovation Grant.[74] Nigeria's National Integrated Electricity Policy recognizes that regulators should create and strengthen incentive structures, particularly in the off-grid space, such as green energy purchase obligations, to attract private investment into the power sector, particularly in underdeveloped and less

[69] Federal Republic of Nigeria, 'National Digital Innovation and Entrepreneurship Policy (2021-2023)' <https://api.apiintelligence.org/upload/a4caf02bbfdfbda7d5b8965e54754fa0.pdf> accessed 15 June 2025.

[70] Nigeria Start Up Act, art 15(3).

[71] ibid art 13.

[72] ibid art 13(4).

[73] art 139(g).

[74] Rural Electrification Agency, 'Unveiling the REF, Research and Innovation Fund' (2021) <https://rea.gov.ng/ref-research-and-innovation-hub.pdf> accessed 15 June 2025.

viable areas.[75] The policy aims to facilitate private sector participation by creating a proper enabling environment and developing market rules, codes, regulations, and orders which prioritize facilitation of private sector investment.[76]

Inverters, rectifying apparatus, fuel cells, photovoltaic modules, regulated and unregulated power supplies, and solar home systems have been covered under item 46 of the Nigerian Investment Promotion Act List of Pioneer Industries which are also recognized under the Startup Act.[77] They are also eligible for tax exemption for a period of three years and an additional two years,[78] full deduction of any expenses on research and development which are wholly incurred in Nigeria and excluded from restrictions placed by the Companies Income Tax Act.[79] Startup companies are exempted from making contributions to the Industrial Training Fund if they provide in-house training to their employees.[80] In addition, startups are eligible for export incentives and financial assistance from the Export Development Fund, Export Expansion Grant, and the Export Adjustment Scheme Fund.[81] They can also access grants and loan facilities administered by the Central Bank of Nigeria and the Bank of Industry which are statutorily empowered to assist small and medium scale enterprises and entrepreneurs.[82]

In 2020, the Nigerian Central Bank established the Framework for Implementation of the Solar Connection Facility.[83] It provides long term low interest credit facilities with a maximum tenure of ten years to the Nigeria Electrification Project (NEP) pre-qualified home solar value chain players including manufacturers and assemblers of solar components and off-grid energy retailers.[84] Importation of fully assembled solar components are prohibited under the facility.[85] In 2024, the Value Added Tax (Modification) Order modified the First Schedule to the Nigerian Value Added Tax Act by inserting electric vehicles; semi-knock-down units for the assembly of electric vehicles; biogas and biofuel equipment; and accessories for clean cooking and transportation. The order has expanded the list of VAT exempt goods under the Fiscal Incentives for the Presidential Gas for Growth Initiative.

In June 2025, the Nigeria Tax Act was signed in to law to introduce sweeping tax reforms. The Act which will come into force in January 2026, amends sections of the Nigeria Start Up Act and exempts small businesses from paying income tax, capital gains tax, and the newly introduced development levy. Under the Nigeria Tax Act, the financial threshold for small companies which are not required to pay tax has increased from an annual gross turnover of NGN25million to NGN100 million and total fixed assets which do not exceed

[75] Nigeria Ministry of Power, 'National Integrated Electricity Policy' 16 <https://power.gov.ng/download/national-integrated-electricity-policy/> accessed 1 November 2025. See also, Nigeria Ministry of Power, 'National Integrated Resource Plan' (2025) <http://power.gov.ng/download/nigeria-integrated-resource-plan-nirp-2024/> accessed 14 July 2025.

[76] ibid 18.

[77] Nigerian Investment Promotion Commission, 'List of Pioneer Industries and Products Approved by the Federal Executive Council' (7 August 2017) https://nipc.gov.ng/wp-content/uploads/2019/01/Gazetted-List-of-Pioneer-Industries-and-Products.pdf accessed 15 June 2025.

[78] Nigeria Start Up Act, art 25(2).

[79] ibid art 25(3).

[80] ibid art 25(5).

[81] ibid art 26.

[82] ibid art 27.

[83] Central Bank of Nigeria, 'Framework for Implementation of the Solar Connection Facility 2020' <https://www.cbn.gov.ng/Out/2020/DFD/Solar%20Connections%20Facility%20Guidelines%201.0.pdf> accessed 15 April 2025.

[84] ibid.

[85] ibid 2.1.

NGN250million.[86] As a result of this reform, small DER entrepreneurs may not have to pay tax until they scale up. Also, renewable energy resources are exempt from a 5 per cent surcharge on fossil-fuel products provided or produced in Nigeria[87] while renewable energy equipment is exempt from value added tax.[88] The Act has introduced the Economic Development Incentive for priority companies to replace pioneer tax incentives which have been operational in Nigeria for decades. The Economic Development Incentive which is a tax credit of 5 per cent per annum for five years covers generation facilities that produce renewable electric energy and manufacturing of renewable energy equipment with a threshold of N100b for a sunset period of twenty years.[89]

As a member of the Economic Community of West Africa (ECOWAS), Nigeria is part of the ECOWAS Regional Standard which covers requirements for off-grid lighting products such as separate fixed (indoor) lighting systems, separate portable lighting systems, and portable in funded-vehicle systems. ECOWAS also aims to establish tariff mechanisms to promote off-grid electricity systems.[90] The ECOWAS Regional Off-grid Electricity Access Project aims to develop a regional market for autonomous solar systems. In July 2023, the ECOWAS Commission adopted two harmonized quality and assurance standards for off-grid solar products to promote renewable energy systems and hybrid systems for rural electrification. ECOWAS has also adopted recommendations for renewable energy and hybrid systems for rural electrification. Action 3 of the ECOWAS Renewable Energy Policy aims to make renewable energy power production an attractive business for private investors/entrepreneurs. The policy recognizes the importance of regional incentives and subsidies.[91]

By replacing pioneer tax incentives with the Economic Development Incentive which expressly refers to renewable energy, the Nigerian government has shown a clear commitment to prioritizing renewable energy. However, in comparison to jurisdictions like India which provide a 100 per cent tax incentive on profits for ten years, during the first fifteen years of a business, the Nigerian can introduce more fiscal and non-fiscal incentives.

IV. Functionalities for a Supportive Business Environment for DER Eco-entrepreneurs in Nigeria

Eco-entrepreneurs experience unique legal barriers in Nigeria. Nigeria has been classified as an electrification headwind due to its low access to electricity which does not directly correspond with its high solar startup activity.[92] Some of the challenges faced by entrepreneurs include limited access to information, lack of adequate financial support, limited expertise, poor energy infrastructure, and incoherent regulatory frameworks.[93] Other

[86] Nigeria Tax Act 2025 (signed 26 June 2025, entered into force 1 January 2026), art 202.
[87] ibid art 162(a).
[88] ibid art 186(2).
[89] ibid schedule X.
[90] 'ECOWAS Energy Policy 2023', Strategic Objective 1 <https://eis.ecowas.int/assets/front/medias/doc/media_20231011104957.pdf> accessed 15 June 2025. On the regional and international law in DERs, see DS Olawuyi, 'International Law on Distributed Energy Resources' in chapter 2 of this book.
[91] ECOWAS Renewable Energy Policy (2015) <https://downloads.unido.org/ot/46/74/4674905/lugmayr_ene_cpn_raf_2015_130007_003.pdf> accessed 15 June 2025.
[92] Fankhauser (n 9).
[93] B Kiraly, 'Best Practices for SMES in the Energy Transition' (European Covenant of Companies for Climate and Energy 2024) <https://www.smeunited.eu/admin/storage/smeunited/ccce-best-practices-for-smes-in-the-energy-transition-final.pdf> accessed 10 April 2025.

challenges are difficulties in reaching consumers, low levels of demand, and exchange rate fluctuations and currency convertibility.[94] According to the UN Conference on Trade and Development, key components of a functional entrepreneurship policy framework are optimization of the regulatory environment, enhancement of entrepreneurship education and skills development, facilitation of technology exchange and innovation, improving access to finance, and promotion of awareness and networking.[95] This section examines five main reforms that can be introduced in Nigeria to promote DER entrepreneurship.

First, it is important for Nigeria to develop a dedicated industrial policy for DER which provides incentives and subsidies for entrepreneurs. While a dedicated legislative instrument on DER has its advantages, an industrial policy in the form of a strategic document should be introduced first. This can be supported later on by DER amendments to the Nigeria Electricity Act. Although the 2023 Mini-Grid Regulations by inference regulate DERs, a more comprehensive policy supported by clauses which clearly define DERs and address the legal challenges identified in the previous sections should be developed with input from relevant stakeholders.

Second, a dedicated department should be established under the Nigerian Ministry of Power for Distributed Energy Resources and Green Startups. This department should provide information on the legal and regulatory support for eco-entrepreneurs online using an UpToDate website. For example, India's Ministry of New and Renewable Energy maintains a National Portal for Rooftop Solar. In Nigeria, the Green Energy Investment Platform which is to act as an online One Stop Investment Platform for Renewable Energy and Energy Efficiency investments in Nigeria does not have a functioning website. The Ministry of Power and the Ministry of Communications, Innovation, and Digital Economy should establish an online register of DER startups. This will ensure that eco-consumers can access information on the services provided by DERs.

The third major reform that should be introduced is dedicated fiscal and non-fiscal incentives for DER startups. Although as examined above, Nigeria has introduced new reforms under the Nigeria Tax Act 2025 which amend the Startup Act, more can be done. Some of the incentives that can be introduced to support entrepreneurs include feed-in tariffs, auctions, and loans.[96] In addition, government can establish industrial free trade zones for the local manufacturing of DERs which will be an appropriate transition to the proposed ban on the importation of solar panels. A good initiative is the introduction of a 10 per cent tax rebate for a period of two to five years to encourage individuals and businesses to purchase DER products. In 2023, South Africa introduced a one-year solar panel tax rebate to encourage installment of solar panels. In Uzbekistan, the government recently reduced the corporate income tax and property tax rates of companies producing solar installations, wind power plants, and small hydroelectric plants for a period of three years by 50 per cent. One of the major economic components of the PESTEL analysis of DER entrepreneurs which affects demand for products is the high cost of installing DERs. Appropriate incentives can increase demand and supply for DERs.

[94] Fankhauser (n 9).

[95] UNCTAD, 'Entrepreneurship Policy and Implementation Guidance' (2012) <https://unctad.org/system/files/official-document/diaeed2012d1_en.pdf> accessed 10 April 2025.

[96] UNCTAD, 'Investment Policies for the Energy Transition: Incentives and Disincentives' (2023) https://unctad.org/system/files/official-document/diaepcbinf2023d8_en.pdf accessed 15 June 2025.

It is important for the government to focus on the specific needs of SMEs. The previous sections of this chapter show that while Nigerian policies refer to start-ups, the main beneficiaries of grants and financial schemes for DERs are large enterprises and subsidiaries of foreign companies. This may largely be due to the focus of the REA on rural communities and the strict conditions imposed under international development bank financing. For example, under the Market Scale-up Challenge Fund (MSCF), REA clearly indicates in the implementation manual that the facility is not aimed at helping grow small and lower potential companies but very few startups.[97]

A third important reform initiative is the regular education and sensitization of eco-citizens and businesses on government policies. The Ministry of Power should also provide regular seminars for entrepreneurs and provide legal aid for small businesses. This will make it easy for startups to receive up to date information at no cost on available incentives and government support. It is also important that there is more transparency in disbursement of government grants and subsidies for DERs. As the previous section of this chapter showed, Nigeria does not have a principal DERs law or green start up policy. This means that participants and stakeholders are usually not well informed about strategic plans and developments. Although government agencies organize workshops, more can be done. In March 2025, as part of the Regional Off-Grid Electricity Access Project, ECOWAS organized a capacity building workshop for over fifty small and medium-sized enterprises in the off-grid solar photovoltaic sector in Nigeria. Regular reports should be made available to the public to educate citizens on DERs on a dedicated website. Recently, the Nigerian Customs Service released a Handbook on Import and Export Procedures for Renewable Energy Equipment to provide clarity on tariff classifications for renewable DERs including solar panels, batteries, and inverters.[98] This is a commendable initiative which will promote predictability for DER entrepreneurs.

A fourth policy initiative is development of a strong trade policy which maintains a balance between Nigeria's green industrial goals and Nigeria's renewable energy objectives. Although tariffs are a key source of revenue for developing countries, by lowering tariffs the government can ease the cost of doing business and encourage DER innovations.[99] The proposed ban on the importation of photovoltaic solar panels has been strongly opposed by entrepreneurs even though this ban has been part of Nigeria's strategic plans for years. The government should postpone the ban on importation of solar panels to reduce the cost of doing business and increase the demand for renewable energy.

Fifthly, while solar renewable energy is extremely important in Nigeria, it is important that other types of DERs are covered in Nigeria's policy and clearly defined. This must be complemented by policy coherence and a shift from purely tech-based and tech-driven eco-entrepreneurship to encompass green startups.

[97] Rural Electrification Agency, 'Project Implementation Manual for the Nigeria Electrification Project' (2020) <https://rea.gov.ng/project-implementation-manual_revised_20200614.pdf> accessed 15 June 2025.

[98] A Iruoghene and L Bushra, 'Customs Unveils Renewable Energy Handbook, Strengthens EU-Germany-GIZ Partnership' Nigerian Customs Newsletter (August 2025) <https://customs.gov.ng/?cat=15> accessed 7 August 2025.

[99] United Nations Conference on Trade and Development, 'Powering Trade: Fine-tuning Trade Policy for Solar and Wind Energy Value Chains' (2024) <https://unctad.org/system/files/official-document/osgttinf2024d3_en.pdf> accessed 15 June 2025.

V. Conclusion

This chapter has examined the regulatory challenges of DER entrepreneurs in Nigeria. It has shown that although Nigeria has myriad policies and incentives for renewable energy startups, implementation challenges remain in place. As captured through the PESTEL analysis, political, economic, technological, environmental, and legal factors contribute to these challenges. Although Nigeria's DER business landscape is diverse and composed of several startups, the main market participants are large enterprises. The newly enacted Nigeria Tax Act of 2025 is a key example of Nigeria's commitment to supporting renewable energy startups. As with many other developing countries, the key challenge remains poor implementation and low political will. To overcome these challenges, it is important to support legal interventions with policy frameworks and institutions which address the barriers faced by the eco-entrepreneurs. This chapter has shown the imperative of establishing an agency for green entrepreneurs and a dedicated DER policy which comprehensively develops a strategic framework for supporting startups in line with Nigeria's sustainable development objectives and commitments.

20

Entrepreneurial Legal Education and Energy Democratization

An Indigenous Renewable Distributed Energy Clinic

Alastair R Lucas and Allan E Ingelson

I. Introduction

Approximately 180 remote Indigenous communities in Canada's northern territories are not connected to North American electricity and natural gas grids. Some 70 per cent of these rely on electricity generated using diesel powered generators for electricity and home heating.[1] This has simply been the way it is done. Water and winter road fuel delivery is expensive and risky. Reliable community electricity services may be problematic. Hydrocarbon release contributes to GHG emissions.

This chapter will focus on distributed renewable energy potential for these communities located in the Canadian Northern Territories (west to east: Yukon, Northwest Territories, and Nunavut)[2] by supporting Indigenous communities with law and business education and advice to ensure informed local choices are made in the sustainable energy transition.

It presents a professional education and Indigenous community renewable energy strategy that will contribute to broader sustainability goals. The analysis and lessons learned will be relevant for other Canadian provinces and other countries in which remote Indigenous communities are located.[3] It will address as a case study establishment of a two-year pilot project proposed by the Faculty of Law, the Haskayne School of Business, and the Canadian Institute of Resources Law at the University of Calgary. The clinic will be a *pro bono* educational program, not a commercial enterprise providing professional advice. It will investigate energy and environmental law and policy relevant to design of an appropriate curriculum for the clinic.

A key literature review thread incorporates Indigenous values and traditions as well as Indigenous community consultation, participation, and accommodation. Consultation is

[1] Canada Energy Regulator, 'Market Snapshot: Clean Energy Projects in Remote Indigenous and Northern Communities' (15 February 2023) <https://www.cer-rec.gc.ca/en/data-analysis/energy-markets/market-snapshots/2023/market-snapshot-clean-energy-projects-remote-indigenous-northern-communities.html>accessed 19 August 2025; L Couves, 'Diesel dependency: the hidden cost of living in remote communities' (20 March 2025) <https://www.pembina.org/reports/diesel-dependency-hidden-cost-living-remote-communities> accessed 19 August 2025.

[2] Unlike the provinces, the territories are not under the Constitution Act, 1867. Rather they have their own constitutive federal statutes (Yukon Act, SC 2002 c 7; Northwest Territories Act, SC 2014 c 2 s 2; and Nunavut Act, SC 1993 c 28). There has been limited devolution of governance and land management powers to elected territorial councils that operate under federally appointed commissioners.

[3] Including Colombia (chapter 11) and Mexico (chapter 9).

Alastair R Lucas and Allan E Ingelson, *Entrepreneurial Legal Education and Energy Democratization* In: *Sustainable Distributed Energy Resources*. Edited by: Íñigo del Guayo, Damilola S. Olawuyi San, Louis de Fontenelle, Milton Fernando Montoya, and José Juan González, Oxford University Press. © International Bar Association 2026. DOI: 10.1093/9780198991946.003.0023

consistent with the University of Calgary Indigenous Strategy[4] and the constitutional legal doctrines concerning consultation and accommodation that Canadian courts have established in interpreting constitutional recognition and affirmation of 'existing Aboriginal and treaty rights' under section 35 of the Constitution Act 1982.[5] The case law includes decisions holding that consultation by energy regulators means real engagement not limited to oral or written announcements. It must include listening to and understanding Indigenous concerns, then responding fully to those concerns and accommodating them if necessary.[6]

A second literature review area is law and policy concerning literature that assesses alternative models for university *pro bono* law and business clinics. The substantive gap addressed is the inability of Canadian Northern Indigenous communities to address the harm resulting from continued community reliance on historic electricity generation and district heating practices.[7] The related gap is lack of preliminary law and business knowledge to participate effectively in project consultation should renewable energy generation projects be proposed in certain communities.

A component of the literature review is articles or in some cases, assessment of the overall legal and business models of established *pro bono* university-based clinics. These will include relevant business clinics, as well as environmental law clinics that are often involved with energy projects. The question will be what lessons can be learned? This supports research on Canadian law and practice in Indigenous energy project consultation and accommodation.[8]

This chapter targets a real and current set of issues faced by Northern Canadian Indigenous communities. Dependence on hydrocarbon energy and wood for electricity generation and home heating affects the quality of community life and health. The proposed clinic has the potential to help communities to address and solve some of these problems.

The remainder of the chapter begins with a look at the nature of the proposed clinic (Section II). Then Section III discusses renewable energy and how this kind of *pro bono* clinic can serve remote Indigenous communities. Sections IV and V assess the significance in a clinic syllabus of environmental rights and constitutional Indigenous rights, as well as the legal status of reserve lands. Relevant federal and territorial renewable distributed energy policy and law is introduced in Section VI. Section VII shows Indigenous business issues and models. Potential barriers to the proposed clinic are reviewed in Section VIII. Finally, Section IX summarizes conclusions and lessons learned.

II. The Nature of the Clinic

This will not be a conventional clinic, such as a public interest environmental law clinic, providing legal research, advice, and representation for clients opposing projects likely

[4] University of Calgary, 'Explore the strategy | Office of Indigenous Engagement | University of Calgary' https://www.ucalgary.ca/indigenous/about-ii-taapohtop/explore-strategy accessed 19 August 2025.

[5] The Constitution Act, 1982, Schedule B to the Canada Act 1982 (UK), 1982, c 11.

[6] *Tsleil-Waututh Nation v Canada* [2018] FCA 153 (*Tsleil-Waututh*).

[7] One example is the Gjoa Haven Clean Energy Implementation Project (2024) involving a range of energy generation and conservation projects aimed at reducing diesel use https://natural-resources.canada.ca/funding-partnerships/gjoa-haven-clean-energy-implementation accessed 19 August 2025.

[8] Section VI below.

to cause environmental and related social harm. The aim for these clients may extend to project cancellation or changes expected to negate or minimize environmental or social damage. Instead, the essence of the proposed clinic is education about renewable energy development in remote off-grid sites including information about the impacts from projects, legal requirements for consultation, and the approvals process for sites on or near Indigenous communities, and different types of business structures and options for project financing. Critically, all of this will be tailored to Indigenous persons and their communities in their physical and cultural context and the relevant body of Aboriginal law.

From the business clinic side, the prospect is similar with emphasis on distributed renewable energy providers, appropriate business structures, finance, government relations, and relevant regulatory matters.

A core focus is knowledge about renewable distributed energy technology and providers, as well as Indigenous community requirements and objectives. In the northern territories wind, solar, and geothermal resources are of particular interest. Significant is information about how distributed renewable energy can address specific remote Indigenous community needs, including social, cultural, and environmental objectives. This requires enough scientific, technical, and Indigenous cultural knowledge to be able to interact with specialist advisors in these areas. An analogy is legal and business information concerning environmental assessment (encompassing social/cultural and economic factors) that may be required in regulatory processes. Of course, the student education objectives must be considered beyond the potential distributed renewable energy benefits and constraints.

Most important is that students and Indigenous communities understand a paradigm that delivers clean energy, environment, and social justice in a clinical context that recognizes the colonial dominance that still characterizes these legal and business relationships. Patricia Barkaskas speaks of 'situated rationality' and emphasizes that Indigenous clinics must be 'rooted within specific First Nations' relationships, language, and ceremony'.[9]

Literature shows development of a public interest law and business clinic model with relatively common philosophy, education values, and clinic structure.[10] This suggests a template for an Indigenous distributed energy clinic. Other literature clarifies the technical and operational side of distributed renewable energy. There are studies that target specifically clinical legal education aimed at Indigenous people and communities.

Concerning workshop curriculum development for this type of clinic, it is necessary to understand the Indigenous rights constitutional context. There is a large body of literature on the constitutional Aboriginal rights and role of Indigenous communities in energy and environmental regulation that is part of the larger literature on the law and policy that supports Canada's federal and provincial regulatory systems. Core democratic principles and energy justice concepts frame all of this.[11]

There are a number of important considerations regarding the content of educational workshops and related education materials on renewable energy (RE) development to be offered to Indigenous partners. Fundamental is recognition of diversity and different

[9] P Barkaskas and others, 'Reflecting on Clinical Legal Education at the Indigenous Community Clinic' (2020) 32 Osgoode Journal of Law and Social Policy 138, 139.

[10] There are law school clinics in eight Canadian jurisdictions that have law schools: Justice Canada, 'Legal Clinics in Canada: Exploring Service Delivery and Legal Outcomes Among Vulnerable Populations in the Context of COVID-19' (November 2021) s 5.2 (Justice Canada).

[11] See *Tsleil-Waututh* (n 6); Constitution Act 1982 (n 5); s 8, Indigenous Rights.

perspectives of Indigenous peoples (First Nations, Metis, and Inuit) in the northern territories, and the different types of lands where RE may be developed only with the consent of Indigenous peoples.

For learner inclusivity the content of workshop materials should reflect the authentic experience of Indigenous peoples in their respective communities and a genuine understanding of the context and lived experience of clinic clients. Context should incorporate different aspects of the relevant cultural, historical, political, legal, and social context and relationships with non-Indigenous peoples and organizations.

In regard to pedagogy, cultural elements such as oral histories, Indigenous history related to the land and energy development, the role of elders, storytelling, the role of knowledge holders and elders in knowledge transfer, multigenerational focus and connection with the land, should be integrated into the pedagogy. Shared learning and problem-solving exercises should be incorporated into clinic workshop pedagogy.

Workshop curricula should be designed to integrate both Indigenous and non-Indigenous learning to avoid the lack of content reflecting the broader Indigenous community to promote shared understanding of the renewable energy development and land use policies, laws, and the associated issues. For delivery of workshop programs consideration should be afforded to traditional opening and learning family sessions and traditional closing ceremonies. Hard copy materials and eBooks should be developed in collaboration with and guidance from First Nations, Métis, and Inuit subject-matter experts, and others.

To promote a safe environment for Indigenous learners and to bridge the Indigenous content, context, and culture, the clinic should attempt to retain an instructor or workshop representative who has the following competencies: (a) a deep understanding of Indigenous ways of knowing, being, and doing; (b) demonstrated Indigenous knowledge, lived experience, and/or meaningful connections to Indigenous communities; and (c) a genuine commitment to inclusive teaching practices that honor Indigenous perspectives and promote workshop participation to support the learners and address the unique educational and professional challenges faced by Indigenous learners.

The above considerations need to be incorporated into workshop contract issues. Contract provisions must recognize the diversity of Indigenous Peoples. Instructional materials should be balanced across the three Indigenous groups: First Nations, Inuit, and Métis. Moreover, the content should reflect the Indigenous population's diversity and include those on First Nations, Métis settlements, and Northern communities. Materials should include references to local Northern communities that may be affected by RE development, treaty areas, or other references to qualify the material as specific to a region and not generalize the Indigenous population. Contracts should provide for the authenticity of Indigenous content—the honest and authentic experience of Indigenous people and their communities. Examples and cases should reflect the subjects of these lived experiences. Content may include types and nature of businesses and organizations, profit and not-for-profit enterprises, organizations, not limited to business enterprises within Indigenous communities. Authenticity is only achieved by partnering with those who have a lived Indigenous experience.

Authenticity of Indigenous context requires that the context reflect Indigenous people's authentic environment and describe Indigenous people in their respective communities. The context, along with the content, creates a sense of actual experience and thus creates a sense of inclusivity for the Indigenous community. Context could include aspects of culture, history, political, legal, and social contexts, as well as relationships with non-Indigenous parties.

Content should employ cultural elements as part of the workshop pedagogy. Specific and common cultural elements should be used in the instruction. These include oral histories and storytelling, the role of Elders and knowledge holders in knowledge transfer, multi-generational focus, and humour and connection with the land should be incorporated into the pedagogy. Participation of local Indigenous knowledge holders or Elders who can provide insight into regional customs and practices is encouraged. Culture, traditional knowledge, practices, and Indigenous history should be incorporated into workshop instruction when complementary. In addition, collective instruction, problem-solving, and shared learning should be a part of workshop pedagogy.

III. Distributed Renewable Energy

Chapter 1 of this book describes distributed energy resources as 'small scale generation, transmission, and storage of electricity [facilities] that are often at or closer to the point off consumption'. It refers to cost and environmental assessment, as well as promotion of 'energy security and sustainable development'. Further, they identify four key vectors of transformation analysis that are adopted and expanded in this chapter. Also helpful is 'Distributed energy systems: A Review of classification, technologies, applications, and policies'.[12] This points to another broad-based vector, namely sustainability. Here it spotlights a key sustainability tool, namely renewable energy. Renewable technology is tailored to distributed energy so that it can be produced by small facilities and scaled to small remote community characteristics and needs. Place definitely matters. So, the focus is not merely on distributed energy but distributed renewable energy.

A. Remote Indigenous Communities

The final vector concerns a singular kind of place—Northern Canadian Indigenous communities. Their small size, Northern hinterland locations, and lack of reliable road access dictate small-scale electricity technology. Historically, this has meant diesel-powered generation. These generators are small and manageable, and critically, diesel fuel can be transported in small loads on barges, and in winter by truck on ice roads. Thus, there is a distributed electricity generation history. But the hydrocarbon fuel is classically unsustainable—both economically and ecologically unreliable, presenting ecosystem and public health risks.[13]

[12] T Bin Nadeem and others, 'Reflecting on Clinical Legal Education at the Indigenous Community Clinic' (2020) 32 Osgoode Journal of Law and Social Policy 138–39.

[13] Intergovernmental Panel on Climate Change, Sixth Assessment Report, 'AR6 Synthesis Report: Climate Change 2023' (March 2023) <https://www.ipcc.ch/report/ar6/syr/> accessed 19 August 2025.

There have been many successful renewable energy projects in Canada's southern provinces and other parts of Canada.[14] A notable example among others[15] is the Fort Chipewyan Solar Generation Project in Northern Alberta.[16]

1. *Pro bono* law and business clinics

These law clinics are common in Canadian law schools.[17] These have two fundamental objectives. First, they are intended to provide law students with hands on professional learning experience. This is mentored client service work that takes them a step beyond the substantive law and legal analysis courses that characterise law school curricula. Second, these clinics are aimed at ameliorating social problems, targeting the prohibitive cost of legal services faced by poor and vulnerable people. These involve criminal or civil justice matters that require legal advice in court or tribunal proceedings that are beyond the means and personal competence of these individuals. Essentially, it is a matter of ensuring that justice is served. Prime issues for these clinics are divorce, spousal and child support, post-separation matters including child custody and access, landlord and tenant, and a range of personal and family issues.[18]

All of these involve personal and justice system issues that require a response. They are not matters that arise when individuals or corporate entities initiate commercial projects or take other actions.

The result is that while designers of *pro bono* Indigenous distributed renewable energy clinics can learn from *the structure and operations of pro bono legal clinics*, this must be done carefully. An Indigenous renewable distributed energy clinic does involve legal information and advice. It also involves technical and business information and advice by knowledgeable citizens and professionals, including lawyers (students and lawyers), renewable energy systems experts, Indigenous community engagement specialists (including community members and elders), and business systems professionals. The clinic should thus have: (1) a law and business (including technical knowledge) provider; (2) a provider of Indigenous traditional knowledge; and (3) a facilitator of legal/regulatory and business collaboration.

2. *Pro bono* business clinics

A good example of a law-based business clinic is the Business Venture Clinic at the University of Calgary Law School. This clinic focuses on business start-ups. According to the website:

> 'The Business Venture Clinic offers you the opportunity to get hands-on work in entrepreneurship and business law. You will be matched with start-up enterprises to work with throughout the year. This is an opportunity for you to gain valuable experience doing a

[14] NRC Canada study, see D Eeles and others, 'Energy Storage: The Regulatory Landscape in Alberta' (2021) 59(2) Alberta Law Review 355 at 359; Alberta Utilities Commission, 'Distribution System Inquiry: Final Report' (February 2021) 144.

[15] Canada Energy Regulator, 'Market Snapshot: Indigenous Ownership of Canadian Renewable Energy Projects is Growing' (21 June 2023) <https://www.cer-rec.gc.ca/en/data-analysis/energy-markets/market-snapshots/2023/market-snapshot-indigenous-ownership-canadian-renewable-energy-projects-growing.html > accessed 19 August 2025.

[16] Three Nations Energy, 'Clean Energy Projects In Fort Chipewyan, AB' https://www.3ne.ca/ accessed 19 August 2025.

[17] Justice Canada (n 10).

[18] ibid, s 5.1.3.

variety of legal work such as drafting agreements and contracts. During this program, you will be matched with a mentor lawyer who practices in the area and who will provide helpful feedback on your work. This helps you learn the ropes before you start practising on your own.

You will also take a class in entrepreneurial law to support your hands-on learning. In this class, you will learn about issues that commonly arise in start-up companies, such as employment issues, trade secrets, dividing up founders' equity, intellectual property licensing, and board of director problems. In most cases, part of the class will be spent in a lecture and part will be spent on practical case study assignments.'[19]

This is a hypothetical file-based type clinic. It is not intended to provide direct legal and business advice to clients.

By contrast, the Aboriginal Law clinic at the University of Toronto provides free legal services to low-income Indigenous people living in Toronto. This is the classic law school *pro bono* legal services model focused on a range of Indigenous people's personal challenges, including: housing problems and tenant rights; complaints about inadequate police protection; and human rights equality issues.

Clearly an Indigenous renewable distributed energy clinic would have a narrower focus. The key difference is that unlike the University of Toronto clinic it would not assist with litigation about Indigenous human rights and citizen entitlements. Further it will not be solely a law clinic but will incorporate relevant business and technical information. The overarching goal will be sustainability—facilitating community action that will contribute to classical sustainability goals[20] while improving community quality of life.

IV. Environmental Rights

There is little doubt that the state of our natural environment is dire, with increasing pollution, ecological and human harm, and unprecedented natural disasters. Climate change has pushed global temperature into the danger zone, Northern ecosystems and Northern residents are particularly at risk.[21]

In considering a syllabus for the clinic, environmental ethical questions abound.[22] Whatever responsive approaches are taken, basic environmental law assumptions need to be examined. For example, attempts have been made by governments and individuals to engage fundamental rights to a healthy environment, including food security, as well as clean air, water, and land.

However, environmental law development has not been equal to the protective actions required. David Boyd has stated:

[19] University of Calgary, 'Business Venture Clinic' <https://law.ucalgary.ca/business-law/learn/business-venture-clinic> accessed 19 August 2025.

[20] UN World Commission on Environment and Development, *Our Common Future* (OUP 1987).

[21] See D Boyd, *Cleaner, Greener, Healthier* (UBC Press 2015) 139.

[22] See S Bagg, 'Ethical Dimensions of Environmental Law' in WA Tilleman and others (eds), *Environmental Law and Policy* (4th edn, Emond 2020) 35.

> 'Contrary to the myth of a pristine green country providing environmental leadership to the world, a huge pile of studies proves beyond a reasonable doubt that Canada lags behind other nations in terms of environmental performance. According to researchers at Simon Fraser University, Canada's environmental performance ranks twenty-fourth out of the twenty-five wealthiest nations in the Organisation for Economic Cooperation and Development (OECD). The OECD has published blistering criticisms of Canada's weak laws and policies, perverse subsidies for unsustainable industries, and poor environmental performance.'[23]

According to Sara Bagg the problems are clear. Environmental laws generally ignore laws of nature, assuming infinite earth resources.

They are made and enforced by governments reliant on revenue produced by environmentally destructive activity, and operate in a context of limitless growth, consumerism, overarching property rights, and powerful corporations.[24]

We would expect to rely on Canada's Constitution for this kind of legal protection. But the constitutional Canadian Charter of Rights and Freedoms, which guarantees rights to life, security of the person, and equality, has been interpreted not to guarantee enforceable environmental rights.[25] There are some statutory rights (subject to removal or amendment by the legislatures) at the provincial level. But these are mainly procedural rights to information, to comment, and a few restricted (requiring prior government consent) to sue.[26] It remains the case that there are no constitutionally targeted environmental rights in Canada,[27] and thus no rights to, for example, federal and provincial financial assistance to support specific Indigenous community renewable energy initiatives such as distributed energy projects. Nor is it clear that constitutionally entrenched Indigenous rights can achieve this kind of objective. This is discussed below.

Weak environmental rights have caused some to look to constitutionalized Aboriginal rights to function as adjunct environmental rights. After all, ecosystem protection is central to the core Aboriginal rights to hunt, fish, and trap. However, Aboriginal rights have been interpreted on a case-by-case basis. The judicially recognized legal 'duty to consult and accommodate' Aboriginal people affected by energy projects may spotlight ecosystem and energy issues. But this duty applies to specific treaty and traditional lands and their energy resources.

The duty to consult, an element of the Crown's protective fiduciary duty to Indigenous people, includes administrative tribunals like the Alberta Energy Regulator (AER), Alberta Utilities Commission (AUC), Alberta Environmental Appeals Board (EAB), Natural Resources Conservation Board (NRCB), and other statutory agencies such as the Alberta government Aboriginal Consultation Office (ACO). But it does not extend to private persons or corporations.

A more basic problem with this approach is that the diverse groups of Indigenous people and First Nations who are parties to different types of treaty agreements are by no means single issue environmental advocates. The primary concerns are food, shelter, and cultural

23 DR Boyd, *The Right to a Healthy Environment: Revitalizing Canada's Constitution* (UBC Press 2012) 6.
24 Bagg (n 22) 44; see below.
25 AR Lucas and A Olugbemi, 'Legal Rights to a Healthy Environment', CIRL Occasional Paper #82, April 2023.
26 See Ontario Environmental Bill of Rights. 1993, SO 1993 c 28.
27 Lucas and Olugbemi (n 25).

resource protection. In these matters, there can be reasonable differences. This may push environmental issues down priority lists and ensure that they do not drive legal strategies.

V. Indigenous Rights, Title, and Legal Status of Reserve Lands

The possibility that constitutional Aboriginal rights and title may have environmental rights traction has been debated.[28] This is based on Section 35(2) of the Constitution Act 1982 which states:

> ... existing aboriginal and treaty rights of the aboriginal peoples of Canada are hereby recognised and affirmed.

Included are rights concerning the protection and use of wildlife and its habitat on traditional lands. Rapidly developing Supreme Court of Canada jurisprudence on the duty to consult and accommodate,[29] Aboriginal title,[30] and Aboriginal governance[31] may present the strongest constitutional support for rights to a healthy environment. The question here is how these rights may work in relation to Indigenous distributed renewable community energy projects.

To establish and protect Aboriginal rights, the Supreme Court of Canada has outlined a two-step formula. This involves first identifying the nature of the Aboriginal right claimed because of, 'the nature of the action which the applicant is claiming was done according to the Aboriginal right, the nature of the government regulation, statute or action being impugned, and the custom or tradition being relied upon to establish the right';[32] and second, proof that the activity or action is an 'element of a practice, custom or tradition integral to the distinctive culture of the Aboriginal group claiming the right'.[33]

The latter requires that this element of distinctive culture has been practised prior to European settlement and that there is reasonable continuity of practice to the present. There must be clear evidence of infringement.

However, once the right is established, infringement is still possible, but it must be justified[34] as in the public interest, with a compelling and substantial objective, consistent with the Crown's fiduciary obligation to the Aboriginal group, a rational connection between the objective and the impugned action, minimum impairment, and benefits that outweigh the negative effects on the Aboriginal group. Further, the Crown's duty to consult and accommodate must be met.

[28] See H Fast and B Gunn, 'Indigenous Peoples' Jurisdiction over the Environment', in W Tilleman and others (eds), *Environmental Law and Policy* (4th edn, Emond 2020) 317, 325–26.

[29] See *Gitxaala Nation v Canada* 2016 FCA 187 (*Gitxaala Nation*).

[30] *Tsilhqot'in Nation v British Columbia* 2014 SCC 44 (*Tsilqot'in Nation*).

[31] See *Reference to the Court of Appeal of Quebec in relation to the Act Respecting First Nations, Inuit, and Metis Children, youth, and families* 2022 QCCA 185 (Quebec Reference).

[32] *R v Van der Peet* [1996] 2 SCR 507 para 53.

[33] ibid para 46.

[34] *R v Sparrow* [1990] 1 SCR 1075.

A. The Duty to Consult and Accommodate

This does not create a free-standing Aboriginal right but is the corollary of fiduciary obligations owed by the Crown to Indigenous peoples of Canada.[35] It arises when the Crown knows of Aboriginal rights and contemplates action that might affect those rights adversely. The Supreme Court of Canada has described a spectrum of Aboriginal rights and impact significance to determine the scope and content of consultation and accommodation required in specific circumstances.[36]

1. Aboriginal title

This is a communal Indigenous land right that extends beyond Aboriginal rights and includes use and occupation of particular land. The Supreme Court of Canada in *Tsilhqot'in Nation v British Columbia*[37] stated that this encompasses economic rights including mineral rights and generally the right to manage the land. Essentially the same prior to European settlement test applies, requiring that clear and obvious use and occupation be shown.

Tsilqot'in Nation, decided in 2014, was the first Supreme Court of Canada declaration of Aboriginal title. The litigation began in 1983.[38] In its reasons, the Supreme Court used what it described as a 'culturally sensitive' approach to sufficiency of territorial occupation. Its approach goes some way towards balancing resource development objectives and environmentally sensitive land use. This is underlined by the statement in the context of justification that,

> . . . incursions into Aboriginal title cannot be justified if they would substantively deprive future generations of the benefit of the land[39]

This is consistent with the intergenerational equity elements of environmental sustainability.[40] Following the Supreme Court's title declaration, the Tsilhqot'in Nation proceeded to use the title area to anchor an adjacent Tribal Park designed to protect the ecological values that support their Aboriginal rights and title.[41]

2. Legal status of reserve lands

This is not straightforward. As a matter of constitutional division of powers, section 91 of the Constitution Act 1867 gives the federal government exclusive legislative power over these lands along with legal title, subject to the beneficial rights of the Aboriginal people concerned ('Indians and lands reserved for the Indians'). However, generally, provincial law made under provincial legislative powers applies to Indians and lands reserved for

[35] *Haida Nation British Columbia* 2004 SCC 73 (Haida Nation).
[36] *Gitxaala Nation* (n 29).
[37] *Tsilqot'in Nation* (n 30).
[38] See N Schabus, '*Tsilhqot'in Nation*: A Territorial Land-Use Concept of Aboriginal Title and Aboriginal Governance' in W Tilleman and others (eds), *Environmental Law and Policy* (4th edn, Emond 2020) Ch 8, 127, 135.
[39] *Tsilhqot'in Nation* (n 30) para 86.
[40] See L Collins, 'Revisiting the Doctrine of Intergenerational Equity in Global Environmental Governance' (2007) 30 Dalhousie Law Joumal 79.
[41] The Firelight Group, 'Dasiqox Tribal Park established by the Tsilhqot'in to protect their land and culture', 31 March 2016, < https://firelight.ca/news/dasiqox-tribal-park-established-by-the-tsilhqot-in-to-protect-their-land-and-culture > accessed 19 August 2025.

the Indians. The result is that Alberta energy regulatory laws administered by the Alberta Energy Regulator (AER) and the Alberta Utilities Commission (AUC) govern renewable energy facilities and their operation on reserves, but uncertainty exists whether provincial facilities reclamation and abandonment provisions apply on reserves.

This body of Aboriginal rights law suggests some support for constitutional environmental rights. However, the qualification is that these may be limited to defined traditional lands, and the goals of environmental rights litigants do not always line up with First Nations aspirations and strategy.

However, it is easy to imagine disputes concerning proposed First Nation community renewable energy projects in which for example, a community council disputes the benefits of a proposed electricity generation, battery storage, or heating facility that focuses on First Nation members, or hereditary chiefs contesting a provincial or federal government regulatory decision-making procedure by which a First Nation Council—private electric utility renewable energy project was approved.

In this scenario, facts about the adequacy of a consultation program are critical. For example, was regulatory agency notice of a public hearing on a project delivered only through the agency's website, social media platforms, and notice in a large city daily newspaper? In applying the Supreme Court's spectrum of Indigenous rights, nature, and impact, facts such as quality and reliability of the proposed electricity service, early community interaction with the relevant government agencies, and communication alternatives available, would be crucial.

Concerning substantive Indigenous impact, what custom or tradition is relied upon? Is there a distinct practice or custom involved? Is there direct physical impact—for example location of solar panel arrays within meters of some residents' homes, nearby wind turbines killing birds, or the safety of proximal geothermal hot water pipelines? This inquiry may lead in another direction, for example, to the established decision role of hereditary chiefs relative to that of the elected council in that First Nation.

VI. Federal and Territorial Indigenous Renewable Energy Policy and Law

There is a Canadian framework of policy and law in which Indigenous renewable distributed energy has developed and can develop further. The federal government has supported renewable energy with various initiatives within a national renewable energy target of 90 per cent by 2030.[42] It has pursued a net-zero fifty GHG emissions goal with five-year targets and carbon pricing legislation. The oil and gas sector's GHG emissions cap (a cap-and-trade system) is in final preparation during August 2025. Overall guidance is in the document Canada's Transition to a Clean Economy,[43] which maps out an electric energy transition path. A key instrument is the Clean Power Regulations issued in late 2025.[44]

[42] Government of Canada, 'Sustainable Development Goal 7: Affordable and clean energy' < https://www.canada.ca/en/employment-social-development/programs/agenda-2030/afforadable-clean-energy.html >accessed 25 July 2025.

[43] Government of Canada, 'Canada's Clean Electricity Future' (2024) <https://www.canada.ca/en/services/environment/weather/climatechange/climate-plan/clean-electricity.html> accessed 19 August 2025.

[44] Clean Power Regulations, SOR/2024-263, 13 December 2024.

Also guiding this federal renewable energy promotional policy is the Cabinet Directive on Regulatory and Permitting Efficiency[45] for Clean Growth Projects.[46] This directive targets key sectors including power/electricity, nuclear, and clean fuels. The objective is to speed up federal project decision making and describe federal roles and responsibilities.[47] Included is increased early engagement with Indigenous peoples affected by projects.[48]

More specifically, the directive clarifies section 35 of the *Constitution Act 1982's* overarching special relationship with Indigenous Peoples.[49] Other recitals include endorsement of the UN Declaration on the Rights of Indigenous People, particularly provision for consultation and co-operation to ensure free, prior, and informed consent' prior to approval of impacting projects. Required support of provinces, territories and municipalities, investors, and proponents to ensure clean growth, is also noted. An 'Action Council' is created to monitor Indigenous consultation and necessary consent.[50]

However, the United States tariff regime unveiled on 4 March 2025 has upended both federal and provincial energy and environmental law and policy.[51] Canadian response[52] to the 25 per cent US tariffs (with 10 per cent on energy products including crude oil, gas, and electricity and uranium) is a 25 per cent surtax applicable to a range of specific US products, with potentially more to be added to the list. Ontario premier Rob Ford imposed a 25 per cent export tax (later 'paused') on Ontario electricity exports to the US and threatened to cut exports.[53] The Alberta government intends to expand oil and gas exports, tariffs notwithstanding. Meanwhile uncertainty reigned in Summer 2025 as financial markets swung, and commerce slowed in both countries.

Even before the US tariffs there were signs in public opinion polls of a 'seismic shift'[54] in Canadian energy and environment policy. 'Carbon tax' became a 'toxic' term.[55] There was emphasis in many contexts of emerging public concern about economic growth and energy security relative to anti-pollution and climate action wrapped in a broad sustainable energy transition.

On the territorial side, the Canadian Territories of Yukon, Northwest Territories, and Nunavut face unique challenges and opportunities in their pursuit of renewable energy frameworks. These vast, sparsely populated northern regions are rich in natural resources yet experience logistical and environmental constraints due to remote locations and extreme weather conditions.[56] Historically reliant on diesel fuel for power generation, the

[45] Government of Canada, Cabinet Directive on Regulatory and Permitting Efficiency (July 2024).

[46] ibid s 4.

[47] ibid s 5, 'Governance'.

[48] ibid s 10, 'Indigenous Partnerships'.

[49] ibid s 1, 'Introduction/Preamble'.

[50] ibid s 5(a) 'Deputy Ministers' Regulatory Efficiency Action Council'.

[51] Executive Order 14193 of 1 February 2025 (Imposing Duties to Address the Flow of Illicit Drugs Across Our Northern Border), as amended by Executive Order 14197 of 3 February 2025 (Progress on the Situation at Our Northern Border).

[52] US United States Surtax Order (2025-1).

[53] R Ferguson, 'Doug Ford slaps 25 per cent tax on electricity to the U.S. in response to Trump tariffs' *Toronto Star* (10 March 2025).

[54] C Varcoe, 'Poll Reveals Seismic Shift on Energy' *Calgary Herald* (14 February 2025) A1.

[55] Toronto Sun, 'Toxic carbon tax losing former allies' (12 September 2024) <https://torontosun.com/opinion/editorials/editorial-toxic-carbon-tax-losing-former-allies> accessed 19 August 2025.

[56] Procido LLP, 'Renewable Energy in the Canadian Territories: Pioneering Clean Energy in Remote Landscapes' <https://procido.com/2024/11/10/renewable-energy-in-the-canadian-territories-pioneering-clean-energy-in-remote-landscapes/> accessed 7 August 2025.

Territories are now increasingly focusing on renewable energy sources to enhance energy security, reduce greenhouse gas emissions, and lower costs.

A. Yukon

Yukon government has two main pieces of legislation governing electrical energy namely the *Yukon Development Corporation Act*[57] and the *Yukon Public Utilities Act.*[58] The *Yukon Development Corporation Act* states that the Yukon Development Corporation (YDC) is to participate with the private sector in the economic development of Yukon and, in particular, with regard to energy systems and generation, production, transmission, and distribution. The Yukon Utilities Board acts as the regulator of electricity utilities. It issues orders setting rates, approving standards, and overseeing utility practices, including for public utilities like Yukon Energy Corporation.

It is noteworthy that the Government of Yukon has fully implemented the Independent Power Production policy, allowing First Nation governments, communities, and entrepreneurs to generate renewable energy and feed it into the electrical grid to help meet local demand.[59] According to the policy, renewable energy sources include hydro, solar, geothermal, and biomass; natural gas does not qualify as a renewable energy source.

The former Minister of Energy, Mines and Resources, Ranj Pillai, had said that this is a true green energy policy that promotes a secure and sustainable energy future for Yukon as it supports the development of local energy infrastructure and will help increase the supply of renewable energy across the territory while reducing the reliance on non-renewable sources in our communities.[60] The YDC was directed to develop and manage the Innovative Renewable Energy Initiative (IREI) with the objective of encouraging and supporting the development of community-led renewable energy generation projects in Yukon.[61]

IREI is open to a broad range of Yukon-based organizations including Municipal Governments, First Nation Governments, First Nation Development Corporations, Community Societies or Associations, and Yukon-based businesses. IREI aims to reduce reliance on fossil fuels and diversify its energy mix by supporting local wind and solar projects. It also seeks to build local capacity and business opportunities in renewable energy by funding at least four generation projects by 2025.[62] Through the IREI, the Government of Yukon has supported the development of independent power producers, including the Vuntut Gwitchin First Nation's solar project in Old Crow and Kluane First Nation's wind project in Destruction Bay.

[57] RSY 2002 c 236.
[58] RSY 2002 c 186.
[59] Yukon, 'Government of Yukon's Independent Power Production policy implemented' <https://yukon.ca/en/news/government-yukons-independent-power-production-policy-implemented> accessed 7 August 2025.
[60] ibid.
[61] Yukon, 'Innovative Renewable Energy Initiative policy'< https://yukon.ca/sites/default/ files/ydc-innovative-renewable-energy-initiative-policy.pdf> accessed 7 August 2025.
[62] ibid.

B. Northwest Territories

The Northwest Territories (NWT) electricity sector is principally governed by the Northwest Territories Power Corporation Act (NTPC[63]) and the Public Utilities Act.[64] The NTPC Act sets up the NTPC to manage energy generation and distribution while the PUA authorizes the Public Utilities Board to act as an independent, quasi-judicial agency that regulates the territory's public utilities to ensure that essential services like electricity are delivered fairly, safely, and at reasonable rates.

The Government of the Northwest Territories (GNWT) is also forward-looking with the 2030 Energy Strategy. The Strategy provides a guide for the development of secure, affordable, and sustainable energy in the NWT for transportation, heat, and electricity. This includes support for energy efficiency and conservation programs, local renewable and alternative energy solutions, and large-scale energy projects.[65]

Along with ensuring that energy in the NWT is secure and affordable, many of the initiatives under the 2030 Energy Strategy are focused on reducing greenhouse gas (GHG) emissions, contributing to the GNWT's Climate Change Strategic Framework's first goal to reduce GHG emissions.[66] By 2030, the NWT aims to reduce GHG emissions from electricity generation in diesel-powered communities by an average of 25 per cent, reduce GHG emissions from road vehicles by 10 per cent per capita, increase the share of renewable energy used for space heating to 40 per cent, and increase residential, commercial, and government building energy efficiency by 15 per cent.[67]

In 2025, a new direction was issued with eleven policy directives focused on improving and expanding net metering and independent power producer programs, implementing rate impact mitigation measures, mandating long-term integrated system planning, and making several necessary regulatory updates. Many of these issues are interrelated. This work responds to long-standing issues raised by utilities, communities, and residents, while creating more space for Northerners to generate their own renewable power and manage their energy costs. It will also better position Indigenous and community governments to advance renewable energy opportunities and support local economic development.[68]

For Indigenous and community-owned power production, the GNWT states that it will introduce a new program named Independent Power Production to 'prioritize Indigenous and community ownership and ensure transparency in power purchase agreements'.[69]

[63] Northwest Territories Power Corporation Act, SNWT 1988 c N-2.

[64] Public Utilities Act, RSNWT 1988, c 24 (Supp).

[65] Government of Northwest Territories, 'Infrastructure' <https://www.inf.gov.nt.ca/en/services/energy/2030-energy-strategy?> accessed 7 August 2025.

[66] ibid.

[67] ibid

[68] Government of Northwest Territories, 'Improving how electricity works for people and communities in the NWT'< https://www.gov.nt.ca/en/newsroom/improving-how-electricity-works-people-and-communities-nwt> accessed 8 August 2025.

[69] Cabin Radio, 'GNWT makes regulatory changes mostly aimed at renewables'<https://cabinradio.ca/232900/news/economy/infrastructure/energy/gnwt-makes-regulatory-changes-mostly-aimed-at-renewables/> accessed 7 August 2025.

C. Nunavut

The Qulliq Energy Corporation Act[70] establishes Qulliq Energy Corporation as the sole electricity utility in Nunavut, responsible for generation, distribution, energy conservation, and alternative generation development. In May 2025, the government launched a Sustainable Energy Support Policy (SESP) to replace the 2021 Renewable Energy Support Policy. This is a testament to the dedication of the government towards cleaner and more sustainable energy solutions for Nunavut.

Building on existing programs for homeowners and cabin owners, the new policy introduces the Independent Power Producer (IPP) subsidy, supporting larger-scale community-led renewable energy projects under the Qulliq Energy Corporation's IPP program.[71] Minister Joanasie said the policy advances locally developed, resilient energy solutions and reinforces Nunavut's commitment to sustainability by supporting Inuit and municipal-led projects.[72]

VII. Indigenous Business Issues and Models.

In the provinces south of the territories, there is evidence that direct Indigenous ownership and participation in energy projects including renewable distributed energy is growing.[73] An example is the Indigenous led Tilley solar project in Alberta. It provides 23.6 MW of power for southern Alberta.[74] Timing was excellent with the project gaining Alberta government approval just before the government's 2023 moratorium on new renewable energy projects. Tilley is one project, though one of the larger ones, among dozens of Indigenous renewable energy projects developed in the last five years.[75] A significant number were located within Indigenous communities.[76]

Public and energy community corporate perception about Aboriginal rights driving reconciliation and the relative ease and benefits of investment in First Nations renewable energy projects is changing. A factor here is the relative degree of public knowledge about legal and government policy developments. These include further judicial clarification of the constitutional duty to consult and accommodate rights-affected Aboriginal people within the broader 'honor of the Crown'[77] concept and better understanding of the concept of Aboriginal title following the 2014 Supreme Court of Canada decision in *Tsilhqot'n Nation v British Columbia*.[78] Metis issues are also relevant following judicial recognition of Metis as 'Aboriginal' people under the Constitution.

[70] Quilliq Energy Corporation Act, RSNWT (Nu) c N-2.

[71] Nunavut's 'New Sustainable Energy Support Policy' <https://climatechangenunavut.ca/news/nunavuts-new-sustainable-energy-support-policy> accessed 13 August 2025.

[72] ibid.

[73] Canada Energy Regulator, 'Market Snapshot: Indigenous Ownership of Canadian Renewable Energy Projects is Growing' (released 21 June 2023) < https://www.cer-rec.gc.ca/en/data-analysis/energy-markets/market-snapshots/2023/market-snapshot-indigenous-ownership-canadian-renewable-energy-projects-growing.html > accessed 19 August 2025.

[74] See D Israelson, 'Solar farm redefines Alberts's energy landscape', *The Globe and Mail* (25 February 2025) B6.

[75] ibid.

[76] ibid.

[77] Above, section VI.

[78] n 34.

Aboriginal self-government rights, supported by recent Supreme Court of Canada decisions is also a factor to consider.[79] This may include elements of natural resources and energy management on reserve and traditional lands. There are numerous examples of negotiation, sometimes with government facilitation, of impact and benefit agreements concerning nonrenewable natural resource extraction and transportation. This is a model equally applicable to renewable energy. First Nation equity ownership has often been a feature of these agreements.[80]

On the financial side, renewable energy investment can be drawn from trust fund sources. These trusts can be created by First Nations from the proceeds of extinguished Aboriginal title land sales and tailored to specific needs and opportunities including renewable energy investments.

Another relevant business development is establishment of Indigenous operated capital markets firms. This has created First Nations' investment success and more broadly, advanced First Nations led reconciliation. Examples include project funding working with the Aboriginal corporation Cedar Leaf Capital.[81] There are also examples of loans and guarantees provided by the Canada Infrastructure Bank, an example being funding investment by a group of First Nations communities in a battery storage project.[82]

VIII. The Uncertainty of Indigenous Rights

The case law on Canadian Aboriginal rights suggests that in any relevant regulatory process, responsive consultation and necessary accommodation is required.[83] But the main difficulty is that the Supreme Court of Canada's Aboriginal rights infringement test is highly fact specific. Litigation based on plausible scenarios could at least seriously delay government actions attacked; but it must be remembered that justifiable rights infringement could be established.

Another possibility is Aboriginal rights challenges based on rights to secure community energy. But arguments based on secure energy rights are again likely to be fact specific and in any event, the Supreme Court of Canada has generally not supported potential rights infringement remedies that require government fiscal action.[84] Yet, Supreme Court recognition of the UN Declaration on the Rights of Indigenous Peoples and its incorporation in Canadian domestic law by the Federal *UNDRIP Act* may strengthen rights claims.[85] The court has indicated that the Declaration can support characterization of a federal government law or action in a way that brings it within an exclusive federal constitutional subject.[86]

A more recent Federal Court decision provides a detailed analysis, concluding that the federal Canadian Nuclear Safety Commission had a duty based on UNDRIP's 'free, prior

[79] Quebec Reference (n 31).

[80] See A Lucas and D Laidlaw, 'Oil and Gas Sector Local Content Decision Processes: Canadian Indigenous Participation' in D Olawuyi (ed), *Local Content and Sustainable Development in Global Energy Markets* (CUP 2021) 343, 352.

[81] See J Berkow, 'Indigenous-run capital markets firms in Canada are hitting their stride', *The Globe and Mail* (15 February 2025) B6.

[82] ibid.

[83] *Gitxaala Nation* (n 29); Québec Reference (n 31).

[84] P Hogg, *Constitutional Law of Canada* (5th edn, Thomson Carswell 2007) Section 35.5 'Remedial Action'.

[85] Quebec Reference (n 31) para 45.

[86] ibid.

and informed consent'. However, the court clarified that while UNDRIP requirements provide interpretive guidance, they do not give the challenging First Nation a veto over the Commission's decision. They guarantee only the right to a 'robust process'[87] that makes the Aboriginal perspective clear.

IX. Conclusions and Lessons Learned

Clinic creation and delivery is feasible, and helpful models are available. Pilot project funding is in place for the proposed University of Calgary clinic. There are promising governmental, private, and Indigenous-owned funding sources for community renewable energy projects. There are significant project constraints or at least uncertainties that must be addressed for each project, including differences among First Nations, federal and territorial governments, and sometimes among community groups. The role of Aboriginal rights in distributed renewable energy project approvals and regulation must be assessed for each project in which the clinic participates.

Changing federal energy and environmental law and policy must be considered, including emergent US tariff and Canadian response issues, as well as gaps such as clear legal site remediation and reclamation obligations on Aboriginal lands. Federal law is needed to empower First Nations councils, as in Alberta, to authorize renewable energy projects with appropriate conditions. Clinic proponents must keep a close watch on the apparent public 'seismic shift' away from green goals towards relatively greater emphasis on energy security and economic integrity. This applies to future remote Indigenous community renewable energy projects and the information and training provided by the proposed clinic.

[87] *Kebaowek First Nation v Canadian Nuclear Laboratories* 2025 FC 319 (CanLII); further appeal may provide the Supreme Court of Canada an opportunity to provide ultimate guidance on these issues.

21

Conclusion

Sustainable Distributed Energy Resources and the Future of Electricity Law

Iñigo del Guayo, Damilola S. Olawuyi, Louis de Fontenelle, Milton Fernando Montoya, and José Juan Gonzalez Márquez

I. Introduction

This volume examines the guiding principles of sustainable distributed energy resources law and policy. With case studies from frontier energy hubs, the book unpacks overlapping legal and policy challenges in the development and implementation of DERs that currently hinder the coherent realization of its decarbonization, decentralization, digitalization, and democratization potentials. The chapters also highlight emerging transformations in energy law and policy aimed at addressing these challenges. In this concluding chapter, we offer some reflections on moving beyond fragmentation in DER implementation and towards holistic and sustainability-aligned governance approaches.

For several decades, the generation and distribution of electricity depended on large, centralized energy systems and power infrastructure, distant from consumption centres, which required the construction of transmission and distribution networks.[1] However, over the last decades, the ongoing global energy transition and the quest for innovative ways to promote energy efficiency; and to reduce the environmental footprints, infrastructure gaps, and costs of electricity from centralized energy systems, have resulted in an exponential rise in initiatives aimed at advancing DERs.[2] As discussed in Chapter 1 of this book, DERs refer to the small-scale generation, transmission, and storage of electricity through microgrids, backup batteries, fuel cells, and virtual power plants that are often at, or closer to, the point of consumption (for example, solar panels in the consumer's home).[3] In addition to reducing the costs and environmental impacts associated with centralized electricity generation, transmission, and storage, DERs aim to promote energy security and sustainable development by increasing the availability, affordability, accessibility, reliability, and energy self-sufficiency of the end-users. DERs also aim to foster energy democracy and diversification by advancing citizen-led energy solutions that address electricity deficits

[1] See Chapter 1 of this book.

[2] ibid; see also I Bantekas, 'The Off-Grid Revolution and the "Promise" of Energy Equality Under the Paris Climate Change Agreement' (2024) 23(1) Northwestern Journal of Human Rights 45–72.

[3] T Bin Nadeem and others, 'Distributed Energy Systems: A Review of Classification, Technologies, Applications, and Policies' (2023) 48 Energy Strategy Reviews 101096.

Iñigo del Guayo, Damilola S. Olawuyi, Louis de Fontenelle, Milton Fernando Montoya, and José Juan Gonzalez Márquez, *Conclusion* In: *Sustainable Distributed Energy Resources*. Edited by: Íñigo del Guayo, Damilola S. Olawuyi San, Louis de Fontenelle, Milton Fernando Montoya, and José Juan González, Oxford University Press.
DOI: 10.1093/9780198991946.003.0024

through innovative renewable energy technologies. With the growing recognition of the critical importance of DERs for energy decarbonization, decentralization, digitalization, and democratization across the world, investments in DER infrastructure and projects are increasing in pace and ambition.[4]

At the same time, as DER implementation and deployment grows in pace and scale, legal challenges are also emerging. As chapters in this book demonstrate, concerns over land use and environmental challenges in the location of DER infrastructure have emerged in many parts of the world, which continue to raise questions on the decarbonization potential of DERs.[5] Similarly, lack of regulatory clarity on the integration and interconnection of DERs with distribution grids have resulted in customer complaints and increased regulatory risks for utilities.[6] Questions also arise about grid integration and balancing, especially how to accelerate investments in smart grids, to ensure that integrating electricity from DERs does not overwhelm and damage existing electricity infrastructure designed for centralized generation and distribution and storage.[7] Furthermore, the high-cost and uneven access to DER technologies raise key concerns on how DERs may lead to exclusion and inequitable access, especially low income and already vulnerable groups including women, youth, and Indigenous communities. Similarly, lack of supportive legal and policy frameworks that promote energy entrepreneurship and citizenship increasingly hinder the abilities of energy citizens and prosumers across the world to benefit from the energy democratization objectives of DERs.[8] This book makes it clear that the expansion of DERs is an expression of the democratization of the just energy transition, not only from a technical perspective but also from a social sustainability perspective. In fact, without a sustainability perspective, the implementation of DERs risks exacerbating environmental pollution, energy poverty, social exclusions, as well as complex human rights, legal, and commercial risks in energy markets worldwide.[9] The lack of a clear integration of sustainability considerations in the design and implementation of DERs pose significant challenges to the just and inclusive penetration and growth of DERs. Such incoherence also sends conflicting signals to investors and entrepreneurs, in terms of regulatory environment, competitiveness, and sustainability of DER investments.[10] As investors seek greater certainty and guidance on what sustainability-aligned DER implementation and contracting practices are, coherent legal frameworks at national, regional, and international levels, as well as bespoke legal advice, are required.

[4] See Frost & Sullivan, 'Nearly $1 Trillion in Investments Estimated by 2030 as Power Sector Transitions to a More Decarbonized and Flexible System' <https://www.frost.com/news/press-releases/nearly-1-trillion-in-investments-estimated-by-2030-as-power-sector-transitions-to-a-more-decarbonized-and-flexible-system/> accessed 10 March 2025; see also European Parliament, 'Will Distributed Energy Resources Change How We Get our Energy?' <https://www.europarl.europa.eu/RegData/etudes/ATAG/2020/651944/EPRS_ATA(2020)651944_EN.pdf> accessed 10 November 2024.

[5] See Chapter 4 of this book.

[6] In 2021 for example, a United States utility was fined one million dollars for exceeding a maximum number of customer service complaints on interconnection delays and failures. See G Brown, 'Minnesota PUC Fines Xcel Energy $1 Million for Interconnection Failures', Inter-state Renewable Energy Council (21 January 2021) https://irecusa.org/blog/regulatory-engagement/minnesota-puc-fines-xcel-energy-1-million-for-interconnection-failures/ accessed 1 November 2025.

[7] See for example G Chazan, 'Bavarian village reaps renewable energy: Success of Bavarian "prosumers" creates dilemma for German electricity system' https://www.ft.com/content/4bd1ad70-793e-11e3-b381-00144feabdc0 accessed 1 November 2025.

[8] See Chapter 19 of this book.

[9] See Chapters 1 and 2 of this book.

[10] See Chapter 19 of this book.

Sustainability-aligned legal frameworks on DERs can support faster, inclusive, and more cost-effective DER development and integration into extant grids. However, as the chapters of this book demonstrate, in several jurisdictions such transformations are not easily forthcoming. Electricity laws and policies that were designed many years ago, based on the centralized model and before the rapid penetration of DERs, are incapable of keeping pace with the legal and sustainability dimensions of integrating DERs into the grid.[11] Taking stock of progress made, as well as innovative approaches for further progress, in addressing social, environmental, and economic aspects of DER design, financing and implementation can help plot out a better path to further reform.

What are the key legal challenges to integrating social, environmental, and economic sustainability in the design and implementation of DERs? What are the required legal and policy approaches to address such gaps? These are what we address in this chapter. After this introduction, section II discusses the drivers of—and sustainability risks—posed by fragmentation in response to the wide range of social, environmental, and economic challenges of DER deployment. Section III discusses the guiding principles of sustainability-aligned DER governance required to advance a just, inclusive, and rights-based deployment of DERs. Sustainability-aligned DER legislation, legal drafting, contract negotiation, policymaking, and education approaches are urgently required to balance and integrate the 4Ds in extant and future DERs, ranging from project due diligence, licensing and approval, to operations, decommissioning, and end-of-lifetime closure. Section IV is the concluding section.

II. Regulatory Fragmentation in DER Implementation: Drivers and Sustainability Risks

Regulatory fragmentation on DERs, meaning the lack of coherent integration of social, economic, and environmental considerations in DER planning and implementation is currently pervasive and has been identified as a threat to sustainable and speedy penetration of DERs.[12] By focusing excessively on decarbonization or environmental objectives, project planners may overlook the social and economic inequalities and risks, which could lead to sub-optimal integration of DERs, or projects that fail to deliver the promised benefits in terms of energy democratization and socio-economic opportunities. On the other hand, excessive focus on the socio-economic aspects of DERs, without examining its environmental and decarbonization dimensions may result in energy policy failure and unjust outcomes.[13] Without clear, comprehensive, and coherent legislation and policies, DER projects will fail

[11] R Valova and G Brown, 'Distributed Energy Resource Interconnection: An Overview of Challenges and Opportunities in the United States' (2022) 2 Solar Compass 100021.

[12] ibid; see also R Cerchione and others, 'Balancing Sustainability and Circular Justice: The Challenge of the Energy Transition' (2025) 494 Journal of Cleaner Production 144942.

[13] Energy policy failure has been defined as 'any energy policy which does not meet local, national, and international energy and climate goals across the activities of the energy life-cycle and where just outcomes are not delivered'. See M Sokołowski and R Heffron, 'Defining and Conceptualising Energy Policy Failure: The When, Where, Why, and How' (2022) 161 Energy Policy 112745; see also P Derwort, N Jager, and J Newig, 'Towards Productive Functions? A Systematic Review of Institutional Failure, its Causes and Consequences' (2019) 52 Policy Sciences 281–98.

to effectively balance and achieve climate change and energy security goals set out in the Paris Agreement and the UN SDGs.

As demonstrated in this book, four key drivers of regulatory fragmentation on sustainable DERs exist. First is the lack of comprehensive global standard on DERs. As discussed in Chapter 2, even though the Paris Agreement recognizes the need for countries to adopt different mitigation and technological approaches, including renewable energy deployment and energy decentralization, the lack of clear and comprehensive guidelines or roadmap at the international level for sustainable DER implementation remains a practical gap.[14] The lack of an international framework on DERs means that countries and sub-regional authorities are implementing energy decentralization programs according to their own local priorities and contexts, especially in the context of their nationally determined contributions (NDCs).[15] While some NDCs include DERs, some do not. Even in the NDCs that include DERs, a wide range of approaches, baselines, and measures have been developed to support DER growth and integration.[16] While some countries express DER implementation in legally binding terms with mandatory renewable performance standards, in some other countries DER projects are voluntary and complimentary. Similarly, the technical standards on DER interoperability and integration in distribution grids that have been announced by industry regulators and standards organizations vary in terms of clarity, scope, and ambition.[17] The result is a proliferation of DER standards with different ambitions, methodology, and scope, with many lacking clear guidelines on addressing socio-economic and human rights aspects of DER implementation. There are still grey areas in terms of consumer protection, transparency in interconnection processes, data ownership, and privacy requirements. As countries continue to clarify the modalities and rules for advancing the goals of the Paris Agreement, there is a need for clear, transparent, and comprehensive standards on the use of DERs for climate mitigation, including the methodology for measuring and reporting progress.

A second driver of regulatory fragmentation is the disconnect between DER strategies and policies, and electricity legislation, contracts, regulations, and risk management frameworks. While investments in DER projects are growing exponentially, electricity laws and policies that were designed many years ago, based on the centralized model remain largely in place, and are increasingly incapable of keeping pace with the legal and sustainability dimensions of integrating DERs into the grid. Chapter 7 of this book for example, evaluates the alignments and contentions between distributed energy developments and legislative efforts to achieve one internal energy market in the European Union (EU). Chapter 8 evaluates legislative efforts on DER in New Zealand, while Chapter 9 unpacks the limitations of China's existing legal framework on decentralization, highlighting how lack of tailored regulation to promote system flexibility and reliability limit the coherent and sustainable

[14] H Carr, 'Distributed Energy Resources: what we learned from regulators about managing the energy transition in Africa' (24 October 2024) <https://crossboundaryenergy.com/regulators-managing-the-energy-transition-in-africa/> accessed 12 March 2025.

[15] See for example, Ontario Energy Board (OEB), *Framework for Energy Innovation: Setting a Path Forward for DER Integration* (30 January 2023) <https://engagewithus.oeb.ca/fei/news_feed/the-oeb-has-released-the-framework-for-energy-innovation-setting-a-path-forward-for-der-integration-2> accessed 12 March 2025.

[16] See UNFCCC, NDC Registry, < https://unfccc.int/NDCREG> accessed 6 August 2025.

[17] See for example, 'IEEE Standard for Interconnection and Interoperability of Distributed Energy Resources with Associated Electric Power Systems Interfaces' in *IEEE Std 1547-2018 (Revision of IEEE Std 1547-2003)*, 1–138, 6 April 2018, doi: 10.1109/IEEESTD.2018.8332112; also Ontario Energy Board (OEB), 'Framework for Energy Innovation: Setting a Path Forward for DER Integration' (30 January 2023).

decentralization of energy systems. Chapter 10 also illustrates how electricity legislation and contracts in Mexico are yet to be infused with human rights provisions needed to address social exclusions in the implementation of DERs projects. Chapter 14 also discusses legislative gaps in terms of addressing data protection and privacy challenges that arise with the digitization of energy networks under a DER model. Lack of comprehensive legislation on DER implementation in many countries limits the coherent integration of social, environmental and economic objectives in a manner that advances the 4Ds. There is an urgent need for sustainability-aligned legislation and electricity contracts to address the disconnect between DER investments and the corresponding legal frameworks to ensure their coherent implementation.

A third driver of regulatory fragmentation is the lack of a supportive legal framework to promote and incentivize citizen-led entrepreneurial energy solutions under the DER framework. As discussed in Chapter 15, DER legislation and contracts should incentivize the producer to generate electricity and ensure transparency in interconnection to distribution grids. There is also a need for governments to provide incentives to utilities and distribution network companies to upgrade their infrastructure and digitalization technologies needed to comply with real-time monitoring and other interconnection requirements.[18] However, in addition to lack of incentives, Chapter 12 of this book discusses how prosumers face significant financing barriers and challenges especially given the high costs of DER technologies and infrastructure in developing countries. Similarly, in the context of Colombia, Chapter 17 highlights legal challenges that energy communities face, especially regulatory barriers, limited access to finance, and the gaps in integration with the country's existing energy infrastructure. As Chapter 19 discusses, simplifying the process of business formalization and registration will go a long way in encouraging individual and private sector participation in DER projects. Furthermore, the wide range of political, environment, social, technological, and legal (PESTEL) gaps and challenges slowing the rapid growth of distributed energy business models and how such barriers in the entrepreneurial ecosystem must be comprehensively dismantled. A reliance on technology transfer alone and solar panel importation will drive costs and will not effectively meet the significant technology requirements of DERs.[19] In addition to addressing the barriers to technology transfer, there is a need to dismantle all barriers to clean technology entrepreneurship at local levels.[20] In addition to barriers that energy citizens face in addressing DER technologies and financing, a gender lens in required to address gender-based exclusions in DER models, particularly the social differences in the production and consumption of energy under a decentralized energy system. As Chapter 18 discusses, it will be difficult to achieve the energy democratization objectives of DERs without addressing the root causes of social exclusions in DER penetration. DER legislation and contracts must integrate risk mitigation strategies including the requirements for human rights due diligence, environmental impact assessments, and gender vulnerability screening to ensure broad-based access to DER technologies and financing opportunities.

[18] See also, Valova (n 11), see also G Brown and S Stanfield, 'Improving Interconnection Timelines: The Need for Data and Enforcement' (*PV Magazine*, 1 February 2022) <https://pv-magazine-usa.com/2022/02/01/improving-interconnection-timelines-the-need-for-data-and-enforcement/> accessed 10 August 2025.

[19] See Chapter 2 of this book.

[20] ibid.

A fourth driver of regulatory fragmentation is the lack of capacity, in terms of resources and training to advance sustainable DER implementation. Chapter 20 discusses the role of entrepreneurial education in promoting energy democratization, especially through clinical legal education that reinforces practical skills on prosumerism. Without adequate awareness and capacity development, local communities may resist or reject DER projects, which may lead to suboptimal deployment and penetration. Entrepreneurship education is required to promote homegrown innovation and technologies needed to advance sustainable DERs. Furthermore, negotiating risk-free renewable electricity contracts will require advanced knowledge of energy law, environmental law, climate change law, and contract law. However, in many countries, the development and implementation of entrepreneurial legal education faces significant resource and capacity constraints. Higher education institutions have key roles to play in designing specialized courses and training programs that promote knowledge and awareness on sustainable DERs.

Coherent solutions to the multiple social, environmental, and economic challenges in advancing sustainable DERs must be based on a holistic transformation of energy and electricity laws. Barriers to systemic integration and coherence must be acknowledged and addressed to succeed in the required transformation. As explained in Chapters 16 and 17, there are still regions where, despite public policies promoting DERs, there remain legal and regulatory barriers to their integration into the grid, and where the majority of energy community cases are still in the experimental phase. The next section discusses the guiding principles of sustainability-aligned DER governance required to advance a just, inclusive, and rights-based deployment of DERs.

III. Roadmap for Sustainability-aligned DER Governance

Sustainability-aligned governance of DER requires a mix of legislative, institutional, and contractual frameworks aimed at addressing sustainability challenges relating to environmental impacts, costs, grid integration and social exclusions in the design and implementation of DER projects and infrastructure. This section explores the institutional and legal futures of sustainable DERs, combining structural analysis with practical priorities. Drawing on all contributions to this volume, it outlines a roadmap for sustainable and inclusive DER governance. Rather than proposing a single model, it identifies key legal challenges and opportunities to shape energy transitions that are just, resilient, and sustainable.

A. Addressing Energy Poverty and Legal Fragmentation

The lack of access to electricity is a major component of energy poverty and signals a failure to meet the commitments set forth in the 2030 Agenda. The chapters in this book demonstrate that DERs are an effective solution for meeting the energy needs of communities located far from traditional power plants.

The concept of DERs is gaining traction globally, yet legal obstacles still impede its full development. In many regions—particularly Latin America—legal frameworks remain fragmented, insufficient, or contradictory. As noted in Chapter 10 (González), Mexico currently lacks specific regulation, although the forthcoming 2025 Energy Sector Law aims to define rules for distributed generation, small systems, and self-supply electricity. Colombia faces similar difficulties, as discussed in Chapter 17 by Montoya and Zapata, where regulation is insufficient and technical and financial barriers persist, even considering that it is a competitive electricity system with public and private companies participating.

This legal instability discourages investment and contributes to DERs' marginal role in national energy mixes. The challenge is to align decentralized electricity generation with sustainability, equity, and the protection of human rights, especially the right to energy access. This is not an insurmountable task. Many jurisdictions are already experimenting with legal reforms, and governance in this sector must be strengthened to support and scale such efforts.

One of the main objectives of DERs should be to primarily benefit communities currently experiencing energy poverty. Legal frameworks must also recognize women as direct beneficiaries, given that they are often most impacted by lack of access to electricity. Legal reform must do more than simply allow DER development. It must proactively structure it to promote high-quality, affordable, clean, and reliable energy. Governments must fulfill their commitments under the Paris Agreement, SDGs, and international human rights law. As Chapter 2 emphasizes, DERs can advance SDG 7 (clean energy), SDG 1 (no poverty), SDG 5 (gender equality), and SDG 10 (reduced inequalities).

DERs are also essential to decarbonization. However, achieving this requires a deliberate roadmap: robust regulation, long-term planning, and coherence across jurisdictions and governance levels. Electricity storage—a vital and underregulated component in developing countries—must be addressed in law to secure system stability. Ultimately, new legislation must establish financial and fiscal mechanisms that support DER implementation. Efficient legal frameworks must avoid fragmentation, define clear institutional responsibilities, and recognize the roles of diverse social actors, including academia and civil society.

In fact, smart metering, predictive software, and decentralized grid management systems are driving the development of DERs at an increasing pace. However, at the same pace as the development of DER-related technology, programs to build and strengthen the technical capacity of communities must also be implemented. Community knowledge, training, and technical skills are also crucial to the democratization of DERs, especially in regions experiencing poverty or without access to electricity, where DERs are the answer to these conditions.[21]

B. Law and Transition: Managing Complexity

The development of DERs is not just a technological shift but a legal one. Traditional legal systems, designed for centralized architectures, are often ill-equipped to govern the distributed nature of modern energy systems. This generates multiple layers of complexity: technological, normative, and institutional.

Regulating DERs requires tailored rules for specific technologies—batteries, smart grids, SMRs, peer-to-peer trading. It also demands legal frameworks capable of integrating sustainability, gender equity, and participatory governance. The diversity of DER

[21] See Olaoye and Ilesanmi, Chapter 19 of this book.

stakeholders—from co-operatives to municipalities—makes this even more pressing. There is also a need for supportive legislation and policies that encourage sustained investments in DERs, especially through incentives that support private sector participation.

Law is increasingly shaped by practice, not doctrine. Regulatory sandboxes, simplified licensing, and community-based energy contracts are essential to adapt to new energy realities. Fragmentation in itself is not necessarily negative, as it may in some cases reflect responsiveness and innovation. But in the absence of clear legal goals, regulatory fragmentation becomes an obstacle to justice, predictability, and systemic coherence. Through comprehensive regulations that clarify how stakeholders should mainstream environmental, social, and economic consideration in the design, financing and implementation of DER projects, and the specific responsibilities of different regulatory actors, coherent DER implementation can be strengthened.

C. Governance, Justice, and the 4Ds

This book has explored how the four vectors of energy transition—decarbonization, decentralization, digitalization, and democratization (4Ds)—interact with the law. These transitions do not occur in harmony. Centralized decarbonization may bypass equity. Digitalization may entrench exclusion. Democratization may clash with market logic.

Justice—particularly energy justice—can provide the normative anchor for these processes. It offers a transversal framework to assess who benefits, who participates, and who is protected.[22] Justice must not remain declaratory. It must be translated into legal norms, rights, and enforcement mechanisms. As shown in Chapters 2, 10, and 17, many jurisdictions still lack the institutional tools to operationalize energy justice. Sustainable DERs require a just and inclusive approach that ensures that all communities, including low income, marginalized, and Indigenous communities have affordable and reliable access to DER technologies and solutions. This includes providing dedicated funding, loans, and incentives that enable low-income households to participate in DERs. Furthermore, industry structure and regulations that limit market participation and access of community microgrids must be carefully addressed to advance the energy democratization objectives of DERs. By integrating energy justice norms, particularly distributive, participatory, reparation (or restorative) justice, and social justice in the design and implementation of DER projects, we can accelerate an advance the 4Ds of DERs in a just, inclusive, and equitable manner that leaves no one behind.

D. A Plural Future Anchored in Law

There is no single future for DERs. Centralized and decentralized models will coexist. The legal challenge is not to impose uniformity, but to articulate plural governance structures with shared normative anchors.

[22] I Del Guayo and others (eds), *Energy Law and Energy Justice* (OUP 2020) 1–10.

To avoid drift, we must reject technocratic deregulation. The energy transition cannot be left to the inertia of markets or path-dependent innovation. As Rimbaud wrote of the 'drunken boat', without direction we risk fragmentation and loss.

Instead, we must build legal frameworks that are imaginative, inclusive, and grounded in rights. Law must serve as translator, integrator, and guardian—making sense of diverse models while preserving coherence, stability, and justice.

Communities also have an important role to play in the future legal design of DERs. Increasingly, communities want to participate in the design of inclusive regulatory frameworks centered on citizens, their concerns, capabilities, and social goals. In other words, the democratization of DER regulation also implies the democratization of energy public policy and law. In this sense, participatory procedures that are much more transparent, less opaque, earlier, inclusive, flexible, and open to diversity are increasingly needed. In times when the decentralization of energy systems is a reality, exclusively technocratic regulation, behind closed doors, without citizen participation, is not viable.

Furthermore, as shown in Chapter 20 of this book, higher education institutions have key roles to play in advancing energy democratization through entrepreneurial legal education that unlock practical skills on innovation, technology commercialization, and minimizing ESG risks. As DER gains increased prominence worldwide, the teaching of energy and electricity law must evolve from a sole focus on centralized electricity generation and distribution. Legal educators should embrace innovative pedagogies such as entrepreneurial legal clinics, simulations, and executive courses that expose students and prosumers to the entire value chain of design, financing, and implementation of DER projects, including drafting and negotiating sustainability-aligned DER contracts.

E. Sustainability-aligned DER Contracts

DER contracts, ie agreements that outline the terms and conditions for implementing DER projects, such as power purchase agreements (PPAs), operations and maintenance (O&M) agreements, and interconnection agreements play key roles in clarifying responsibilities and obligations of the project planners, utilities, and other parties in the DER value chain. To advance sustainable DERs, it is essential for such contracts to integrate sustainability standards that clearly allocate responsibilities and risks for managing adverse social, environmental, and economic impacts, while maintaining the benefits and viability of distributed generation. Sustainable DER contracts should clearly integrate requirements for environmental impacts assessments and human rights due diligence to better anticipate and manage adverse human rights impacts, especially land rights, environment, health, and community participation rights. With the rise of environment, social, and governance (ESG) norms, especially the adoption of the United Nations Guiding Principles on Business and Human Rights in 2011, EU Corporate Sustainability Due diligence Directives, and other national actions plans on business and human rights across the world, having ESG specific requirements in DER contracts can help clarify expectations and responsibilities.[23] DER planners, financing parties, and other stakeholders in the DER value chain could

[23] DS Olawuyi and EI Athwal, 'Net Zero, Human Rights, and ESG Risk Management', in D Olawuyi and others, *Net Zero and Natural Resources Law* (OUP 2024) 175–92.

avoid costly ESG risks and legal lability that may result from DER projects by ensuring their contracts clearly integrate ESG-related due diligence, representations, warranties, and indemnity clauses.[24]

This book also makes it clear that one of the cornerstones of democratizing DERs is the financial conditions that make the projects viable, not only for their installation but also for their operation, maintenance, and technical assistance. These aspects, of course, must be addressed in regulations, but also in contract clauses, so that the development of DER projects is also financially sustainable and provides legal certainty for all parties involved.

Furthermore, given the inevitability of disputes in DER projects, as it is with many contractual relations, alternative dispute resolution (ADR) mechanisms, particularly arbitration, that could enhance the speedy, timely, and cost-effective resolution will be crucial, especially given the time sensitive nature of DER investments.[25] Given that arbitration is a consensual process, investors, and their lawyers alike, should integrate arbitration clauses into DER contracts to ensure their application for the resolution of disputes that may arise. Similarly, regulatory challenges that limit the widescale application of ADR in the context of DER projects must be carefully addressed. For example, in some countries where legislation limits the application of arbitration to commercial disputes only, such restrictions may limit the use of ADR for land and environment related disputes that may arise from DER projects.[26] It is crucial to provide supportive legislation, as well as institutional frameworks that advance the use of arbitration and other ADR mechanisms in the timely and effective resolution of DER-related disputes.

IV. Conclusion and Future Directions

This book has analysed DERs from the perspective of sustainability, in its triple environmental, economic, and social dimensions. Almost a century ago, our ancestors believed that progress and well-being rested on the construction of national electricity grids and large power generation units—in other words, on centralization. Society today understands that progress and well-being lie in the opposite process. The fundamental lever of change is decarbonization, which, together with DERs driven, in turn, by digitalization, leads to the decentralization and democratization of the energy system.

DERs are linked to renewable energy. It is true that there are non-renewable DERs, such as, for example, a small diesel engine capable of generating electricity for domestic use. In this example, we are dealing with energy consumed by the person who produces it; therefore, it is distributed generation. However, in this book, we have left that example and others like it out of our analysis because our perspective has been one of sustainability, particularly environmental sustainability, which today means nothing other than decarbonization and, therefore, the progressive reduction in the use of fossil fuels.

[24] International Bar Association, 'Report on Use of ESG Contractual Obligations and Related Disputes' (IBA 2023) https://www.ibanet.org/document?id=report-on-use-of-ESG-contractual-obligations accessed 1 November 2025.

[25] ibid.

[26] See D Olawuyi and O Olujobi, 'Recent Trends in African Energy and Mining Arbitration in Light of the Paris Agreement: Current Opportunities and Future Directions' in V Nalule and DS Olawuyi (eds), *The Palgrave Handbook of Arbitration in the African Energy and Mining Sectors* (Palgrave Macmillan 2024).

As with any change, governments and regulators must act as leaders, but the change must be led by society. A lack of public involvement in the process is just as detrimental as excessive public involvement. Regulatory inaction can lead to the failure of the emergence of a new energy paradigm based on DERs. Excessive involvement, in the form of, for example, large, sustained subsidies, can lead to an inefficient, inequitable, and uncompetitive energy system, and some consumers may be financing the change to the benefit of others. Excessive public leadership hinders the social acceptance of new energy facilities, which generates considerable suspicion, particularly among the communities where these facilities are located.

The widespread use of DERs has obvious advantages, but it is not without new problems. First, as the International Energy Agency has warned, DERs can jeopardize security if they are not regulatory aligned with grid needs. This does not mean that DERs must conform to grid needs, but rather that the current and future state of the grid must be considered so that their gradual replacement with off-grid energy resources is possible and safe. From an economic perspective, DERs should contribute to alleviating the problem of energy poverty, since, by definition, consumers generate the electricity they need at a much lower cost. However, it cannot be forgotten that this advantageous economic situation for vulnerable consumers is the goal of a long road, one that requires overcoming technological and systemic difficulties (such as, where applicable, the charges these consumers must pay for distribution network maintenance). Socially, distributed energy resources generate rejection and bring new and old problems associated with the NIMBY rule.

In this context, the law relating to DERs must facilitate their emergence and development. This is a double facilitation: on the one hand, the law must facilitate DERs' ability to gain an ever-increasing foothold in an economic system based on the use of fossil fuels, both for electricity generation and transportation, and therefore compete with the subsidies established for such fuels; secondly, the law must facilitate their expansion into electrical systems based on extensive transmission and distribution networks, which connect some production centres with millions of users connected to those networks.

The law of distributed energy resources is a law that must respond to demand and act on demand. Consumers are the protagonists in a decentralized energy system. It is necessary to recognize that in such a system, the law becomes extremely complex. When law had to be limited to ensuring that there was sufficient supply to satisfy demand, and that this supply had to be as cheap as possible and as environmentally sustainable as possible, the law was relatively simple. Demand- and consumption-oriented law is like a big puzzle, as numerous solutions are required.

A law oriented towards demand and consumption is like a large puzzle, as numerous regulatory pieces are required, all of which must exist simultaneously and harmoniously: rules on self-consumption, storage, demand aggregation, data control of millions of users—all these rules are important in themselves, but also in relation to the energy system we intend to abandon (the system based exclusively on grids). The law and regulation of distributed energy resources thus takes on a granular nature. Broad principles are not so much of use (though they are still valid to properly construe the law), as the regulator must delve into numerous details, which will vary with the evolution of technology itself. Let's look at three examples. First, it is necessary to regulate when batteries can be charged from the grid and when stored electricity can be fed into the grid, while also establishing the necessary technical specifications. Second, if we want to allow consumers to actively participate in the market and sell their surplus electricity, it is necessary to regulate the role of aggregators and establish a fair system of economic relations with marketers. Thirdly, if we want to expand the use of electric vehicles, also as potential energy storage devices, numerous urban planning and municipal regulations must be implemented.

It is a kind of law which must be drafted carefully, balancing the sometimes opposite exigencies of economic efficiency, sustainability, and social acceptance. As in any other area of law, regulators must take on board, when discussing a future new regulation, the technological components involved, the economic costs (be them of the transactions themselves or the costs of an equitable distribution of the regulatory burdens), the environmental benefits, as well as the environmental externalities. A sustainable system based on DERs is decentralized and democratic, and so should be the law related to such a system, decentralized insofar it involves infra-state legal competences and democratic insofar consumers should participate in drafting the law via new participation channels.

The democratization of renewable energy sources and DERs is also a challenge for all countries where the technology has not yet become widespread. In other words, it is not only important for regulations to eliminate barriers and for financial mechanisms to exist to make projects viable. It's also important for developing countries to be able to incorporate the technology widely in communities that lack permanent access to energy. In other words, one of the strongest messages of this book is that in developing countries, DERs are also a solution to energy poverty.

We have written this book in the hope it offers readers (be they lawyers, regulators, academics, governments, community leaders, amongst others) both an analysis of the disparate legal problems around DERS, and a set of answers to the heterogeneous questions posed in current times by them to the energy system and to society. Through supportive regulations, policies, and contracts that ensure equitable access to DER technologies and innovation, we can maximize the full potential of DERs to advance a resilient, secure, and sustainable energy future.

Index

For the benefit of digital users, indexed terms that span two pages (e.g., 52–53) may, on occasion, appear on only one of those pages.